RF SYSTEM DESIGN OF TRANSCEIVERS FOR WIRELESS COMMUNICATIONS

RF SYSTEM DESIGN OF TRANSCEIVERS FOR WIRELESS COMMUNICATIONS

Qizheng Gu
Nokia Mobile Phones, Inc.

Springer

Qizheng Gu

RF System Design of Transceivers for Wireless Communications

Library of Congress Control Number: 2005049760

ISBN 0-387-24161-2 e-ISBN 0-387-24162-0
ISBN 978-0387-24161-6

Printed on acid-free paper.

Printed in the United States of America.

9 8 7 6 5 4 3

springer.com

To my wife, Lixian,

and

to our family's younger generations, Ye and Ethan

Contents

Preface

This book is about radio frequency (RF) transceiver system design for wireless communication systems. Most digital communications texts focus on the system design of the digital base-band rather than the RF section. The text is written for RF system design engineers as well as RFIC design engineers involved in the design of radios for digital communication systems. It is also appropriate for senior undergraduates and graduate students in electrical engineering.

The text develops systematic design methods of RF receivers and transmitters along with a corresponding set of comprehensive design formulas. Attention is given equally to the analysis of the RF systems. The book is focused on mobile communication systems implemented in RF application specific integrated circuits (ASICs) but it is applicable to other wireless systems such as, for examples, WLAN, Bluetooth and GPS. It covers a wide range of topics from general principles of communication theory, as it applies to digital radio designs, to specific examples on the implementation of multimode mobile systems. It is assumed that the reader has a good RF background, basic knowledge of signal and communication theory, and fundamentals of analog and mixed signal circuits.

Completion of this book is the result of helps and encouragement from many individuals, to all of whom I express my sincerest thanks and appreciation. To Dr. Bjorn Bjerede, I wish to pay tribute for his constant encouragement, many valuable inputs and technical discussions during the course of writing this book. I am deeply indebted to Professor Peter Asbeck of UCSD, for his time to review the whole manuscript, and his very useful comments and helpful suggestions. Much of my RF knowledge and analysis skill were learned and developed when I was engaged in research at the Center for Electromagnetic Wave Theory and Application headed by Professor Kong, in the Research Laboratory of Electronics at MIT. I would like to express my sincerest gratitude to Professor Jin Au Kong of MIT for his creating and providing an academic environment in which there is much knowledge sharing, learning, and searching. My knowledge of wireless communication systems and RF transceivers was learned and accumulated from practical designs when working for wireless companies, such as, PCSI,

Rockwell Semiconductor Systems and Torrey Communications, and especially for Nokia. I would like specially to thank my ex-colleague Dr. Leon Lin for initiating some Matlab programs for nonlinear system simulation. Within Nokia I wish to gratefully acknowledge managers for providing the opportunity for me to work on RF system design of multiple R&D projects, and to thank colleagues for sharing their knowledge and lab results. Special thanks are given to Mr. Greg Sutton and Dr. George Cunningham for their time to review some of the text.

In addition, I would like to express my gratitude to technical reviewers, Dr. Sule Ozev, Prof. Osama Wadie Ata, Mr. Paul D. Ewing, Dr. Rolf Vogt, and Dr. Jaber Khoja for their time and very useful comments. I also wish to thank the staff of Springer, especially Editorial Director, Mr. Alex Greene and his assistant Ms. Melissa Guasch for their effective supports and advices. Finally yet importantly, I would like to thank my wife and family for understanding, patience, and unwavering support.

Qizheng Gu
April, 2005

Chapter 1

Introduction

1.1. Wireless Systems

In wireless systems, the connection between equipments, such as a mobile station and a base station, is by means of electromagnetic waves instead of a cable or a wire. In this sense, optical and infrared communications systems are also wireless systems. However, the wireless systems discussed in this book are only those based on the connection medium of *radio frequency* (*RF*) electromagnetic waves. The present practicable limits of radio frequency are roughly 10 kHz to 100 GHz [1]. Wireless systems at present commonly operate in hundreds MHz or a few GHz frequency. Electromagnetic waves with a frequency in these regions have a propagation distance with an acceptable attenuation and a good penetrating capability through buildings and vehicles and are able to carry wide-band signals.

1.1.1. Mobile Communications Systems

Mobile communications started in 1920s, but the real mobile communication era began in the early 1980s. The cellular mobile system, *advanced mobile phone service* (*AMPS*), first operated for commercial telecommunication service in the United States in 1983. All the first-generation mobile communications systems — including AMPS, *TACS* (*total access communications system*) in the U.K., *NTT* (*Nippon Telephone and Telegraph*) systems in Japan, and *NMT* (*Nordic Mobile Telephones*) in Europe — are analog systems.

The second-generation mobile systems — digital mobile systems — were introduced in late 1980s. There are several competing worldwide digital standards, such as the *global system for mobile communications* (*GSM*), IS-95/98 *code-division multiple-access system* (*CDMA*), IS-136 *time-division multiple-access system* (called as U.S. *TDMA* or *D-AMPS*), and the *personal digital cellular* (*PDC*) system in Japan. GSM systems normally operate in the 900 MHz frequency band. The GSM specification translated to higher frequency 1800 MHz band is known as the *digital communication system* (*DCS*) but also called as 1800 MHz GSM. IS-95/98 CDMA systems and IS-136 DAMPS usually operate in either 800 MHZ cellular band or 1900 MHz PCS band. The second-generation mobile systems, as the first-generation systems, are still used mainly for voice communications. However, the efforts to bring the data capability to existing mobile systems in the wireless industry have not slowed down. The *general package radio system* (*GPRS*) and the *enhanced data rate for GSM evaluation* (*EDGE*), so-called 2.5-generation technologies were then born one after another. An EDGE up to 384 kbps is an evolutionary technology for GSM systems. It allows these systems to provide voice, data, Internet, and other connectivity solutions.

The need for more voice capacity to accommodate more callers reliably and to generate more revenue through increased billable minutes of use of high-speed data services, 384 kbps mobile and 2 MHz fixed, to cater to the needs of an increasing mobile user community, corporate local network access, wireless Internet access, and global roaming have pushed the limits of existing 2G networks [2]. In late 1990s, the third-generation systems, cdma2000_1x and *wide-band code division multiple access* (*WCDMA*) systems, emerged. The new-generation systems are able to support all these needs and to provide video and multimedia services as well as voice communications.

1.1.2. Wireless Local Area Network (WLAN)

Wireless LAN is seen as the technology that will enable the most convenient link between existing wired networks and portable computing and communications equipments, such as laptop computers and *personal digital assistants* (*PDAs*), at the office, hotel, company, or campus level. Building company- or campuswide data communications through the WLAN can reduce the need for wiring among several buildings. In general, the applications of the WLAN systems can be simply between two computers or between a computer and a wired network, all the way up to a complete network with many users and a great number of data paths.

In mid 1997, the Institute of Electrical and Electronics Engineers (IEEE) finalized the initial standard for WLANs, IEEE 802.11. This standard specifies a license-free (ISM band) 2.4 GHz operating frequency with data rates of 1 and 2 Mbps using a direct sequence or a frequency-hopping spread spectrum. The 802.11 is not a single but a family of standards addressing WLAN. The IEEE 802.11a standard defines a WLAN system based on the *orthogonal frequency division multiplexing* (*OFDM*) technology that splits an information signal across 52 separate subcarriers to provide a transmission-of-data rate up to 54 Mbps and throughputs over 24.3 Mpbs, and the operating frequency of the system is in the license-free (UNII band) 5.15 to 5.35 GHz and 5.725 to 5.875 GHz bands. The 802.11a devices, as the other 802.11 based devices, share the channel on a *time-division multiple-access* (*TDMA*) basis. The 802.11b specifies a WLAN system using *direct sequence spread spectrum* (*DSSS*) to achieve the maximum 11 Mbps transmission-of-data rate in 2.4 GHz band. To benefit from the high data rates in 2.4 GHz band, a new standard, 802.11g, that is based on the OFDM and fully backward compatible with the 802.11b standard was developed in early 2000. The high-data-rate WLANs like the 802.11a and 802.11g systems satisfy requirements of multimedia applications including streaming HDTV-quality video in the home, high-speed Internet, and file transfer.

The supplement standards, 802.11e, 802.11f, 802.11h, and 802.11i are defined to enhance the capability of IEEE 802.11-based WLANs. The 802.11e is for enhancements of the *quality of service* (*QOS*). The 802.11f provides a recommended practice for an interaccess-point protocol. The 802.11h extends the spectrum and transmit power management at 5 GHz for European operation. The 802.11i enhances MAC layer security.

The wireless local area networking emerged from the communication market in late 1990s and now has become one of major forces driving wireless communications technologies.

1.1.3. Bluetooth

Bluetooth is a *wireless personal area network* (*WPAN*) and was intended for the short distance (~ 10 m) connectivity and communications between devices in a room where cables had been used. Bluetooth operates in the unlicensed 2.4 GHz *industrial, scientific, and medical* (*ISM*) band. This band is 83.5 MHz wide, beginning at 2.4 GHz and ending at 2.4835 GHz. The *frequency-hopping spectrum spread* (*FHSS*) technique is used in Bluetooth. Signals of Bluetooth devices hop in a pseudo random manner among 79 defined channels spaced at 1 MHz. In this sense, Bluetooth occupies the entire ISM band, but at any instant only a small portion of the band (~1 MHz) is ever occupied. The frequency-hopping rate of the Bluetooth signal is 1600 hops per second.

The Bluetooth device should have low cost and low current consumption. It supports only up to a data rate of 720 kbps. To maintain its low current consumption and cost characteristics, its typical transmission power is around 1 mW or 0 dBm. The Bluetooth receiver has sensitivity –70 dBm or better. The minimum performance of a Bluetooth radio is defined in [2].

The main application of the Bluetooth is pervasive connectivity among a myriad of small office- or home-located wireless devices, including Bluetooth-enabled computers, telephones, home environment, security and small-ffice or home-management systems. Furthermore, some of these devices will have Bluetooth-enabled subdevices, such as wireless cameras, keyboards, mice, headset, and microphones. All these devices are wirelessly interconnected using the protocols of the Bluetooth.

A wireless personal area network (WPAN) standard, IEEE 802.15.1, has been developed based on the Bluetooth protocol. Since the Bluetooth, WPAN and the 802.11b/g WLAN operate in the same 2.4 GHz ISM frequency band, to alleviate potential problems with the coexistence interference, the coexistence solutions for these systems have been extensively discussed in the industry and proposed by the IEEE 802. 15.2 coexistence task group, Bluetooth special interest group and 802.11 working group [3-4].

1.1.4. Global Positioning System (GPS)

The *global positioning system (GPS)* is made up of 24 satellites, which orbit 12,000 miles above the earth, constantly transmitting the precise time and their position in spaces. GPS receivers use triangulation of signals from the satellites to determine their precise locations on earth. GPS satellites know their location in space and GPS receivers can determine their distance from a satellite by using the travel time of a radio message from the satellite to the receiver.

Commercial GPS operates at the frequency of 1575.42 MHz and a 1.023 Mchip/sec *pseudo noise* (*PN*) sequence spread spectrum is used to spread its ranging signal resulting in a bandwidth 2.046 MHz. The sensitivity of a stand-alone GPS receiver is –130 dBm or better. The commercial GPS is commonly used in vehicle-tracking systems and navigation systems for automobile driving or boat sailing. Embedding GPS capability in mobile stations has become a Federal Communications Commision (FCC) requirement. In the mid-1990s, the FCC required that wireless mobile stations should embed GPS capability and the wireless service providers include features that would make the *Enhanced 911* (*E911*) available to their customers. In this application, the wireless network assistant GPS receiver must have a close to –150 dBm sensitivity [5]. The GPS may become a technology for everyday convenience. GPS-capable mobile computers or PDAs and portable GPS devices will advise the user of the nearest hotel, restaurant, point of interest, or gas station [6].

1.1.5. Ultra Wide-band Communications

The *ultra wide-band* (*UWB*) technology will be used in communications for the wireless personal area networks (WPAN) typically operating within ~10 meters [7]. The USB was defined as any signal that occupies more than 500 MHz in the 3.1 to 10.6 GHz band. The Federal Communications Commission's ruling limits UWB EIRP power to –41.25 dBm/MHz over the above band, and the emission limits vary according to frequency bands, such as rolling-down to –75 dBm/MHz in the GPS band.

In the UWB communications, two technical approaches are commonly used. One is based on 'impulses,' actually extremely short digital pulses in the subnanosecond range (1 to 1000 picosecond). To carry information, impulse trains can be modulated with either position, amplitude, or phase. The pulses can be modulated directly by the base-band signal instead of using a high-frequency carrier. Another is the multiband approach. The concept is to utilize multiple-frequency bands (say, 528 MHz for each band [8]) to efficiently use the UWB spectrum by frequency hopping over multiple UWB bands. These signals do not interfere with each other because they operate at different frequency bands within the UWB spectrum. Each of the UWB signals can be modulated to achieve a very high data rate. Several digital modulation techniques commonly employed in wireless communications, such as OFDM, can be used on each individual UWB signal. The advantage of the multibands approach is that they are extremely adaptive and scalable. The number of bands used in communications can vary. The bands can dynamically be adjusted to remove the affected bands with interference or to avoid using the band already utilized by another service.

The ultra wide bandwidth of UWB signals potentially has very high data rates of up to 480 Mpbs. It is worth knowing that based on Shannon's information theory, the information-carrying capacity is linearly proportional to bandwidth and logarithmically with power, making it very attractive to increase the bandwidth for achieving a high data rate. Limiting UWB to a very short distance is the key as it isolates the wide-band transmission to a very local area. The UWB is being targeted at WPANs, audio and video distribution within the home, as well as a cable replacement option for USB and FireWire. The standard of the WPAN based on the UWB technology, IEEE P802.15.3a, requires a data rate of 110 Mpbs at 10 meters with up to 4 bands and a power consumption of less than 100 mW.

The rates will be scaled up to 200 Mpbs at 5 m, with rates increasing down past 4 m [9].

1.2. System Design Convergence

Convergence in the wireless worlds of communications, computing, global positioning systems, and consumer electronic devices is an irresistible trend today. The growing complexity of today's wireless products results largely from the rapid convergence of traditionally separate technologies. One inevitable consequence of the convergence is the impending integration of voice, data, image, video, music, the Internet, instant messaging, home automation, and GPS.

The design task in this converging world is tough to do because of the merging of once-distinct systems. The best way to overcome the challenges caused by the converging technologies is to successfully incorporate RF analog-digital, hardware-software, *system-on-a-chip* (*SoC*), and *printed-circuit-board* (*PCB*) designs. RF analog systems must be designed in sequence with digital systems, not as a separate exercise. The blurring of distinctions and proper partitioning between hardware and software make it critical to architect the system properly before deciding on the implementation. Increased complexity and the shrinking size of wireless products have resulted in designing and using highly integrated SoCs to reduce number of discrete devices on the PCBs as possible [10–11].

The age in which each of us lived in our own cozy design environment, caring little about what was going on in others, has passed [11]. Today, the key to success in our wireless system and product designs is knowledge proliferation across technical disciplines. There is no exception for the RF system design either. The design of the RF system in wireless products must cross technical boundaries to incorporate digital, DSP, and SoC designs.

The early cellular phone looked like and felt like a brick. Today's multiband and multimode mobile phone can be kept in a small pocket and has multiple functions — such as GPS-based E911, Bluetooth-enabled short distance connectivity, and digital camera — other than the voice and data communications. It is impossible to shrink the multi-function mobile phone into a pocket size without using highly *integrated circuits* (*ICs*), i.e., SoC solutions. The highly integrated RF analog chips that can be considered as SoCs came to the market later than digital BB ICs in the middle of 1990's. For instance, the so-called α chip set consisting of one digital base-band chip and four RF analog chips for the *personal handy-phone system* (*PHS*) mobile stations was developed by Pacific Communication Science Inc. in

1994. Actually, the SoC not only makes a complex wireless product able to fit in a small-form factor but also is a unique enabler for implementation of the *direct conversion architecture receiver*. Since a transceiver is now built as a system of an IC chip, RF system design of the transceiver is tightly associated with the development of relevant RF analog ICs. The relevant ICs must be designed and developed according to the selected radio architecture, given system partitioning and the defined specifications of individual stages that resulted from the RF system design.

The RF system is actually a subsystem of the overall wireless digital transceiver. It is apparent that the performance of the digital base-band and the RF analog systems of the transceiver will affect each other. The digital base-band demodulator performance and processing gain determine the maximum allowed *noise figure* of the RF receiver for a defined sensitivity. The group delay distortion of channel filtering in the RF receiver chain impacts minimum requirement of the *signal-to-noise/interference ratio* (*SNIR*) to achieve a certain data *bit error rate* (*BER*). The RF receiver *automatic gain control* (*AGC*) loop is closed in the digital *base-band* (*BB*) since the received signal strength is measured in the digital domain instead of using analog power detector as in old days. On the transmitter side, the transmission power level is entirely controlled by the digital base-band, and the DC offset and the imbalance between *in-phase* (*I*) and *quadrature* (*Q*) channels in the transmitter analog block are compensated by means of adjusting BB digital signal level in the I or Q channel. The interfaces between the RF and digital BB are the *analog-to-digital converter* (*ADC*) in the receiver and the *digital-to-analog converter* (*DAC*) in the transmitter, respectively. The dynamic range or resolution of the ADC and the DAC impacts the gain control range and filtering requirements of the RF receiver and transmitter. The RF system design, thus, together with the digital BB design (or what is referred to as RF-BB co-design), becomes necessary, and to be competent for this a qualified RF designer must possess enough knowledge in the digital base-band system and modern communications theory.

Another area that needs to be seriously considered in the RF system design is the *digital signal processing* (*DSP*) and associated software. Today, RF design demands more and more supports from the DSP. The RF analog block gain controls of transceiver AGC loops are usually executed by means of the DSP embedded in the digital base-band. Operating mode change, frequency band switch, and/or channel selection of a multimode and multiband wireless transceiver are carried out through software. In the wireless mobile station, the DSP is also heavily used in power management to dynamically control the power amplifier and IC biases according to transmission power. Some radio architectures, such as the direct conversion and low IF receivers, need more supports from the DSP than ever requested by the superheterodyne receiver. These architecture receivers without the DSP cooperation to compensate DC offset, to correct I and Q channel imbalance, and/or to run more complicated AGC system cannot perform properly as expected in the best case or even may not work in the worst case. The DSP and software will play a much more important role in the future *software-defined radio* (*SDR*), in which the ADC and DAC have been pushed as close to the antenna as possible and the RF front-end including RF filters and antennas are programmable as well as the digital base-band.

We are facing all these challenges in RF system-design convergence. In this book, the RF system of wireless transceivers will be developed based on highly integrated RF analog integrated circuits. The fundamental approaches and formulas presented in this book, however, can be definitely applied in discrete RF system design. To cope with the design convergence, fundamentals of the digital base-band system are introduced in the next chapter. The design incorporating the DSP will be described wherever it is needed.

1.3. Organization of This Book

This book consists of six chapters. It systematically describes RF system design methodology for transceivers used in various wireless systems, particularly for those in mobile stations. Practically used radio architectures are discussed in detail. RF receiver and transmitter system designs are firmly backed by relevant formulas, and simple but efficient design tools for RF receiver and transmitter can be then developed in terms of the Excel spreadsheet or the Matlab program. Almost all examples in this book are based on mobile stations of different protocol wireless systems.

Fundamentals of the general system and the digital base-band are introduced in Chapter 2. Understanding them is very significant in today's RF system design. Linear system theory is the foundation for analysis and design of RF and digital base-band systems. The system analysis and design can be carried out either in the time domain or in the frequency domain in terms of Fourier transformations. Nonlinear modeling and simulation approaches, which are important for evaluating performance of power amplifier and other nonlinear devices in the transmitter, are addressed. Noise and random processes play substantial roles in communications, and they are extensively discussed in Chapter 2. The last section of this chapter presents digital base-band system related topics including the sampling theorem and process, jitter and quantizing noises, commonly used modulation schemes, pulse shaping technology and intersymbol interference, error probability of detection and carrier-to-noise ratio estimation for achieving certain bit or frame error rate.

Chapter 3 mainly discusses different radio architectures and their design considerations. Selection of radio architecture is the first thing to do when developing an RF transceiver. The architecture discussions in this chapter are based on mobile stations adopted in wireless communication systems, such as GSM, CDMA, etc., but they are generally also applicable to other wireless transceiver. The most popularly used radio architecture was the superheterodyne transceivers. It has outstanding performance in receiver sensitivity, selectivity, and current consumption, and therefore it was commonly employed in all kinds of wireless systems. The direct-conversion or so-called zero IF, architecture transceiver is currently becoming more popularly used in wireless than the superheterodyne radio. Its extraordinary advantages and technical challenges are discussed. The modern RFIC and DSP technologies make the direct-conversion architecture implementation feasible. The third radio architecture addressed in this chapter is the low IF one. This architecture overcomes the technical issues of the direct conversion radio at a cost of more difficulty in obtaining needed image rejection. Today most GPS receivers are using low IF architecture, and some GSM mobile station receivers employ this architecture too under certain DSP supports. The last radio architecture is based on band-pass sampling, which steps further forward to the software-defined radio (SDR). In this architecture, the ADC and the DAC play the RF quadrature down- and up-converter roles, and most of analog function blocks are moved to the digital domain except the RF front-end, i.e., RF filters, *low noise amplifiers* (*LNA*) in the receiver, and power amplifiers in the transmitter.

Receiver RF system analysis and design are discussed in Chapter 4. Formulas for receiver performance evaluation are derived and presented in an individual section of the receiver key parameter analysis. They are the foundation of the receiver system design. Noise figure of a receiver is first analyzed since it determines one of the most important receiver parameters, sensitivity. Impacts of the transmitter power leakage and antenna mismatch on the receiver sensitivity are also analyzed. The relationship between the receiver linearity and the intermodulation spurious response attenuation is discussed in a generic way. In the CDMA mobile station receiver, there is a special requirement referred to as single-tone desensitization, which results from the AM transmission leakage cross-modulating a strong interference tone caused by the third-order nonlinearity of the receiver LNA. An approximate estimation approach of the single-tone desensitization is presented in the fourth subsection. The selectivity of adjacent and alternate channels is a commonly used specification in the wireless systems with channelization. They are determined mainly by the phase noise of local oscillators at the adjacent and the alternate channel frequencies and at the receiver filtering characteristics as analyzed in Section 4.5. The automatic gain control (AGC) system has substantial importance for any configuration receiver. The basic design principles of a receiver AGC loop are discussed. The last section describes the receiver system design method and performance evaluation approach. A Matlab program for the receiver performance evaluation is provided in the appendix of this chapter.

Chapter 5 discusses transmitter system analysis and design. Similar to the previous chapter, formulas for transmitter performance evaluation are provided in individual sections of transmitter key parameter analysis. However, in the transmitter case, simulation instead of calculation may be necessary to evaluating nonlinear effects of devices in the transmitter chain on some transmitter parameters. Transmission power and spectrum are first discussed. In the successive sections, calculation approaches of modulation accuracy or so called waveform quality factor of a transmission signal are presented, and main factors that degrade the modulation accuracy are analyzed. Another key parameter of a transmitter is the adjacent/alternate channel power. The main contributor to the adjacent/alternate channel power is the nonlinearity of the transmitter chain. Simulation methods and approximate formulas for the adjacent/alternate channel power calculation are provided in the fourth section of this chapter. Out-of-channel band noise and spurious emissions of a transmitter are usually interference sources to adjacent channel mobile stations and other system equipments, and their level is tightly restricted. The calculation of noise and spurious emission of the transmitter is introduced in the fifth section. The transmitter AGC is usually incorporated with power-management controls through adjusting device bias or switching the device on or off. In Section 5.6, a preliminary discussion of the transmitter AGC and power management is presented. The final section of this chapter describes the transmitter system design considerations including architecture comparison and system block partitioning. A Matlab program for calculating adjacent/alternate channel power can be found in Appendix 5C.

The last chapter of this book provides RF system design examples of wireless mobile transceivers. A system design of multi-mode and multi-band superheterodyne transceiver is first discussed. The designed mobile transceiver is able to operate in GSM (GPRS), TDMA, and AMPS systems and to run in the 800 MHz Cellular band and 1900 MHz PCS band. The second application example is an RF system design dedicated to the CDMA direct conversion transceiver. Among wireless mobile communication systems, the CDMA system has the toughest performance requirements on its mobile stations as a result of its complexity and operation bandwidth allocations. It discusses in detail how the challenges of CDMA direct-conversion transceiver can be efficiently overcome through a proper RF system design.

References

[1]. F. Jay, ed., *IEEE Standard Directory of Electrical and Electronics Terms*, Fourth Edition, IEEE, Inc. New York, NY, Nov. 1988.

[2]. Bluetooth Special Interest Group (SIG), 'Specification of the Bluetooth System version 1.2, Part A: Radio Specification,' pp. 28 - 46, May 2003

[3]. Bob Heile, "Living in Harmony: Co-Existence at 2.4-GHZ," *Communication Systems Design*, vol. 7, no. 2, pp. 70–73, Feb. 2001.

[4]. O. Eliezer and M. Shoemake, "Bluetooth and Wi-Fi Coexistence Schemes Strive to Avoid Chaos," *RF Design*, pp. 55–72, Nov. 2001.

[5]. TIA-916, *Recommended Minimum Performance Specification for TIA/EIA/IS-801-1 Spread Spectrum Mobile Stations*, April 2002.

[6]. R. Lesser, "GPS: The Next VCR or Microwave," *RF Design*, p. 57, Mar. 1998.

[7]. S. Roy et al., "Ultrawideband Radio Design: The Promise of High-Speed, Short-Range Wireless Connectivity," *Proceedings of IEEE*, vol. 92, no. 2, pp. 295-311, Feb. 2004.

[8]. A. Batra, "Multi-Band OFDM Physical Layer Proposal," *IEEE P80.15 Working Group for WPANs*, Sept. 2003.

[9]. P. Mannion, "Ultrawideband Radio Set to Redefine Wireless Signaling," *EE Times*, Sept. 2002.

[10]. J. Blyler, "Get a High-Level View of Wireless Design," *Wireless Systems Design*, vol. 7, no. 2, pp. 21–28, Feb. 2002.

[11]. R. Bingham, "Managing Design-Chain Convergence Is a Must," *Wireless Systems Design*, vol. 7, no. 3, p. 17, Mar. 2002.

Associated References

[1]. M. J. Riezenmam, "The Rebirth of Radio," *IEEE Spectrum*, pp. 62–64, January 2001.

[2]. J. Tomas et al., "A New Industrial Approach Compatible with Microelectronics Education: Application to an RF System Design," *1999 IEEE International Conference on Microelectronic Systems Education*, pp. 37 – 38, July 1999.

[3]. B. Nair, "3G and Beyond: The Future of Wireless Technologies," *RF Design*, pp. 56–70, Feb. 2001.
[4]. K. Hansen, "Wireless RF Design Challenges," *2003 IEEE Radio Frequency Integrated Circuits Symposium*, pp. 3–7, June 2003.
[5]. A.A. Abidi, "RF CMOS Comes of Age," *IEEE Microwave Magazine*, pp. 47–60, Dec. 2003.
[6]. K. Lim, S. Pinel et al., "RF-System-On-Package (SOP) for Wireless Communications," *IEEE Microwave Magazine*, pp. 88–99, March 2002.
[7]. V. Loukusa et al., "Systems on Chips Design: System Manufacturer Point of View," *Proceedings of Design, Automation and Test in Europe Conference and Exhibition*, vol. 3, pp. 3–4, Feb. 2004.
[8]. J. Lodge and V. Szwarc, "The Digital Implementation of Radio," *1992 IEEE Global Telecommunications Conference*, vol. 1, pp. 462 – 466, Dec. 1992.
[9]. G. Miller, " Adding GPS Applications to an Existing Design," *RF Design*, pp. 50–57, March 1998.
[10]. T. Rao, "High-Speed Packet Service Kick-Starts Migration to 3G," *Communication Systems Design*, vol. 9, no. 9, Sept. 2003.
[11]. H. Honkasalo et al., "WCDMA and WLNA for 3G and Beyond," *IEEE Wireless Communications*, pp. 14–18, April 2002.
[12]. R. Steele, "Beyond 3G," *2000 International Zurich Seminar on Broadband Communications*, pp. 1 – 7, Feb. 2000.
[13]. J. Craninckx and S. Donnay, "4G Terminals: How Are We Going to Design Them?" *Proceedings of 2003 Design Automation Conference*, pp. 79–84, June 2002.
[14]. J. Klein, "RF Planning for Broadband Wireless Access Network," *RF Design*, pp. 54–62, Sept. 2000.
[15]. D. M. Pearson, "SDR (System Defined Radio): How Do We Get There from Here?" *2001 Military Communication Conference*, vol. 1, pp. 571–575, Oct. 2001.
[16]. J. Sifi and N. Kanaglekar, "Simulation Tools Converge on Large RFICs," *Communication Systems Design*, vol. 8, no. 6, June 2002.

Chapter 2

Fundamentals of System Design

2.1. Linear Systems and Transformations

2.1.1. Linear System

A *system* refers to any entity that produces a unique output in response to a legitimate input. In wireless communications, communication channels, base stations, mobile stations, receivers, transmitters, frequency synthesizers, and even filters all are physical systems with different complexity. Mathematically, a system can be described in

$$y(t) = \boldsymbol{T}[x(t)], \tag{2.1.1}$$

where $x(t)$ is the input (or excitation), $y(t)$ is the output (or response), t is an independent variable usually representing time, and $\boldsymbol{T}$ is the operation performed by the system. Then the system is also viewed as a transformation (or mapping) of $x(t)$ into $y(t)$.

A system is linear if and only if the principle of superposition holds — i.e., its output can be expressed as a linear combination of responses to individual inputs:

$$\boldsymbol{T}[\alpha_1 x_1(t) + \alpha_2 x_2(t)] = \alpha_1 \boldsymbol{T}[x_1(t)] + \alpha_2 \boldsymbol{T}[x_2(t)], \tag{2.1.2}$$

where α_1 and α_2 are arbitrary scalars. A system that does not satisfy the superposition relationship (2.1.2) is classified as nonlinear.

There is a class of linear systems called *linear time-invariant* (*LTI*) systems that play particularly important role in communication system theory and design. A system is referred to as time-invariant if a time shift, τ, in the input signal $x(t-\tau)$ causes the same time shift in the output signal $y(t)$. This is expressed as

$$y(t - \tau) = \mathbf{L}[x(t - \tau)], \tag{2.1.3}$$

where $\boldsymbol{L}$ is LTI system operator. LTI systems provide accurate models for a large number of building blocks in a communication system. Important examples of LTI systems include many of the basic devices used in base and mobile stations, such as filters, isolators, duplexers, and amplifiers operating in their linear region.

The LTI system can be completely characterized by what is referred to as the impulse response of the system, which is defined as the response of the system when the input is an impulse signal $\delta(t)$ — i.e.,

$$h(t) = \boldsymbol{L}[\delta(t)] . \tag{2.1.4}$$

The impulse signal $\delta(t)$ also known as impulse function or Dirac delta function, is defined by its effect on a test function $\phi(t)$ as follows:

$$\int_{-\infty}^{\infty} \phi(t)\delta(t)dt = \phi(0) , \tag{2.1.5a}$$

and in general, a delayed delta function $\delta(t\text{-}t_o)$ is defined by

$$\int_{-\infty}^{\infty} \phi(t)\delta(t-t_o)dt = \phi(t_o) . \tag{2.1.5b}$$

The $\delta(t)$ function possesses the following properties:

$$\delta(t) = \begin{cases} 0 & t \neq 0 \\ \infty & t = 0 \end{cases} \tag{2.1.6}$$

and

$$\int_{\substack{-\varepsilon \\ \varepsilon \to 0}}^{\varepsilon} \delta(t)dt = 1 , \tag{2.1.7}$$

where ε is a small value approaching zero.

The time response of LTI systems to their inputs can be simply derived by performing convolution of the input and the impulse response of the system. From (2.1.5b), the input $x(t)$ can be expressed as

$$x(t) = \int_{-\infty}^{\infty} x(\tau)\delta(t-\tau)d\tau . \tag{2.1.8}$$

The response $y(t)$ of a LTI system to an input $x(t)$ is

$$y(t) = \boldsymbol{L}[x(t)] = \boldsymbol{L}\left[\int_{-\infty}^{\infty} x(\tau)\delta(t-\tau)d\tau\right] = \int_{-\infty}^{\infty} x(\tau)\boldsymbol{L}[\delta(t-\tau)]d\tau$$

$$= \int_{-\infty}^{\infty} x(\tau)h(t-\tau)d\tau = x(t) * h(t). \tag{2.1.9}$$

In (2.1.9), $*$ denotes the convolution.

It is worth while to note an example of the response of a LTI system with an impulse response $h(t)$ to a complex exponential input signal $x(t) = e^{j\omega t}$, where $j = \sqrt{-1}$ and $\omega = 2\pi f$ is referred to as *angular frequency*. From (2.1.9), we have

$$\begin{aligned} y(t) = \mathbf{L}[e^{j\omega t}] &= \int_{-\infty}^{\infty} h(\tau)e^{j2\pi f(t-\tau)}d\tau \\ &= e^{j2\pi ft}\int_{-\infty}^{\infty} h(\tau)e^{-j2\pi f\tau}d\tau = H(f)e^{j2\pi ft}, \end{aligned} \tag{2.1.10}$$

where

$$H(f) = |H(f)|e^{j\angle H(f)} = \int_{-\infty}^{\infty} h(\tau)e^{-j2\pi f\tau}d\tau. \tag{2.1.11}$$

In (2.1.11), $|H(f)|$ and $\angle H(f)$ are amplitude and phase of function $H(f)$, respectively. (2.1.10) tells us that the response of a LTI system to a complex exponential signal with a frequency $f = \omega/2\pi$ is still a complex exponential signal with the same frequency. Thus, $e^{j\omega t}$ is the eigenfunction of the LTI system and $H(f)$ is the eigenvalue of $\boldsymbol{L}$ associated with $e^{j\omega t}$. This result provides a foundation of Fourier analysis.

2.1.2. Fourier Series and Transformation

Fourier series and transformation convert time-domain signals to frequency domain (or spectral) representations. This conversion, in many cases, provides better insight into the behavior of certain types of systems. Based on Fourier series and transformation, a frequency-domain approach of analyzing LTI systems called Fourier analysis has been developed. This approach makes analyzing some LTI systems in the frequency domain easier than directly in the time domain.

A periodic signal $x(t)$ with a period T_o can be expanded into a complex exponential Fourier series as follows if it satisfies the Dirichlet conditions:§

$$x(t) = \sum_{k=-\infty}^{\infty} x_k e^{jk\cdot 2\pi f_o t}, \qquad f_o = \frac{1}{T_o}, \tag{2.1.12}$$

where x_k are known as the *complex Fourier coefficients* and are given by

$$x_k = \frac{1}{T_o}\int_{T_o} x(t)e^{-jk2\pi f_o t}\,dt \tag{2.1.13}$$

where $\int_{T_o}$ denotes the integral over any one period. The complex coefficients x_k can be expressed as

$$x_k = |x_k| e^{j\varphi_k}. \tag{2.1.14}$$

where $|x_k|$ and φ_k are amplitude and phase of the kth harmonic of $x(t)$. A plot of $|x_k|$ versus the frequency is called *amplitude spectrum* of the periodic signal $x(t)$, and a plot of φ_k versus frequency is referred to phase spectrum of $x(t)$. When $x(t)$ is real, from (2.1.1.3) we have

$$x_{-k} = x_k^* \tag{2.1.15}$$

In (2.1.14), x_k^* means the complex conjugate of x_k.

The average power of a periodic signal $x(t)$ over any period is defined as

$$P = \frac{1}{T_o}\int_{T_o} |x(t)|^2\,dt \tag{2.1.16}$$

---

§ 1. $x(t)$ is absolutely integrable over its period — i.e., $\int_{T_o} |x(t)|dt < \infty$; 2. The number of maximum and minimum of $x(t)$ is finite and 3. The number of discontinuities of $x(t)$ in each period is finite.

Substituting (2.1.12) into the above equation, we obtain the following expression:

$$\frac{1}{T_o}\int_{T_o}|x(t)|^2 dt = \sum_{k=-\infty}^{\infty}|x_k|^2 . \tag{2.1.17}$$

This is called *Parseval's identity*. It says that the average power of a period signal $x(t)$ is equal to the sum of the power contents of its harmonics or its spectral lines.

For a nonperiod signal $x(t)$, we can image that the signal has a $T_o \rightarrow \infty$. Assuming $x(t)$ satisfies the Dirichlet conditions given in the foot note of the previous page, we can introduce new variables $df = 1/T_o$ and $2\pi f = \omega = k\omega_o$ to (2.1.12) and (2.1.13) and obtain

$$x(t) = \int_{-\infty}^{\infty}\left(\int_{-\infty}^{\infty} x(\tau)e^{-j2\pi f\tau}d\tau\right)e^{j2\pi ft}df . \tag{2.1.18}$$

Let us define the Fourier transform of $x(t)$ as

$$X(f) = \int_{-\infty}^{\infty} x(t)e^{-j2\pi ft}dt = \boldsymbol{F}[x(t)]. \tag{2.1.19}$$

Then (2.1.18) becomes the inverse Fourier transform of $X(f)$:

$$x(t) = \int_{-\infty}^{\infty} X(f)e^{j2\pi ft}df = \boldsymbol{F}^{-1}[X(f)], \tag{2.1.20}$$

where notations $\boldsymbol{F}$ and $\boldsymbol{F}^1$ denote the Fourier transform and inverse Fourier transform, respectively. Equations (2.1.19) and (2.1.20) are referred to as a *Fourier transform pair*. In general, $X(f)$ is a complex function, and it is also referred to as voltage spectrum of the signal $x(t)$.

Comparing equations (2.1.8) and (2.1.18), we derive that

$$\delta(t-\tau) = \int_{-\infty}^{\infty} e^{j2\pi f(t-\tau)}df \tag{2.1.21a}$$

or in general

$$\delta(t) = \int_{-\infty}^{\infty} e^{j2\pi f \cdot t} df . \qquad (2.1.21b)$$

Equation (2.1.21) expresses that the impulse signal $x(t) = \delta(t)$ can be resulted from the inverse Fourier transformation of a spectrum $X(f) = 1$ — i.e., the spectrum of the impulse signal $\delta(t)$ is a constant equal to 1, as shown in Fig. 2.1.

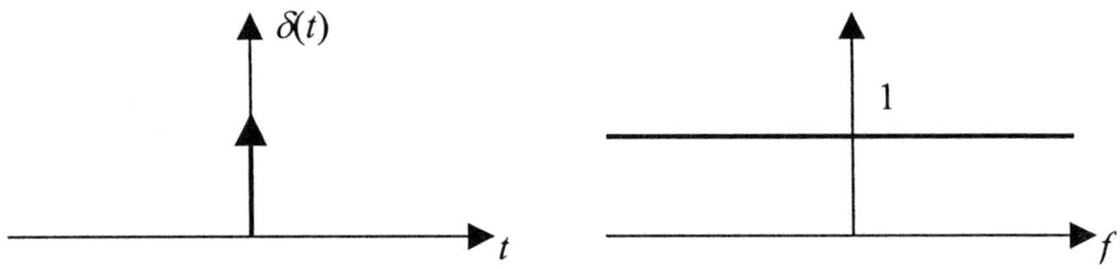

Figure 2.1. Impulse signal and its spectrum

In the following, we present some Fourier transformation properties, which are useful in the RF system design, and the other properties can be found in many textbooks [1–2].

- *Linearity.* The Fourier transform operation is linear. If $x_1(t)$ and $x_2(t)$ possesses Fourier transforms $X_1(f)$ and $X_2(f)$, respectively, then for any two scalars α and β, results in

$$\boldsymbol{F}[\alpha x_1(t) + \beta x_2(t)] = \alpha X_1(f) + \beta X_2(f) . \qquad (2.1.22)$$

- *Convolution.* If the time functions $x(t)$ and $h(t)$ both have Fourier transforms, then

$$\boldsymbol{F}[h(t) * x(t)] = \boldsymbol{F}[h(t)] \cdot \boldsymbol{F}[x(t)] = H(f) \cdot X(f) . \qquad (2.1.23)$$

  The Fourier transform of the convolution of two time functions is the product of their Fourier transforms in the frequency domain.

- *Modulation.* The Fourier transform of a modulated carrier, $x(t)e^{j2\pi f_o t}$, is

$$\boldsymbol{F}[x(t)e^{j2\pi f_o t}] = X(f - f_o) . \qquad (2.1.24)$$

The Fourier transform of an amplitude modulated signal, $x(t)\cos(2\pi f_o t)$, has a form

$$\boldsymbol{F}[x(t)\cos(2\pi f_o t)] = \frac{1}{2}\boldsymbol{F}\left[x(t)(e^{j2\pi f_o t} + e^{-j2\pi f_o t})\right]$$
$$= \frac{1}{2}\left[X(f+f_o) + X(f-f_o)\right].$$

- *Time shifting*. A time shift of t_o from its origin results in a phase shift of $-2\pi f\, t_o$ in the frequency domain — i.e.,

$$\boldsymbol{F}[x(t \quad t_o)] = X(f)e^{-j2\pi f t_o} \tag{2.1.25}$$

- *Parseval's relation*. If $X(f) = \boldsymbol{F}[x(t)]$ and $Y(f) = \boldsymbol{F}[y(t)]$, then

$$\int_{-\infty}^{\infty} x(t)y^*(t)dt = \int_{-\infty}^{\infty} X(f)Y^*(f)df\,. \tag{2.1.26}$$

Note that if $y(t) = x(t)$, we have Parseval's identity for the Fourier transform

$$\int_{-\infty}^{\infty} |x(t)|^2\, dt = \int_{-\infty}^{\infty} |X(f)|^2\, df\,. \tag{2.1.27}$$

This expression is similar to (2.1.17) of the periodic signal. It explains that the energy content of the signal $x(t)$ can be computed by integrating the energy spectral density $|X(f)|^2$ of $x(t)$ over all the frequency.

2.1.3. Frequency Response of LTI Systems

Comparing (2.1.11) with the Fourier transform (2.1.19), $H(f)$ is the Fourier transform of the impulse response $h(t)$ of a LTI system. The Fourier transform of the time response $y(t)$ for a LTI system with an input $x(t)$ is the product of $X(f) = \boldsymbol{F}[x(t)]$ and $H(f)$ as given in (2.1.23) — i.e.,

$$Y(f) = \boldsymbol{F}[h(t) * x(t)] = H(f)X(f)\,. \tag{2.1.28}$$

The function *H(f)* is defined as the transfer function of the LTI system, and it is expressed as the system output frequency response *Y(f)* divided by its input response *X(f)* or the Fourier transform of the impulse response *y*(*t*).

$$H(f) = |H(f)|e^{j\angle H(f)} = \frac{Y(f)}{X(f)} = \int_{-\infty}^{\infty} h(t)e^{-j2\pi ft}\,dt\,. \tag{2.1.29}$$

The transfer function *H(f)* uniquely characterizes a LTI system. $|H(f)|$ is called the *amplitude response* or *gain* of the system, and $\angle H(f)$ the *phase response* in radians. The gain is usually expressed in decibels (dB) by using the definition

$$G(f) = 20\log_{10}|H(f)|\,. \tag{2.1.30}$$

The *bandwidth* of a system is defined as the spectral region of constancy of the amplitude response $|H(f)|$ or the gain $G(f)$ of the system. There are many different bandwidth definitions used in communications, such as the noise bandwidth of a receiver, the equal ripple bandwidth of a Chebyshev filter, etc. A commonly used bandwidth is called the 3 dB bandwidth that is defined as the frequency range over which the amplitude response $|H(f)|$ remains within $1/\sqrt{2}$ times its maximum values — i.e., the frequency range over which the gain drops by less than or equal to 3 dB.

A real LTI system, such as a filter or communication channel, is usually a *dispersive* (frequency-selective) device. The phase response $\angle H(f)$ of the dispersive system is usually a nonlinear function of the frequency. Some delay is introduced into the output in relation to the input of the system when a signal passes through it. Assuming that an input signal consists of a narrow-band signal defined as

$$x(t) = m(t)\cos(2\pi f_o t)\,,$$

where *m(t)* is a low-pass information-bearing signal with its spectrum limited to the frequency interval $|f| \le W$, and $f_o >> W$, we may approximately express $\angle H(f)$ as

$$\angle H(f) \cong \angle H(f_o) + (f - f_o)\frac{\partial \angle H(f)}{\partial f}\Big|_{f=f_o}\,. \tag{2.1.31}$$

Define phase delay τ_p and group delay, τ_g, respectively, as

$$\tau_p = -\frac{\angle H(f_o)}{2\pi f_o} \tag{2.1.32}$$

and

$$\tau_g = -\frac{1}{2\pi}\frac{\partial \angle H(f)}{\partial f}\bigg|_{f=f_o}. \tag{2.1.33}$$

Then (2.1.31) turns into

$$\angle H(f) \cong -2\pi\tau_p - 2\pi(f - f_c)\tau_g. \tag{2.1.34}$$

If the device has a constant gain g_o, its transfer function is

$$H(f) = g_o e^{-j[2\pi f_o \tau_p + 2\pi(f-f_o)\tau_g]}. \tag{2.1.35}$$

It can be proved that the output response of the device to the input $x(t) = m(t)\cos(2\pi f_o t)$ has the form

$$y(t) = g_o m(t-\tau_g)\cos[2\pi f_o(t-\tau_p)].$$

The carrier $\cos(2\pi f_o t)$ is delayed by τ_p seconds, and therefore τ_p represents the phase delay. The information-bearing envelope $m(t)$ is delayed by τ_g seconds, and hence τ_g represents group delay.

2.1.4. Band-Pass to Low-Pass Equivalent Mapping and Hilbert Transform

The signals that we deal with in RF system design are usually modulated carriers, which are processed in band-pass systems. An amplitude- and phase-modulated narrow-band signal can always be represented by means of in-phase and quadrature components as

$$x(t) = a_I(t)\cos 2\pi f_o t - a_Q(t)\sin 2\pi f_o t, \tag{2.1.36}$$

where $a_I(t)$ is the in-phase component, $a_Q(t)$ is the quadrature component, and f_o is the carrier frequency. The spectrum of the signal $x(t)$ is limited to frequencies within a bandwidth $\pm B$ of the carrier frequency f_o, and usually $f_o >> 2B$. For example, among the wireless mobile communication systems, the wide-band CDMA (WCDMA) signal has the broadest bandwidth, $2B$ = 3.78 MHz, but the carrier frequency f_o is within the band 1920 to 2170 MHz in the cellular frequency band.

The systems used in the wireless communications are often band-pass systems with impulse response $h(t)$ and transfer function $H(f)$. The impulse response $h(t)$ of a LTI system, which has a frequency response limited to frequencies within $\pm B$ of the carrier frequency f_o, can also be represented in terms of two quadrature components as

$$h(t) = h_I(t)\cos 2\pi f_o t - h_Q(t)\sin 2\pi f_o t, \tag{2.1.37}$$

where $h_I(t)$ and $h_Q(t)$ are in-phase and quadrature components, respectively.

It is usually too much involved if we directly analyze modulated carriers or directly perform simulation of band-pass systems. This is due to computer-based modern analysis and simulation by using sampled signals and discrete system models. From sampling theorem [3], we know that a discrete model uniquely represents a continuous signal only if the sampling frequency is at least twice the highest frequency in the signal spectrum. In the previous example, the spectrum of the band-pass signal $x(t)$ has a band within $f_o - B \leq f \leq f_o + B$. Thus the sampling frequency has to be $2f_o + 2B$ or higher to obtain an accurate analysis or simulation. This is apparently not an efficient way even if possible.

In fact, the analysis and simulation of the modulated carriers and band-pass systems can be simplified by utilizing their low-pass equivalents. The modulated carrier signal (2.1.36) can be written as

$$x(t) = a(t)\cos[2\pi f_o t + \phi(t)] = \mathrm{Re}[a(t)e^{j\phi(t)}e^{j2\pi f_o t}], \tag{2.1.38}$$

where $a(t) = \sqrt{a_I^2 + a_Q^2}$ is the amplitude modulation, and $\phi(t) = \tan^{-1}\left[a_Q(t)/a_I(t)\right]$ is phase modulation of the signal. The modulation signal on the right side of (2.1.38)

$$\tilde{m}(t) = a(t)\cdot e^{j\phi(t)} \tag{2.1.39}$$

contains all the message information, and is of a low-pass nature. It is referred to as the *complex low-pass equivalent* or the *complex envelope* of the signal. The bandwidth of $\widetilde{m}(t)$ is $2B$, and the narrow-band condition $f_o >> 2B$ is usually satisfied. Under the same condition, the band-pass system with an impulse response $h(t)$ can be evaluated through an equivalent low-pass impulse response

$$\widetilde{h}_L(t) = h_I(t) + jh_Q(t)\,. \qquad (2.1.40)$$

The sampling rate in the analysis or simulation of the low-pass signals and/or systems should be on the order of $2B$.

As will be seen it is possible that the analysis or simulation of the band-pass signals and/or systems is greatly simplified by utilizing their low-pass equivalents. To form a low-pass equivalent of a band pass-signal like $x(t)$ given in (2.1.36), we need a signal $\hat{x}(t)$ similar to $x(t)$ but having a -90° phase shift — i.e.,

$$\hat{x}(t) = a_I(t)\sin 2\pi f_o t + a_Q(t)\cos 2\pi f_o t\,. \qquad (2.1.41)$$

A complex function referred to as *analytic signal* of $x(t)$ or as *pre-envelope* of $x(t)$ is defined as [1]

$$x_+(t) = x(t) + j\hat{x}(t)\,. \qquad (2.1.42)$$

Substituting (2.1.36) and (2.1.41) into (2.1.42), we obtain the pre-envelope having a form

$$x_+(t) = [a_I(t) + ja_Q(t)]e^{j2\pi f_o t} = a(t)e^{j\phi(t)}e^{j2\pi f_o t}\,. \qquad (2.1.43)$$

It is the same as that in the square brackets on the right side of (2.1.38). Taking the carrier $e^{j2\pi f_o t}$ off from (2.1.43), we obtain the complex low-pass equivalent signal possessing the same form as (2.1.39). The spectrum of $x_+(t)$ is

$$X_+(f) = A_I(f - f_o) + jA_Q(f - f_o)\,. \qquad (2.1.44)$$

The *Hilbert transform* of a time function results in ± 90° phase shift of the function. The Hilbert transform of $x(t)$, which is denoted as $\hat{x}(t)$, is defined by

$$\hat{x}(t) = \boldsymbol{H}[x(t)] = \frac{1}{\pi}\int_{-\infty}^{\infty}\frac{x(\tau)}{t-\tau}d\tau = x(t) * \frac{1}{\pi \cdot t}. \tag{2.1.45}$$

The *inverse Hilbert transform*, by means of which the original function $x(t)$ is recovered from $\hat{x}(t)$, is defined as

$$x(t) = \boldsymbol{H}^{-1}[x(t)] = -\frac{1}{\pi}\int_{-\infty}^{\infty}\frac{\hat{x}(\tau)}{t-\tau}d\tau \tag{2.1.46}$$

Hilbert transform has the following useful properties:

- *Fourier transform.* If $X(f)$ is the Fourier transform $x(t)$, the Fourier transform of the Hilbert transform is

$$\boldsymbol{F}[\hat{x}(t)] = \hat{X}(f) = -jX(f)\,\mathrm{sgn}(f), \tag{2.1.47}$$

where sgn(f) is a sign function

$$\mathrm{sgn}(f) = \begin{cases} -1, & f < 0 \\ 1, & f \geq 0 \end{cases}. \tag{2.1.48}$$

- *Quadrature filtering.* The Hilbert transform of $x(t) = \cos 2\pi f_o t$ is

$$\hat{x}(t) = \boldsymbol{H}[\cos 2\pi f_o t] = \sin 2\pi f_o t\,. \tag{2.1.49}$$

- *Double Hilbert transform.* The Hilbert transform of the Hilbert transform of $x(t)$ is - $x(t)$ — i.e.,

$$\boldsymbol{H}[\hat{x}(t)] = \hat{\hat{x}}(t) = -x(t)\,. \tag{2.1.50}$$

Therefore, from (2.1.49) we have

$$\boldsymbol{H}[\sin 2\pi f_o t] = -\cos 2\pi f_o t\,. \tag{2.1.51}$$

- *Band-limited modulation.* If modulation signals $a_I(t)$ and $a_Q(t)$ have band-limited spectrum — i.e.,

$$A_I(f) = A_Q(f) = 0 \qquad \text{for } |f| \geq 2B$$

— then the Hilbert transform of the modulated signal $x(t) = a_I(t)\cos 2\pi f_o t - a_Q(t)\sin 2\pi f_o t$ is given by (2.1.41)

$$\hat{x}(t) = a_I(t)\sin 2\pi f_o t + a_Q(t)\cos 2\pi f_o t\,, \tag{2.1.41}$$

provided that $f_o \geq 2B$.

- *Orthogonality*. $x(t)$ and its Hilbert transform $\hat{x}(t)$ are orthogonal:

$$x(t) = \int_{-\infty}^{\infty} x(t)\hat{x}(t)dt = 0 \tag{2.1.52}$$

Using the pre-envelope (2.1.43), the modulated signal (2.1.36) can be expressed as

$$x(t) = \mathrm{Re}[x_+(t)] = \mathrm{Re}[\tilde{x}_L(t)e^{j2\pi f_o t}]. \tag{2.1.53}$$

The low-pass equivalent signal or complex envelope is formally defined as

$$\tilde{x}_L(t) = x_+(t)e^{-j2\pi f_o t} = a_I(t) + ja_Q(t) = a(t)e^{j\phi(t)}. \tag{2.1.54}$$

Comparing (2.1.54) with (2.1.39), we know that they are exactly the same. The low-pass equivalent signal is actually the modulation signal.

The spectrum of the low-pass equivalent signal $\tilde{x}_L(t)$ can be derived from the spectrum of the pre-envelope $x_+(t)$. Using (2.1.42) and (2.1.47), we obtain the spectrum of the pre-envelope of $x(t)$ is

$$X_+(f) = X(f) + j\hat{X}(f) = X(f) + j[-j\,\mathrm{sgn}(f)]X(f) = \begin{cases} 2X(f), & f \geq 0 \\ 0, & f < 0 \end{cases} \tag{2.1.55}$$

Therefore, the low-pass equivalent spectrum of the band-pass signal $x(t)$ can be derived to have the following form

$$X_L(f) = A_I(f) + jA_Q(f) = X_+(f + f_o) = \begin{cases} 2X(f + f_o), & f \geq -f_o \\ 0, & f < -f_o\,. \end{cases} \tag{2.1.56}$$

Assuming that the band-pass signal $x(t)$ has a spectrum as shown in Fig. 2.2a, the spectra of $x_+(t)$ and $\widetilde{x}_L(t)$ are depicted in Fig. 2.2b and Fig. 2.2c, respectively.

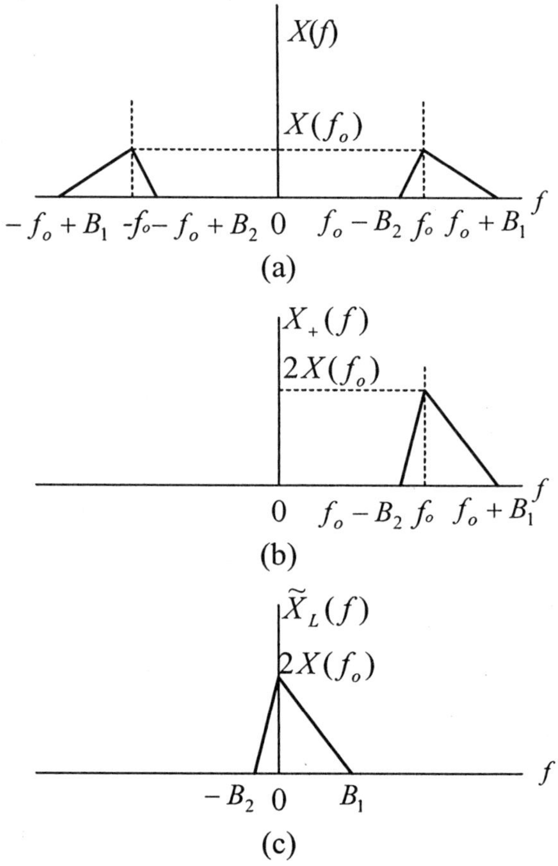

Figure 2.2. (a) Amplitude spectrum of band-pass signal $x(t)$;(b) amplitude spectrum of pre-envelope $x_+(t)$; and (c) amplitude spectrum of low-pass equivalent $\widetilde{x}_L(t)$

Using the low-pass equivalent definition (2.1.54), we can easily derive the low-pass equivalent of the band-pass system (2.1.37) to be

$$\widetilde{h}_L(t) = h_+(t)e^{-j2\pi f_o t} = h_I(t) + jh_Q(t), \tag{2.1.57}$$

where $h_+(t)$ is the pre-envelope of the system impulse response $h(t)$ having the form

$$h_+(t) = h(t) + j\hat{h}(t) . \tag{2.1.58}$$

It is similar to the pre-envelope $x_+(t)$ of band-pass signal. The spectrum of $h_+(t)$ can be expressed as

$$H_+(f) = H(f) + j\hat{H}(f) = \begin{cases} 2H(f) & f \geq 0 \\ 0 & f < 0 \end{cases} \tag{2.1.59}$$

The spectrum (transfer function) of the low-pass equivalent impulse response $\widetilde{h}_L(t)$ should have a similar expression as (2.1.56) for the $\widetilde{x}_L(t)$ spectrum. However, for simplifying the expression of the low-pass equivalent convolution, the transfer function of the $\tilde{h}_L(t)$ is defined as

$$H_L(f) = H_I(f) + jH_Q(f) = \frac{1}{2}H_+(f + f_o) = \begin{cases} H(f + f_o), & f \geq -f_o \\ 0, & f < -f_o . \end{cases} \tag{2.1.60}$$

In Fig. 2.3. the transfer functions of the band-pass LTI system $H(f)$ and its low-pass equivalent system $H_L(f)$ are presented. In practice, the low-pass equivalent transfer function is obtained by truncating the negative frequency component of the band-pass system $H(f)$ and shifting the positive part to the zero axis by an amount equal to the carrier frequency f_o.

When the band-pass system $h(t)$ has a real input $x(t)$, the output $y(t)$ is equal to the convolution of $h(t)$ and $x(t)$ as given in (2.1.9)

$$y(t) = h(t) * x(t) = \int h(t - \tau)x(\tau)d\tau . \tag{2.1.9}$$

and the corresponding expression in the frequency domain is (2.1.28)

$$Y(f) = H(f) \cdot X(f) . \tag{2.1.28}$$

It can be imagined that the spectrum $Y_+(f)$ of the analytic signal $y_+(t)$ of the output response will have a relationship with $Y(f)$ same as the relationship (2.1.55) between $X_+(f)$ and $X(f)$. Thus from (2.1.55), (2.1.59), and (2.1.28) we obtain

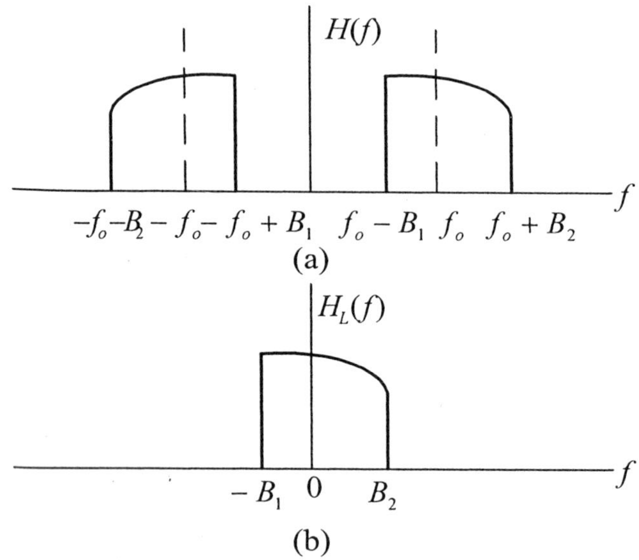

Figure 2.3. (a) Spectrum of band-pass system and (b) spectrum of low-pass equivalent system

$$Y_+(f) = \frac{1}{2} H_+(f) X_+(f) . \tag{2.1.61}$$

Referring to (2.1.56), the spectrum of the low-pass equivalent output response $\widetilde{y}_L(t)$ should have the form

$$Y_L(f) = Y_+(f + f_o) = \begin{cases} 2Y(f + f_o) & f \geq -f_o \\ 0 & f < -f_o . \end{cases} \tag{2.1.62}$$

Substituting (2.1.56), (2.1.60), and (2.1.62) into (2.1.61), we obtain

$$Y_L(f - f_o) = H_L(f - f_o) X_L(f - f_o) \tag{2.1.63}$$

In the time-domain, the above equation has the following form

$$\widetilde{y}_L(t) e^{j2\pi f_o t} = e^{j2\pi f_o t} \int_{-\infty}^{\infty} \widetilde{h}_L(t-\tau) \widetilde{x}_L(\tau) d\tau = e^{j2\pi f_o t} \widetilde{h}_L(t) * \widetilde{x}_L(t) ,$$

or the low-pass equivalent response of the LTI system is

$$\tilde{y}_L(t) = \int_{-\infty}^{\infty} \tilde{h}_L(t-\tau)\tilde{x}_L(\tau)d\tau = \tilde{h}_L(t) * \tilde{x}_L(t) . \quad (2.1.64)$$

The output response of the band-pass system can be easily derived from this low-pass equivalent output by

$$y(t) = \text{Re}[\tilde{y}_L(t)e^{j2\pi f_o t}]$$

The significance of the result is that the analysis or simulation of a complicated band-pass system, which is due to the presence of the multiplying factor $e^{j2\pi f_o t}$, is replaced by an equivalent but much simpler low-pass analysis or simulation. The low-pass equivalent essentially gives us the same outcome as its band-pass counterpart.

2.2 Nonlinear System Representation and Analysis Approaches

A system that does not satisfy the superposition principle as given by (2.1.2) is nonlinear. In reality, all physical systems are nonlinear. A linear system is actually an approximation of a nonlinear system under certain conditions, such as signal strength below certain level. An RF receiver is usually considered as a linear system for a weak desired signal, but it will appear to have nonlinear behavior to strong interference signals when the receiver is in its high gain mode operation. A typical nonlinear device in a mobile station is the power amplifier in its transmitter. Transmitter power amplifiers for mobile stations are usually designed to have good power efficiency while maintaining the required linearity for meeting overall transmitter performance.

An analytical treatment of nonlinear systems is much more difficult than that for linear systems, but it is still possible with some simplification and idealization especially for mildly nonlinear systems such as RF receiver systems. However, for other types of nonlinear systems like the transmitter of mobile stations, it is better to use a simulation approach.

In this book, we address only the nonlinear systems that are memoryless. This is not only for simplification, but more important the nonlinear systems, which we are going to design and analyze, can be well approximated as memoryless. Analysis of nonlinear systems with memory will use a Volterra series that is characterized by a set of time-domain functions called *nonlinear impulse response*, while in the frequency domain, it is characterized by a set of frequency-domain functions called *nonlinear transfer functions*. However, this topic is not within the scope of this book, and readers who are interested in this topic can find it in references [4–7].

2.2.1. Representation of Memoryless Nonlinear Systems

A memoryless system is one whose output $y(t)$ depends only on its input present value $x(t)$. For a memoryless nonlinear system, its output can be generally expressed in a functional form of its input as

$$y(t) = F[x(t)]. \tag{2.2.1}$$

In handling memoryless nonlinear systems, it is quite common to expand the function on the right side of (2.2.1) into a power series:

$$y(t) = F[x(t)] = \sum_{n=1}^{\infty} a_n x^n(t) \cong \sum_{n=1}^{N} a_n x^n(t). \tag{2.2.2}$$

Based on these expansions, we can develop analytical approaches and obtain useful insights into important nonlinear phenomena, such as intermodulation, crossmodulation, compression, and desensitization.

2.2.2. Multiple Input Effects in Nonlinear Systems

In the wireless systems and other communication systems, the input of a receiver used in these systems is almost always the sum of a desired signal and one or more interfering signals (also called *interferers*). These signals interact with the nonlinearities of the receiver to produce additional spectral components not contained in the original input signals. The extra spectral components or responses result from various types of nonlinear effects that are classified as intermodulation, crossmodulation, compression, desensitization, and blocking.

Assuming that the input signal $x(t)$ consists of a desired signal $s_d(t)$ and two interferers $s_{I1}(t)$ and $s_{I2}(t)$, respectively, the output $y(t)$ of a nonlinear system is then

$$y(t) = a_1\left[s_d(t) + s_{I1}(t) + s_{I2}(t)\right] + a_2\left[s_d(t) + s_{I1}(t) + s_{I2}(t)\right]^2 + a_3\left[s_d(t) + s_{I1}(t) + s_{I2}(t)\right]^3 + \cdots \tag{2.2.3}$$

Gain Compression and Desensitization Let us start with a simpler case that the input $x(t)$ contains an unmodulated desired signal $s_d(t)$ and a single tone interference $s_{I1}(t)$— i.e.,

$$\left.\begin{aligned} s_d(t) &= A_d \cos 2\pi f_o t \\ s_{I1}(t) &= A_{I1} \cos 2\pi f_1 t \\ s_{I2}(t) &= 0 \end{aligned}\right\}. \tag{2.2.4}$$

Substituting (2.2.4) into (2.2.3), wc obtain

$$\begin{aligned} y(t) &= a_1\left[A_d \cos 2\pi f_o t + A_{I1} \cos 2\pi f_1 t\right] \\ &\quad + a_2\left[A_d^2 \cos^2 2\pi f_o t + A_{I1}^2 \cos^2 2\pi f_1 t + 2A_d A_{I1} \cos 2\pi f_o t \cos 2\pi f_1 t\right] \\ &\quad + a_3\left[A_d^3 \cos^3 2\pi f_o t + A_{I1}^3 \cos^3 2\pi f_1 t + 3A_d^2 A_{I1} \cos^2 2\pi f_o t \cos 2\pi f_1 t \right. \\ &\quad \left. + 3A_d A_{I1}^2 \cos 2\pi f_o t \cos^2 2\pi f_1 t\right] + \cdots \\ &= a_1 A_d\left[1 + \frac{3a_3}{4a_1} A_d^2 + \frac{3a_3}{2a_1} A_{I1}^2\right] \cos 2\pi f_o t + \cdots \end{aligned} \tag{2.2.5}$$

When the interference is not present and the desired signal is small $A_d << 1$, the system *small signal gain* is equal to a_1 since all terms on the right side of (2.2.5) will be negligibly small for a mildly nonlinear system if compared with the first term $a_1 A_d \cos\omega_o t$, the desired linear response. However, as the amplitude of the desired signal increases, the gain of the system will vary with the input signal level. The second term $\frac{3a_3}{4a_1} A_d^2$ in the brackets on the right side of (2.2.5) gradually becomes significant. If the sign of a_3 is opposite to that of a_1, then the output will be smaller than that predicated by linear theory. This phenomenon is called *gain compression*, and the compressive gain G_c in decibel (dB) is

$$G_c = 20\log\left|a_1\left(1 + \frac{3a_3}{4a_1} A_d^2\right)\right|. \tag{2.2.6}$$

The compressive gain drops with increasing A_d, and the signal level $A_d = A_{-1}$, where the gain is 1 dB lower than the small signal gain, is referred to as *1 dB compression point*. From the above compressive gain expression, the 1 dB compression point A_{-1} should satisfy the following equation:

$$20\log\left|\left(1+\frac{3a_3}{4a_1}A_{-1}^2\right)\right| = -1$$

Therefore, A_{-1} can be expressed as

$$A_{-1} = \sqrt{(1-10^{-1/20})\frac{4}{3}\left|\frac{a_1}{a_3}\right|} = \sqrt{0.145\left|\frac{a_1}{a_3}\right|}\,. \tag{2.2.7}$$

In the case of interference presence, the desired signal output will also reduce when the amplitude A_{I1} of the interferer increases if $a_3 < 0$. From (2.2.5), we can see that the interference causes the gain drop of amount $3a_3 A_{I1}^2/2$. This nonlinear effect is called *desensitization*. In this case, the weak desired signal suffers from experiencing very small gain. The gain drop with the strong interference is twice as fast compared to the gain compression case. Thus the desired signal will be completely blocked when the gain goes down to zero if the interference is strong enough.

Crossmodulation When the input $x(t)$ consists of a weak desired signal $s_d(t)$ and a strong interferer $s_{I1}(t)$ with an amplitude modulation $1+m(t)$ as follows

$$\left.\begin{aligned} s_d(t) &= A_d \cos 2\pi f_o t \\ s_{I1}(t) &= A_{I1}\left[1+m(t)\right]\cos 2\pi f_1 t \\ s_{I2}(t) &= 0 \end{aligned}\right\}, \tag{2.2.8}$$

the output $y(t)$ of a nonlinear system is then

$$y(t) = a_1 A_d \left\{1+\frac{3a_3}{4a_1}A_d^2 + \frac{3a_3}{2a_1}A_{I1}^2\left[1+m^2(t)+2m(t)\right]\right\}\cos 2\pi f_o t + \cdots \tag{2.2.9}$$

Expression (2.2.9) contains the desensitization and compression terms of (2.2.5) plus new terms, $(3a_3/2a_1)A_{I1}^2 m^2(t)$ and $(3a_3/a_1)A_{I1}^2 m(t)$. These new terms mean that the amplitude modulation on the strong interferer is now transferred to the desired signal through the interaction with the system nonlinearity. This nonlinear phenomenon is referred to as *crossmodulation*.

In the CDMA mobile stations the crossmodulation is an issue. The transmitted signal from the mobile station is not only an OQPSK phase modulated signal, but it is also amplitude modulated. The leakage of the transmission signal and a strong single tone interferer may generate crossmodulation components in the receiver front-end due to its nonlinear behavior under the strong interference attack. If the crossmodulation components are located in the receiver band, they will severely impact the receiver performance.

Intermodulation Now considering the case of more than one interferer, suppose there are interferers $s_{I1}(t)$ and $s_{I2}(t)$ accompanying a desired signal $s_d(t)$ in the input $x(t)$. Assuming that all of them are single tone signals, they have the forms

$$\left.\begin{aligned} s_d(t) &= A_d \cos 2\pi f_o t \\ s_{I1}(t) &= A_{I1} \cos 2\pi f_1 t \\ s_{I2}(t) &= A_{I2} \cos 2\pi f_2 t \end{aligned}\right\}. \tag{2.2.10}$$

After rearrangement the output resulting from (2.2.4), we obtain

$$\begin{aligned} y(t) = a_1 A_d &\left[1 + \frac{3a_3}{4a_1} A_d^2 + \frac{3a_3}{2a_2}\left(A_{I1}^2 + A_{I2}^2\right)\right] \cos 2\pi f_o t \\ &+ a_2 A_{I1} A_{I2} \left[\cos 2\pi (f_1 + f_2) t + \cos 2\pi (f_1 - f_2)\cdot t\right] \\ &+ \frac{3}{4} a_3 \left[A_{I1}^2 A_{I2} \cos 2\pi (2f_1 \pm f_2) t + A_{I1} A_{I2}^2 \cos 2\pi (2f_2 \pm f_1)\cdot t\right] + \cdots \end{aligned} \tag{2.2.11}$$

In (2.2.11), the terms with frequencies $(f_1 \pm f_2)$ are called the second-order intermodulation products, and the terms having frequencies $(2f_1 \pm f_2)$ and $(2f_2 \pm f_1)$ are the *third-order intermodulation products*. Actually, the output (2.2.11) also contains $(n+m)$th ($n, m = 2, 3, \cdots$) order intermodulation products with frequencies $|nf_1 \pm mf_2|$. These terms are usually much smaller than the second- and third-order intermodulation products, and for mild nonlinear systems they usually do not need to be considered. The second- and third-order intermodulation products may cause serious problems in a receiver if the frequencies of the interferers are close to the receiver operating frequency and have a proper spacing.

2.2.3. Memoryless Band-Pass Nonlinearities and Their Low-Pass Equivalents

The nonlinear devices, such as power amplifiers, used in wireless communication systems usually have a band-pass frequency response. These nonlinear devices are commonly considered as memoryless, and they can be represented by a nonlinear gain (AM-AM), and a phase distortion also called amplitude-to-phase conversion (AM-PM). These representations are generally valid to band-pass signals if the bandwidth of the signals is narrow enough, and over the signal bandwidth the device characteristic is essentially frequency independent.

In most cases, a memoryless band-pass nonlinear device to a narrow-band signal $x(t)$ with a carrier at f_o

$$x(t) = A(t)\cos\left[2\pi f_o t + \phi(t)\right] \tag{2.2.12}$$

has a response of the following form:

$$y(t) = f\left[A(t)\right]\cos\left\{2\pi f_o t + \phi(t) + g\left[A(t)\right]\right\}, \tag{2.2.13}$$

where $f\left[A(t)\right]$ is the nonlinear gain (AM-AM conversion), and $g\left[A(t)\right]$ is the amplitude-to-phase conversion (AM-PM conversion). We can see from (2.2.13) that the nonlinear part of the device output depends only on the amplitude $A(t)$of the input $x(t)$. This nonlinear characteristic is referred to as *envelope nonlinearity*. The typical AM-AM (upper) and AM-PM (lower) characteristics of a solid-state power amplifier are shown in Fig. 2.4.

Usually, it is convenient to use the complex form or the analytic form of the signals [8]. In this case, the input signal (2.2.12) turns into

$$x_+(t) = A(t)e^{j[2\pi f_o t + \phi(t)]}. \tag{2.2.14}$$

The analytic output of the nonlinear memoryless system can be derived from (2.2.13) in a similar way, and it has the form

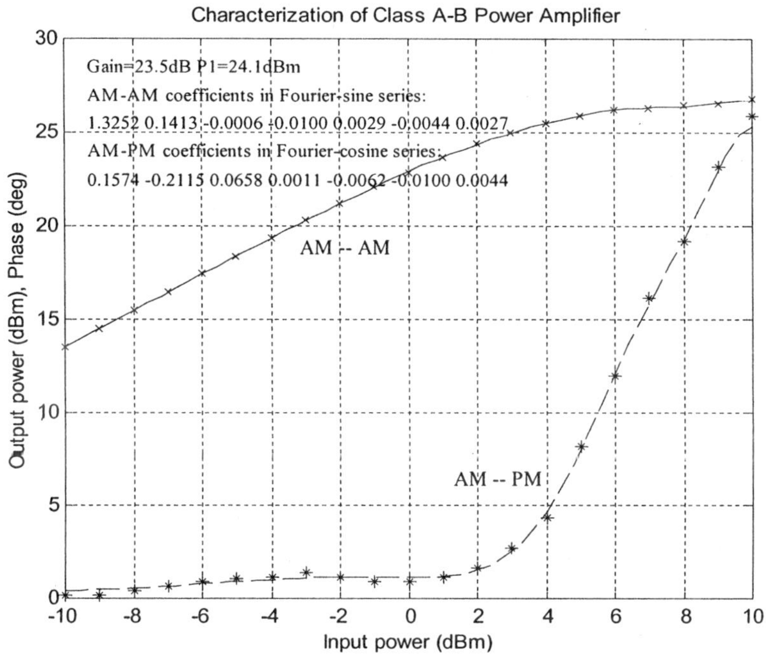

Figure 2.4. Typical magnitude and phase characteristics of a power amplifier

$$y_+(t) = f[A(t)]e^{j\{2\pi f_o t+\phi(t)+g[A(t)]\}}. \qquad (2.2.15)$$

Based on (2.2.15), the nonlinear memoryless system can be represented by means of a block diagram model as depicted in Fig. 2.5 [8]. The AM-AM and AM-PM nonlinearities shown in (2.2.15) can also be modeled in terms of two instantaneous amplitude nonlinearities. Rearranging the right side of (2.2.15), we obtain

$$y_+(t) = \widetilde{s}(t)e^{j[2\pi f_o t+\phi(t)]} = [s_I(t) + js_Q(t)]e^{j[2\pi f_o t+\phi(t)]}, \qquad (2.2.16)$$

where $\widetilde{s}(t)$ is defined as

$$\widetilde{s}(t) = f[A(t)]e^{jg[A(t)]} \qquad (2.2.17)$$

and $s_I(t)$ and $s_Q(t)$ are, respectively,

$$s_I(t) = f[A(t)]\cos\{g[A(t)]\} \tag{2.2.18a}$$

and

$$s_Q(t) = f[A(t)]\sin\{g[A(t)]\}. \tag{2.2.18b}$$

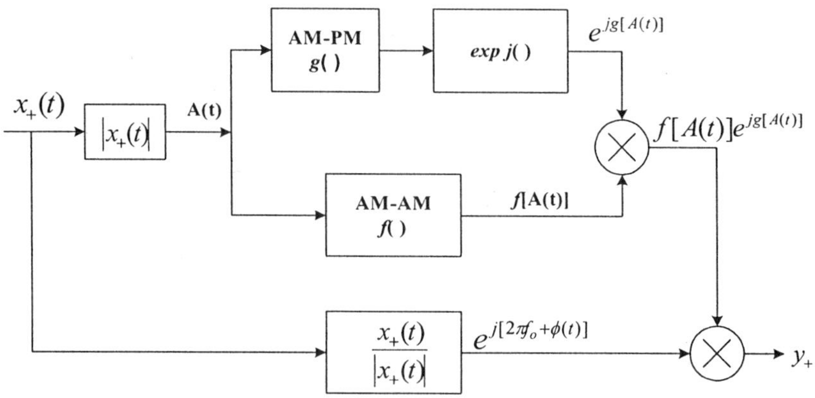

Figure 2.5. Block diagram representation of nonlinear memoryless system

The block diagram of a quadrature representation of the nonlinear memoryless model is given in Fig. 2.6. The output of the envelope nonlinearity is the real part of $y_+(t)$, and its quadrature form is

$$y_+(t) = s_I(t)\cos[2\pi f_o t + \phi(t)] - s_Q(t)\sin[2\pi f_o t + \phi(t)]. \tag{2.2.19}$$

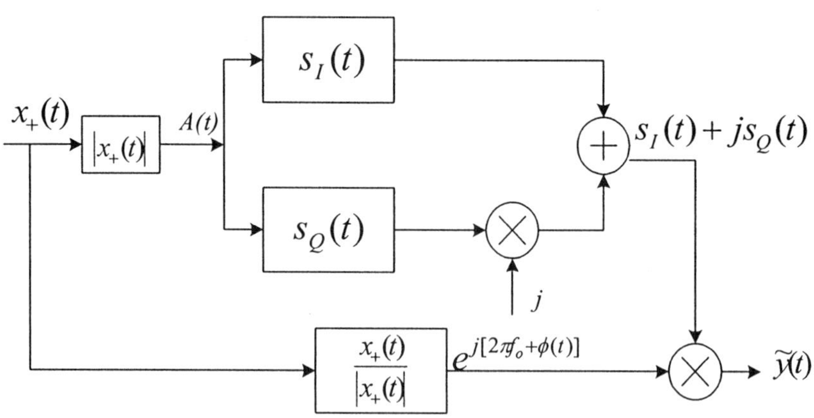

Figure 2.6. Block diagram of quadrature representation of nonlinear memoryless system

As mentioned in the previous section, in system simulation it is more efficient to use the low-pass equivalents of the band-pass systems. The complex envelope of the analytic input signal $x_+(t)$ (2.2.14) is

$$x_L(t) = \widetilde{x}(t)e^{-j2\pi f_o} = A(t)e^{j\phi(t)}. \tag{2.2.20}$$

The low-pass equivalent of the nonlinear system output from (2.2.15) and (2.2.16) is

$$\begin{aligned}\widetilde{y}_L(t) &= y_+(t)e^{-j2\pi f_o t} = f[A(t)]e^{j\{\phi(t)+g[A(t)]\}} \\ &= \widetilde{s}(t)e^{j\phi(t)} = [s_I(t) + js_Q(t)]e^{j\phi(t)}.\end{aligned} \tag{2.2.21}$$

The block diagram representations of the low-pass equivalent memoryless nonlinearity (2.2.21) are identical in form to Fig. 2.5 and Fig. 2.6, respectively, if $x_+(t)$ and $y_+(t)$ in these figures are correspondingly replaced by $\widetilde{x}_L(t)$ and $\widetilde{y}_L(t)$. In these cases, the output from the block $x_+(t)/|x_+(t)|$ in Figs. 2.5 and 2.6 will be $e^{j\phi(t)}$ without the carrier term $e^{j2\pi f_o t}$.

We should know that the output of a nonlinear system with a band-pass input has a wider spectral bandwidth than that of its input signal and also exhibits extra spectral components centered around the frequencies $\pm nf_o$, $n = 0, 1, 2, \ldots$. The spectral bandwidths at frequencies $\pm nf_o$ are n times wider than those at the carrier frequencies $\pm f_o$. In most cases, only the output spectral components around the carrier frequencies $\pm f_o$ are of interest. The low-pass equivalent output of a band-pass nonlinearity is valid only for the output spectral components around the carrier frequency $\pm f_o$. We shall see that the low-pass equivalent model for the nonlinear memoryless systems will be used in the adjacent channel power calculation in the transmitter system design and analysis.

2.3. Noise and Random Process

The most important noise in communication systems that can be treated in a systematic way is *thermal noise*. Thermal noise arises from random currents due to Brown motion of electrons in transceiver components, such as resistors and other lossy elements. Thermal noise has a flat power spectral density over a very wide frequency range and is said to be white since all frequencies are equally represented. Thermal noise depends only on the temperature.

Other types of noise are as follows. *Flicker noise* is observed at low frequencies in semiconductors and has a spectral density proportional to $1/f$. *Impulse noise* is short, randomly occurring electrical spikes. *Quantization noise* occurs when a continuous waveform is sampled and converted into a discrete time signal, such as in a analog-to-digital converter (ADC). *Shot noise* is caused by the variations of electron emission in vacuum tubes and semiconductor depletion regions or due to the fluctuations of the arrival rate of photons in optical devices.

A physical system including the communication systems is always accompanied by noise, and therefore it becomes essential to study and understand noise behavior and their influence to the system performance. In this section, we shall briefly review noise and random process to the extent required for the RF system analysis and design.

2.3.1. Noise Power and Spectral Representation

It is important to quantitatively determine the effect of systems on noise and the influence of noise on system performance. To do so, we need know the ways to represent and measure the noise. It will be seen that a spectral representation of the noise plays an important role in the analysis of a system with noise. The frequency analysis of noise and random signals is different from that of deterministic signals. It will be found that the analysis is based on the spectral distribution of the power in the noise and random wave instead of the amplitude spectrum.

Noise function $n(t)$ is a random time-varying wave, and this random time-varying function is referred to as a *random process*. A typical noise appearance is shown in Fig. 2.7. A sample of $n(t)$ taken at an arbitrary time is a random variable with some *probability density function* $p_n(n)$. The *probability density function* (*pdf*) determines the probability of the sampled value locating in $n < n_x < n + dn$ to be $p_n(n_x)dn$:

$$P(n < n_x < n + dn) = p_n(n_x)dn . \tag{2.3.1}$$

In communication systems, a commonly used probability density function for additive noise is the Gaussian pdf, which has the following form:

$$p_n(n) = \frac{1}{\sigma\sqrt{2\pi}} \exp \frac{-(n-m)^2}{2\sigma^2}, \tag{2.3.2}$$

where m is the average $E[n(t)]$ or called the *mean* of the distribution, σ^2 and σ are called variance and standard deviation of the random distribution, respectively. The physical meaning of the variance σ^2 will be explored in the following paragraphs.

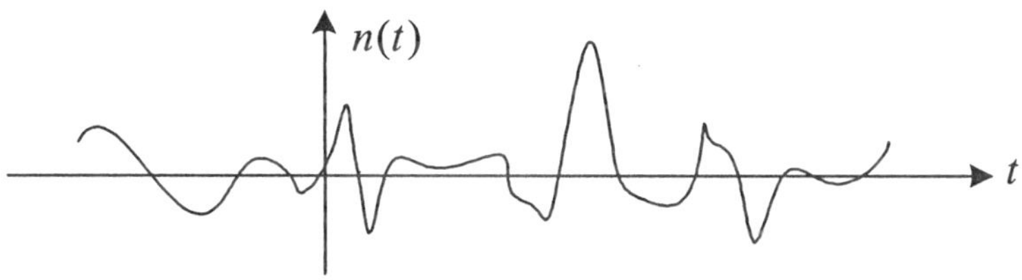

Figure 2.7. Noise-random waveform

The noise-time waveform will be different when measured in different intervals. A full characterization of a random process, therefore, requires a collection of an infinite length of time functions $n_k(t)$ (k =1, 2, ...), where $n_k(t)$ is the kth measured time waveform of the noise-random process $n(t)$. A very large set $\{n_k(t)\}$ of time waveforms is referred to as an *ensemble*, and each of the waveforms is called as *a sample function*.

If one measures the noise for a time interval of T, the average value of the noise waveform is calculated as

$$\overline{n(t)} = \lim_{T\to\infty} \frac{1}{T} \int_{-T/2}^{T/2} n(t)dt \, . \tag{2.3.3}$$

This average is performed over the time interval of T, and it therefore is called *time average*. In general, $\overline{n(t)}$ is a random variable since $n(t)$ varies randomly. Depending on when the average is performed, we may obtain different values of $\overline{n(t)}$. For a stationary random process (e.g., the thermal noise at a constant temperature), its statistical properties are invariant with a time shift. In this case, $\overline{n(t)}$ is independent of when the averaging is performed. In RF systems, most random processes that we shall deal with are stationary.

On the other hand, a noise-random process can be fully characterized in terms of an ensemble of the noise waveforms as mentioned above. Simultaneously taking samples from all the noise waveforms in the ensemble at the same time instant t_1 (as shown in Fig. 2.8), we obtain an ensemble of samples that is a set of random variables with the same probability density function $p_n(n)$ as that of the sampled value of the noise time function $n(t)$ at an arbitrary time t. The sum of all the sample values divided by the number of samples is called the *ensemble average* or *statistical average* of the noise-random process, and it is noted as $E[n(t_1)]$. The average value of the samples taken in the ensemble at time t_2 or at any other time t will be the same as $E[n(t_1)]$. The ensemble average thus can be formally defined based on the probability density function $p_n(n)$ as

$$E[n(t)] = \int_{-\infty}^{\infty} n(t) p_n(n) dn \,. \tag{2.3.4}$$

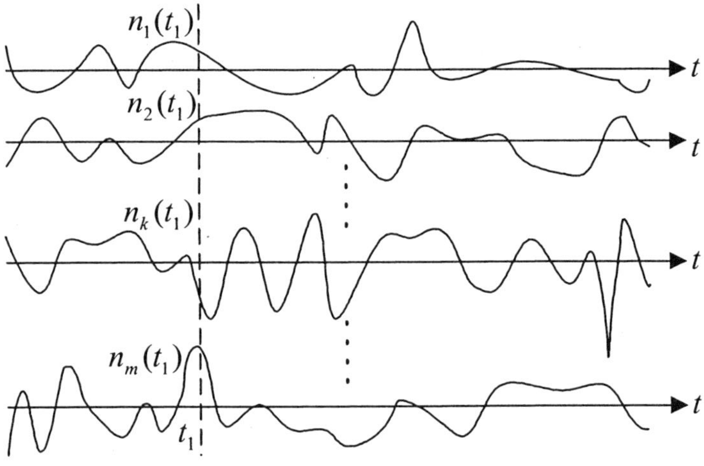

Figure 2.8. An ensemble containing a set of noise-random waveforms

Most of random processes $n(t)$ in RF systems have the following property:

$$\overline{n(t)} = \lim_{T\to\infty} \frac{1}{T} \int_{-T/2}^{T/2} n(t) dt = E[n(t)] \,. \tag{2.3.5}$$

This kind of random process is referred to as an *ergodic process*.

The average power of $n(t)$ in the time domain is defined as

$$P_N = \overline{n^2(t)} = \lim_{T\to\infty} \frac{1}{T} \int_{-T/2}^{T/2} n^2(t)dt\,. \tag{2.3.6}$$

The ensemble average of $n^2(t)$ is

$$E[n^2(t)] = \int_{-\infty}^{\infty} n^2(t) p_n(n) dn\,. \tag{2.3.7}$$

For an ergodic-random process (which is always assumed in this book), we have

$$P_N = E[n^2(t)]\,. \tag{2.3.8}$$

The DC power of the noise-random process is

$$\overline{n(t)}^2 = E^2[n(t)]\,. \tag{2.3.9}$$

The time average fluctuation power denoted as N is equal to the variance σ^2 of the ensemble average of $n^2(t)$ — i.e.,

$$\begin{aligned} P_N - \overline{n(t)}^2 &= \lim_{T\to\infty} \frac{1}{T} \int_{-T/2}^{T/2} \left[n(t) - \overline{n(t)}\right]^2 dt \\ &= E[n^2(t)] - E^2[n(t)] = \sigma^2 = N. \end{aligned} \tag{2.3.10}$$

A measure of the degree of dependence between random variables $n(t_1)$ and $n(t_2)$ of a noise process (Fig. 2.9) is the autocorrelation function, which is defined as

$$R_n(t_1, t_2) = E[n(t_1)n(t_2)]\,. \tag{2.3.11}$$

It is apparent that if $t_2 \to t_1$, then $R_n \to E(n^2)$. If variables $n(t_1)$ and $n(t_2)$ tend to become statistical independent when the spacing $t_2 - t_1$ is beyond certain value, such as greater than 1/(bandwidth of the process), then $R_n \to E^2(n)$, or 0 if $E(n) = 0$.

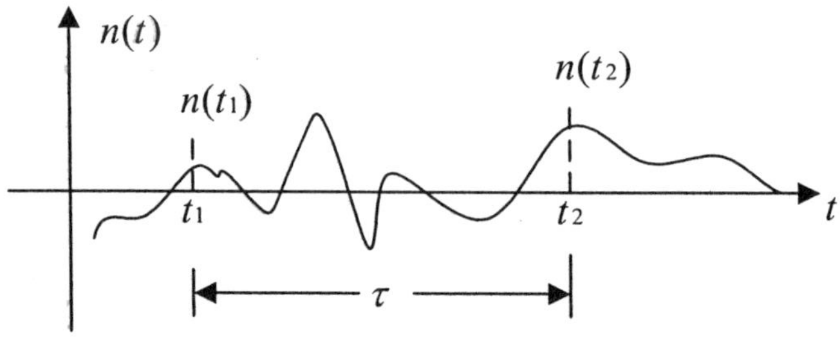

Figure 2.9. Autocorrelation

For a stationary process, the auto-correlation function is independent of time, and it has a form

$$R_n(\tau) = E[n(t)n(t+\tau)] . \tag{2.3.12}$$

The time-domain expression of the autocorrelation function is

$$R_n(\tau) = \lim_{T\to\infty} \frac{1}{T} \int_{-T/2}^{T/2} n(t)n(t+\tau)dt . \tag{2.3.13}$$

The Fourier transform of $R_n(\tau)$ is called the *power spectral density*, or frequently *power spectrum*. The power spectrum is denoted by $S_n(f)$:

$$S_n(f) = \int_{-\infty}^{\infty} R_n(\tau)e^{-j2\pi f\cdot\tau}d\tau . \tag{2.3.14}$$

The inverse Fourier transform of $S_n(f)$ exists, and $R_n(\tau)$ may be formally obtained from $S_n(f)$ by writing

$$R_n(\tau) = \int_{-\infty}^{\infty} S_n(f)e^{j2\pi f\cdot\tau}df . \tag{2.3.15}$$

But if $\tau = 0$, from (2.3.15) we have

$$R_n(0) = E(n^2) = \int_{-\infty}^{\infty} S_n(f)df . \tag{2.3.16}$$

From the above expression, we can see that $R_n(0)$ is equal to the total power in the noise process, and (2.3.16) explains why $S_n(f)$ is called power spectral density. In the special case where the noise $n(t)$ has a zero mean $E(n) = 0$ and thus the noise power is equal to variance N,

$$N = \int_{-\infty}^{\infty} S_n(f)df \,. \tag{2.3.17}$$

One particular spectral density that plays an extremely important role in communications is that in which the spectral density $S_n(f)$ is flat or constant — say, equal to $n_o/2$ — over all frequencies:

$$S_n(f) = \frac{n_o}{2} \,. \tag{2.3.18}$$

Strictly speaking, this is physically inadmissible since it implies infinite noise power. However, this is a very good model for many typical situations in which the noise bandwidth is larger than the frequencies of interest. Noise $n(t)$ with a flat spectral density $n_o/2$, as given in (2.3.18), is referred to as *white noise*.

Note the white noise can be seen as a band-limited noise by letting the bandwidth, $B \to \infty$, as shown in Fig. 2.10. The transform pair, the auto-correlation and the spectral-density functions, of the band-limited noise are, respectively:

$$R_n(\tau) = N \frac{\sin 2\pi B \tau}{2\pi B \tau} \tag{2.3.19}$$

and

$$S_n(f) = \begin{cases} n_o/2 & |f| \le B \\ 0 & |f| > B. \end{cases} \tag{2.3.20}$$

The noise of Fig. 2.10 is often called *band-limited white noise* since its spectral density is flat within $|f| \le B$. The spectral density of (2.3.20) is also referred to as a two-sided power spectral density since the spectrum equally appears in the negative frequency band $-B$. It is apparent from this transform pair (2.3.19) and (2.3.20) that the autocorrelation of a white noise is just an impulse function centered at the origin:

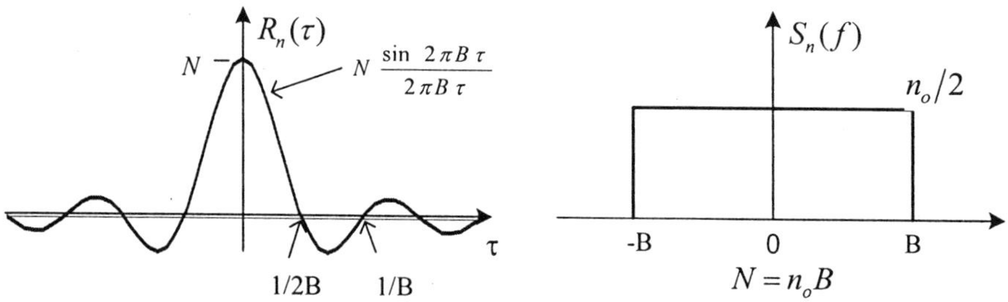

Figure 2.10. Correlation function and spectral density of band-limited noise

$$R_n(\tau) = \frac{n_o}{2}\delta(\tau) \tag{2.3.21}$$

The spectral-density and autocorrelation functions of the white noise are shown in Fig. 2.11.

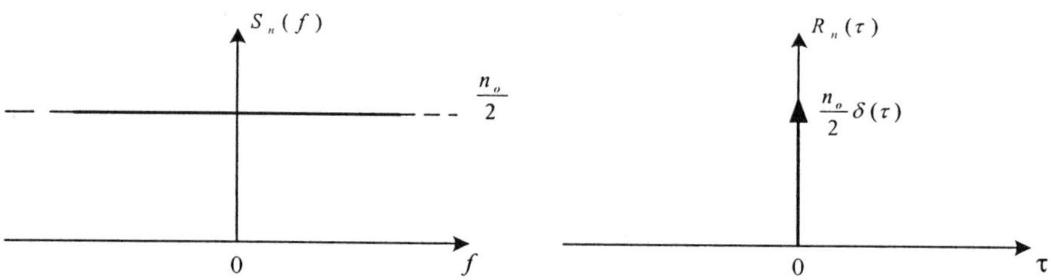

Figure 2.11. White-noise spectrum and auto-correlation function

The thermal noise appearing at the input of a receiving antenna is white noise. Although the white-noise model may appear to be physically inadmissible, in practice we shall consider an actual physical-noise process with a bandwidth B as white noise if the system in which the noise is propagating has a time response >> $1/B$. Band-limited white noise looks to us for all practical purposes like white noise if its bandwidth B is much greater than the frequency range of significant frequency response of our systems.

As an example, the white-noise power spectral density $S_n(f)$ across a resistor of value R will be calculated. The mean square noise voltage generated by the resistor R in a bandwidth B is

$$\overline{v_n^2} = 4kTR \cdot B, \tag{2.3.22}$$

where T is the absolute temperature of the resistor, in ^oK (Kelvin), R is in ohms, and k is the Boltzman constant equal to 1.38×10^{-23} J/ ^oK . The white noise $\overline{v_n^2}$ in the bandwidth B can be also expressed as N_oB with a two-sided spectral density $n_o/2$ in volts squared. Therefore, from (2.3.22) the thermal noise spectral density for the two-sided spectrum is

$$S_n(f) = \frac{\overline{v_n^2}}{2B} = 2kTR . \tag{2.3.23}$$

In most cases of practical interest, there will be conjugate matched between a white-noise source and a load, such as a mobile station antenna with equilibrium temperature T and an input impedance of a mobile receiver (Fig. 2.12). In this case, the noise power delivered to the load R_2 by the equivalent voltage source $\overline{v_n^2}$ in series with R_1 is

$$\frac{\overline{v_n^2}R_2}{(R_1 + R_2)^2 + (X_1 + X_2)^2} = \frac{\overline{v_n^2}}{4R} = kT \cdot B . \tag{2.3.24}$$

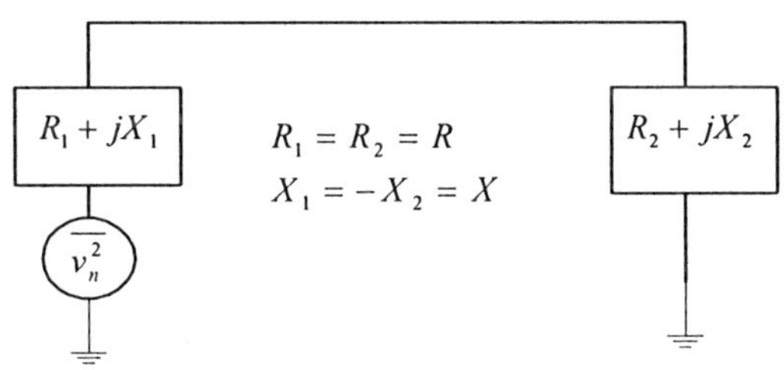

Figure 2.12. Thermal-noise circuit model and conjugate matched load

From the above expression, we can obtain the two-sided power spectral density of thermal noise to be

$$S_n(f) = \frac{kT}{2} . \tag{2.3.25}$$

In reality, the one-sided power spectral density for thermal noise is more often used in RF system designs. The spectrum appearing at negative frequencies results from Fourier analysis. There are no negative frequencies in the real physical world. The one-sided spectrum is obtained from folding the negative frequency part of the spectrum around the zero frequency vertical axis and adding it to positive frequency part. The one-sided thermal noise power density in W/Hz is

$$S_{n\_oneside} = n_o = kT. \tag{2.3.26}$$

If substituting $k = 1.38 \times 10^{-23}$ J/ ^oK, and $T = 290$ ^oK, we obtain $n_o = 4.002 \times 10^{-21}$ W = 4.002×10^{-18} mW. In RF system designs, we would like to use dBW or dBm as power units. Therefore, one-sided thermal noise spectral density in dBm/Hz is equal to

$$N_o = 10 \log n_o = 10 \log kT = -173.98 \cong -174 \text{ dBm/Hz}. \tag{2.3.27}$$

This is the thermal-noise spectral-density level at $T = 290$ ^oK (nominal room temperature).

2.3.2. Noise and Random Process Through Linear Systems

Noise $n_i(t)$ with a spectral density $S_{n_i}(f)$ and a correlation function $R_{n_i}(\tau)$ passes through a linear system with a transfer function $H(f)$ and impulse response $h(t)$. The output response $n_o(t)$ of the linear system is the convolution of the impulse response and the input noise:

$$n_o(t) = \int_{-\infty}^{\infty} h(t-\tau) n_i(\tau) d\tau \tag{2.3.28}$$

The spectral densities between the output and input noises have the following relation:

$$S_{n_o}(f) = |H(f)|^2 S_{n_i}(f) \tag{2.3.29}$$

If the bandwidth *B* of the input noise $n_i(t)$ is much smaller than the system bandwidth *W*, the spectrum of the output noise $n_o(t)$ will be very little different from that of the input noise $n_i(t)$ except the magnitude. On the other hand, if $W << B$, the system will not respond rapidly enough to fast fluctuations of the input noise $n_i(t)$, and the output $n_o(t)$ can vary only at roughly the rate of W.

Now look at an example of the application of (2.3.29). White noise of spectral density $n_o/2$ is applied to the input of an RC low-pass with a transfer function

$$|H(f)|^2 = \frac{1}{1+(2\pi RCf)^2}. \tag{2.3.30}$$

The spectral density of the output noise from the filter is

$$S_{n_o} = \frac{n_o/2}{1+(2\pi RCf)^2} = \frac{n_o/2}{1+(f/f_1)^2}, \tag{2.3.31}$$

where $f_1 = 1/2\pi RC$. The spectrum of the output noise from the RC filter is depicted in Fig. 2.13. The bandwidth of the output noise is inversely proportional to the filter time constant RC. The correlation function of the filtered noise $n_o(t)$ results from the Fourier transform of (2.3.31),

$$R_{n_o} = \frac{n_o}{4RC} e^{-|\tau|/RC}. \tag{2.3.32}$$

The correlation function will drop to 1/10 of its maximum value $n_o/4RC$ when $\tau = 2.3RC$. The correlation time, therefore, is in the order of 2.3 RC.

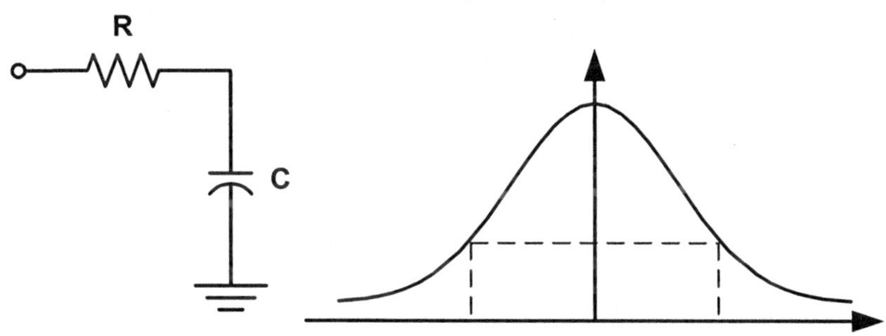

Figure 2.13. Output spectrum of white noise passing through an RC filter

For low-pass linear systems, a *noise equivalent bandwidth* is defined as follows. Suppose that we have a white-noise source of zero mean and power spectral density $n_o/2$ connected to he input of a low-pass system with a transfer function $H(f)$. The resulting average output noise power is then given by

$$N_{out} = \frac{n_o}{2} \int_{-\infty}^{\infty} |H(f)|^2 df = n_o \int_{0}^{\infty} |H(f)|^2 df, \tag{2.3.33}$$

where in (2.3.33) we have made use of the fact that the transfer function $|H(f)|$ has even symmetry about the frequency origin.

Now consider that the same white-noise source is connected to the input of an *ideal low-pass filter* with a zero-frequency response $H(0)$ and a bandwidth *B*. In this case, the average output noise power is

$$N_{out} = n_o B H^2(0) . \tag{2.3.34}$$

Assuming that this output noise power is equal to that in (2.3.33), we then have the equivalent noise bandwidth defined as

$$B = \frac{\int_{0}^{\infty} |H(f)|^2 df}{H^2(0)} . \tag{2.3.35}$$

The equivalent noise bandwidth calculation is illustrated in Fig. 2.14.

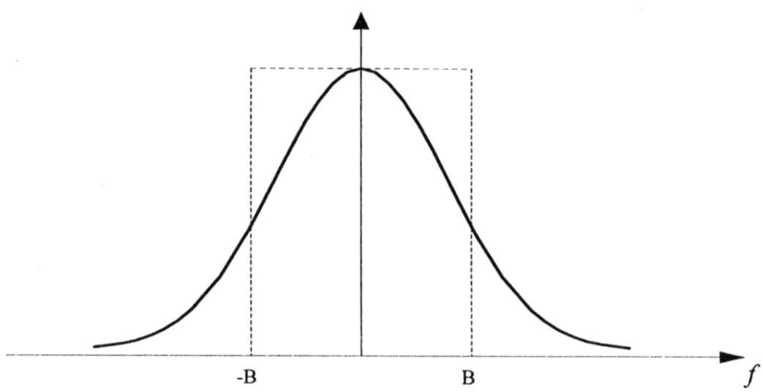

Figure 2.14. The equivalent noise bandwidth

2.3.3. Narrow-Band Noise Representation

The noise that we deal with in communication systems is usually *narrow-band noise*. The RF receiver in a communication system typically has a bandwidth just large enough to pass the desired signal essentially undistorted but not so large as to admit excessive noise through the receiver. In the RF receiver, narrow-band band-pass filters are used to control and restrict the noise bandwidth. Although the noise at the receiver input is possibly wide-band, the noise $n(t)$ appearing at the output of the narrow-band band-pass filter in the receiver is narrow-band. Assuming the transfer function of the filter is denoted as $H(f)$ and input white noise has a zero mean and unit power spectral density, the power spectral density $S_n(f)$ of narrow-band noise $n(t)$ has an expression as

$$S_n(f) = |H(f)|^2 . \tag{2.3.36}$$

The spectral components of narrow-band noise are usually concentrated about the center frequency $\pm f_o$ as shown in Fig. 2.15.

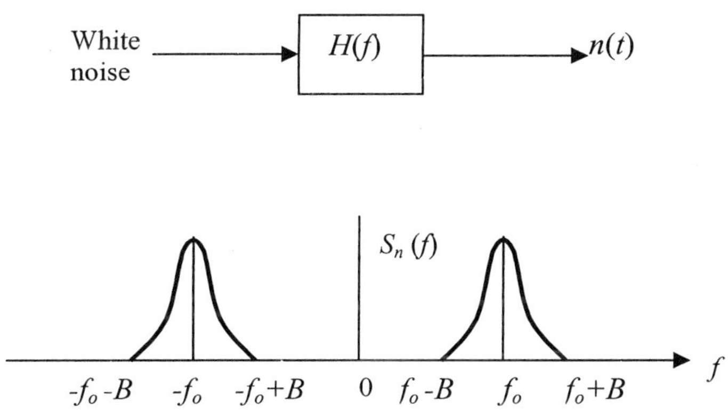

Figure 2.15. Generation of narrow-band noise and spectral density of narrow-band noise

Utilizing the pre-envelope and complex envelope concepts used in Section 2.1.4, we can represent narrow-band noise $n(t)$ with a center frequency f_o as follows. The analytic form of narrow-band noise $n_+(t)$ can be written in the form

$$n_+(t) = n(t) + j\hat{n}(t), \tag{2.3.37}$$

where $\hat{n}(t)$ is the Hilbert transform of $n(t)$. The complex low-pass equivalent or complex envelope $\tilde{n}(t)$ can be obtained from $n_+(t)$ by multiplying $\exp(-j2\pi f_o t)$ — i.e.,

$$\tilde{n}(t) = n_+(t)e^{-j2\pi f_o t}. \tag{2.3.38}$$

The complex low-pass equivalent can also be expressed as the sum of an in-phase component $n_I(t)$ and a quadrature component $n_Q(t)$:

$$\tilde{n}(t) = n_I(t) + jn_Q(t). \tag{2.3.39}$$

From (2.3.37) to (2.3.39), the in-phase component $n_I(t)$ and the quadrature component $n_Q(t)$ are related to the narrow-band noise $n(t)$ and its Hilbert transformation $\hat{n}(t)$ as follows, respectively,

$$n_I(t) = n(t)\cos(2\pi f_o t) + \hat{n}(t)\sin(2\pi f_o t) \tag{2.3.40}$$

and

$$n_Q(t) = \hat{n}(t)\cos(2\pi f_o t) - n(t)\sin(2\pi f_o t). \tag{2.3.41}$$

From the last two equations, eliminating $\hat{n}(t)$ we obtain the canonical representation of the narrow-band noise $n(t)$ in terms of $n_I(t)$ and $n_Q(t)$,

$$n(t) = n_I^{;}(t)\cos(2\pi f_o t) - n_Q(t)\sin(2\pi f_o t). \tag{2.3.42}$$

The in-phase component $n_I(t)$ and the quadrature component $n_Q(t)$ of narrow-band noise $n(t)$ have zero mean. The components $n_I(t)$ and $n_Q(t)$ have the same variance σ^2 as the narrow-band noise $n(t)$:

$$E[n_I^2(t)] = E[n_Q^2(t)] = E[n^2(t)] = \sigma^2. \tag{2.3.43}$$

Comparing (2.3.38) and (2.3.39) with (2.3.37), it is apparent that the power spectral densities of the quadrature components $n_I(t)$ and $n_Q(t)$

may be visualized as those of the original noise $n(t)$ shifted down to zero frequency. The power spectra of the in-phase and quadrature noises with a bandwidth B are both centered at DC unlike that of $n(t)$ centered about the frequency f_o. It can be shown that the power spectral densities $S_{n_I}(f)$ and $S_{n_Q}(f)$ of the quadrature components are equal, and they are related to the power density $S_n(f)$ of the original narrow-band noise $n(t)$ as follows:

$$S_{n_I}(f) = S_{n_Q}(f) = \begin{cases} S_n(f - f_o) + S_n(f + f_o), & -B \le f \le B \\ 0, & \text{elsewhere}, \end{cases} \qquad (2.3.44)$$

The low-pass power spectral density of the in-phase noise $n_I(t)$ and quadrature noise $n_Q(t)$ and the power spectrum of the corresponding band-pass narrow-band noise $n(t)$ with a carrier frequency f_o are shown in Fig. 2.16.

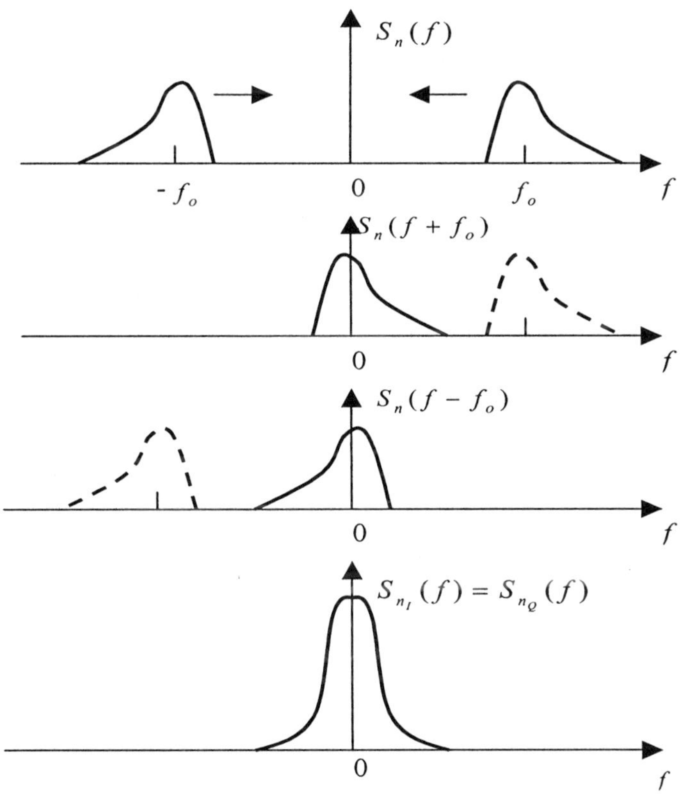

Figure 2.16. Band-pass power spectral density of narrow-band noise and low-pass power spectrum of quadrature components

In a special case, $S_n(f)$ is symmetrical about the carrier frequency f_o, and then $S_{n_I}(f)$ and $S_{n_Q}(f)$ have a form of

$$S_{n_I}(f) = S_{n_Q}(f) = 2S_n(f + f_o) = 2S_n(f - f_o) . \qquad (2.3.45)$$

If the narrow-band noise $n(t)$ with the symmetrical power spectral density is Gaussian with a zero mean, then the in-phase noise $n_I(t)$ and the quadrature noise $n_Q(t)$ are statistically independent. In this case, the joint probability density function of the random variables $n_I(t_k + \tau)$ and $n_Q(t_k)$ is equal to the product of their individual probability density functions — i.e.,

$$f_{n_I n_Q}(n_I, n_Q) = \frac{1}{2\pi\sigma^2} e^{-\frac{n_I^2 + n_Q^2}{2\sigma^2}} , \qquad (2.3.46)$$

where σ^2 is variance of the original noise $n(t)$ as given in (2.3.43).

The narrow-band noise $n(t)$ can alternatively be represented in terms of its envelope $r(t)$ and phase $\varphi(t)$ as follows:

$$n(t) = r(t)\cos[2\pi f_c t + \varphi(t)], \qquad (2.3.47)$$

where

$$r(t) = \left[n_I^2 + n_Q^2\right]^{1/2} \qquad (2.3.48)$$

and

$$\varphi(t) = \tan^{-1}\left[\frac{n_Q(t)}{n_I(t)}\right]. \qquad (2.3.49)$$

Expanding (2.3.47) and comparing it with (2.3.42), we obtain

$$n_I(t) = r(t)\cos\varphi(t) \qquad (2.3.50)$$

and

$$n_Q(t) = r(t)\sin\varphi(t) . \qquad (2.3.51)$$

The probability density function of the random process $r(t)$ has a Rayleigh distribution as

$$f_r(r) = \begin{cases} \dfrac{r}{\sigma^2} e^{-\frac{r^2}{2\sigma^2}}, & r \geq 0 \\ 0 & \text{elsewhere,} \end{cases} \tag{2.3.52}$$

where σ^2 is variance of the original noise $n(t)$. The phase noise $\varphi(t)$ has a uniformly distributed probability density function inside the range 0 to 2π as

$$f_\varphi(\varphi) = \begin{cases} \dfrac{1}{2\pi}, & 0 \leq \varphi \leq 2\pi \\ 0, & \text{elsewhere.} \end{cases} \tag{2.3.53}$$

The random processes $r(t)$ and $\varphi(t)$ are statistically independent. The joint probability density function of $r(t)$ and $\varphi(t)$ is

$$f_{r\varphi}(r,\varphi) = \frac{r}{2\pi\sigma^2} e^{-\frac{r^2}{2\sigma^2}}. \tag{2.3.54}$$

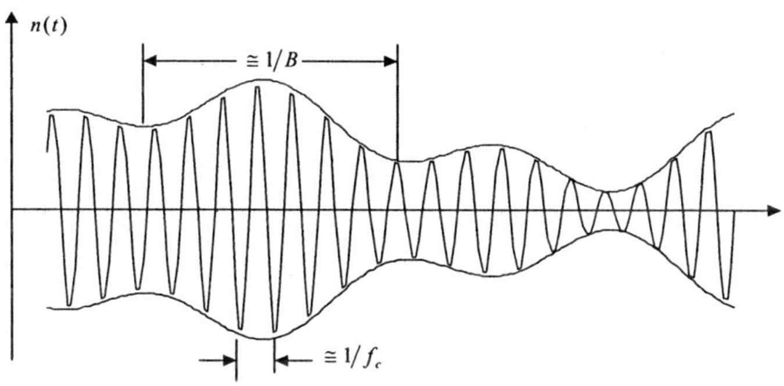

Figure 2.17. Narrow-band noise

We would expect that envelope $r(t)$ and phase $\varphi(t)$ of the narrow-band noise vary roughly at the rate B Hz in a random fashion as depicted in Fig. 2.17.

2.3.4. Noise Figure and Noise Temperature

Any physical system has inherent noise. The inherent noise of a system, such as a receiver, can be measured in different ways. For communication systems, noise figure or *noise temperature* is used to characterize their inherent noise effect.

The *noise figure* (*NF*) is defined as the ratio in decibels (dB) of the total system noise to the noise that would be present if the system is noiseless but excited by input noise from a source at temperature $T = 297^{\circ}\mathrm{K}$. The ratio is sometimes called the *noise factor* (*F*). The noise factor is also equivalently defined as the ratio of input signal-to-noise power ratio, $SNR_i = P_{S_i}/P_{N_i}$, to the output signal-to-noise power ratio, $SNR_o = P_{S_o}/P_{N_o}$ — i.e.,

$$F = \frac{SNR_i}{SNR_o}. \tag{2.3.55}$$

The noise factor of a system can also be expressed in the ratio of the system output noise P_{N_o} converted into the input equivalent to the input noise ratio. The input equivalence of the system output noise conversion is equal to

$$P_{N_t} = \frac{P_{N_o}}{g} = \frac{g(P_{N_i} + P_{N\_system})}{g} = P_{N_i} + P_{N\_system}, \tag{2.3.56}$$

where g is the system power gain, and P_{N_system} is the system internal noise converted to its input. Therefore, the noise factor of the system has the expression

$$F = \frac{P_{N_t}}{P_{N_i}} = \frac{P_{N_i} + P_{N\_system}}{P_{N_i}} = 1 + \frac{P_{N\_system}}{P_{N_i}}. \tag{2.3.57}$$

The noise figure of the system is the noise factor expressed in decibels, and it has the form

$$NF = 10\log\frac{SNR_i}{SNR_o} = 10\log\left(1 + \frac{P_{N\_system}}{P_{N_i}}\right) \text{ dB}. \tag{2.3.58}$$

The physical meaning of the noise figure or the noise factor is a measure of the signal-to-noise ratio degradation as the signal passes through a system.

Most RF systems, whether active or passive, are a two-port system — i.e., a system has an input port and an output port — or they can be defined as a two-port system. A linear two-port system can be modeled by two input noise sources — a series voltage source E_n and a parallel current source I_n, followed by a noiseless system with gain and output impedance as determined by the two-port parameters and the impedance at the input port [9]. In general, these two noise sources are correlated. The equivalent representation of the two-port system for characterizing its inherent noise effect is depicted in Fig. 2.18. Where it is assumed that E_g is a noise source at input port, and Z_g is the impedance of the noise source.

In the case of no relation between E_n and I_n, the total noise power input to the noiseless system is proportional to $[E_g^2 + E_n^2 + (I_n Z_g)^2]$, whereas the source noise power is proportional to E_g^2. From (2.3.58), we obtain the noise factor of the two-port system to be [9]

$$F = 1 + \frac{E_n^2 + (I_n Z_g)^2}{E_g^2}. \tag{2.3.59}$$

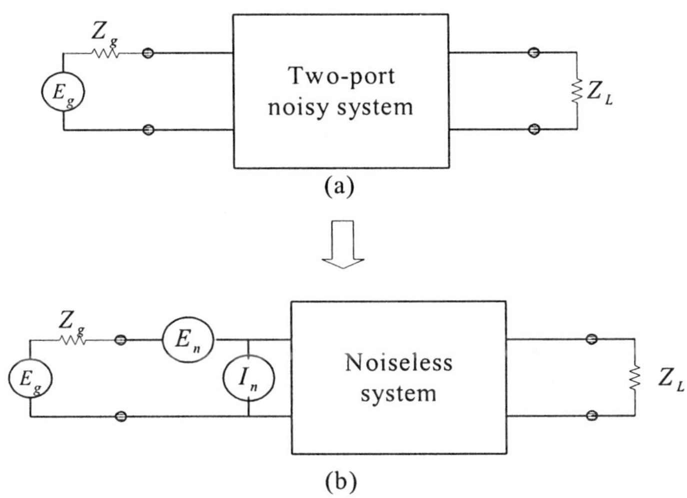

Figure 2.18. Equivalent represent of two-port noisy system

Considering $Z_g = R_g + jX_g$ and $E_g^2 = 4kTBR_g$, we have

$$F = 1 + \frac{E_n^2}{4kTBR_g} + I_n^2 \frac{R_g^2 + X_g^2}{4kTBR_g}. \tag{2.3.60}$$

In general, the source reactance X_g can be tuned out through a matching circuit. It is easy to prove that F will have its minimum value when the source resistance R_g equals the ratio of the system equivalent noise voltage to the noise current or

$$R_{go}^2 = \frac{E_n^2}{I_n^2} = \frac{R_n}{G_n}, \tag{2.3.61}$$

where the following two expresses, $E_n^2 = 4kTBR_n$ and $I_n^2 = 4kTBG_n$, are used. The minimized noise factor is

$$F_o = 1 + 2\sqrt{R_n G_n}. \tag{2.3.62}$$

In the case of existing correlation between E_n and I_n, the noise factor expression becomes [9]

$$F = 1 + \frac{(E_n + I_n Z_g)(E_n^* + I_n^* Z_g^*)}{E_g^2}. \tag{2.3.63}$$

The cross products of E_n and I_n is not equal to zero since they are correlated. The products $E_n I_n^*$ and $E_n^* I_n$, when averaged over time, and the ensemble of noise functions produce complex values, $C_n = C_{n,r} + jC_{n,x}$ and $C_n^* = C_{n,r} - jC_{n,x}$, respectively. Then the noise factor turns into

$$F = 1 + \frac{E_n^2 + I_n^2(R_g^2 + X_g^2) + 2(C_n Z_g^* + C_n^* Z_g)}{4kTBR_g}. \tag{2.3.64}$$

An alternative noise factor expression is in terms of E_n being broken into an uncorrelated with I_n part E_u and a correlated part E_c — i.e., $E_n = E_u + E_c = E_u + I_n Z_c$, where Z_c is a complex constant referred to as correlation impedance. Using this assumption and the noise resistance and conductance equivalents as used in (2.3.61), we obtain

$$F = 1 + \frac{R_u + G_n\left[(R_c + R_g)^2 + (X_c + X_g)^2\right]}{R_g}. \tag{2.3.65}$$

If the source impedance Z_g has the following relations with the system input noise impedance, $X_{go} = -X_c$ and $R_{go}^2 = R_u / G_n + R_C^2$, the noise factor F will be optimized, and it is

$$F = 1 + 2R_c G_n + 2\sqrt{R_u G_n + R_c^2 G_n^2}. \tag{2.3.66}$$

In the satellite communication systems and other low-noise systems, engineers like to use the *noise temperature* instead of the noise figure to measure the thermal noise produced by the systems. A low noisy system with a gain G can be represented by a noiseless system with an equivalent noise source at temperature T_e, as shown in Fig. 2.19. Assuming the system internal noise at its output is $P_{N\_internal} = kT_e BG$, where B is the system-noise bandwidth, then the system-noise temperature T_e is defined as [13]

$$T_e = \frac{P_{N\_internal}}{kBG}. \tag{2.3.67}$$

The relationship between the noise temperature and noise factor can be derived as follows. From (2.3.57), we know that the system-noise factor is determined by the ratio of the system noise P_{N_system} to the input noise P_{N_i}. In measuring the noise factor, the input is a standard noise source with power kT_oB, where $T_o = 290\ ^\circ\mathrm{K}$. The system noise is

$$P_{N\_system} = P_{N\_internal} / G = kT_e B.$$

Thus, the noise factor can be expressed by using the system equivalent noise temperature as

$$F = 1 + \frac{P_{N\_system}}{P_{N_i}} = 1 + \frac{T_e}{T_o}, \tag{2.3.68}$$

or the system noise temperature is expressed in

$$T_e = (F-1)T_o\,. \tag{2.3.69}$$

For a system with $T_e = 0$ or $F = 1$, it is an ideal noiseless system.

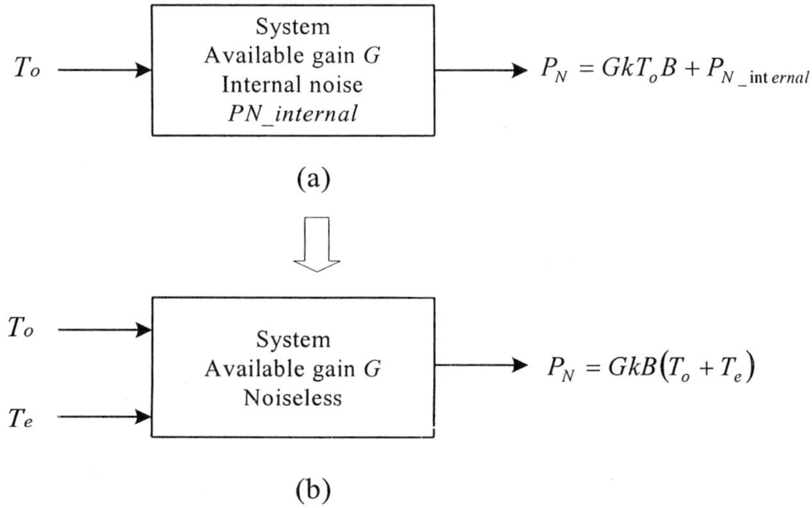

Figure 2.19. Definition of effective noise temperature

2.4. Elements of Digital Base-Band System

In this section, we present a broad range of topics that are essential for understanding and analyzing digital base-band system and important for transceiver RF system designs. The digital base-band system in a transceiver used for a wireless communication system is the portion of a transceiver from an input port of source information, voice and/or data, to a digital-to-analog converter (DAC) in the transmitter and from an analog-to-digital converter (ADC) to an output port of estimated information, voice and/or data, in the receiver.

Source information, either voice or data, is usually a base-band low-pass signal from DC to a few megahertz. To transmit information in a wireless digital communication system, the base-band signal must be transformed into digital symbols, the symbols are then converted to digital waveforms, and finally the digital waveforms modulate an RF carrier for transmitting. In the receiver chain, a reverse conversion and transformation process is carried out to finally detect received messages.

We discuss sampling theorem, quantization effects, pulse shaping, intersymbol interference, detection error probability, and signal-to-noise ratio or carrier-to-noise ratio.

2.4.1. Sampling Theorem and Sampling Process

In digital communication systems, the analog information must first be transformed into a digital format. The transform process starts with the *sampling process*. This process can be implemented by means of sampling and holding operations. The output of the sampling process is a sequence of pulses with amplitudes derived from the input waveform samples. The following discussion on the sampling process uses an intuitive approach similar to that presented in reference [10].

A band-limited signal without spectral component beyond frequency f_{max} can be completely reconstructed from a set of uniformly spaced discrete-time samples if the samples are obtained with a sampling rate f_s,

$$f_s \geq 2 f_{\max} . \tag{2.4.1}$$

This particular statement is known as the *uniform sampling theorem*. The sampling rate $f_s = 2 f_{\max}$ is also referred as the *Nyquist rate*.

Assume that an analog waveform $x(t)$ with a Fourier transform $X(f)$, which is equal to zero, while $|f| \geq f_{\max}$ as shown in Fig. 2.20(a) and (b).

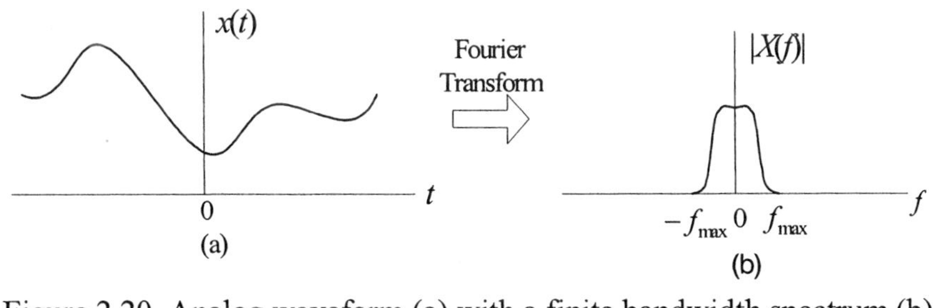

Figure 2.20. Analog waveform (a) with a finite bandwidth spectrum (b)

An *ideal sampling* of $x(t)$ can be viewed as the product of $x(t)$ with a periodic train of impulse functions $x_\delta(t)$, defined as

$$x_\delta(t) = \sum_{n=-\infty}^{\infty} \delta(t - nT_s), \tag{2.4.2}$$

where $T_s = 1/f_s$ is the sampling period and $\delta(t)$ is the unit impulse or Dirac delta function. The sampled version of $x(t)$ as depicted in Fig. 2.21(a) is denoted as $x_s(t)$ and can be expressed as

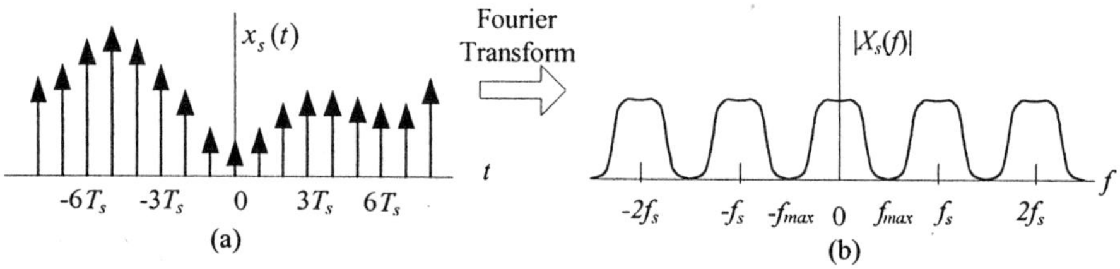

Figure 2.21. Impulse sampled signal and its spectrum

$$x_s(t) = x(t)x_\delta(t) = \sum_{n=-\infty}^{\infty} x(t)\delta(t - nT_s) = \sum_{n=-\infty}^{\infty} x(nT_s)\delta(t - nT_s), \tag{2.4.3}$$

where the following shifting property of the impulse function is used:

$$x(t)\delta(t - t_o) = x(t_o)\delta(t - t_o). \tag{2.4.4}$$

The spectrum of the sampled signal $x_s(t)$ can be obtained from the Fourier transform of (2.4.3). Considering the convolution property of the Fourier transform (see Section 2.1), the Fourier transform of the sampled signal, $X_s(f)$ can be expressed as the convolution of $X(f)$ and the Fourier transform of the impulse train $x_\delta(t)$, $X_\delta(f)$ —i.e.,

$$X_s(f) = X(f) * X_\delta(f) = X(f) * \left[\frac{1}{T}\sum_{n=-\infty}^{\infty} \delta(f - nf_s)\right] = \frac{1}{T}\sum_{n=-\infty}^{\infty} X(f - nf_s), \tag{2.4.5}$$

where the following frequency domain form of the impulse train is used:

$$X_\delta(f) = \frac{1}{T_s} \sum_{n=-\infty}^{\infty} \delta(f - nf_s). \tag{2.4.6}$$

The spectrum $X_s(f)$ of the sampled signal $x_s(t)$ is, to within a constant factor ($1/T_s$), exactly the same as $X(f)$ of the original signal $x(t)$. In addition, the spectrum repeats itself periodically in frequency every f_s hertz as shown in Fig. 2.21(b).

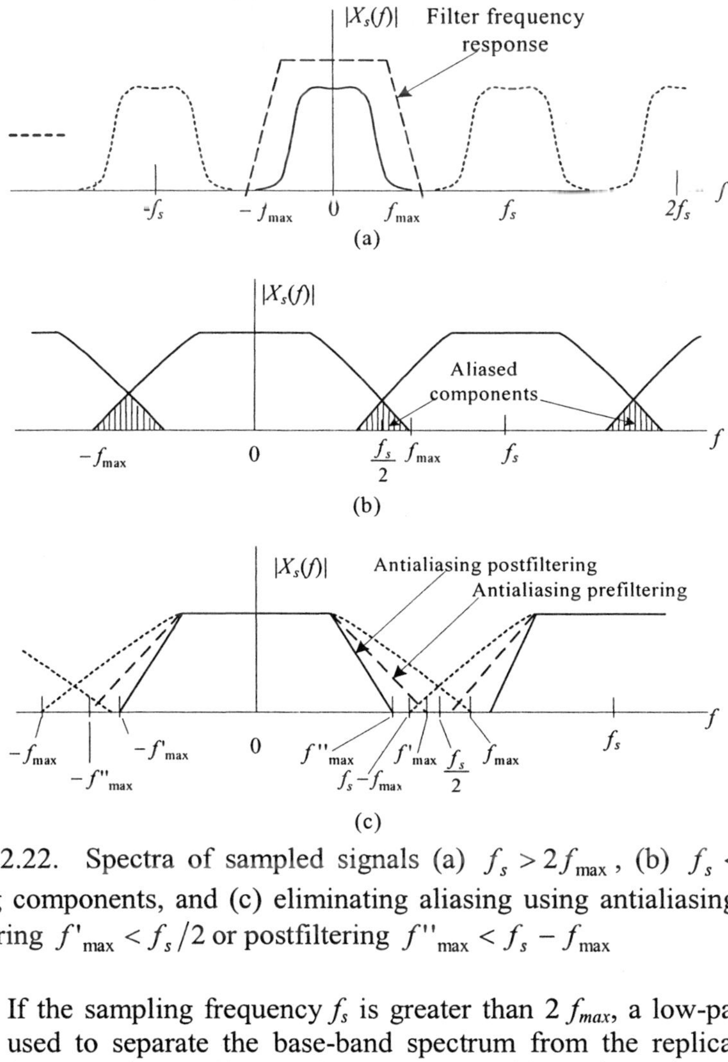

Figure 2.22. Spectra of sampled signals (a) $f_s > 2f_{\max}$, (b) $f_s < 2f_{\max}$, aliasing components, and (c) eliminating aliasing using antialiasing filters: prefiltering $f'_{\max} < f_s/2$ or postfiltering $f''_{\max} < f_s - f_{\max}$

If the sampling frequency f_s is greater than $2\,f_{max}$, a low-pass filter can be used to separate the base-band spectrum from the replications at higher frequencies as illustrated in Fig. 2.22(a). When the sampling frequency is less than the Nyquist rate, $f_s < 2f_{\max}$ — i.e., in the *undersampling* case, the replications will overlap as depicted in Fig. 2.22(b). This spectral overlap of the replications is called *aliasing phenomenon*. Part of the information contained in the original signal will be lost when aliasing occurs. Using antialiasing filters in the undersampling case, we can eliminate the aliasing phenomenon as shown in Fig. 2.22(c). The analog signal is prefiltered so the new maximum frequency f'_{max} is reduced to equal to or less than $f_s/2$. The aliased components can also be removed by postfiltering after sampling; the filter cutoff frequency f'''_{max} should be less than $f_s - f_{\max}$. However, using antialiasing filters it will still result in a loss of some of the signal information. For an engineering application, it is usually best to choose the lowest sampling rate as

$$f_s \geq 2.2 f_{\max}. \tag{2.4.7}$$

The instantaneous impulse sampling is a theoretically convenient model. A more practical approach of accomplishing the sampling is to use the rectangular pulse train or switching waveform, $x_p(t)$, as shown in Fig. 2.23(a). Each rectangular pulse in $x_p(t)$ has a width T and an amplitude $1/T$. The Fourier series of the pulse train with a repeated rate f_s has a form

$$x_p(t) = \sum_{n=-\infty}^{\infty} a_n e^{j2\pi n f_s t}, \tag{2.4.8}$$

where a_n is a sinc function in the form

$$a_n = \frac{1}{T_s} \operatorname{sinc}\left(\frac{\pi \cdot nT}{T_s}\right) = \frac{1}{T_s} \frac{Sin(\pi \cdot nT/T_s)}{\pi \cdot nT/T_s} \tag{2.4.9}$$

and $T_s = 1/f_s$. The magnitude spectrum of the periodic pulse train is given in Fig. 2.23(b), and the envelope of the spectrum has the sinc function shape. The sampled sequence $x_s(t)$ of a band-limited analog signal $x(t)$ (Fig. 2.20(a)) can be expressed as

$$x_s(t) = x(t)x_p(t) = x(t) \sum_{n=-\infty}^{\infty} a_n e^{j2\pi n f_s t}. \tag{2.4.10}$$

This kind of sampling is referred to as *natural sampling*, since the top of each pulse in $x_s(t)$ retains the same shape of its corresponding analog waveform segment in the pulse interval as illustrated in Fig. 2.23(c.). The Fourier transform $X_s(f)$ of the sampled signal is as follows:

$$X_s(f) = \boldsymbol{F}\left\{x(t)\sum_{n=-\infty}^{\infty} a_n e^{j2\pi n f_s t}\right\} = \sum_{n=-\infty}^{\infty} a_n X(f - f_s). \qquad (2.4.11)$$

The spectrum of the natural sampled signal is depicted in Fig. 2.23(d).

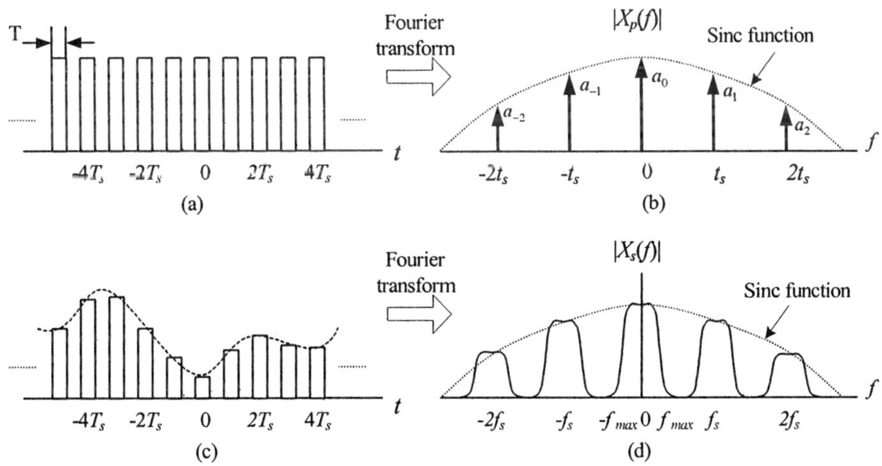

Figure 2.23. Finite pulse sampled signal and its spectrum

In reality, the most popular sampling method is the *sampling and hold*. It can be represented by the convolution of the sampled impulse train (2.4.3), $[x(t)x_\delta(t)]$, with a unity amplitude rectangular pulse $p(t)$ of pulse width T_s:

$$x_s(t) = p(t) * [x(t)x_\delta(t)] = p(t) * \left[x(t)\sum_{n=-\infty}^{\infty}\delta(t - nT_s)\right]. \qquad (2.4.12)$$

This convolution results in a flat-top sampled sequence. Its Fourier transform is the product of the Fourier transform $P(f)$ of the rectangular pulse and the spectrum (2.4.5) of the impulse sampled sequence — i.e.,

$$X_s(f) = P(f)\frac{1}{T_s}\sum_{n=-\infty}^{\infty} X(f - nf_s), \qquad (2.4.13)$$

where $P(f) = T_s \text{sinc}(fT_s)$. The spectrum of the sample and hold sequence is similar to that presented in Fig. 2.23(d). The hold operation will significantly attenuate the high-frequency replicates. Usually postfiltering is necessary to further suppress the residual spectral components of the replicates at the multiples of the sampling frequency.

A typical sample and hold circuit is shown in Fig. 2.24. In this circuit, the sampled analog voltage is held on capacitor C_H during the analog to digital conversion. After the sample and hold circuit, usually an analog-to-digital converter follows.

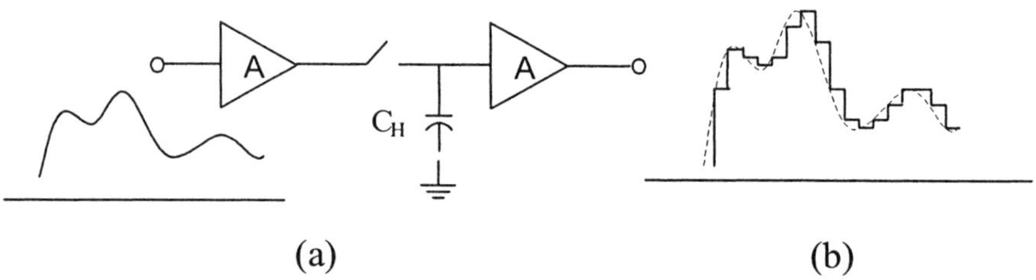

Figure 2. 24. Typical sample and hold circuit (a) and its output (b)

The pulses presented in Fig. 2.23(c) are called quantized samples, and the output waveform of sample and hold circuit is shown in Fig. 2.24(b). These formats can be interfaced with a digital system when the sample values are quantized to a finite set. After quantization, the analog waveform can still be recovered to certain precision, but the reconstruction fidelity can be improved by increasing the number of the quantization level.

2.4.2. Jitter Effect of Sampling and Quantizing Noise

The sampling theorem predicted precise reconstruction of the signal is based on uniformly spaced samples of the signal. The sampling becomes no longer uniform if there is a slight jitter in the position of samples. The jitter is usually a random process, and the position of the samples is not exactly known. The jitter effect is equivalent to FM modulation of the base-band signal. If the jitter is random, noise with a low-level wide-band spectrum is induced. If the jitter appears in a periodic manner, an FM disturbance with low-level discrete spectral lines is generated.

The jitter effect can be measured by means of *signal-to-noise ratio* (*SNR*). If the input signal is

$$v(t) = A\sin(2\pi ft), \tag{2.4.14}$$

where A and f are amplitude and frequency of the signal, respectively. The time derivative of the signal is

$$\frac{dv(t)}{dt} = 2\pi fA\cos 2\pi ft\ . \tag{2.4.15}$$

The *root mean square* (*rms*) value of the derivative can be expressed as

$$\frac{\Delta v_{j\_rms}}{\Delta t_{j\_rms}} = 2\pi fA\sqrt{\int_{-1/2f}^{1/2f}\cos^2(2\pi ft)dt} = \sqrt{2}\pi fA, \tag{2.4.16}$$

where Δv_{j_rms} is the rms jitter noise, and Δt_{j_rms} represents the rms value of the jitter time. If the jitter has a normal distribution with a zero mean and a variance of $\sigma_j^2 = \Delta t_{j\_rms}^2$, the rms noise voltage can be then expressed as [11]

$$\Delta v_{j\_rms} = \sqrt{2}A\pi f\sigma_j \tag{2.4.17}$$

The signal-to-jitter noise ratio then can be written as

$$\begin{aligned} SNR_{jitter} &= 20\log\left(\frac{A/\sqrt{2}}{\Delta v_{j\_rms}}\right) \\ &= 20\log\left(\frac{1}{2\pi f\cdot\Delta t_{j\_rms}}\right) = 20\log\left(\frac{1}{2\pi f\cdot\sigma_{j\_rms}}\right). \end{aligned} \tag{2.4.18}$$

In the quantization process, this inherent distortion results from the truncation error. This distortion is referred to as *quantization noise*. The step size between the quantization levels is called the *quantile interval* and is denoted as Δq volts. When the quantization levels are uniformly distributed over the full range, the quantizer is called a *uniform quantizer*. Each quantized sample value of the analog waveform is an approximation of the true value. The error between the approximation and the true value will be within $\pm\Delta q/2$. We assume that the quantization error is uniformly distributed over a single quantile interval Δq. Thus, the probability density function of the quantization error is $1/\Delta q$. The quantization noise or error variance is obtained from

$$P_{N_q} = \sigma_q^2 = \int_{-\Delta q/2}^{\Delta q/2} \frac{x^2}{\Delta q} dx = \frac{\Delta q^2}{12}. \tag{2.4.19}$$

If the peak-to-peak maximum voltage swing $V_{p\text{-}p}$ of the analog signal can be quantized into L_q levels — i.e., $V_{p-p} = L_q \cdot \Delta q$ — the peak power for a sine wave signal can be expressed as

$$P_{peak} = \frac{1}{2Z}\left(\frac{V_{p-p}}{2}\right)^2 = \frac{1}{2Z}\left(\frac{L_q \Delta q}{2}\right)^2 = \frac{L_q^2 \cdot \Delta q^2}{8 \cdot Z}, \tag{2.4.20}$$

where Z is the input impedance of the quantizer. The ratio of peak signal power to quantization noise SNR_q then is

$$SNR_q = \frac{L_q^2 \Delta q^2 / 8}{\Delta q^2 / 12} = \frac{3L_q^2}{2}. \tag{2.4.21}$$

When we consider the jitter noise and quantization noise both, the total noise from a quantizer is

$$P_{N_{jq}} = \frac{\Delta q^2}{12} + 2(A\pi f \sigma_j)^2. \tag{2.4.22}$$

For the sinusoidal signal with amplitude A, the signal-to-noise ratio is

$$SNR_{jq} = 10\log\left[\frac{A^2}{2\left(\frac{\Delta q^2}{12} + 2(\pi f A \sigma_j)^2\right)}\right] = 10\log\left[\frac{3L_q^2}{2 + 3(2\pi f L_q \sigma_j)^2}\right]. \tag{2.4.23}$$

The quantization levels L_q are usually represented by bits — i.e.,

$$L_q = 2^b, \tag{2.4.24}$$

where b is the bit number of a quantizer. For example, a 4-bit quantizer has 16 quantization levels.

2.4.3. Commonly Used Modulation Schemes

To transmit a voice or data message through a wireless system, the message information or message source must be encoded into a manner suitable for transmission. The translation process from a base-band message source to a band-pass signal, which is suitable for transmission, is referred to as *modulation*. The modulation may be implemented by means of amplitude, frequency, or phase variation of the transmission carrier. The carrier frequency for the wireless systems can be hundreds or thousands of MHz. *Demodulation* is a reverse process of the modulation. In a receiver, the base-band message is extracted or demodulated from the modulated carrier.

In this section, only the most commonly used modulation schemes in the wireless mobile systems are addressed. The FM modulation was the most popular modulation employed in the analog wireless mobile systems, such as the AMPS system. However, all the second- and third-generation mobile systems employ digital modulation. The digital modulation has many advantages including greater noise immunity, more robustness to channel impairments, easier multiplexing of voice, data, and image information, and better security. M-ary phase shift keying (BPSK, QPSK, OQPSK, etc.), minimum shift keying (MSK), and M-ary quadrature amplitude (16 QAM and 64 QAM) are the most popular digital modulations adopted by different protocol wireless systems.

Power efficiency and *bandwidth efficiency* are two main criteria of selecting modulation scheme. The *power efficiency* η_P is often defined as the ratio of the signal energy per bit E_b to noise power spectral density N_o — i.e., E_b/N_o required at the receiver input for a certain probability of error. The *bandwidth efficiency* η_{BW} is expressed as

$$\eta_{BW} = \frac{R}{BW} \quad \text{bps/Hz}, \tag{2.4.25}$$

where R is data rate in bits per second (bps), and BW is the bandwidth of the modulated RF signal.

Shannon capacity theorem [12] limits the maximum bandwidth efficiency in a channel contaminated by additive white Gaussian noise (AWGN). The channel-capacity relationship and therefore the limitation of the bandwidth efficiency are presented by

$$\eta_{BW\_max} = \frac{C}{BW} = \log_2\left(1 + \frac{S}{N}\right) \text{ bps/Hz}, \tag{2.4.26}$$

where C is the channel capacity in bps, and S/N is the signal-to-noise ratio. For the case where the data rate equals the channel capacity — i.e., $R = C$ — considering $S/C = E_b$ the signal energy per bit, (2.4.26) can be rewritten in the following form

$$\eta_{BW\_max} = \log_2\left(1 + \frac{E_b}{N_o}\eta_{BW\_max}\right), \tag{2.4.27}$$

where N_o is noise power spectral density. It is apparent from (2.4.27) that E_b/N_o can be expressed as a function of η_{BW_max}

$$\frac{E_b}{N_o} = \frac{(2^{\eta_{BW\_max}} - 1)}{\eta_{BW\_max}}. \tag{2.4.28}$$

From (2.4.28), we can derive the limiting value of E_b/N_o equal to 0.639 or –1.59 dB for an error-free communication by letting $\eta_{BW\_max} \to 0$.

In wireless communication system design, there is always a need for tradeoffs between bandwidth efficiency and power efficiency when choosing modulation schemes. For example, the GSM system employs a Gaussian pulse shaping minimum shift keying (GMSK) modulation, which results in a carrier with *constant envelope*. This modulation allows a transmitter power amplifier operating in high efficiency. The GSM system has a raw data rate of 22.8 kbps while occupying a bandwidth of approximate 200 kHz. To raise the data rate to over 300 kpbs but to keep the same occupying bandwidth as the GSM signal, the EDGE system was developed. The EDGE system uses an 8-ary PSK modulation to achieve a better bandwidth efficiency. However, in exchange for the high bandwidth efficiency, we obtain a lower power efficiency since the carrier envelope now varies with a swing possessing approximate a 3 dB peak-to-average ratio. This means that we need use a low efficiency linear (or close to linear) power amplifier in the transmitter of the EDGE system.

2.4.3.1. Analog Frequency Modulation (FM)

Frequency modulation (*FM*) is commonly employed in the analog mobile communication systems such as AMPS in the Americas and TACS in Europe. FM is a form of angle modulation in which the instantaneous carrier frequency is varied linearly with the base-band message signal $m(t)$. The FM signal is expressed as

$$v_{FM}(t) = A_o \cos\left(2\pi f_o t + 2\pi k_f \int_{-\infty}^{t} m(x)dx + \theta\right), \tag{2.4.29}$$

where A_o is the amplitude of the carrier, f_o is the carrier frequency, k_f is the frequency deviation constant, and θ is an arbitrary phase shift. The quadrature form of the FM modulation is as the following:

$$v_{FM}(t) = I(t)\cos(2\pi f_o t + \theta) - Q(t)\sin(2\pi f_o t + \theta), \tag{2.4.30}$$

where

$$I(t) = A_o \cos\left(2\pi k_f \int_{-\infty}^{t} m(x)dx\right) \text{ and } Q(t) = A_o \sin\left(2\pi k_f \int_{-\infty}^{t} m(x)dx\right). \tag{2.4.31}$$

FM is a modulation with constant envelope since the amplitude of its carrier is kept constant. This modulation has high power efficiency. In the analog mobile systems, a sinusoidal signal is usually used as the test signal to measure the performance of a receiver in a mobile station. For example, in the AMPS a 1 kHz sinusoidal signal with amplitude providing 8 to 12 kHz maximum frequency deviation is employed as the test signal. If the test tone has an amplitude A_m, and a frequency f_m — i.e., $m(t) = A_m \cos(2\pi f_m t)$ — the FM signal is expressed as [13]

$$v_{FM}(t) = A_o \cos\left[2\pi f_o t + \beta_f \sin(2\pi f_m t) + \theta\right], \tag{2.4.32}$$

where β_f is the frequency modulation index defined as the ratio of the peak frequency deviation Δf_p to the modulation frequency f_m or

$$\beta_f = \frac{A_m k_f}{f_m} = \frac{\Delta f_p}{f_m}. \tag{2.4.33}$$

The rule-of-thumb relation (2.4.34) approximately gives the RF bandwidth of an FM:

$$BW = 2\Delta f_p + 2B, \tag{2.4.34}$$

where B is the base-band bandwidth of the modulation message, and $B = f_m$ in the sinusoidal modulation signal case. The test tone used in the AMPS has a 1 kHz frequency and provides 8 to 12 kHz peak frequency deviation. The bandwidth is approximately in the range of 18 to 26 kHz.

2.4.3.2. Digital Modulation

In digital communications, voice and data message are translated into a binary-source data stream $\{m_i\}$, and then the source data stream is converted into in-phase and quadrature binary bit streams $\{m_{I,i}\}$and $\{m_{Q,i}\}$. They are the digital modulation signals. In digital modulation, the modulating signal may be represented as time sequence of symbols. Each symbol of an M-ary keying modulation has M finite states, and it consists of $n_b = \log_2 M$ bits of source data stream.

Waveforms of digital modulation schemes commonly used in the important wireless mobile systems can be expressed in complex envelope form

$$M(t) = I(t) + jQ(t) = A(t)e^{j\varphi(t)}, \tag{2.4.35}$$

where $I(t)$ and $Q(t)$ are in-phase and quadrature envelope waveforms, respectively, and they have forms as follows:

$$I(t) = \sum_k I_k p_I(t - kT_s - \tau) \tag{2.4.36a}$$

and

$$Q(t) = \sum_k Q_k p_Q(t - kT_s - \tau). \tag{2.4.36b}$$

In (2.4.36), I_k and Q_k are sequences of discrete variables mapped from message data with a symbol rate of $1/T_s$, $p_I(t)$ and $p_Q(t)$ are finite energy pulses (such as rectangular, filtered rectangular, or Gaussian), τ is possible delay, and the envelope amplitude $A(t)$ and the phase $\varphi(t)$ are, respectively,

$$A(t) = \sqrt{I^2(t) + Q^2(t)} \tag{2.4.37}$$

and

$$\varphi(t) = \tan^{-1} \frac{Q(t)}{I(t)}. \tag{2.4.38}$$

The I_k and Q_k sequences of modulations of practical interest, such as MPSK, QPSK, OQPSK, π/4QPSK, MSK, and MQAM, are presented in Table 2.1 [8]. Most of the finite energy pulses $p_I(t)$ and $p_Q(t)$ given in Table 2.1 are rectangular except the MSK modulation, and the pulse shaping techniques (other than rectangular pulse shaping for the bandwidth efficiency) are discussed in the next section. Usually, the I_k and Q_k sequences have a symbol rate $1/T_s$ or symbol duration T_s. The symbol duration T_s of an M-ary keying modulation is related to the bit duration T_b of originally binary data stream as

$$T_s = \log_2 M \cdot T_b. \tag{2.4.39}$$

Thus, $T_s = 2T_b$ for the QPSK, OQPSK, π/4QPSK, and MSK modulation.

Table 2.1. Parameters in (2.4.36) of modulation schemes commonly used in mobile wireless systems

Modulation Scheme	(I_k, Q_k)	$p_I(t), p_Q(t)$
M-PKS	$I_k + jQ_k = e^{j\varphi_k}$ $\varphi_k = 2n_k\pi/M,\ n_k = 0, 1, \ldots M\text{-}1$	$p_I(t) = 1,\ 0 \le t \le T_s$ $p_I(t) = p_Q(t)$
QPSK	$(I_k, Q_k) = (\pm 1, \pm 1)$ or $\varphi_k = \pm\pi/4, \pm 3\pi/4$	$P_I(t) = 1,\ 0 \le t \le T_s$ $p_I(t) = p_Q(t)$
OQPSK	$(I_k, Q_k) = (\pm 1, \pm 1)$ or $\varphi_k = \pm\pi/4, \pm 3\pi/4$	$p_I(t) = 1,\ 0 \le t \le T_s$ $p_Q(t) = 1,\ T_s/2 \le t \le 3T_s/2$
π/4 DQPSK	$I_k + jQ_k = e^{j\theta_k}$ $\theta_k = \theta_{k-1} + \varphi_k$ $\varphi_k = \pm\pi/4, \pm 3\pi/4$	$p_I(t) = 1,\ 0 \le t \le T_s$ $p_I(t) = p_Q(t)$
MSK	$(I_k, Q_k) = (\pm 1, \pm 1)$	$P_I(T) = COS(\pi T/T_S), 0 \le T \le T_s$ $p_Q(t) = \sin(\pi t/T_s), 0 \le t \le T_s$
M-QAM	$(I_k, Q_k) = [\pm 1, \pm 3, \ldots,$ $\pm(\sqrt{M} - 1)]$	$p_I(t) = 1,\ 0 \le t \le T_s$ $p_I(t) = p_Q(t)$

The in-phase and quadrature base-band waveforms of MPSK have forms

$$I_{MPSK}(t) = \sum_k \cos\left(\frac{n_k 2\pi}{M}\right) \tag{2.4.40a}$$

and

$$Q_{MPSK}(t) = \sum_k \sin\left(\frac{n_k 2\pi}{M}\right), \tag{2.4.40b}$$

where $n_k = 0, 1, 2, \ldots,$ or $M-1$ depending on the values of $m_{I,}$ and m_Q streams during the corresponding interval. The *I(t)* and *Q(t)* of QPSK and OQPSK are the special case of (2.4.40) with M = 4, and they can be expressed in 45° phase shifted as

$$I_{QPSK}(t) = \sum_k \cos\left(\frac{n_k \pi}{2} + \frac{\pi}{4}\right) \tag{2.4.41a}$$

and

$$Q_{MPSK}(t) = \sum_k \sin\left(\frac{n_k \pi}{2} + \frac{\pi}{4}\right). \tag{2.4.41b}$$

Although OQPSK has the same *I(t)* and *Q(t)* expressions as QPSK, in OQPSK *I(t)* and *Q(t)* are offset in their alignment by half a symbol period. The phase switching in OQPSK is more often than in QPSK, and thus OQPSK eliminate 180° phase transition that may occur in QPSK. Nonlinear amplification of OQPSK signal does not regenerate high frequencies as much as in QPSK, and therefore OQPSK is very attractive for mobile communication systems where bandwidth efficiency and low power consumption are critical. These digital modulation schemes are used in CDMA wireless communication system.

The most attractive feature of the π/4QPSK modulation is its noncoherent detection that greatly simplifies the receiver design. It is employed in TDMA (IS-54/136) and PHS mobile communication systems. The base-band *I(t)* and *Q(t)* waveforms are represented by

$$I_{\pi/4QPSK}(t) = \sum_k I_k = \sum_k \left(I_{k-1}\cos\varphi_k - Q_{k-1}\sin\varphi_k\right) \tag{2.4.42a}$$

and

$$Q_{\pi/4QPSK}(t) = \sum_k Q_k = \sum_k \left(I_{k-1}\sin\varphi_k - Q_{k-1}\cos\varphi_k\right), \tag{2.4.42b}$$

where $\varphi_k = \pi/4$, $-\pi/4$, $3\pi/4$, or $-3\pi/4$, depending on the values of message data stream during the corresponding symbol interval.

Modified minimum shift keying (MSK) modulation is applied in a GSM mobile communication system. MSK is continuous phase frequency shift keying (FSK) with a modulation index 0.5. The modulation index is defined as $k_{FSK} = 2\Delta F_{max} / R_b$, where ΔF_{max} is the peak frequency deviation and R_b is the bit rate. In MSK, the base-band *I(t)* and *Q(t)* waveforms have the following forms:

$$I_{MSK}(t) = \sum_k I_k \cos\left(\frac{\pi(t - k \cdot 2T_b)}{2T_b}\right) \tag{2.4.43a}$$

and

$$Q_{MSK}(t) = \sum_k Q_k \sin\left(\frac{\pi(t - k \cdot 2T_b)}{2T_b}\right), \tag{2.4.43b}$$

where I_k and Q_k are message data-dependent and equal to +1 or –1, and they change only every $2T_b$ seconds.

Based on expressions (2.4.35) and (2.4.36), a digital modulation signal can be represented in terms of a two-dimensional constellation diagram. For examples, constellation diagrams of QPSK, π/4QPSK, and 16QAM are depicted in Fig. 2.25.

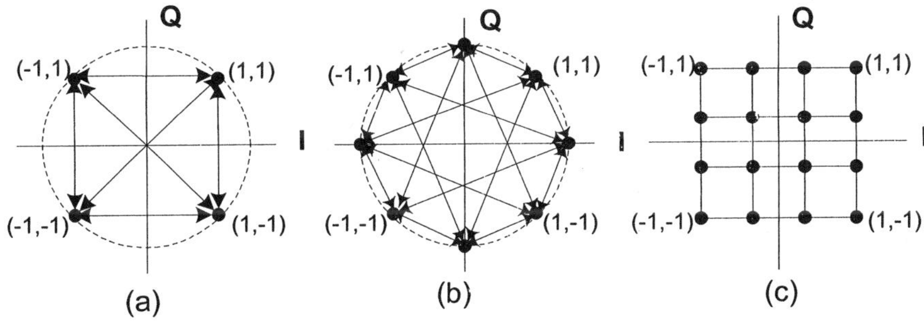

Figure 2.25. Constellation diagrams of (a) QPSK, (b) π/4QPSK, and (c) 16 QAM

An RF carrier modulated by a digital modulation scheme can be expressed in the following form similar to (2.4.30):

$$v_{RF}(t) = I(t)\cos(2\pi f_o t + \theta) - Q(t)\sin(2\pi f_o t + \theta), \tag{2.4.44}$$

where *I*(*t*) and *Q*(*t*) have the form of (2.4.36), and depending on the modulation scheme they can be (2.4.40), (2.4.41), (2.4.42), or (2.4.43), f_o is the RF carrier frequency, and θ is an arbitrary phase shift of the RF carrier.

2.4.3.3. Spread-Spectrum Modulation

Spread-spectrum techniques employ a transmission bandwidth much greater than the minimum bandwidth required for transmitting the message data. The advantage of spread spectrum is that many users can share the same bandwidth without significantly interfering with each other. Spread-spectrum modulation has inherent interference rejection capability. A spread-spectrum signal can be identified from a number of other spread-spectrum signals. It can also be recovered from other types of interference, such as narrow-band jammers. Another feature of the spread-spectrum signal is that it is not sensitive to multipath fading. All these features make the spread-spectrum modulation very suitable for application in the wireless mobile communications.

The most commonly used techniques for spectrum spreading are *direct sequence* (DS) and *frequency hopping* (FH). In these two spread-spectrum techniques, a *pseudo noise* (PN) sequence is used for spreading and despreading. The pseudo noise is not a truly random signal, and it is a deterministic and periodic signal. However, it has noise-like properties when compared with digital message data.

A PN sequence is a periodic binary (1 and 0) sequence, which can usually be generated by means of a *feedback shift register*. A feedback shift register consisting of a k-stage register for storage and a logic circuit that is connected to form a multiloop feedback are as depicted in Fig. 2.26. The binary PN sequence can be mapped to a two-level PN waveform, and it may be represented by

$$c(t) = \sum_{n=-\infty}^{\infty} c_n p(t - nT_c) \tag{2.4.45}$$

where $\{c_n\}$ is the binary PN sequence of ± 1s and $p(t)$ is a rectangular pulse of duration T_c. Each single pulse of the PN waveform is called a *chip*.

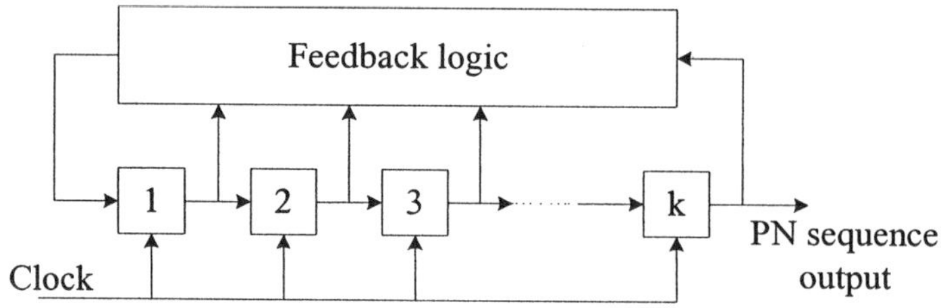

Figure 2. 26. Block diagram of generalized feedback shift register with k stages

When the feedback logic is entirely made up of modulo-2 adders, the feedback shift register is linear. A PN sequence is called as *a maximum length sequence* if the period of the PN sequence generated by a linear feedback register is $N = 2^k - 1$. The maximum length sequence has broad applications, and many important properties. In each period of a maximum length sequence, the number of 1s is always one more than the number of 0s. Its autocorrelation is periodic and binary valued. The autocorrelation function of a maximum length PN sequence with period $T_{PN} = NT_c$ is expressed as

$$R_p(\tau) = \begin{cases} 1 - \dfrac{N+1}{NT_c}|\tau|, & |\tau| \le T_c \\ -\dfrac{1}{N}, & \text{for the remainder of the period} \end{cases} \tag{2.4.46}$$

where τ is the time shift between two cross-correlating PN sequences. The plot of (2.4.46) is shown in Fig. 2.27. This kind of autocorrelation function is one of the most properties of the PN sequence.

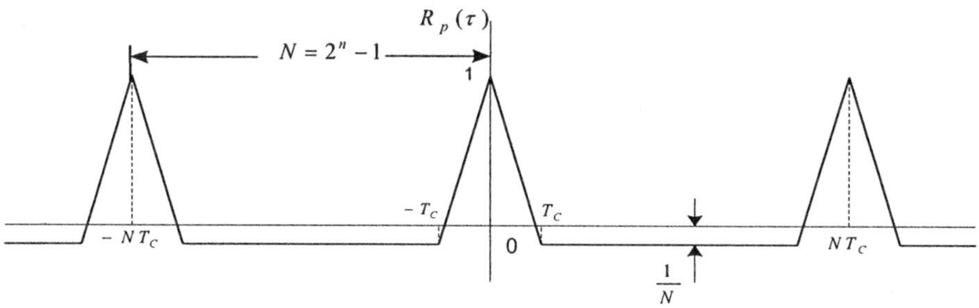

Figure 2.27. Autocorrelation function of maximum length PN waveform

A *direct-sequence spread-spectrum* (DS-SS) system spreads the modulated base-band data signal by directly multiplying this base-band data with a binary PN waveform. The base-band data signal $m(t)$ is denoted by

$$m(t) = \sum_{m=-\infty}^{\infty} a_m p_m(t - mT_b), \tag{2.4.47}$$

where $\{a_m\}$ represents the data sequence of ± 1, and $p_m(t)$ is a rectangular pulse of duration T_b. The spreading PN sequence $c(t)$ has the form as given in (2.4.45). The spread-spectrum signal can then be expressed as

$$v_{DS}(t) = A \cdot m(t)c(t) \cdot \cos(2\pi f_o t + \varphi), \tag{2.4.48}$$

where $A \cdot \cos(2\pi f_o t + \varphi)$ is an RF carrier, and φ is an arbitrary phase shift.

Processing gain of the direct sequence spread spectrum is defined as

$$PG_{DS} = \frac{T_b}{T_c} = \frac{R_c}{R_b}. \tag{2.4.49}$$

It represents the gain achieved by processing a spread-spectrum signal over the unspread signal. On the other hand, the ratio of R_c / R_b is also an approximate measure of the interference rejection capability.

Many DS spread-spectrum signals are able to share the same channel bandwidth provided that each signal has its own PN (signature) sequence. Thus, it is possible to have multiple users transmit messages simultaneously over the same channel bandwidth. The digital communication system in which each user has its own signature code for transmitting message data over a common channel bandwidth is called *code division multiple access* (*CDMA*). The spreading code used in the CDMA IS-2000 and WCDMA systems is not only the PN sequence but also the 64 orthogonal Walsh codes [14–15].

The *frequency-hopping spread-spectrum* (FH-SS) system is another commonly used spectrum-spreading system in the wireless communications. Frequency hopping means a periodic change of transmission frequency over an available bandwidth. The available frequency band is divided into a large number of nonoverlapping frequency slots called the *hopset*. Each frequency slot has a bandwidth large enough to contain most of the power in a narrow-band modulation burst (usually MFSK), and its central frequency is the carrier frequency of the modulated data bursts. The frequency slot or the carrier for each modulation burst is pseudo-randomly selected according to a PN sequence. Therefore, a frequency-hopping signal may be considered as a sequence of modulated data bursts with pseudo-randomly changing carrier frequencies. A block diagram of FH signal generation is shown in Fig. 2.28. An example of a frequency-hopping pattern is presented in Fig. 2.29.

The bandwidth of a frequency slot in the hopset is referred as the *instantaneous bandwidth B*. The bandwidth over which the frequency hopping occurs is called *total hopping bandwidth W*. The processing gain for FH systems is defined as

$$PG_{FH} = \frac{W}{B}. \tag{2.4.50}$$

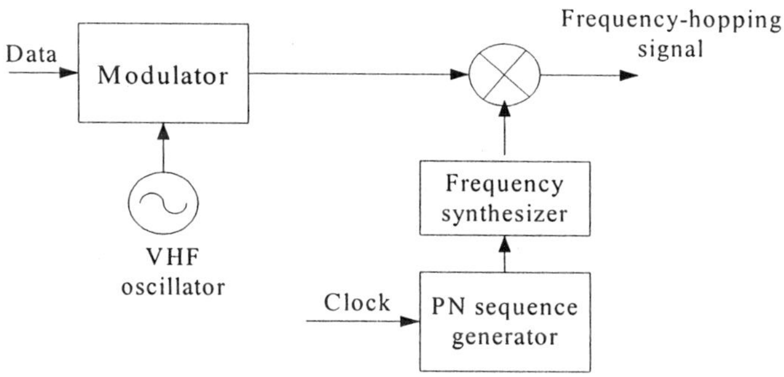

Figure 2.28. Block diagram of FH signal generation

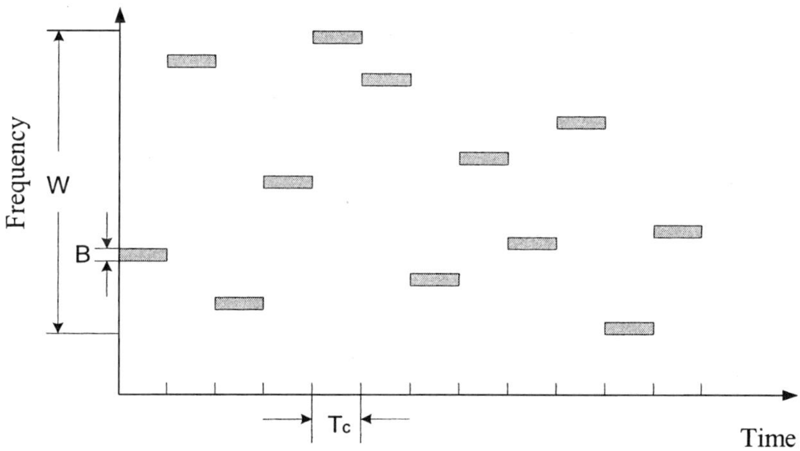

Figure 2.29. Example of a frequency hopping pattern

Frequency hopping may be classified as fast or slow. If the frequency-hopping rate R_{FH} is equal to or less than symbol rate R_S, the frequency-hopping is called *slow hopping*. If there are multiple hops per symbol — i.e., $R_{FH} > R_S$, the frequency hopping is called *fast hopping*.

The frequency-hopping technique is employed by the GSM mobile system and the Bluetooth wireless short-distance connecting system. The frequency hopping used in both systems are slow hopping.

2.4.4. Pulse-Shaping Techniques and Intersymbol Interference (ISI)

The available frequency bandwidth for mobile communication systems is limited. It is always desirable to make the modulated digital signal occupying small bandwidth, and thus the available bandwidth can be more efficiently utilized. In the previous Section 2.4.3, it is assumed that the individual pulse of each symbol in the digital modulation waveform is rectangular for simplicity. In reality, the rectangular pulse is rarely employed in wireless communication systems except in a GPS system since its spectrum is a sinc function, which has relatively high and slowly decaying side lobes. The rectangular pulses in the modulation waveform need be shaped to reduce the modulation bandwidth and suppress radiation in the adjacent channels. In the pulse-shaping techniques, band-limiting filters are usually utilized to shape the rectangular pulse. After pulse shaping, however, the individual pulse of each symbol will spread in time, and its tail will smear into the intervals of adjacent symbols to interfere with them. This interference is termed as *intersymbol interference* (*ISI*).

In pulse shaping- techniques, two important criteria are used to reduce the spectral bandwidth of a modulated digital signal and to simultaneously minimize the ISI effect. Nyquist [16] showed that the ISI could be zero if the overall response of a system (including transmitter, channel, and receiver) is designed so that at every sampling instant, the response due to all symbols except the current symbol is equal to zero. Assuming that the impulse response of the overall system is $h_{sys}(t)$, then the mathematical represent of the above statement is

$$h_{sys}(kT_s) = \begin{cases} A & k = 0 \\ 0 & k \neq 0, \end{cases} \tag{2.4.51}$$

where T_s is the symbol period, k is an integer, and A is a nonzero constant. The impulse response given in (2.4.52) meets the condition of (2.4.51):

$$h_{sys}(t) = \frac{\sin(\pi t/T_s)}{\pi t/T_s}. \tag{2.4.52}$$

This is a sinc function with zero crossings at kT_s ($k = \pm 1, \pm 2, \ldots, \pm n, \ldots$). However, the sinc impulse response is not physically realizable since it implies an infinite time delay and a frequency response of rectangular brick wall as

$$H_{sys}(f) = \frac{1}{R_s}\Pi\left(\frac{f}{R_s}\right), \tag{2.4.53}$$

where $\Pi(f/R_s)$ is a rectangular function and its value equals 1 when $f \le R_s/2$ and equals 0 when $f > R_s/2$.

The following impulse response, which also satisfies conditions given in (2.4.51) — i.e., without ISI — can usually be implemented in the physical world.

$$h_{sys}(t) = \frac{\sin(t/T_o)}{t} \cdot z(t) \tag{2.4.54}$$

where $z(t)$ is an even function and its frequency response has zero magnitude outside the rectangular spectrum of the Fourier transform of $sinc(t/T_o)$, and $1/(2T_o) \ge 1/(2T_s)$ is the bandwidth of the rectangular spectrum. If the overall system can be modeled as a pulse-shaping filter, then (2.4.54) is the impulse response of the filter and its Fourier transform is the transfer function or frequency response of the filter.

2.4.4.1. Raised Cosine Pulse Shaping

One of the pulse-shaping filters most popularly used in mobile communication systems is the raised cosine filter. A raised cosine filter meets the criteria of (2.4.51), and it has a (2.4.54) type of impulse response. The corresponding impulse response has a form as

$$h_{RC}(t,\alpha) = \frac{\sin(\pi t/T_s)}{\pi t/T_s} \cdot \frac{\cos(\pi\alpha t/T_s)}{1-(2\alpha t/T_s)^2}, \tag{2.4.55}$$

where α is a roll-off factor with a value between 0 and 1. The impulse responses of the raised cosine filters with $\alpha = 1$, 0.5, and 0.2 are shown in Fig. 2.30.

The frequency response or the transfer function of the raised cosine filter can be obtained from the Fourier transform of (2.4.55), and it is expressed as

$$H_{RC}(f,\alpha)=\begin{cases}1 & 0\le|f|\le(1-\alpha)/2T_S\\ \dfrac{1}{2}\left[1+\cos\left(\dfrac{\pi\left(2T_S|f|-1+\alpha\right)}{2\alpha}\right)\right] & (1-\alpha)/2T_S<|f|\le(1+\alpha)/2T_S\\ 0 & |f|>(1+\alpha)/2T_S\end{cases}$$

(2.4.56)

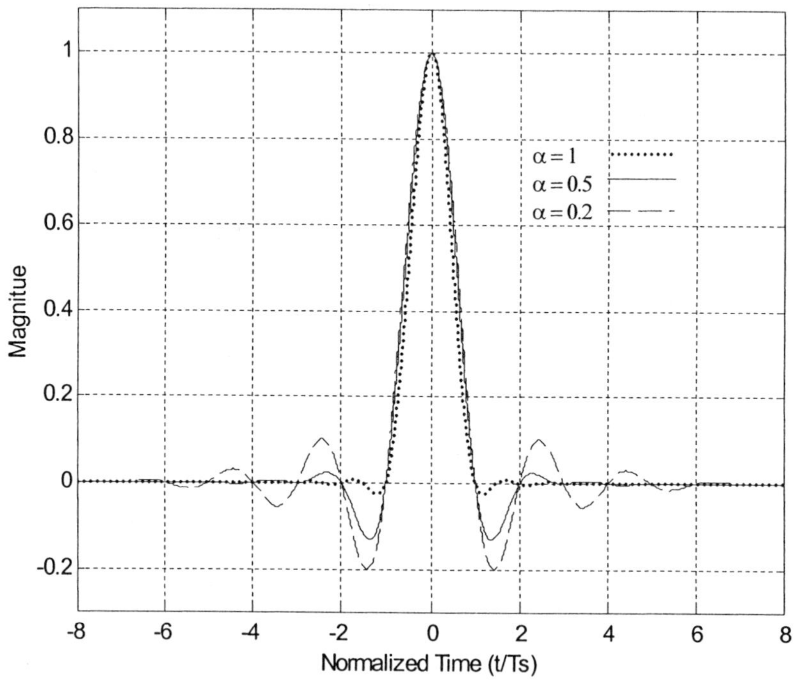

Figure 2.30. Impulse responses of raise cosine filters

Fig. 2.31 presents three power spectra of raised cosine filtered signals when the roll-off factor α equals 1, 0.5, and 0.2, respectively. The spectrum is a "brick wall" rectangular when $\alpha = 0$. The absolute bandwidth B of a base-band raised cosine filter with a roll-off factor α is associated with symbol rate R_s as

$$B=\frac{1}{2}(1+\alpha)R_s \tag{2.4.57}$$

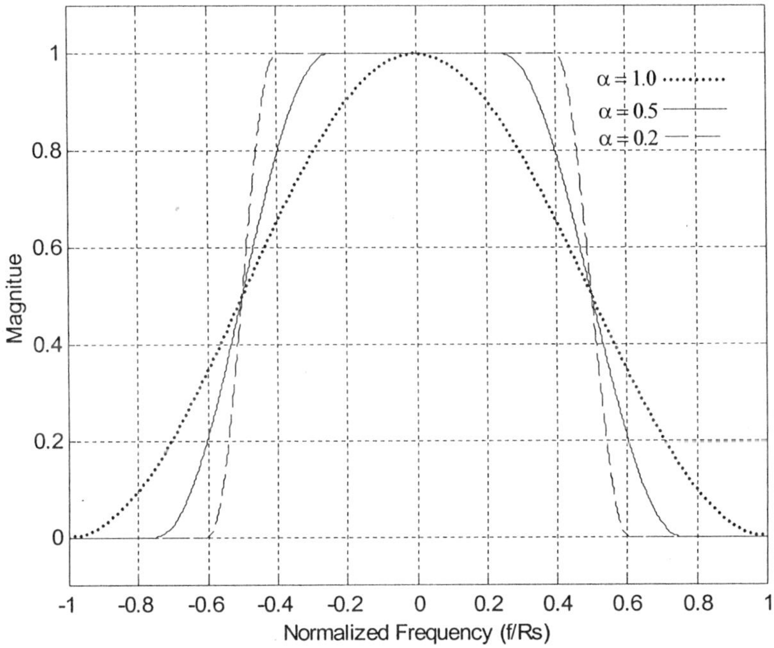

Figure 2. 31. Frequency responses of raised cosine filters

For a band-pass RF signal, which is shaped by a raised cosine filter, the bandwidth BW is twice of the base-band bandwidth, and thus it can be expressed by R_s as

$$BW = (1 + \alpha)R_s . \quad (2.4.58)$$

When α = 0.5, to transmit an RF signal with a symbol rate R_s the raised cosine filter should have a base-band bandwidth $B = 3R_s/4$ and an RF bandwidth $BW = 1.5\ R_s$.

The impulse response and the frequency response presented in (2.4.55) and (2.4.56) should be the responses of the overall system. In a mobile communication system usually consisting of transmitter, channel, and receiver, a root raised cosine filter will be used in the transmitter and receiver each side. The composition characteristic of two root raised cosine filters with the same α is equal to that of a raised cosine filter. The impulse response of a root raised cosine filter is expressed as

$$h_{RRC}(t,\alpha)=\begin{cases} 1-\alpha+4\dfrac{\alpha}{\pi} & \text{for } t=0 \\ \dfrac{1}{1-16\alpha^2(t/T_S)^2}\left[\dfrac{\sin[(1-\alpha)\pi t/T_S]}{\pi t/T_S}+4\alpha\dfrac{\cos[(1+\alpha)\pi t/T_s]}{\pi}\right] & \\ \dfrac{\alpha}{\sqrt{2}}\left[\left(1+\dfrac{2}{\pi}\right)\sin\dfrac{\pi}{4\alpha}+\left(1-\dfrac{2}{\pi}\right)\cos\dfrac{\pi}{4\alpha}\right] & \text{for } t=\dfrac{T_S}{4\alpha} \end{cases}$$

(2.4.59)

The impulse response of the root raised cosine filter has non-zero crossing at the peak of adjacent symbol pulses as depicted in Fig. 2. 32, but in a communication system the root raised cosine filter is usually used in a pair. Thus the composite time response of the overall system has no ISI issue.

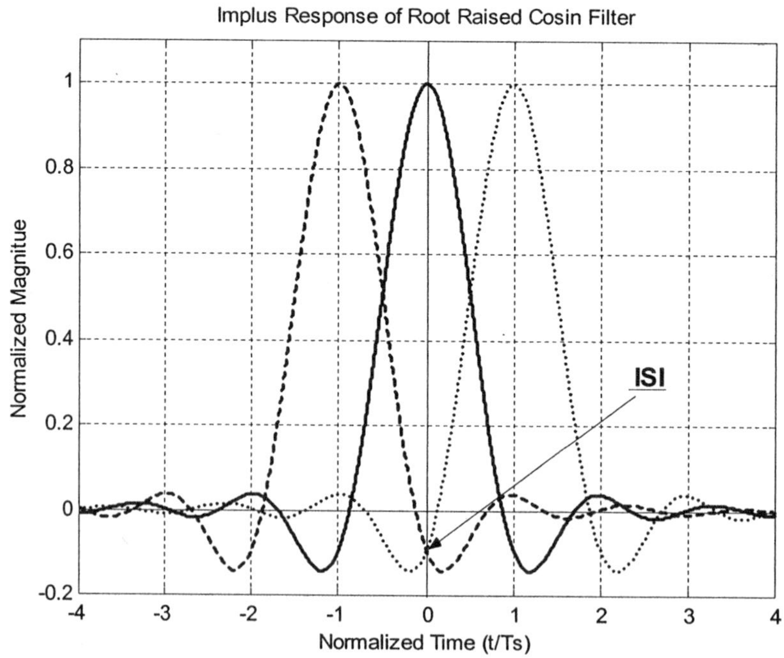

Figure 2. 32. Impulse response of root raised cosine filter with $\alpha = 0.5$ and showing nonzero crossing at an adjacent symbol pulse peak

The frequency response of a root raised cosine filter has a form as

$$H_{RRC}(f,\alpha)=\begin{cases}1 & 0\le|f|<(1-\alpha)/2T_s\\ \cos\left[\dfrac{T_s}{4\alpha}\left(2\pi|f|-\dfrac{\pi(1-\alpha)}{T_s}\right)\right] & (1-\alpha)/2T_s\le|f|\le(1+\alpha)/2T_s\\ 0 & (1+\alpha)/2T_s\le|f|\end{cases} \tag{2.4.60}$$

2.4.4.2. Gaussian Pulse Shaping

A Gaussian pulse shaping filter is another commonly used base-band shaping filter in the communication systems. The Gaussian filter has a smooth transfer function with no zero crossings. The impulse response of the Gaussian filter is expressed as

$$h_G(t,\sigma)=\frac{1}{\sqrt{2\pi}\sigma}\exp\left(-\frac{t^2}{2\sigma^2}\right), \tag{2.4.61}$$

where the parameter σ is related to 3 db bandwidth and bit duration product BT_b and it is expressed as

$$\sigma=\frac{\sqrt{\ln 2}}{2\pi BT_b}. \tag{2.4.62}$$

The corresponding frequency response of the Gaussian impulse is still a Gaussian function, and it is represented by

$$H_G(f,\sigma)=\exp\left(-2\pi^2\sigma^2 f^2\right) \tag{2.4.63}$$

The Gaussian pulse-shaping filter is particularly used for the base-band pulse shaping of the MSK modulation. The MSK modulation with base-band Gaussain pulse data is referred to as GMSK. The most attractive features of the GMSK are its excellent power efficiency and excellent spectral efficiency. These are due to its constant envelope and the fact that Gaussian pulse shaping significantly reduces the side lobes of the MSK signal spectrum. However, Gaussian pulse shaping introduces ISI between symbols, but it is not so severe especially when $BT_b \ge 0.5$. The signal-to-noise ratio (E_b/N_o) degradation is approximately 0.14 dB when BT_b =

0.5887. The ISI increases with as the BT_b value decreases although the spectrum becomes more compact.

In the GSM mobile system, the GSMK modulation with a $BT_b = 0.3$ is employed. The power spectral density of a GSM signal is shown in Fig. 2.33. The bit duration T_b and the parameter σ of the GSM system are, respectively,

$$T_b = 48/13 = 3.69 \quad \mu\text{sec} \quad \text{and} \quad \sigma = 0.4417.$$

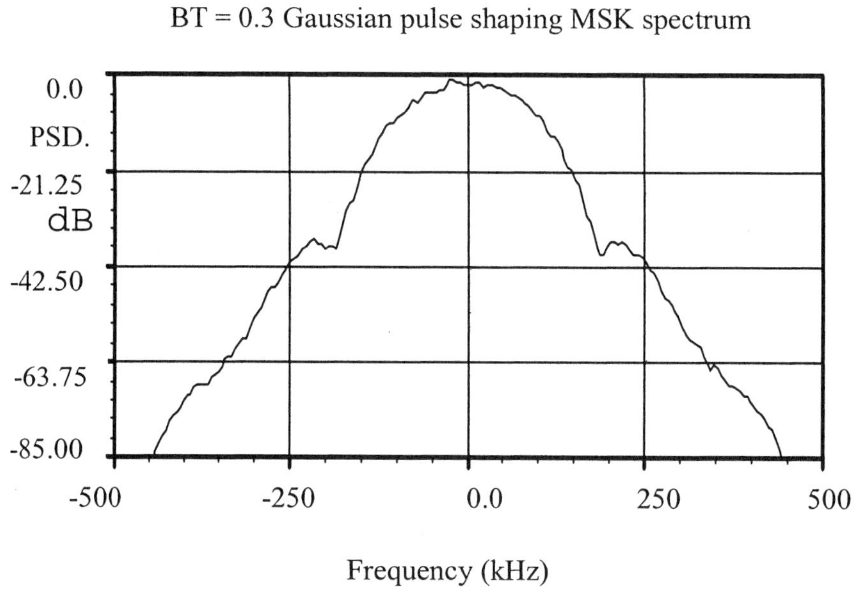

Figure 2. 33. Power spectral density of GMS signal

2.4.4.3. Pulse Shaping Used in the CDMA System

In the CDMA system, the base-band spreading PN sequence in the transmitter is pulse-shaped by means of an impulse response $h_c(t)$, which should satisfy the following mean squared error equation [17]:

$$\sum_{k=0}^{\infty} \left[\alpha \cdot h_c(kT_c/4) - h(k)\right]^2 \leq 0.03, \tag{2.4.64}$$

where the constants α and τ are used to minimize the mean square error, T_c is the chip duration of the PN sequence, and the coefficients $h(k)$ for $k < 48$ are listed in Table 2.2 and $h(k) = 0$ for $k \geq 48$.

Table 2.2. Coefficients of $h(k)$

k	$h(k)$	k	$h(k)$	k	$h(k)$
0, 47	-0.025288315	8, 39	0.037071157	16, 31	-0.012839661
1, 46	-0.034167931	9, 38	-0.021998074	17, 30	-0.143477028
2, 45	-0.035752323	10, 37	-0.060716277	18, 29	-0.211829088
3, 44	-0.016733702	11, 36	-0.051178658	19, 28	-0.140513128
4, 43	-0.021602514	12, 35	0.007874526	20, 27	0.094601918
5, 42	0.021602514	13, 34	0.084368728	21, 26	0.441387140
6, 41	0.091002137	14, 33	0.126869306	22, 25	0.785875640
7, 40	0.081894974	15, 32	0.094528345	23, 24	1.000000000

The impulse responses of $h_c(t)$ are shown in Fig. 2.34. From this plot we can see that after the rectangular pulses of the spreading PN sequence pass through the base-band pulse shaping filter, the tails of the shaped pulse is no longer crossing zero at the peak of the adjacent chip pulses — i.e., there exists *interchip interference* (ICI).

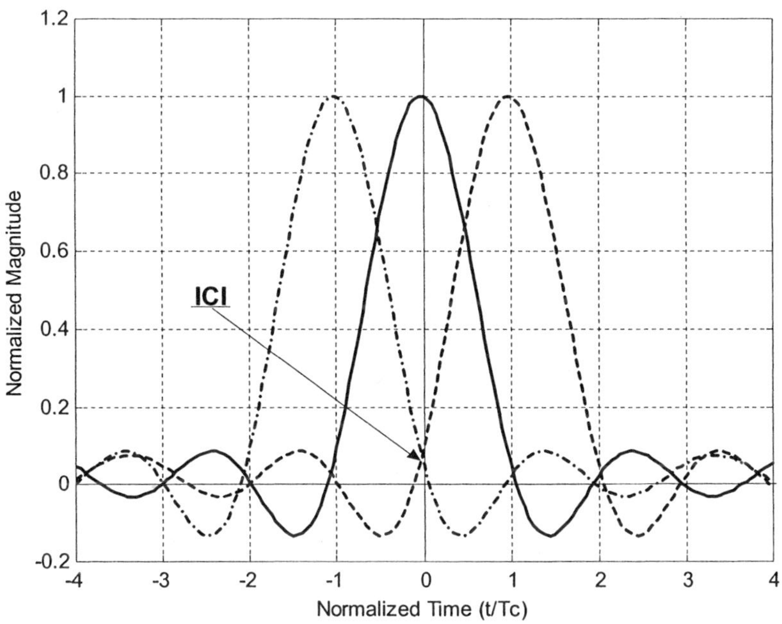

Figure 2.34. Impulse response of $h_c(t)$ used in CDMA system

The power spectral density of the base-band filtered spreading PN sequence is depicted in Fig. 2.35. It has a very low side lobe or a high spectral efficiency at the cost of creating the ICI. However, in the receiver of the CDMA system a complementary filter is used to remove or minimize the

ICI introduced by the transmitter base-band filter. The impulse response of the composite filter consisting of the transmitter base-band filter and the receiver complementary filter connected in cascade shall approximately satisfy Nyquist criterion of zero ICI, or in the other words, the nonzero crossing value shall be at least 50 dB below the peak value.

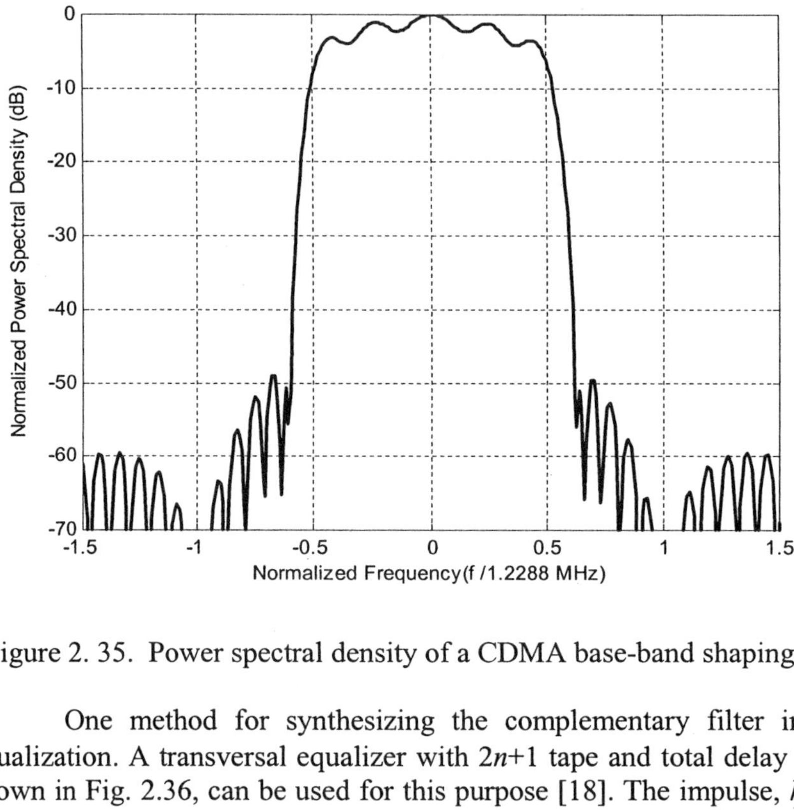

Figure 2. 35. Power spectral density of a CDMA base-band shaping pulse

One method for synthesizing the complementary filter involves equalization. A transversal equalizer with $2n+1$ tape and total delay $2n\tau$, as shown in Fig. 2.36, can be used for this purpose [18]. The impulse, $h_c(t)$, of the CDMA base-band pulse-shaping filter at the input of the complementary filter is assumed to have its peak at $t = 0$ and ICI on both sides. The output pulse of the complementary filter can be expressed as

$$h_{Nq}(t) = \sum_{m=-n}^{n} c_m h_c(t - mT_c - nT_c) \tag{2.4.65}$$

or in a discrete form and sampling at $t_k = kT_c + nT_c$:

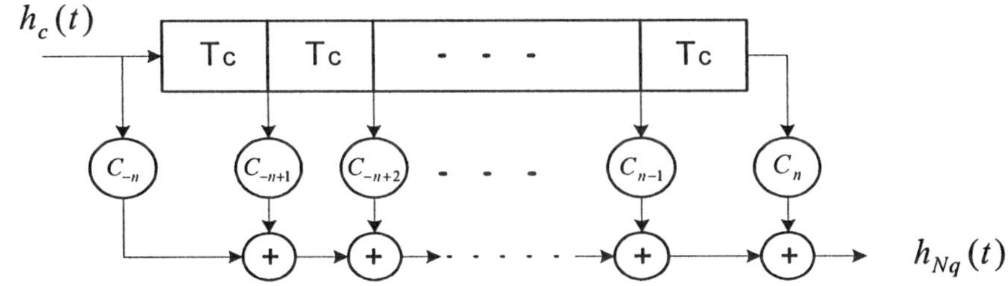

Figure 2.36. A complementary filter of a CDMA base-band shaping filter

$$h_{sys}(t_k) = \sum_{m=-n}^{n} c_m h_c(kT_c - mT_c) = \sum_{m=-n}^{n} c_m h_{c,k-m} , \tag{2.4.66}$$

where $h_{c,k-m} = h_c(kT_c - mT_c)$, and c_m ($m = 0, \pm1, \pm2, \ldots\pm n$) are called tap gains.

To minimize the ICI, the tap gains are chosen to satisfy the following equation — i.e., to force n zero values on each side of the $h_{sys}(t)$ peak:

$$h_{sys}(t_k) = \begin{cases} 1 & k = 0 \\ 0 & k = \pm1, \pm2, \cdots, \pm n. \end{cases} \tag{2.4.67}$$

The corresponding tap gains can be obtained by solving the matrix equation (2.4.68):

$$\begin{bmatrix} h_{c,0} & \cdots & h_{c,-2n} \\ \vdots & & \vdots \\ h_{c,n-1} & \cdots & h_{c,-n-1} \\ h_{c,n} & \cdots & h_{c,-n} \\ h_{c,n+1} & \cdots & h_{c,-n+1} \\ \vdots & & \vdots \\ h_{c,2n} & \cdots & h_{c,0} \end{bmatrix} \begin{bmatrix} c_{-n} \\ \vdots \\ c_{-1} \\ c_0 \\ c_1 \\ \vdots \\ c_n \end{bmatrix} = \begin{bmatrix} 0 \\ \vdots \\ 0 \\ 1 \\ 0 \\ \vdots \\ 0 \end{bmatrix}, \tag{2.4.68}$$

where $h_{c,i}$ ($i = 0, \pm1, \ldots, \pm2n$) is from (2.4.66).

In reality, a equalizer may be used in the CDMA receiver not only for removing the ICI caused by the transmitter base-band filter but also for minimizing the ICI generated by the IF SAW filter, analog base-band low-pass filter, and DC blocking circuit in the receiver chain. In this case, the impulse response on the right side of (2.4.66) should be replaced by the convolution of $h_c(t)$ and the filter impulse responses

$$h(t) = h_c(t) * h_{SAW}(t) * h_{A\_BBF}(t) * h_{DC\_block}(t), \qquad (2.4.69)$$

where $h_{SAW}(t)$, $h_{A\_BBF}(t)$, and $h_{DC\_block}(t)$ are impulse responses of the IF SAW filter, analog base-band filter, and DC block circuit, respectively.

2.4.5. Error Probability of Detection, Signal-to-Noise Ratio (SNR), and Carrier-to-Noise Ratio (CNR)

Error-producing behavior of a digital communication system is an important measure of system performance. A commonly used measure is the average error of symbol or bit detection in the receiver over the total number of the received symbols or bits, and it is referred to as *symbol error rate* (*SER*) or *bit error rate* (*BER*).

Signals before the detection in a receiver of a communication system are usually perturbed or contaminated by noise or interference in their propagation path and channel. Assuming that a data sequence of N symbols, which consist of $M = 2^k$ symbol waveforms, $s_i(t)$ (i = 1, ...M) with a duration T_s, corresponding to MSK modulation values, is transmitted and noise in the signal path and channel is denoted as $n(t)$, then the received signal $r(t)$ can be represented by the symbol waveforms plus noise as

$$r(t) = \sum_{k=1}^{N} s_i(t - kT_s) + n(t - kT_s) \qquad i = (1,2,\cdots M). \qquad (2.4.70)$$

If the receiver detection is based on correlation and there exist M correlators in the detector, the decision rule for the detector is to choose the maximum one from the set of the M correlator outputs $\{x_i(T_s)\}$ — i.e., to select

$$\max\{x_i(T_s)\} = \max\left\{ \int_{kT_s}^{(k+1)T_s} r(t)s_i(t)dt \right\} \quad (i = 1,2,\cdots M) \qquad (2.4.71)$$

The decision of the receiver detection is not always made correctly since the received signal was corrupted by the noise or interference in its propagation path and channel. The wrong decisions generate detection errors. If $n_e(N_s)$ is the number of decision errors of N_s transmitted symbols, the error probability is usually defined as $P_e = \lim_{N\to\infty} n_e(N_s)/N_s$. In the case of system using M-ary modulation, if the transmission probability of M symbols, $s_i(t)$ (i = 1, …, M), is equal, the average probability of symbol errors P_e over all possible transmitted symbols can be expressed as

$$\begin{aligned} P_e &= \frac{1}{M}\sum_{i=1}^{M} P\left(\text{wrong decision based on } \max\{x_i\} \middle| s_i \text{ sent}\right) \\ &= 1 - \frac{1}{M}\sum_{i=1}^{M} P\left(\text{correct decision based on } \max\{x_i\} \middle| s_i \text{ sent}\right) \end{aligned} \tag{2.4.72}$$

It is apparent that the bit error probability P_b is less than the symbol error probability P_e since a symbol usually consists of multiple bits. For M ary orthogonal modulation, such as MFSK, the relationship between P_b and P_e is [19]

$$\frac{P_b}{P_e} = \frac{2^{k-1}}{2^k - 1} = \frac{M/2}{M-1}. \tag{2.4.73}$$

For nonorthogonal modulation, such as MPSK, Gray code [21] mapping the binary to M-ary code is used such that binary sequences corresponding adjacent symbols differ only one bit position. Thus when a symbol error occurs, it will be more likely to mistake one of the nearest symbols. The bit error probability of the Gray code M-ary modulation is approximately related to its symbol error probability as follows:

$$P_b = \frac{P_e}{\log_2 M} = \frac{P_e}{k}. \tag{2.4.74}$$

The probability of symbol or bit error is directly determined by the signal-to-noise ratio (*SNR*) of the received signal. The *SNR* is usually defined as average signal power over average noise power:

$$SNR = \frac{\text{Signal Power}}{\text{Noise Power}} = \frac{S}{N}. \tag{2.4.75}$$

However, in expressing the error probability, P_b or P_e, in AWGN channel, the ratio of signal energy per bit E_b to single-sided thermal noise density n_o (W/Hz) — i.e., E_b/n_o is used generally instead of the *SNR*. The relationship between E_b/n_o and *SNR* is

$$\frac{E_b}{n_o} = \frac{ST_b}{n_o} = \frac{SB}{R_b n_o B} = \frac{S}{N}\left(\frac{B}{R_b}\right), \tag{2.4.76}$$

where T_b is bit time duration, $R_b = 1/\,T_b$ is bit rate, B is signal bandwidth in base-band, and $N = n_o B$.

2.4.5.1. Error-Probability Formulas of Popular Modulation Signals

In this section, expressions of bit or symbol error probability for FSK, PSK, DPSK, MSK, and QAM modulation signals in Gaussian noise are presented with no derivation [8, 10, 20]. The following two functions, *erfc*(x) and $Q(x)$, are often used in the error-probability expressions:

$$erfc(x) = \frac{2}{\sqrt{\pi}} \int_x^{\infty} e^{-u^2}\, du \tag{2.4.77}$$

and

$$Q(x) = \frac{1}{\sqrt{2\pi}} \int_x^{\infty} e^{-u^2/2}\, du\,. \tag{2.4.78}$$

These two functions have a relationship

$$Q(x) = \frac{1}{2}\, erfc\left(\frac{x}{\sqrt{2}}\right) \tag{2.4.79}$$

or

$$\frac{1}{2}\, erfc(x) = Q\left(\sqrt{2}x\right). \tag{2.4.80}$$

In the following formulas of bit or symbol error probabilities for different modulation signals, the $Q(x)$ function will be used, but it is easy to convert these formulas to expressions with *erfc*(x) function.

- *M-ary coherent PSK*

$M = 2$:

$$P_b = Q\left(\sqrt{\frac{2E_b}{n_o}}\right) \tag{2.4.81}$$

$M = 4$: $$P_e(4) = 2Q\left(\sqrt{\frac{2E_b}{n_o}}\right)\left[1 - \frac{1}{2}Q\left(\sqrt{\frac{2E_b}{n_o}}\right)\right] \quad (2.4.82)$$

$M > 4$: $$P_e(M) \cong 2Q\left(\sqrt{\frac{2E_b \log_2 M}{n_o}} \sin\frac{\pi}{M}\right) \quad (2.4.83)$$

- *M-ary Non-coherent DPSK*

$M = 2$: $$P_b = \frac{1}{2}\exp\left(-\frac{E_b}{2n_o}\right) \quad (2.4.84)$$

$M = 4$: $$P_b = \left[\sum_{k=0}^{\infty}\left(\frac{b}{a}\right)^k I_k(ab) - \frac{1}{2}I_o(ab)\right]\exp\left(-\frac{a^2 + b^2}{2}\right) \quad (2.4.85)$$

$$a = \sqrt{2\left(1 - \sqrt{\frac{1}{2}}\right)\frac{2E_b}{n_o}} \quad \text{and} \quad b = \sqrt{2\left(1 + \sqrt{\frac{1}{2}}\right)\frac{2E_b}{n_o}}$$

I_k — the modified Bessel function of order k

$M > 4$: $$P_e(M) \cong 2Q\left(\sqrt{\frac{2E_b \log_2 M}{n_o}} \sin\frac{\pi}{\sqrt{2}M}\right) \quad (2.4.86)$$

- *M-ary Orthogonal FSK*

$M = 2$: $$P_b = \frac{1}{2}\exp\left(-\frac{1}{2}\frac{E_b}{n_o}\right) \quad (2.4.87)$$

$M > 2$: $$P_e(M) = \frac{1}{M}\sum_{i=2}^{M}(-1)^i \frac{M!}{i!(M-i)!}\exp\left(\frac{(i-1)E_s}{i \cdot n_o}\right) \quad (2.4.88)$$

$$E_s = E_b \cdot \log_2 M$$

- *GMSK*

$$P_e = Q\left(\sqrt{\frac{2\alpha E_b}{n_o}}\right) \quad (2.4.89)$$

$\alpha = 0.68$ for $BT = 0.25$

$\alpha = 0.85$ for $BT = \infty$ (MSK)

o *M-ary QAM*

$$P_e(M) = 2P_{e\_\sqrt{M}}\left(1 - \frac{1}{2}P_{e\_\sqrt{M}}\right) \tag{2.4.90}$$

$$P_{e\_\sqrt{M}} \cong \frac{2\left(\sqrt{M}-1\right)}{\sqrt{M}} Q\left(\sqrt{\frac{3\log_2 M}{M-1}\cdot\frac{E_b}{n_o}}\right)$$

$$P_b = \frac{P_e}{\log_2(M)}$$

The above formulas can be used to estimate the bit or symbol error rates of modulated signals in AWGN channel. The bit error probabilities versus E_b/n_o of the modulation signals popularly used in wireless communication systems are presented in Fig. 2.37.

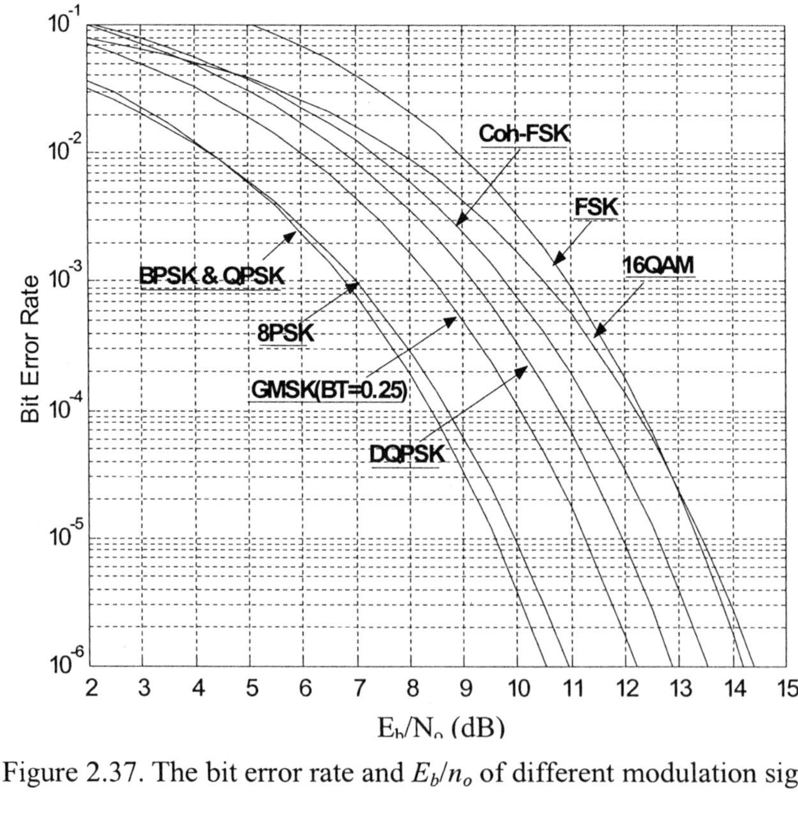

Figure 2.37. The bit error rate and E_b/n_o of different modulation signals

2.4.5.2. Degradation of Error-Rate Performance in Practical Systems

The bit or symbol error rates for different modulation schemes presented in the previous are estimates based on the modulated signal passing through AWGN channel and path with no other interference or perturbation. In wireless systems, the signal channel is possibly a multipath fading channel, and the signal path in a receiver is usually a frequency selective path since certain band-pass and low-pass filters are employed in the receiver to suppress unwanted signal and interference. Either the fading channel or the frequency selective path will degrade the error rate performance of modulated signal, and this means the bit or symbol error rate of a real system is higher than that calculated from the formulas presented in the previous section. For the mobile stations in different protocol wireless systems, the minimum performance requirements of the receiver in most cases are specified in static case or in AWGN, and thus in the RF system design the frequency-selective path impact on the receiver performance is more important than the multipath fading channel influence. In this section, we are going to discuss how the analog filters and other frequency-selective devices in a receiver will affect the error rate performance of a modulated signal through certain practical examples.

Analog frequency filters, especially band-pass and low-pass filters, are commonly used in communication receivers to extract the desired signal from noise or interference. The frequency responses of analog filters, such as *surface acoustic wave* (SAW) filters and ceramic filters, are usually not so ideal. Their magnitude can vary over 2 dB, and their group delay may be severely distorted depending on bandwidth of the filter. The direct consequence of these filters on the information data sequences passing through is to make their intersymbol interference (ISI) increase. Thus, the symbol error rate is raised at the same signal-to-noise ratio. Let us first discuss an example of the BER degradation of a signal with the π/4 DQPSK modulation passing through band-pass filters since there is a closed-form formula for the π/4 DQPSK signal to describe the BER degradation due to the ISI [22–23].

A noise-corrupted DQPSK signal in a narrowband receiver can be expressed as

$$\begin{aligned} s(t) &= A(t)\cos[2\pi f_o t + \phi(t)] + n_c(t)\cos(2\pi f_o t) - n_s(2\pi f_o t) \\ &= R(t)\cos[2\pi f_o t + \theta(t)], \end{aligned} \tag{2.4.91}$$

where $n(t)$ is wide-sense stationary narrow-band Gaussian noise with a single-sided noise spectral density n_o and in-phase and quadrature components $n_c(t)$ and $n_s(t)$, respectively, and

$$R(t) = \sqrt{[A(t)\cos\phi(t) + n_c(t)]^2 + [A(t)\sin\phi(t) + n_s(t)]^2}$$

and

$$\theta(t) = \tan^{-1}\frac{A(t)\sin\phi(t) + n_s(t)}{A(t)\cos\phi(t) + n_c(t)}.$$

For the mth symbol, the phase of the π/4DQPSK signal over a symbol interval, $(m\text{-}1)T_s < t \le mT_s$ is

$$\phi(t) = \phi_m = \phi_{m-1} + \phi_o + 2k_m \cdot \pi/4 \tag{2.4.92}$$

where k_m = 0, 2, …, 3, and ϕ_o = $\pi/4$. The signal amplitude $A(t)$ may fluctuate because of interference, multipath fading, or narrow-band filtering, but here we discuss only the amplitude fluctuation caused by the frequency filters and other devices, such as DC block circuit and analog to digital converter, used in a receiver. The amplitude may vary from symbol to symbol due to the ISI between symbols since the in-band *group delay distortion* and magnitude response ripple of the narrow-band filters in the signal path distort the symbol pulse waveform. In the following analysis the ISI caused by the previous and successive symbol pulses is considered, and the normalized ISI $\Delta I_{p,,s}$ is defined as

$$\Delta I_{p,s} = \frac{\int_{t_o-\delta t}^{t_o+\delta t} s(t \pm T_s)dt}{\int_{t_o-\delta t}^{t_o+\delta t} s(t)dt}, \tag{2.4.93}$$

where subscripts p and s denote the previous and successive symbols, respectively, the plus and minus signs within the integrand in the numerator are corresponding to ΔI_p, and ΔI_s, respectively, t_o is the sampling instant, and $2\delta t$ is the sampling duration.

The base-band signal of the mth symbol can also be expressed in a vector form as

$$\begin{aligned}\bar{r}_m(t) &= r_{mI}(t) + j \cdot r_{mQ}(t) \\ &= \bar{a}_m(t) + \bar{n}_m(t),\end{aligned} \tag{2.4.94}$$

where $\bar{a}_m(t)$ is the mth symbol base-band vector, $n_m(t)$ is noise vector for the mth symbol, and

$$r_{mI}(t) = a_{mI}(t) + n_c(t) \quad (2.4.95a)$$
$$r_{mQ}(t) = a_{mQ}(t) + n_s(t) \quad (2.4.95b)$$

and

$$a_{mI}(t) = A(t)\cos\phi_m + \Delta I_p \cos\phi_{m+1} + \Delta I_s \cos\phi_{m-1} \quad (2.4.96a)$$
$$a_{mQ}(t) = A(t)\sin\phi_m + \Delta I_p \sin\phi_{m+1} + \Delta I_s \sin\phi_{m-1}. \quad 2.4.96b)$$

The phase detection statistic for the DQPSK signal is

$$\psi = \angle\bar{r}_m - \angle\bar{r}_{m-1}, \quad (2.4.97)$$

where $\angle$ denotes the angle of the vector. The decision rule is defined by

$$\psi_{o(k_m-1,k_m)} < \psi - \phi_o \le \psi_{o(k_m,k_{m+1})}, \qquad k_m \text{ transmitted}. \quad (2.4.98)$$

In (2.4.98), $\psi_{o(k_m-1,k_m)}$ (= $\pi/4$, $3\pi/4$, $5\pi/4$, $7\pi/4$) is the decision boundary angle between the m-1-th and mth symbols.

The symbol error rate of the mth data symbol for a given symbol sequence is given by

$$P_e(m|\text{symbol sequence}) = F(\psi_{o(k_m,k_{m+1})} + \phi_o|\Delta\phi_m) - F(\psi_{o(k_{m-1},k_m)} + \phi_o|\Delta\phi_m), \quad (2.4.99)$$

where

$$F(\psi|\Delta\phi_m) = \frac{W\sin(\Delta\phi_m - \psi)}{4\pi}\int_{-\pi/2}^{\pi/2}\frac{e^{-E_1}dt}{U - V\sin t - W\cos(\Delta\phi_m - \psi)\cos t} + \frac{r\sin\psi}{4\pi}\int_{-\pi/2}^{\pi/2}\frac{e^{-E_1}dt}{1 - r\cos\psi\cos t} \quad (2.4.100)$$

$$E_1 = \frac{U - V\sin t - W\cos(\Delta\phi_m - \psi)}{1 - r\cos\psi\cos t} \quad (2.4.101)$$

$$\Delta\phi_m = \angle\bar{a}_m - \angle\bar{a}_{m-1} \quad (2.4.102)$$

$$U = \frac{1}{2}[\rho(t + T_s) + \rho(t)] \quad (2.4.103)$$

$$V = \frac{1}{2}\left[\rho(t+T_s) - \rho(t)\right] \tag{2.4.104}$$

$$W = \sqrt{\rho(t+T_s) + \rho(t)} = \sqrt{U^2 - V^2} \tag{2.4.105}$$

$$\rho(t) = \frac{\left|\bar{a}_m\right|^2}{2n_o} \tag{2.4.106}$$

and

$$r = \frac{E\left[\bar{n}_m(t)\bar{n}_{m-1}(t)\right]}{n_o}, \tag{2.4.107}$$

where E[·]denotes expected value.

Consider all symbol sequences of length L and denote the kth sequence by S_k^L. The average symbol error probability $P_e(M)$ can be expressed as

$$P_e(M) = \frac{1}{M^L}\sum_{k=1}^{M^L} P_e\left(M\left|S_k^L\right.\right). \tag{2.4.108}$$

In the case of π/4 DQPSK and considering ISI from adjacent symbols only, we have $M = 4$ and $L = 3$. In fact, (2.4.92) to (2.4.108) can be used not only for the π/4 DQPSK symbol error-rate calculation but also for other M-ary DPSK error-probability evaluation considering ISI and correlation-phase noise between adjacent symbols.

Using (2.4.92) to (2.4.108), we analyze the degradation of BER performance for a raised cosine filtered π/4 DQPSK signal with a symbol rate 192 ksps (thousand symbols per second) passes through a SAW and a ceramic IF filters with a cascaded impulse response and frequency responses as shown in Fig. 2.38.

The noise bandwidth of this composite filter is approximately 206 kHz. The ISI caused by the composite filter can be calculated by using the impulse response of the raised cosine filter presented in (2.4.54) with a roll-off parameter $\alpha = 0.5$ and $T_s = (1/192)\times10^{-3}$ sec and the impulse response of the cascaded filters. The distorted symbol pulse can be obtained by taking the convolution of these two impulse responses. Then utilizing (2.4.93), we evaluate the ISI. The normalized ISI caused by previous and successive symbols is approximately 0.075 and 0.066. The BER versus E_b/n_o curve with the ISI is depicted in Fig. 2.38. The E_b/n_o has the following relationship with the symbol energy to noise density ratio ρ given in (2.4.106):

$$\frac{E_b}{n_o} = \frac{\rho}{\log_2 M} = \frac{\rho}{2} \tag{2.4.109}$$

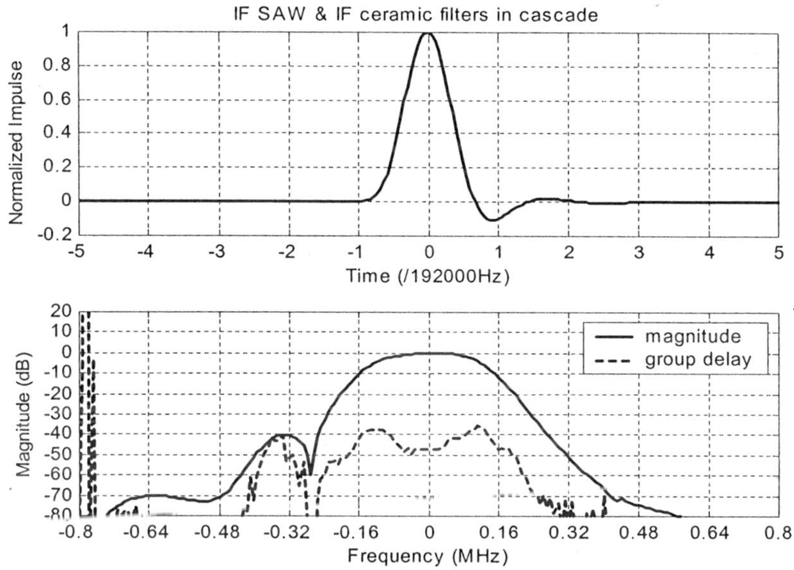

Figure 2.38. Impulse and frequency responses of cascaded SAW and ceramic IF band-pass filters

From Fig. 2.39, we can see that it needs 0.55 dB more E_b/n_o to obtain the same 1% BER in the case of having ISI.

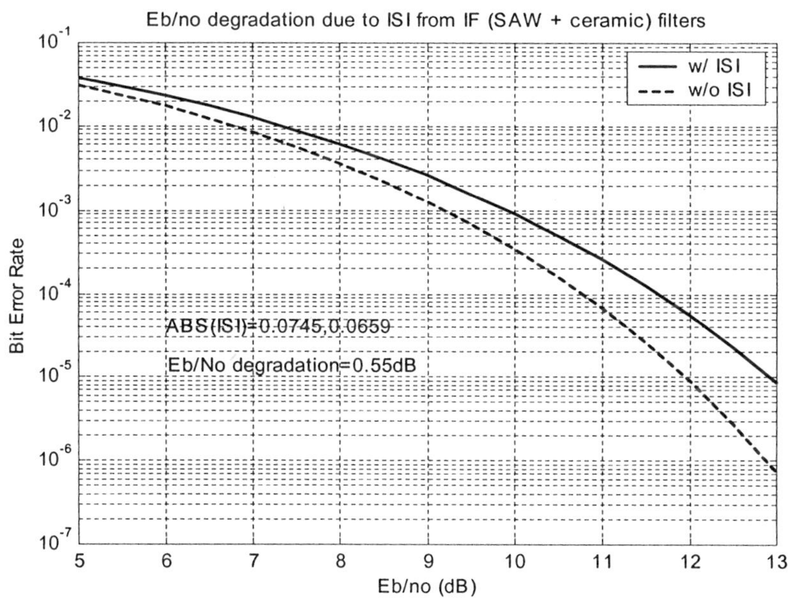

Figure 2.39. BER degradation of π/4 DQPS due to ISI

A spread spectrum QPSK signal is used for the forward link from the base station to the mobile station in the CDMA wireless system. This spread spectrum signal in the mobile station receiver is usually filtered by means of an IF SAW band pass filter and/or an analog base-band low-pass filter in the frequency conversion path of the receiver. The CDMA RF signal with a carrier in frequency band 869 to 894 MHz or 1930 to 1990 MHz is finally converted to an I/Q base-band signal with a low-pass bandwidth approximately 615 kHz. The in-band group delay distortion of the IF SAW is around 1 μsec, and the group delay fluctuation of the analog base-band filter in the signal bandwidth is usually 2.0 to 4.0 μsec depending on the type and the order of the filter. In addition, an AC coupling (DC blocking) with a cutoff frequency approximately 1 kHz between the analog base-band and the digital base-band units may be employed to remove DC offset and simplify the DC offset compensation circuit. The composite group delay distortion of the filters and the AC coupling both cause interchip interference (ICI) increase and then degrade the performance of detection error probability of the CDMA signal. The frame error rate (FER) is used in the CDMA system to measure the receiver detection capability.

In Fig. 2.40, the FER simulation result of the CDMA forward link fundamental channel signal in AWGN passing through the filters, and AC coupling circuit is presented. For comparison, the FER of the same signal is also presented in this figure when using a match filter to replace the SAW and analog base-band filters, and employing DC coupling instead of AC coupling. In these simulations, 9.6 kpbs data rate and the radio configuration (RC) 3 [17] are used. We should notice the effective noise density at the receiver input, N_t, is used there. From Fig. 2.40, we can see that the filters and the AC coupling in the receiving path degrade the FER performance, and in this case 0.4 dB more signal-to-noise ratio, E_b/N_t is needed to achieve 1% FER if comparing with the FER of using match filter.

In general, other than the filters and AC coupling, I and Q channel output magnitude and phase imbalances, the DC offset at the ADC input, etc., in a receiver will also cause decrease of signal-to-noise ratio, and thus the error-rate performance is impacted. Possible degradation of E_b/N_t due to different factors for a CDMA mobile station receiver is summarized in Table 2.3. The DC offset contribution should not be considered if AC coupling is used. The E_b/N_t degraded values shown in this table are presented just as a reference for initial design of an RF receiver system, and they may vary with the receiver configuration, subsystem design/implementation, and particular devices chosen in the receiver.

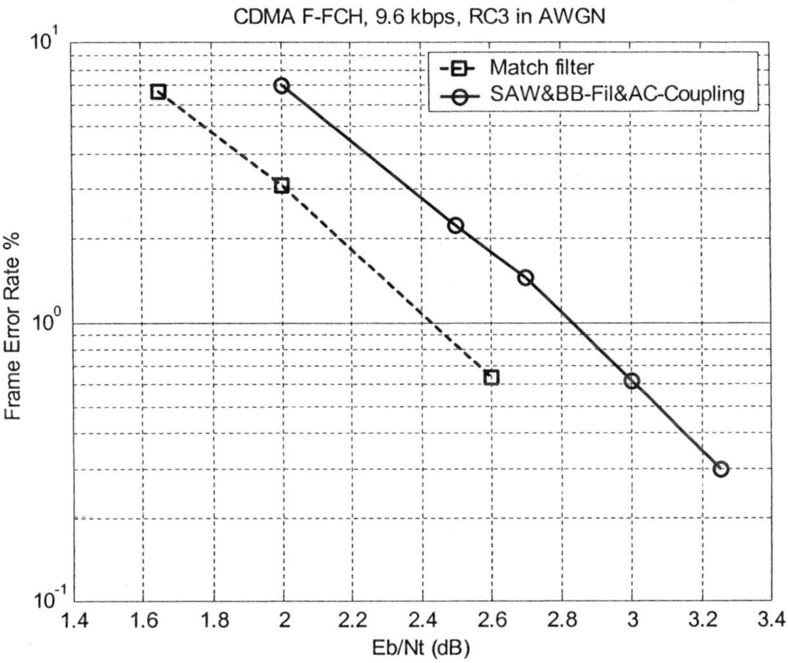

Figure 2 -40 FER of CDMA forward link fundamental channel signal with a data rate 9.6 kbps and RC3 in AWGN

Table 2.3. Main contributions to FER performance degradation in CDMA Receiver

Contribution Factor	Amount	Eb/Nt Degraded (dB)
SAW+BBFIL Group delay distortion (μs)	2.5 – 4	0.3 – 0.4
DC blocking (cutoff frequency in kHz)	1.5 – 3	0.1 – 0.15
DC offset (4 bit ADC LSB)	0.25	0.1
I/Q imbalance (dB, Degree)	1, 10	0.1 – 0.2

2.4.5.3. Carrier-to-Noise Ratio (CNR) and Base-Band Signal-to-Noise Ratio

The *carrier-to-noise ratio* (*CNR*) is an average power ratio of a signal *C* with a carrier to the noise *N* within the signal bandwidth *BW*. The *C*/*N* and *CNR* have the following relationships with E_b/n_o.

$$\frac{C}{N}=\frac{C\cdot T_b}{BW\cdot n_o\cdot T_b}=\frac{R_b}{BW}\frac{E_b}{n_o} \tag{2.4.110a}$$

and

$$CNR=10\log(C/N)=10\log(E_b/n_o)+10\log(R_b/BW)\quad \text{dB}, \tag{2.4.110b}$$

where T_b is data bit duration, $R_b = 1/T_b$ is bit rate, and n_o is noise density within signal bandwidth. Usually, the signal bandwidth in the C/N is specified the symbol rate bandwidth (the double-sided Nyquist bandwidth) — i.e.,

$$BW=R_s=\frac{R_b}{\log_2 M} \tag{2.4.111}$$

where R_s is the symbol rate, and M is the number of bits in each symbol. It should be noted that the noise bandwidth of a practical receiver is determined by the characteristics of the band-pass or low-pass filter employed in the receiver, and it is usually not equal to the symbol rate bandwidth.

For a π/4DQPSK signal with $M = 4$, we have

$$CNR=10\log(E_b/n_o)+3\qquad \text{dB}.$$

In the case of a CDMA signal (such as IS-98D), the signal bandwidth is equal to the spectrum spreading chip rate R_c, and the forward link desired signal C includes not only the traffic data with $1/\alpha$ of total desired signal power but also other signals, such as pilot. Thus (2.4.110) turns into

$$\frac{C}{N}=\frac{R_b}{R_c}\cdot\frac{E_{b\_traffic}\cdot\alpha}{n_t}=\frac{E_{b\_traffic}}{n_t}\cdot\frac{\alpha}{PG_{DS}} \tag{2.4.112a}$$

and

$$CNR=10\log\left(\frac{E_{b\_traffic}}{n_t}\right)+10\log\alpha-10\log(PG_{DS}), \tag{2.4.112b}$$

where $E_{b\_traffic}$ is the bit energy of the traffic channel, n_t is effective noise density since noise in CDMA system usually includes noiselike CDMA signals transmitted from other mobile stations, and PG_{DS} is the processing gain as given in (2.4.48).

For the IS-98D CDMA mobile station system, it has a chip rate of 1.2288×10^6 cps. The receiver sensitivity in AWGN (it is also called *static sensitivity*) is defined under the following conditions: the traffic data rate is 9600 bps, the data signal is $10\log(1/\alpha) = -15.6$ dB lower than overall desired signal power, and the minimum $E_{b\_traffic} / n_t$ for 0.5% FER shall be 4.5 dB or lower. It is also defined that total effective noise within 1.23 MHz is defined as –54 dBm. From the spreading chip rate 1.2288×10^6 cps and the traffic data rate 9600 bps, we obtain the processing gain, PG_{DS} = 128. Thus the CNR can be calculated from (2.4.112b), and it is

$$CNR_{CDMA} = 4.5 + 15.6 - 21.1 = 20.1 - 21.1 = -1 \quad \text{dB}.$$

In practical receivers of CDMA mobile stations, $E_{b\_traffic} / n_t$ for 0.5% FER is in the range of 3.0 to 4.0 dB. The corresponding *CNR* is in between –2.5 dB and –1.5 dB for the receiver sensitivity of the CDMA mobile stations.

The receiver sensitivity of the WCDMA is defined as the received signal strength for a user having a data rate of 12.2 kpbs and a spread chip rate of 3.84 Mpbs at the BER = 0.001. In this case, the processing gain of the system GP_{DS} is $3.84 \times 10^6 / 12.2 \times 10^3 \cong 314.8$ or 25 dB approximately. The ratio of bit energy of the dedicated physical channel to the total noise E_{b_DPCH} / n_t for a 0.001 BER is 7.2 dB including 2 dB implementation margin [24]. The dedicated physical channel data signal is 10.3 dB lower than the total received signal — i.e., $10\log(1/\alpha) = -10.3$ dB. The minimum *CNR* for the WCDMA receiver sensitivity from (2.4.112b) should be

$$CNR_{WCDMA} = 7.2 + 10.3 - 25 = -7.5 \quad \text{dB}.$$

In the analog wireless system AMPS, the modulation scheme is FM, and the receiver performance measure is based *SINAD* instead of data-detection error rate. The *SINAD* in dB is defined by the following equation:

$$SINAD = 10\log\frac{S + N + D}{N + D}, \tag{2.4.113}$$

where *S* is signal, *N* is noise, and *D* is distortion. All of them are variables in base-band. The conversion between the *SINAD* and *CNR* can be obtained from (2.4.113) and the following equations of *S*/*N* and *D*/*S*.

For a FM signal modulated by a sinusoidal tone with a maximum frequency deviation Δf, its signal-to-noise ratio in the base-band is related to the RF or IF carrier-to-noise ratio C/N by [25]

$$(S/N)_o = \frac{\frac{3}{2}\left(\frac{\Delta f}{B}\right)^2\left(\frac{BW}{B}\right)(C/N)}{1+\sqrt{\frac{3}{\pi}}\,\frac{(BW/B)^2\sqrt{(C/N)}e^{-(C/N)}[1+12(\Delta f/BW)^2(C/N)]}{\left(1-e^{-(C/N)}\right)^2}}. \tag{2.4.114}$$

In (2.4.114), B is the base-band bandwidth, BW is the bandwidth of the modulated RF or IF signal. The voice channel final output signal-to-noise ratio SNR therefore in dB scale is

$$SNR = 10\log(S/N) = 10\log(S/N)_o + G_p, \tag{2.4.115}$$

where G_p is the overall voice signal processing gain of the output signal-to-noise ratio developed from the deemphasis, expanding, and C-message weight.

An empirical formula expresses the distortion-to-signal ratio in terms of the signal-to-noise ratio:

$$D/S = 10^{-\frac{SNR-[6-(G_p-SNR)/2]}{20}}. \tag{2.4.116}$$

Considering (2.4.115) and (2.4.116), the $SINAD$ can be rewritten into

$$SINAD = 10\log\frac{1+10^{-[SNR-6+(G_p-SNR)/2]/20}+10^{-SNR/10}}{10^{-[SNR-6+(G_p-SNR)/2]/20}+10^{-SNR/10}}. \tag{2.4.117}$$

We should notice that the SNR in (2.4.117) is a function of CNR as presented in (2.4.114). Therefore, (2.4.117) is an implicit relationship between the $SINAD$ and CNR. It is not so straightforward to find out the CNR value for a given $SINAD$ from equation (2.4.117), but the graphic approach as shown in Fig. 2.41 can be utilized to determine the CNR from a given $SINAD$.

The $SINAD$ versus CNR curve in Fig. 2.41 is obtained from (2.4.117) by means of the following parameters: $\Delta f = 8$ kHz, $B = 3$ KHz, $BW = 30$ kHz, and $G_p = 24.5$ dB. From this figure we can see that CNR_{AMPS} is 2.48 while $SINAD = 12$ dB. In practical receivers of AMPS mobile stations, the CNR_{AMPS} at $SINAD = 12$ dB has a value in 2.0 to 3.5 dB.

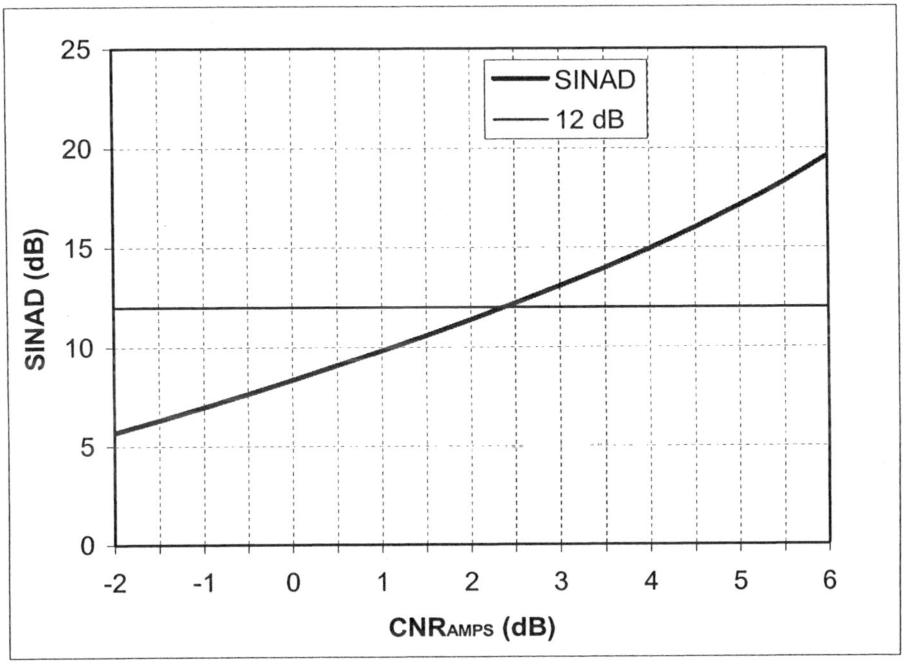

Figure 2.41 SINAD versus CNR of an AMPS mobile station receiver

The required maximum *CNR* values for the receiver static sensitivity of mobile stations for different protocol wireless systems are presented in Table 2.4. In practical receivers of these mobile stations, the corresponding *CNR* for the reference sensitivity in AWGN is lower than the value given in Table 2.4. Carrier-to-noise of a receiver system is mainly determined by the modulation scheme, symbol rate and coding used in the receiver, and the required bit error rate as well. However, the group delay distortion of IF band-pass channel filter and analog base-band low-pass filter also has noticeable impact on the *CNR*. The *CNR* is one of the most important parameters for the RF receiver design and performance analysis. It is used not only for the receiver sensitivity calculation but also for other performance evaluations, such as adjacent channel selectivity, intermodulation spurious attenuation, etc.

Table 2.4. Estimated maximum CNR for reference static sensitivity of different protocol mobile station receivers

System	Maximum CNR (dB)	Reference Sensitivity (dBm)	Modulation	Symbol Rate (ksps)	Maximum Error Rate (%)	Note
AMPS	3.5	-116	FM			
TDMA	10	-110	π/4DQPSK	24.3	3 (BER)	
PHS	11	-97	π/4DQPSK	192	1 (BER)	
GSM	8	-102	GMSK	270.833	4 (RBER)	TCH/FS Class II
CDMA	-1	-104	QPSK	1,228.8 (kcps)	0.5 (FER)	9.6 kbps voice data rate
GPRS	10	-102/-99	GMSK	270.833	10 (BLER)	CS1 code/ Uncoded
Edge	12	-98	8PSK	271	10 (BLER)	
WCDMA	-7.5	-106.7	QPSK	3,840.0 (kcps)	0.1 (BER)	12.2 kbps voice data rate

2.4.6. RAKE Receiver

In spread spectrum systems, such as the CDMA and the GPS systems, a RAKE receiver is used to separately detect multipath signals or signals from different basestations or satellites. Here we discuss only the basics of the RAKE receiver used in CDMA mobile stations. More details cab be found in [26–28] and [20].

CDMA spreading codes have a very low correlation between successive chips. The multipath propagation delay spread in the radio channel creates multiple time-delayed duplications of the transmitted CDMA signal at the receiver. These multipath components are uncorrelated from one another if their relative delays in time are more than a chip duration. Since these multipath components carry useful information, CDMA receivers may in a certain way combine these delayed versions of the original signal to increase the signal-to-noise ratio and then improve the receiver performance. The RAKE receiver is specially designed for achieving this purpose.

A simplified block diagram of the RAKE receiver with elements to detect the L strongest multipath components is shown in Fig. 2.42(a). Each multipath component detector is called a finger of the RAKE receiver. The basic configuration of the kth (k = 1, 2, ..., L) finger for detecting the kth delayed version of the transmitted CDMA signal is depicted in Fig. 2.42(b).

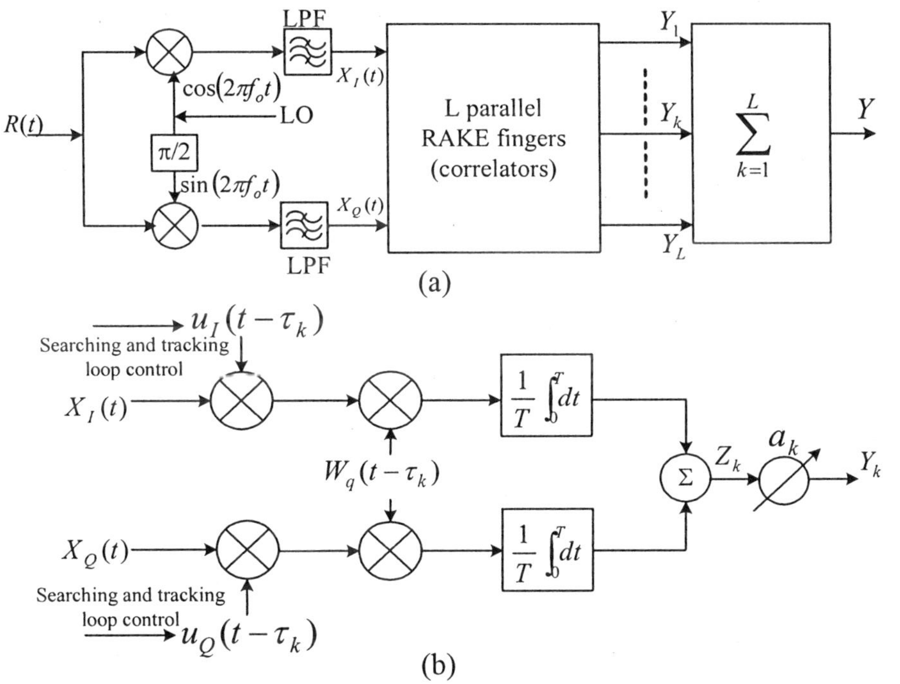

Figure 2.42. Simplified RAKE receiver block diagram and the kth (k = 1, 2,..., L) RAKE finger configuration

The kth finger searches the kth delayed pilot pseudorandom (PN) sequences through correlating the receiver local PN sequences $u_I(t-\tau_k)$ and $u_Q(t-\tau_k)$ with the kth delayed pilot PN sequences in the received signals, $X_I(t)$ and $X_Q(t)$, respectively. After the kth delayed PN sequences have been acquired, an early-late phase lock loop is used to track any time delay or phase change. The finger in the RAKE receiver of the qth (q = 1, 2, ..., M) mobile station is also capable to identify the corresponding data signals by means of correlating with the local Walsh code, $W_q(t-\tau_k)$. The output of the kth finger Y_k is expressed as

$$Y_k = \frac{a_k}{T}\int_0^T \left[X_I(t)u_I(t-\tau_k) - X_Q(t)u_Q(t-\tau_k)\right]\cdot W_q(t-\tau_k)dt = a_k \cdot Z_k, \tag{2.4.118}$$

where a_k is a weighting coefficient. The kth multipath signal voltage V_k is the expected value of Z_k, and the interference power caused by other delayed components and CDMA signals for other mobile stations I_k is equal to the variance of Z_k. They are expressed as, respectively,

$$V_k = E\{Z_k\} \quad \text{and} \quad I_k = Var\{Z_k\}. \tag{2.4.119}$$

The outputs of the L RAKE fingers are combined based on certain ways, which determine how the weighting coefficients a_k (k = 1, 2, …, L) are generated. When combining the individual outputs with weighting coefficients $a_k = V_k / I_k$, which result in the maximum signal-to-noise ratio, the combined output Y is expressed as

$$Y = \sum_{k=1}^{L} Y_k = \sum_{k=1}^{L} a_k Z_k = \sum_{k=1}^{L} \frac{V_k}{I_k} Z_k. \tag{2.4.120}$$

The total output signal-to-interference ratio, S/I_o, is now

$$S/I_o = E^2\{Y\}/Var\{Y\} = \sum_{k=1}^{L} V_k^2 / I_k = \sum_{k=1}^{L} S_k / I_k. \tag{2.4.121}$$

The total output signal-to-interference ratio S/I_o is equal to the sum of the individual signal-to-interference ratio S_k/I_k of the multipath components. The combined ratio may still be large enough for reliable reception even if each individual ratio S_k/I_k is too small for the receiver to work reliably.

RAKE fingers with strong multipath amplitudes may not necessarily provide strong output after correlation due to multiple access interference. To have better RAKE receiver performance, weighing coefficients should be generated based the actual outputs of individual fingers. A large weight is assigned to a finger with large output and vice versa.

Each finger of the RAKE receiver tracks its corresponding components independently and can be reassigned to different components at a given time by adjusting the delay of the □ispreading sequence. The number of fingers in a RAKE receiver may be approximately determined by the signal bandwidth BW and the rms delay spread σ_τ, and it has an [27–28]

$$n_{finger} \cong 1 + \sigma_\tau \cdot BW\,, \tag{2.4.122}$$

where the delay spread σ_τ is expressed as

$$\sigma_\tau = \sqrt{\overline{\tau^2} - \mu_\tau}\,. \tag{2.4.123}$$

In the above expression, the mean excess delay μ_τ and the mean squared excess delay $\overline{\tau^2}$ are, respectively,

$$\mu_\tau = \frac{\sum_{i=1}^{J} \alpha_i \cdot \tau}{\sum_{i=1}^{J} \alpha_l} \quad \text{and} \quad \overline{\tau^2} = \frac{\sum_{i=1}^{J} \alpha_i \cdot \tau^2 \sum_i}{\sum_{i=1}^{J} \alpha_l}\,, \tag{2.4.124}$$

where α_i (i = 1, 2, …, J) are channel weights with $\alpha_1 \geq 1 \geq \alpha_2 \geq \ldots \geq \alpha_J$, and τ_i (i = 1, 2, …, J) are the excess delay as depicted in Fig. 2.43.

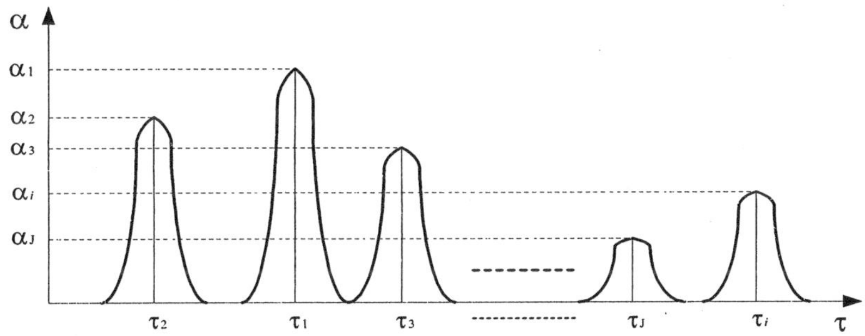

Figure 2.43. Relative amplitude and excess delay of multipath components

Since the working principle of a RAKE receiver is based on the fact that the multipath components are practically uncorrelated each other, it is essentially a diversity receiver. The RAKE receiver in the CDMA mobile stations is also employed for the soft handoff. The RAKE receiver detects the signals from different sectors of the same cell and combines the resulting signals to provide a more reliable reception. It may select signals if detected signals come from different cells.

References

[1]. S. *Haykin, Communication System*, 3rd ed., John Wiley & Sons, Inc., 1994.

[2]. H. P. Hsu, *Signals and Systems*, Schaum's Outline Series, McGraw-Hill, 1995.

[3]. H. S. Black, *Modulation Theory*, D. Van Nostrand Company, Princeton, N.J., 1953.

[4]. M. C. Jeruchim, P. Balaban, and K. S. Shanmugan, *Simulation of Communication System*, Kulwer Academic/Plenum Publishers, New York, 1992.

[5]. D. D. Siljak, *Nonlinear Systems*, Wiley, New York, 1969.

[6]. A. A. M. Saleh, *Frequency Independent and Frequency-dependent Nonlinear Models of TWT Amplifiers*, IEEE Trans. Commun. COM-29(11), 1715–1720 (1981).

[7]. S. A. Maas, *Microwave Mixers*, Artech House, Inc., 1993.

[8]. M. C. Jeruchim, P. Balaban, and K. S. Shanmugan, *Simulation of Communication Systems*, Kluwer Academic/Plenum Publishers, 1992.

[9]. U. L. Rohde, J. Whitaker, and T.T.N. Bucher, *Communications Receivers*, 2nd ed., McGraw-Hill, 1996.

[10]. B. Sklar, *Digital Communications Fundamentals and Applications*, PTR Prentice Hall, 1988.

[11]. T. Tsui, *Digital Techniques for Wideband Receivers*, Artech House,

[12]. C. E. Shannon, "A Mathematical Theory of Communication," *Bell Syst. Tech. J.*, vol. 27, 1948, pp. 379–423.

[13]. M. Schwartz, *Information Transmission, Modulation, and Noise*, 4th ed., McGraw-Hill Publishing Co., 1990

[14]. IS-2000.2, *Physical Layer Standard for cdma2000 Spread Spectrum Systems*, July 1999.

[15]. WCDMA, *Specifications of Air-Interface for 3 G Mobile System*, ARIB, Jan. 1999.

[16]. H. Nyquist, "Certain Topics in Telegraph Transmission Theory," *AIEE Transaction of AIEE*, vol. 47, pp. 617–644, Feb. 1928.

[17]. IS-98D, *Recommended Minimum Performance Standards for cdma2000 Spread Spectrum Mobile stations*, Release A, March 2001.

[18]. A. B. Carlson, *Communication Systems*, McGraw-Hill Book Company, 1986.

[19]. A. J. Viterbi, *Principle of Coherent Communications*, McGraw-Hill Book Company, New York, 1966.

[20]. T. S. Rappaport, *Wireless Communications Principles and Practice*, Printice Hall PTR, 1996.

[21]. I. Korn, *Digital Communications*, Van Nostand Reinhold Company, Inc., New York, 1985.
[22]. R. F. Pawula, "On M-ary DPSK Transmission Over Terrestrial and Satellite Channels," *IEEE Trans. Commun*. Vol. COM-32(7), pp. 752–761, July 1984.
[23]. J. H. Winters, "On Differential Detection of M-ary DPSK with Intersymbol Interference and Noise Correlation," *IEEE Trans. Commun*. Vol. COM-35(1), pp. 117 – 120, Jan., 1987.
[24]. T. Gee, "Suppressing Error in W-CDMA Mobile Devices,' *Communication System Design*," vol. 7, no. 3, pp. 25–34, March 2001.
[25]. A. Mehrotra, *Cellular Radio Performance Engineering*, Artech House, Inc., 1994.
[26]. A. J. Viterbi, *CDMA Principles of Spread Spectrum Communication*, Addison Wesley Longman, Inc, 1995.
[27]. D. V. Nicholson, *CDMA IS-95: Communication System Performance and Optimization*, Continuing Engineering Education Program, George Washington University, 1996.
[28]. *cdma2000 notes* by CDMA Wireless Academy, Inc.

Associated References

[1]. B. Lindoff and P. Malm, "BER Performance Analysis of a Direct Conversion Receiver," *IEEE Trans. Communications*, vol. 50, no.5, pp. 856–865, May 2002.
[2]. M. Juntti and M. Latva-aho, "Bit Error Probability Analysis of Linear Receivers for CDMA Systems," *1999 IEEE International Conference on Communications*, vol. 1, pp. 51–56, June 1999.
[3]. B. Debaillie, B. Come, et al., "Impact of Front-End Filters on Bit Error Rate Performances in WLAN-OFDM Transceiver," *2001 IEEE Radio and Wireless Conference*, pp. 193–196, Aug. 2001.
[4]. J. S. Pattavina, "Estimating BER in Broadband Design," *Communication System Design*, vol. 7, no. 2, Feb. 2001.
[5]. A. Smokvarski, J. S. Thompson, and B. Popovski, "BER Performance of a Receiver Diversity Scheme with Channel Estimation," *ICECom 2003 17th International Conference on Application Electromagnetic and Communications*, pp.87–90, Oct. 2003.
[6]. Y. Zhao and S. G. Hagman, "BER Analysis of OFDM Communication Systems with Intercarrier Interference," *1998 International Conference on Communication Technology*, pp. S38-02-1–S38-02-5, Oct. 1998.

[7]. H. G. Ryu et al., "BER Analysis of Clipping Process in the Forward Link of the OFDM-FDMA Communication System," *IEEE Trans. on Consumer Electronics*, vol. 50, no. 4, pp. 1058–1064, Nov. 2004.
[8]. E. Biglieri et al., "How Fading Affects CDMA: An Asymptotic Analysis with Linear Receiver," *IEEE Journal on Selected Area in Communications*, vol. 19, no. 2, pp. 191–201, Feb. 2001.
[9]. T. Gee, "Suppressing Errors in W-CDMA Mobile Devices," *Communication Systems Design*, vol. 7, no. 3, March 2001.
[10]. K. B. Huang et al., "A Novel DS-CDMA Rake Receiver: Architecture and Performance," *2004 IEEE International Conference on Communications*, vol. 5, pp. 2904–2908, June 2004.
[11]. T. W. Dittmer, "Advances in Digitally Modulated RF Systems," *1997 IEE International Broadcasting Convention*, pp. 427–435, Sept. 1997.
[12]. K. Matis, "Improving Accuracy in Edge-Based Designs," *Communication Systems Design*, vol. 7, no. 4, April 2001.
[13]. M. LeFevre and P. Okrah, "Making the Leap to 4G Wireless," *Communication Systems Design*, vol. 7, no. 7, July 2001.
[14]. J. Yang, "Diversity Receiver Scheme and System Performance Evaluation for a CDMA System," IEEE Trans. on Communications, vol. 47, no. 2, pp. 272–280, Feb. 1999.
[15]. S. N. Diggavi, "On Achievable Performance of Spatial Diversity Fading Channels," *IEEE Trans. on Information Theory*, vol. 47, no. 1, pp. 308–325, Jan. 2001.
[16]. L. Litwin and M. Pugel, "The Principle of OFDM," *RF Design*, pp. 30–48, Jan. 2002.
[17]. H. G. Ryu and Y. S. Lee, "Phase Noise Analusis of the OFDM Communication System by the Standard Frequency Deviation," *IEEE Trans. on Consumer Electronics*, vol. 49, no. 1, pp. 41–47, Feb. 2003.
[18]. M. Speth et al., "Optimum Receiver Design for Wireless Broad-Band Systems Using OFDM — Part I," *IEEE Trans. Communications*, vol. 47, no. 11, pp. 1668–1677, Nov. 1999.
[19]. M. Speth et al., "Optimum Receiver Design for OFDM-Based Broad-Band Transmission — Part II, A Case Study," *IEEE Tran. Communications*, vol. 49, no. 4, pp. 571–578, April 2001.
[20]. P. Banelli and S. Caeopardi, "Theoretical Analysis and Performance of OFDM Signals in Nonlinear AWGN Channels," *IEEE Trans. Communications*, vol. 48, no. 3, pp. 430–441, March 2000.
[21]. L. Piazzo and P. Mandarini, "Analysis of Phase Noise Effects in OFDM Modems," *IEEE Trans. Communications*, vol. 50, no.10, pp. 1696–1705, Oct. 2002.

[22]. M. S. Baek et al., "Semi-Blind Channel Estimation and PAR Reduction for MIMO-OFDM System with Multiple Antennas," IEEE Trans. on Broadcasting, vol. 50, no. 4, pp. 414–424, no. 4, Jan. 2004.

[23]. D. J. Love et al., "What Is the Value of Limited Feedback for MIMO Channels?" *IEEE Communications Magazine*, pp. 54–59, Oct. 2004.

[24]. J. McCorkle, "Why Such Uproar Over Ultrawideband?" *Communication Systems Design*, vol. 8, no. 3, March 2002.

[25]. B. Kull and S. Zeisberg, "UWB Receiver Performance Comparison," *2004 International Workshop on Ultra Wideband Systems Joint with Conference on Ultrawideband Systems and Technologies*, pp. 21–25, May 2004.

[26]. M. Weisenhorn and W. Hirt, "Robust Noncoherent Receiver Exploiting UWB Channel Properties," *2004 International Workshop on Ultra Wideband Systems Joint with Conference on Ultrawideband Systems and Technologies*, pp. 156–160, May 2004.

[27]. E. Saberinia and A. H. Tewfik, "Receiver Structures for Multi-Carrier UWB Systems," *2003 Proceedings of 7th International Symposium on Signal Processing an Its Applications*, vol. 1, pp. 313–316, July 2003.

[28]. S. Gezici et al., "Optimal and Suboptimal Linear Receivers for Time-Hopping Impulse Radio Systems," *2004 International Workshop on Ultra Wideband Systems Joint with Conference on Ultrawideband Systems and Technologies*, pp. 11–15, May 2004.

[29]. E. Grayver and Daneshrad, "A Low-Power All-Digital FSK Receiver for Space Applications," *IEEE Trans. Communications*, vol. 49, no. 5, pp. 911–921, May 2001.

[30]. M. S. Braasch and A. J. Van Dierendonck, "GPS Receiver Architectures and Measurements," Proceedings of the IEEE, vol. 87, no. 1, Jan. 1999.

Chapter 3

Radio Architectures and Design Considerations

In designing and developing an RF transceiver for a wireless mobile communication system, we shall first determine what kind of architecture will be employed based on considerations of performance, cost, power consumption, and robust implementation. In this chapter we are going to present the architectures of the RF receiver and transmitter that are practically applicable to the mobile stations of wireless communication systems.

In general, an RF receiver is defined from the port connected to a receiver antenna to an *analog-to-digital converter* (*ADC*), and an RF transmitter is defined from a *digital-to-analog converter* (*DAC*) to the port connected to a transmitter antenna. The RF receiver and transmitter are usually formed not only by RF circuitry and devices but also by *intermediate frequency* (*IF*) and analog base band circuitry and devices. The ADC and DCA are often used as a boundary between the RF transceiver and its digital counterpart. However, this boundary is getting ambiguous with the state of the art of the ADC/DAC running at higher and higher sampling since the ADC/DAC and the corresponding digital signal processors now become IF or even RF devices. In these cases, the ADC/DAC and some digital signal processing will be also considered as part of the RF transceivers.

The main building blocks of an RF transceiver can be functionally classified into following categories: frequency filters, amplifiers, frequency converters, modulator/demodulators, oscillators, synthesizers, ADC/DAC, signal coupler/divider/combiner/attenuators, switches, power/voltage detectors, etc. An RF transceiver will use most of these function blocks but may not be all of them. The characterization and specification of these function blocks are addressed in the last section of this chapter.

At present most RF transceivers in wireless communication systems are using *superheterodyne architecture*. This architecture has the best performance if compared with the others, and therefore it has been the most popular transceiver architecture since it was invented in 1918. We first discuss the superheterodyne architecture in this chapter. To obtain a great cost saving and to take the advantage of multimode operation without increasing extra parts, the *direct conversion* or *homodyne architecture* has now become a very hot radio architecture for wireless mobile communication systems since the radio-paging receiver using direct conversion was revived in 1980. The direct conversion architecture including its key issues and the corresponding solutions are described in the second section. To overcome some issues of direct conversion architecture, a modified architecture referred to as low IF architecture is then created. Some wireless communication receivers especially those based on the CMOS technology started to employ this architecture to cope with the flicker noise and the DC offset problems of the direct conversion. The low IF architecture is presented in the third subsection. Finally, a radio architecture based on the IF bandpass sampling technique is introduced since the sampling rate and resolution of the modern ADC/DAC with an acceptable power consumption has been improved enough to provide a foundation for practically utilizing this radio architecture.

Readers may be also interested in the radio architecture directly sampling at RF and something called *software defined radio*, but these are not discussed here since a true software radio is not so mature enough, at present, for wireless mobile station applications.

3.1. Superheterodyne Architecture

This is the most popular architecture used in communication transceivers. It is based on the heterodyne process of mixing an incoming signal with an offset frequency *local oscillator* (*LO*) in a nonlinear device to generate an intermediate frequency (IF) signal in the receiver or to produce an RF signal from its IF version in the transmitter. The nonlinear device executing the heterodyne process is called a *frequency mixer* or *frequency converter*. In a superheterodyne transceiver, the frequency translation processes may be performed more than once, and thus it may have multiple intermediate frequencies and multiple IF blocks.

It is apparent that the same IF can be generated by an incoming signal with a frequency either above or below the LO frequency. Between these two frequencies, the one corresponding to the undesired signal is referred to as an *image frequency*, and the signal with this frequency is called as an *image*. The frequency difference between the desired signal and its image is twice the IF. To prevent the possible image interfering with the desired signal and other strong unwanted signals jamming the superheterodyne receiver, sufficient filtering before the frequency converter is usually needed. The bandwidth of this preselection filter is quite broad, and it usually covers the overall reception frequency band in a wireless mobile transceiver. The channel filtering of a superheterodyne receiver is performed in the IF blocks by means of passive filters with high selectivity. The reception channel tuning is often carried out through programming an RF synthesizer, and thus the frequency of each IF block can remain fixed.

In the superheterodyne transceiver, most of the desired signal gain is provided by the IF blocks. At fixed intermediate frequencies, it is relatively easier to obtain high and stable gains. The power consumption for achieving high gain at IF is significantly lower than that if the same gain is developed at RF. This is due to the fact that the channel filters effectively suppress strong unwanted signals or interferers before they are substantially amplified, and therefore a high dynamic range of IF amplifiers is not demanded. In addition, IF amplifiers and circuits are often designed with higher impedances. The high channel selectivity also helps to achieve higher receiver sensitivity since adequate gain prior to the channel filtering can be set for obtaining the best sensitivity but still not to saturate the later stage amplifiers. The unwanted signals or interferers can be further filtered in the analog base-band by using active low-pass filters.

Since this architecture is most commonly used in the wireless communication systems, its detailed configuration will be described through a full duplex transceiver. The multiple intermediate frequencies used in a system will cause a *spurious response* problem, as we expected. It is necessary to have a good frequency plan for a superheterodyne transceiver to successfully operate over a specified frequency band. The frequency planning will be discussed in the second section. A general consideration of system design for a superheterodyne transceiver will be presented in the last section.

3.1.1. Configuration of Superheterodyne Radio

The wireless systems, such as, CDMA, WCDMA, and AMPS, use *full-duplex* transceivers. In these transceivers, the transmitter and the receiver simultaneously operate at offset frequencies. The configuration of a full duplex transceiver is usually more complicated than that of a *half-duplex* one since some means are needed to protect the receiver from the high power transmission and potential spurious emissions of the transmitter. A block diagram of a typical superheterodyne transceiver is depicted in Fig. 3.1.

There are two frequency conversions from RF to based-band and one IF block in both the receiver and transmitter. This is a typical configuration of superheterodyne transceivers employed in mobile stations of different protocol wireless systems. In Fig. 3.1, the upper portion is the

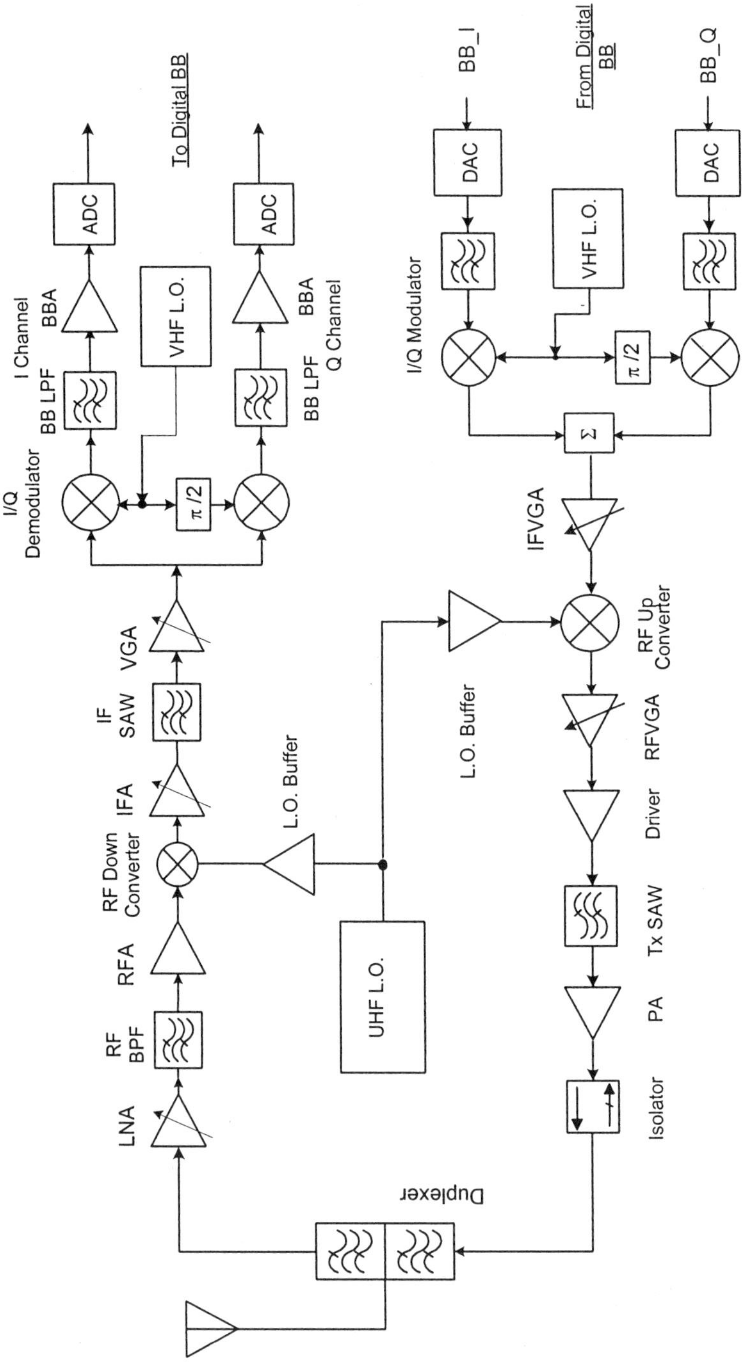

Figure 3.1. Block diagram of a superheterodyne full-duplxer transceiver

receiver block diagram, and the lower portion corresponds to the transmitter. The duplexer and the synthesizer local oscillator (LO) operating at the *ultrahigh frequency* (UHF) band are shared by both the receiver and the transmitter.

The duplexer consists of two band-pass filters with a common input port and two output ports. One filter is centered at receiver frequency band. It is used as the receiver preselection filter and to suppress transmission power leaking to the receiver. Another one is a transmitter filter that is employed to suppress out of transmission band noise and spurious emissions. The duplexer is necessary for a full duplex transceiver only if its receiver and transmitter use a common antenna. Sharing an UHF synthesizer is not necessary for a full duplex transceiver, but it can lower the current consumption and cost of the overall transceiver. The UHF synthesizer provides not only the LO power to the RF converters in the receiver and transmitter but also plays the role of channel tuning for the transceiver.

In a superheterodyne receiver, it usually contains three sections — the RF, IF, and BB. The IF section may have multiple blocks, which operate at different intermediate frequencies, but there is only one IF block in Fig. 3.1 as most receivers used in the wireless mobile stations. The RF section of the receiver includes part of the duplexer as the frequency preselector, a *low noise amplifier* (LNA), an RF *band-pass filter* (BPF), an RF amplifier as the preamplifier of the mixer, and an RF-to-IF down-converter (mixer). The LNA plays an important role in achieving good reception sensitivity. Its gain can be stepped-controlled to cope with the receiver dynamic range. The RF BPF is usually a SAW filter. The function of this filter is to further suppress the transmission leakage, the image, and other interference. Not all superheterodyne receivers will use this SAW filter if the preselector has a high enough rejection to the transmission power or the receiver is one in a half-duplex system. The RF amplifier (RFA) or preamplifier of the mixer provides enough gain to the receiver chain and thus the noise figure of the down-converter and the later stages has only slight influence to the receiver overall noise figure and sensitivity. The RFA is definitely needed when a passive mixer is chosen as the RF down-converter. The down-converter performs the signal frequency translation from RF to IF. Following the down-converter is an IF amplifier (IFA) and then an IF BPF for channel selection and suppressing unwanted mixing products. At present, IF SAW or crystal filters with a high selectivity are often used for the channel filtering. As described above, the IF block provides most of the gain for the overall receiver chain, and the IF *variable gain amplifier* (VGA), which is actually formed by multiple amplifier stages, is the main gain block of the IF section. The I/Q demodulator is the second frequency converter, which down-converts the signal frequency from IF to BB. The demodulator contains two mixers, and it converts the IF signal into I and Q signals — i.e., two 90° phase shifted BB signals. The 90° phase shift is implemented through a polyphase filter shifting the phase between *very-high frequency* (VHF) LO signals going to the mixers in the I and Q channels or by using a VHF VCO with a frequency of the twice IF and a by 2 frequency divider to generate two IF LO signals with a 90° phase difference for the I and Q channel mixing. A *low-pass filter* (LPF) follows the mixer in I and Q each channel to filter out the unwanted mixing products and to further suppress interferers. The filtered I and Q BB signals are amplified by BB amplifiers, and then the ADC converts the amplified BB signals into digital signals for further processing in the digital base-band.

It is similar to the superheterodyne receiver that a superheterodyne transmitter also consists of BB, IF, and BB three sections. The lower portion of Fig. 3.1 is a typical superheterodyne transmitter used in wireless mobile stations. I and Q digital BB signals that bear transmission information are converted to the corresponding analog BB signals by the DAC in the I and Q channels of the transmitter BB section. After BB filtering, the I and Q BB signals are up-converted into IF signals, and the Q channel IF signal obtains 90° more phase shift than that in the I channel during the frequency up-conversion in the I/Q modulator. The output of the I/Q modulator is the sum of the I and Q IF signals. The composite IF signal is amplified by a VGA, which usually consists of multistage amplifiers. An up-converter follows the IF VGA, and the amplified IF signal, then, is up-converted to an RF signal. The RF signal is further amplified by an RF VGA and then by a driver amplifier to a power level that is enough to drive the *power amplifier* (PA). A RF BPF (SAW filter) is inserted in between the driver and the PA to select the desired RF signal and suppress other mixing products generated by the RF up-converter. It is better to place the RF BPF just after the up-converter, but this arrangement may not be convenient since the whole block from the DAC to the driver may be integrated on a single semiconductor die or chip. The power amplifier boosts the desired RF signal to a power level that is high enough to make the transmission power at antenna port being still greater than the minimum requirement after deducting the insertion losses of the isolator and the duplexer. The PA may be a class AB amplifier as used in the TDMA, CDMA, and WCDMA mobile systems or a class C amplifier as employed in the GSM and AMPS systems. In either operating condition, the PA appearances certain nonlinear characteristics. The class C PA has higher power efficiency than the class AB amplifier, but it can be used only for the system with constant envelope modulation schemes, such as FM and GMSK modulations. The PA performance including the gain and linearity is quite sensitive to its loading. An *isolator* need be used in between the PA and the antenna to reduce the influence of the mobile station antenna input impedance variation with its position to the PA. It is apparent the duplexer will further suppress the out of transmitter band noise and spurious emissions and of course reduce the transmission power leaking to the receiver.

From Fig. 3.1 we can see that the gain control of the superheterodyne transceiver takes place in the IF and RF sections. The gain control in the IF section probably occupies about 75% of the overall gain control range or more. It is rare to see that the gain control is implemented in the analog BB section for this radio architecture. The reason for this is that the BB section either in the receiver or in the transmitter has I and Q two channels, and it is hard to keep the I and Q channel magnitude imbalance within an allowable tolerance over the BB gain variation range.

The configuration of a half-duplex superheterodyne transceiver may have some difference from that of the full-duplex one. In this case, it is possible to replace the duplexer in Fig. 3.1 by an antenna switch since the transmitter and the receiver in the half-duplex system do not work simultaneously. The UHF synthesizer LO is switched back and forth between the transmitter and receiver.

3.1.2. Frequency Planning

In the wireless communication systems, the typical frequency band assignment for the *down-link* (or *forward link*) communications from base-stations to mobile stations and for the *up-link* (or called as *reverse link*) communications from mobile stations to base stations is shown in Fig. 3.2. Usually, the bandwidth of the down- and up-link frequency bands is equal — i.e., $B_u = B_d = B_a$ — and both bands are channelized with an even channel spacing. It is apparent the channels in the down-link frequency band are for the base station transmitter and the mobile station receiver, and the channels in the up-link frequency band are for the mobile station transmitter and the base station receiver. The channels in down- and up-link frequency bands are used in pairs as depicted in Fig. 3.2. For example, if channel #Ch1_u in the up-link frequency band is assigned for the transmission of a mobile station, the corresponding channel #Ch1_d in the down-link frequency band will be automatically allocated for the receiver use of the mobile station. The center frequency spacing between any pair of channels is fixed and equal to $B_a + B_s$. For example, the PCS band in theUnited States is allocated a 60 MHz band from 1850 MHz to 1910 MHz for the up-link communications, and another 60 MHz band from 1930 MHz to 1990 MHz for the down-link communications, and the separation between these

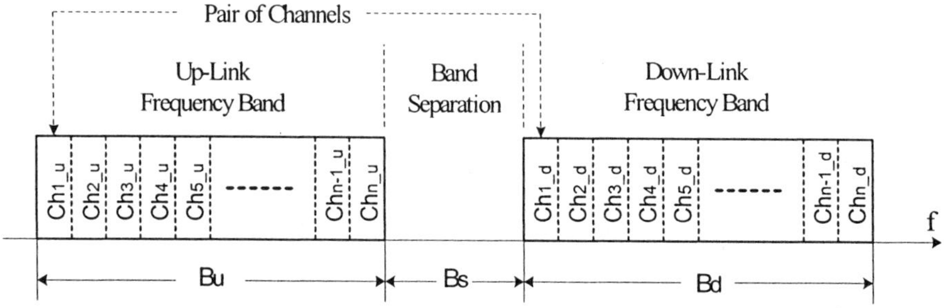

Figure 3.2. Down- and up-link frequency band configuration and channelization

two bands is 20 MHz. The other commonly used frequency bands for wireless communication systems are listed in Table 3.1. We should notice that the bandwidth of a desired RF signal might be greater than the channel spacing. For instance, the RF signal bandwidth of a CDMA signal is approximately 1.25 MHz, but the channel spacing is only 30 kHz in the cellular band and 50 kHz in the PCS band as given in Table 3.1.

Table 3.1. Frequency band allocations of wireless communication systems

Frequency Band/System	Up-Link Frequency Band (MHz)	Down-Link Frequency Band (MHz)	Band Separation (MHz)	Channel Spacing (kHz)
Cellular	824 – 849	869 – 894	20	30 (CDMA)
GSM 900	890 – 915	935 – 960	20	200
E-GSM 900	880 – 915	925 – 960	10	200
DCS 1800	1710 – 1785	1805 – 1889	20	200
PCS	1850 – 1910	1930 – 1990	20	50 (CDMA)
WCDMA	1920 – 1980	2110 – 2170	130	200
802.11b	2400 – 2484	2400 – 2484	—	13000
802.11a	5150 – 5350 5725 – 5825	5150 – 5350 5725 – 5825	— —	20000 20000

The frequency band allocation for a wireless mobile system will have substantial influence to the frequency planning of a superheterodyne transceiver used in this system. The frequency planning is mainly to search and select intermediate frequencies, which should minimize the spurious response problem of the superheterodyne transceiver and should provide the possibility of developing a transceiver with excellent performance. However, this is a very tedious task even if the transmitter and receiver each has only one IF block as presented in Fig. 3.1. It is more difficult to develop a frequency plan for a full-duplex transceiver than for a half duplex one. In a full-duplex transceiver, the receiver and the transmitter operate simultaneously, and the signals in both sides and their harmonics and mixing products need be taken into account in the frequency planning. In this case, it becomes really essential to avoid that any low-order transmitter spurious appears in the receiver band.

A superheterodyne transceiver may contain following fundamental signals: an UHF LO signal, a reference oscillator signal, two or multiple VHF LO signals, two or multiple IF signals, a weak RF reception signal and a powerful transmission signal. The real problem is that these signals may produce a tremendous amount of mixing products and harmonics because most of the devices compromising the transceiver have certain nonlinear characteristics. In the frequency planning, we must analyze these potential mixing products and harmonics — i.e., locate their position and determine their strength. It is not too hard to locate spurs of different order, but it is really difficult to determine the strength of the mixing products and harmonics without proper nonlinear models of corresponding devices. Fortunately, we are able to roughly estimate spurious levels based on the orders of the mixing products and harmonics and based on the order being an even or odd number as well.

3.1.2.1. Selection of Intermediate Frequency (IF)

In frequency planning, it is apparent that the first step of frequency planning is to select the intermediate frequencies. The basic criteria for choosing IFs for a full duplex mobile transceiver are presented as follows:

1. Choose the receiver IF when the receiver and the transmitter share an UHF LO as the configuration given in Fig. 3.1. This is when a high selective IF BPF (SAW filter) is used in the receiver for channel selection. The filter performance, such as insertion loss and selectivity, may be frequency dependent, and in general the performance is better when its center frequency is lower. On the other hand, the transmitter usually need not employ an IF SAW filter, and its IF is automatically determined by the chosen receiver IF, and the frequency separation between the transmitter channel and the corresponding receiver channel — i.e. $\Delta F_{Rx-Tx} = B_a + B_s$ (see Fig. 3.2). It will depend upon the LO frequency F_{LO} being

higher or lower than the receiver and transmitter operation frequencies F_{Rx} and F_{Tx} and also on the receiver frequency F_{Rx} being higher or lower than the transmitter frequency F_{Tx}. The transmitter intermediate frequency IF_{Tx} can be calculated from the receiver intermediate frequency IF_{Rx} as

$$IF_{Tx} = IF_{Rx} + \Delta F_{Rx-Tx}, \quad F_{LO} > F_{Rx} > F_{Tx} \text{ or } F_{Tx} > F_{Rx} > F_{LO} \quad (3.1.1)$$

or

$$IF_{Tx} = IF_{Rx} - \Delta F_{Rx-Tx}, \quad F_{LO} > F_{Tx} > F_{Rx} \text{ or } F_{Rx} > F_{Tx} > F_{LO}. \quad (3.1.2)$$

For example, a commonly used IF_{Rx} in the cellular band CDMA mobile stations is 85.36 MHz, the center frequency difference ΔF_{Rx-Tx} of the receiver and transmitter channel pair in this band is 45 MHz (see Table 3.1), and the IF_{Tx} thus is equal to 130.36 MHz from (3.1.1) when a high injection UHF LO — i.e., $F_{LO} > F_{Rx} > F_{Tx}$, is used.

2. To prevent potential receiver in-band jamming, the receiver IF should be chosen to meet the following inequality:

$$IF_{Rx} > B_{Tx} + B_s + B_{Rx} = 2B_a + B_s \quad (3.1.3)$$

or

$$IF_{Rx} < B_s, \quad (3.1.4)$$

where $B_{Tx} = B_u$, $B_{Rx} = B_d$, and $B_{Tx} = B_{Rx} = B_a$ are assumed. The mechanism of the potential receiver in-band interference if the chosen IF'_{Rx} does not meet (3.1.3) or (3.1.4) can be explained in terms of Fig. 3.3. From this figure we can see that the frequency difference between the channels within ΔF_{T_A} in the mobile transmitter band B_{Tx} and the down-link signals or interferers within ΔF_{R_B} at the end area of the receiver frequency band B_{Rx} can be equal to or very close to $IF'_{Rx} < B_{Tx} + B_s + B_{Rx}$. When a mobile station uses a transmitter and receiver channel pair within frequency bands ΔF_{T_A} and ΔF_{R_A}, the mobile station receiver is possibly jammed by the mixing product of its transmission leakage and a strong interferer or another base station down-link signal within the frequency band ΔF_{R_B} if their frequency offset is equal to or very close to the receiver intermediate frequency, IF'_{Rx}, as depicted in Fig. 3.3. It is apparent that this kind of jamming problem will not happen if the $IF_{Rx} < B_s$.

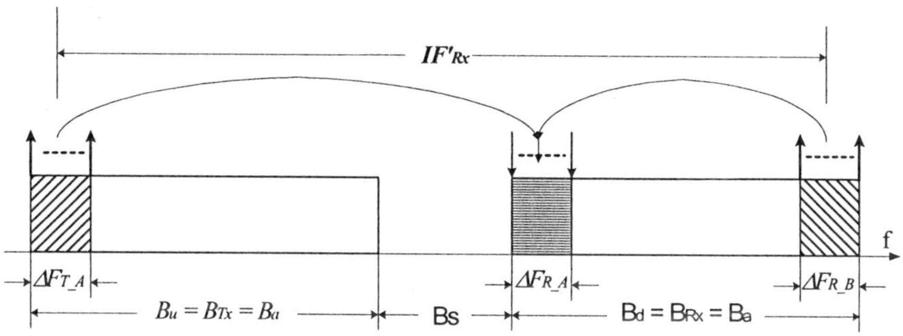

Figure 3.3. Potential in-band interference causing

In the Cellular band, the overall frequency band span is

$$B_{Tx} + B_s + B_{Rx} = 25 + 20 + 25 = 70 \text{ MHz}.$$

For the CDMA mobile stations, the most popular IF_{Rx} is 85.36 MHz, and it is greater than $B_{Tx} + B_s + B_{Rx} = 70$ MHz.

3. The IF/2 issue is that an interference product with a receiver IF will be generated due to second-order harmonic mixing if the frequency of the interferer has a frequency IF/2 offset from the UHF LO frequency. To suppress the possible IF/2 interferer, the receiver intermediate frequency IF_{Rx} should meet the following inequality:

$$F_{Rx\_lowest} + \frac{IF_{Rx}}{2} >> F_{Rx\_highest}, \quad \begin{matrix} \text{if } F_{LO} > F_{Rx} > F_{Tx} \\ \text{or } F_{LO} < F_{Rx} < F_{Tx} \end{matrix} \tag{3.1.5a}$$

or

$$IF_{Rx} >> 2(F_{Rx\_highet} - F_{Rx\_lowest}) = 2B_a, \quad \begin{matrix} \text{if } F_{LO} > F_{Rx} > F_{Tx} \\ \text{or } F_{LO} < F_{Rx} < F_{Tx} \end{matrix} \tag{3.1.5b}$$

Where the double greater than signs just mean that the interferer frequency is away from the boundaries of the receiver frequency band, and the interferer will be suppressed to quite low level by the receiver preselector.

In the previous examples, the 85.36 MHz IF used in the Cellular band CDMA receiver also meets (3.1.5). In this case, the potential

IF/2 interferer is 17.68 MHz offset from the receiver frequency band boundary, 894 Hz.
The selected IF_{Rx} based on (3.1.3), in most cases, also meets the criterion of (3.1.5) since the skirt of the receiver preselector is usually quite steep and the strength of the potential IF/2 interferer will be significantly suppressed.

4. In a multiband transceiver, the receivers operating at different frequency bands may employ the same IF and thus only one IF SAW filter will be used as the channel filter for all band receivers if they work for the same protocol wireless system. However, the receiver IF selection should be based on the frequency band with the largest operation band span — i.e., the largest $B_{Tx} + B_s + B_{Rx}$.
For example, in the CDMA mobile station another commonly used IF_{Rx} is 183.6 MHz in addition to 85.36. The high IF meets the criteria of (3.1.3) and (3.1.5) even in the PCS band since the operation frequency band span, $B_{Tx} + B_s + B_{Rx}$, in the PCS band is 140 MHz. In a cellular and PCS dual band transceiver of a CDMA mobile station, the frequency 183.6 MHz should be chosen as the common IF_{Rx} for both band receivers.
5. Avoid the low order mixing products of the transmitter IF signals with the transmission or the UHF LO signal locating in other wireless system bands. A modern mobile station may have a *GPS* receiver and a *Bluetooth* transceiver (other than a standard protocol wireless system, such as GSM, CDMA, or TDMA) for the voice or data communications. The GPS receiver or the Bluetooth transceiver may simultaneously operate with the CDMA or the other protocol transceiver. To avoid interfering with the GPS receiver or the Bluetooth transceiver, the possible selection ranges of the receiver IF have some restraints.
Using IF selection of a PCS CDMA transceiver as an example, the frequency of the GPS receiver is centered at 1575.42 MHz with a bandwidth of 2.046 MHz, the lowest transmission frequency of a PCS mobile station is 1850 MHz, and thus their frequency difference is 274.58 MHz. It is not desirable to make the IF of the PC CDMA transmitter between 274.58 MHZ and 274.58+60 = 344.58 MHz since the transmitter IF signal with a frequency in this frequency range and the CDMA transmission signal within the band from 1850 MHz to 1910 MHz may potentially generate a mixing product locating in the GPS receiver band 1575.42 ± 1.023 MHz. A simple calculation can verify this. If the IF_{Tx} is chosen equal to 304 MHz, and the transmission frequency is 1880 MHz, one of the possible mixing products of the transmission and the IF

signals has a frequency equal to 1880 – 304 = 1576 MHz within the GPS receiver bandwidth. Therefore, in this case, the IF_{Rx} should be less than 274.58 – 80 = 194.58 MHz or greater than 344.58 – 80 = 254.58 MHz when high LO injection is used, and the receiver frequency is higher than the transmitter frequency — i.e.,

$$IF_{Rx} < 194 \text{ MHz or } IF_{Rx} > 255 \text{ MHz}, \quad \begin{matrix} \text{if } F_{LO} > F_{Rx} > F_{Tx} \\ \text{or } F_{Tx} > F_{Rx} > F_{LO}. \end{matrix} \tag{3.1.6}$$

However, the operation band of the Bluetooth transceiver is from 2400 MHz to 2484 MHz. To avoid the Bluetooth, the signal becomes an image interferer of the PCS CDMA receiver when the high LO is employed. The IF_{Rx} should be less than (2400 – 1990)/2 = 205 MHz or it can be expressed as

$$IF_{Rx} < 205 \quad \text{MHz}. \tag{3.1.7}$$

From (3.1.7) and (3.1.6), we can come to the conclusion that the IF_{Rx} of the PCS CDMA receiver should be less than 194 MHz.

6. In a multiband transceiver the common IF_{Rx} selection may also consider the possibility of using only one UHF VCO and frequency divider for the multiple-band operation. For instance, if the transceiver operates over the cellular and PCS bands, from Table 3.1 we know the cellular receiver band from $F_{Cell\_Rx\_L}$ = 869 MHz to $F_{Cell\_Rx\_H}$ = 894 MHz; the PCS receiver band from $F_{PCS\_Rx\_L}$ = 1930 MHz to $F_{PCS\_Rx\_H}$ = 1990 MHz; and then the IF_{Rx} is better to be able to make the following frequency ranges overlapping as large as possible:

$$\text{Cellular: } 2\times(F_{Cell\_Rx\_L} + IF_{Rx}) : 2\times(F_{Cell\_Rx\_H} + IF_{Rx}),$$

and

$$\text{PCS:} \quad (F_{PCS\_Rx\_L} + IF_{Rx}) : (F_{PCS\_Rx\_H} + IF_{Rx}),$$

and to make the overall frequency variation range Δf_V,

$$\Delta f_V = \left(F_{PCS\_Rx\_H} + IF_{Rx}\right) - Min\left[\left(F_{PCS\_Rx\_L} + IF_{Rx}\right), 2\left(F_{Cell\_Rx\_L} + IF_{Rx}\right)\right], \tag{3.1.8}$$

within an UHF VCO tuning range, which is 5 to 7 % of the VCO center frequency. In (3.1.8), *Min*[*A*, *B*] is a function to choose the one with the minimum value between *A* and *B*.
In the above example, a 2 GHz VCO will be probably used, and its tuning range, thus, is approximately 100 to 140 MHz. If the receiver IF is 183.6 MHz, the overall frequency variation range for the cellular and PCS both band operation is

$$\Delta f_V = (1990 + 183.6) - 2 \times (869 + 183.6) = 68.4\,\mathrm{MHz}.$$

It is well within the tuning range of a 2 GHz VCO.

7. The channel selectivity of the receiver is also dependent upon the receiver IF selection in a certain extension. As a general rule, the lower IF is easier for obtaining higher selectivity than the higher IF. Especially for narrow-band wireless systems, such as AMPS and TDMA systems, it is difficult to develop an IF SAW with good performance if the required center frequency is over 150 MHz since the channel bandwidth of these systems is less than 25 kHz. Therefore, not only the selectivity but also the availability of the IF filter impacts the IF selection.

We have discussed how to select the receiver IF based on the full-duplex transceiver with a single IF block. The same criteria of IF selection can also be applied for the half-duplex transceiver, and in this case the situation will be better than the full-duplex one since its receiver and transmitter do not run at the same time. The spurious generated by the transmitter does not directly interfere with the receiver but may affect another mobile station that operates at the same channel frequency. For a transceiver with multiple IF blocks, its first IF can be certainly selected based on the criteria presented in this section.

3.1.2.2. Spurious Analysis

After the receiver IF has been selected, the transmitter IF can be easily determined in terms of UHF LO frequency if the LO is shared by the receiver and the transmitter. Otherwise, the transmitter IF is chosen independently if the LO is not shared with the receiver. Fundamental signals in a full-duplex transceiver usually contain a strong RF transmission signal, a weak RF received signal, a frequency programmable UHF LO, a frequency fixed transmitter VHF VCO, a frequency fixed receiver VHF VCO, and a reference signal in addition to the receiver IF and the transmitter

IF signals. The frequency of the transmitter or the receiver VHF VCO can be same as or twice of the frequency of the corresponding IF signal depending on how the $\pi/2$ phase shift is implemented. There is no doubt that nonlinearities of the transceiver generate a great amount of spurs consisting of the harmonics and mixing products of the fundamental signals. In spurious analysis the most important things are to locate spurs and to estimate their strength. Based on the location and strength data, the frequency plan can be then evaluated to judge whether it is applicable or not.

It depends on the operation frequency band of the transceiver and other frequency bands of interest to determine the orders of harmonics and mixing products of the fundamental signals, which need be considered. Up to 8th to 12th orders of RF or UHF signals, up to 14th to 20th of IF or VHF signals, and up to 30th to 40th orders of the reference signal are probably enough for using spurious analysis to measure a frequency plan. The harmonics frequency calculation of the fundamental signals is straightforward. The mixing products may be generated by multiple signals, and their frequencies can be expressed as $m{\times}F_{Tx} \pm n{\times}IF_{Tx} \pm p{\times}F_{Ref} \ldots \pm q{\times}F_{Rx\_VCO}$, where m, n, p, ..., and q are integers. However, these mixing products are less important since they are usually high-order spurs with very low strength. In reality, the potentially dangerous spurs are often relatively low-order mixing products of two fundamental signals with frequencies f_A and f_B, and their mixing products have frequencies

$$f_S = m \times f_A \pm n \times f_B , \tag{3.1.9}$$

where m and n are integers equal to 0, 1, 2, 3, $\cdots$, respectively. It represents the harmonic frequency of the fundamental signal A or B when n or m equals zero. These two signals could be any two of the six fundamentals as described in the previous paragraph. The composition frequency of different-order mixing products of two signals is easily calculated. This means that it is always possible to precisely predicate the location of spurs even if the spurs are resulted from multiple signals while their strength may be not so easy to be precisely predicated.

In the frequency planning or selecting IF, it is necessary to minimize the number of the spurious frequencies, especially those generated by the transmitter, in the following frequency bands:

- The receiver operation frequency band:
  (F_{Rx_L}, F_{Rx_H}) (see Table 3.1)
- The image band of the receiver operation frequencies:

$\left(F_{Rx\_L} + 2 \cdot IF_{Rx}, F_{Rx\_H} + 2 \cdot IF_{Rx}\right)$ for high LO injection

or

$\left(F_{Rx\_L} - 2 \cdot IF_{Rx}, F_{Rx\_H} - 2 \cdot IF_{Rx}\right)$ for low LO injection

- IF/2 band:

$$\left.\begin{array}{l}\left(F_{Rx\_L} + IF_{Rx}/2, F_{Rx\_H} + IF_{Rx}/2\right) \ \& \\ \left(F_{Rx\_L} + 3 \cdot IF_{Rx}/2, F_{Rx\_H} + 3 \cdot IF_{Rx}/2\right)\end{array}\right\} \text{for high LO injection}$$

or

$$\left.\begin{array}{l}\left(F_{Rx\_L} - IF_{Rx}/2, F_{Rx\_H} - IF_{Rx}/2\right) \ \& \\ \left(F_{Rx\_L} - 3 \cdot IF_{Rx}/2, F_{Rx\_H} - 3 \cdot IF_{Rx}/2\right)\end{array}\right\} \text{for low LO injection}$$

- The transmitter operation frequency band:

$$(F_{Tx\_L}, F_{Tx\_H}) \text{ (see Table 3.1)}$$

- The UHF LO frequency operation band:

$\left(F_{Tx\_L} - IF_{Rx}, F_{Tx\_H} - IF_{Rx}\right)$ for low LO injection

or

$\left(F_{Tx\_L} + IF_{Rx}, F_{Tx\_H} + IF_{Rx}\right)$ for high LO injection

- Other bands of interest: such as, GPS band: (1575.42 ± 2) MHz. Assume that a GPS receiver transceiver incorporates with the transceiver whose frequency plan is under development.

In reality, it may be not possible to avoid any spur falling into the above bands. It is necessary to examine the spurs within these bands whether in the operation frequency channel of the mobile station transceiver or not. The best way to do this is to utilize spurious response charts.

The following example shows how to use the spurious response chart. A cellular band full-duplex transceiver with an IF_{Rx} = 183.6 MHz, and the transceiver is using a high LO injection configuration. In this case, the transmitter IF is IF_{Tx} = 183.6 + 45 = 228.6 MHz, the UHF LO tuning rang is from 1052.6 MHz to 1077.6 MHz, and the frequencies of the VHF receiver and transmitter VCOs are twice IF_{Rx} and IF_{Tx}, respectively. The reference frequency, F_{Ref}, is assumed 19.2 MHz. Based on these fundamental signal frequencies, the frequencies of the potential spurs can be calculated by using (3.1.9).

When the transceiver is tuned over the cellular frequency band, some mixing products of $3 \times F_{Tx}$ and $7 \times IF_{Tx}$, $3 \times F_{UHF\_LO}$ and $5 \times F_{VHF\_LO}$, and $4 \times F_{Tx}$ and $11 \times IF_{Tx}$ fall in the receiver band as shown in Fig. 3.4. The spurious response frequency lines of ($3 \times F_{Tx}$, $7 \times IF_{Tx}$) and ($3 \times F_{UHF\_LO}$, 5 ×

F_{VHF_LO}) entirely overlap together. In this figure and the next two, the *x*-axis is the transmitter tuning frequency. It can be clearly seen that the spurious response frequency lines have no any intersection with the receiver tuning line. This means there is no lower than 15th order mixing spurs of the transmitter RF and IF signals directly locating within any channel over the receiver operating band.

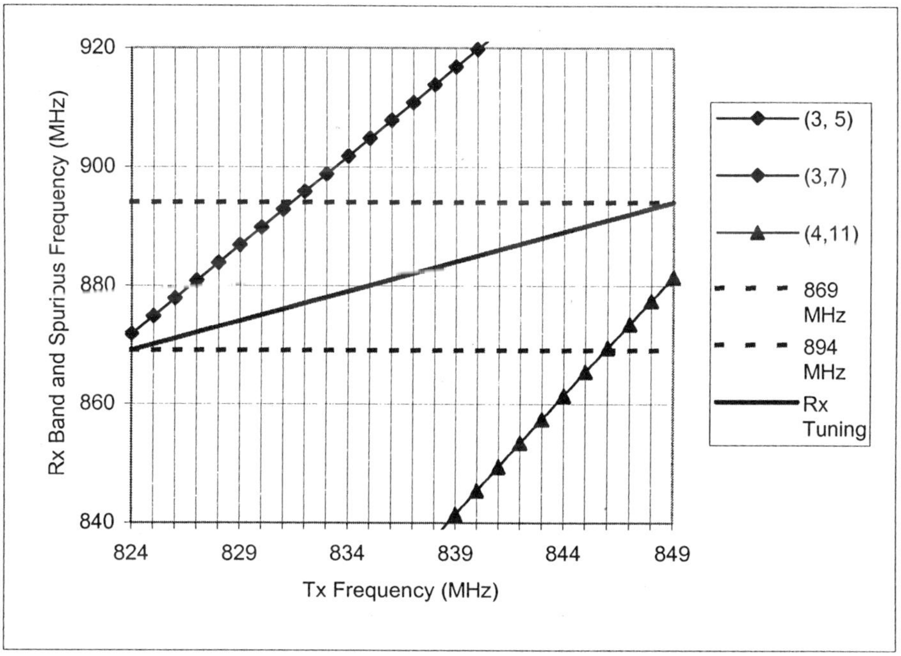

Figure 3.4. Spurious responses in receiver band

The mixing spurious responses of $4 \times F_{Tx}$ and $10 \times IF_{Tx}$ (or $4 \times F_{UHF\_LO}$ and $7 \times F_{Tx\_VHF\_VCO}$), and $2 \times F_{UHF\_LO}$ and $7 \times F_{Tx\_VHF\_VCO}$ (or $2 \times F_{Tx}$ and $12 \times IF_{Tx}$) cross over the UHF LO frequency band, and these two spurious response lines intersect with the UHF LO tuning line at the same point, (838.2, 1066.8) MHz as depicted in Fig. 3.5. The corresponding receiver channel frequency is 883.2 MHz. These two spurs with a frequency near or equal to 1066.8 MHz mix with the receiver signal to generate in-channel interferers with a frequency close to or equal to IF_{Tx} = 183.6 MHz. However, their interference effect can be minimized if carefully to do the circuit board layout and shielding since their order is relatively high, higher than 9th order, and the low-order spurs are generated by UHF and VHF VCOs, which have lower power than the transmitter RF and IF signals that produce high-order spurs in the UHF LO band.

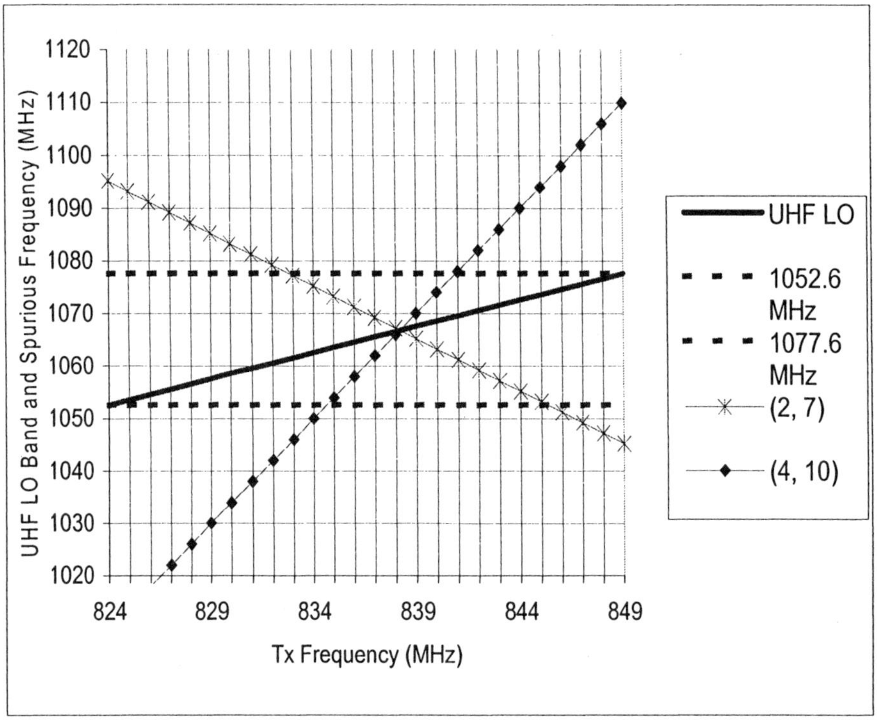

Figure 3.5. Spurious response in UHF LO tuning band

There exist two IF/2 bands, (960.8, 985.8) MHz and (1144.4, 1169.4) MHz. The spurious response lines in these two IF/2 bands do not intersect with the IF/2 tuning lines. Fig. 3.6 presents the mixing spurious response of $2 \times F_{Tx}$ and $3 \times IF_{Tx}$ and the IF/2 tuning lines in the lower IF/2 band. The spurious response line is close to the IF/2 tuning line at the lower-frequency edge, but they do not touch within this frequency band. No lower than 15th order spurs fall into the image band and in the GPS band. No harmonics of the transmitter IF and VHF VCO signals and no harmonics of the receiver IF and VHF VCO locate in all the bands of interest as described above. However, the 46th order harmonic of 19.2 MHz reference signal are exactly at 883.2 MHz channel. This spur may desensitize a narrow-band system, such as AMPS, receiver, but it does not have much impact on a wide-band system, such as CDMA, receiver. Even in the narrow-band system, this high-order reference harmonic spurious problem is easy to suppress in implementation.

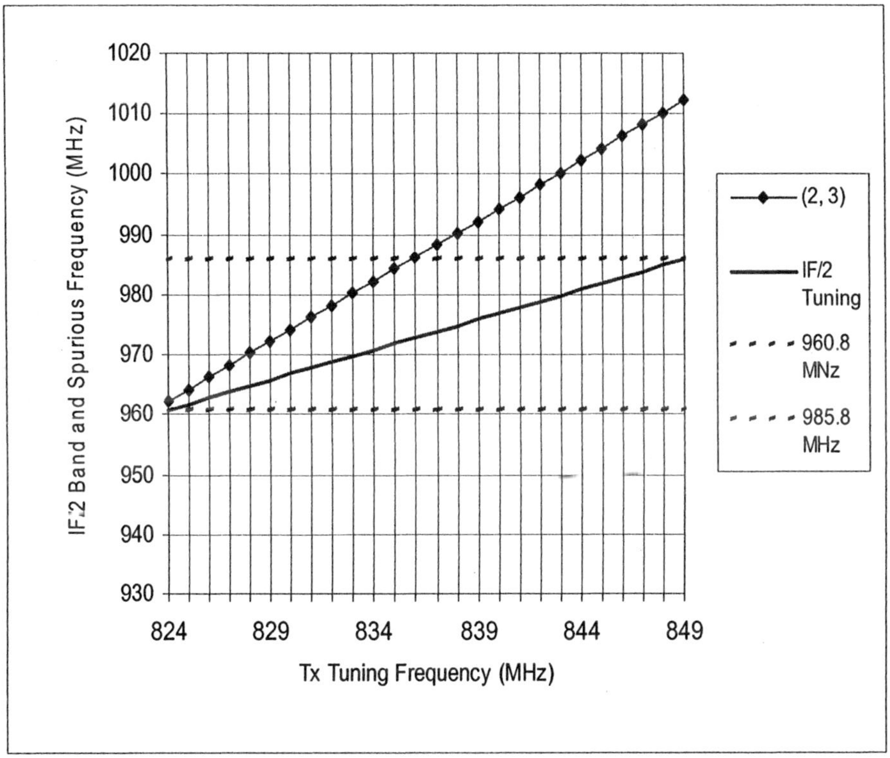

Figure 3.6. Spurious response in IF/2 frequency band

In the transmitter band, there is a spurious response of $1 \times F_{UHF\_LO}$ and $12 \times F_{Ref}$. This spurious response line running in parallel with the transmitter tuning line, and their separation is only 1.8 MHz as depicted in Fig. 3.7. Thus a spur always accompanies with the desired transmission signal for all the channels. This spur may affect transmitter emission specification if its level is significant. However, it is usually not difficult to keep this spur in an insignificant level by means of using differential circuit design, proper filtering, and careful chip and circuit board layouts since this is a mixing product resulted from the 12th-order harmonic of the reference signal, which is an even-order nonlinearity usually weaker than odd-order ones, and the 12th-order harmonic can be sufficiently suppressed by filtering.

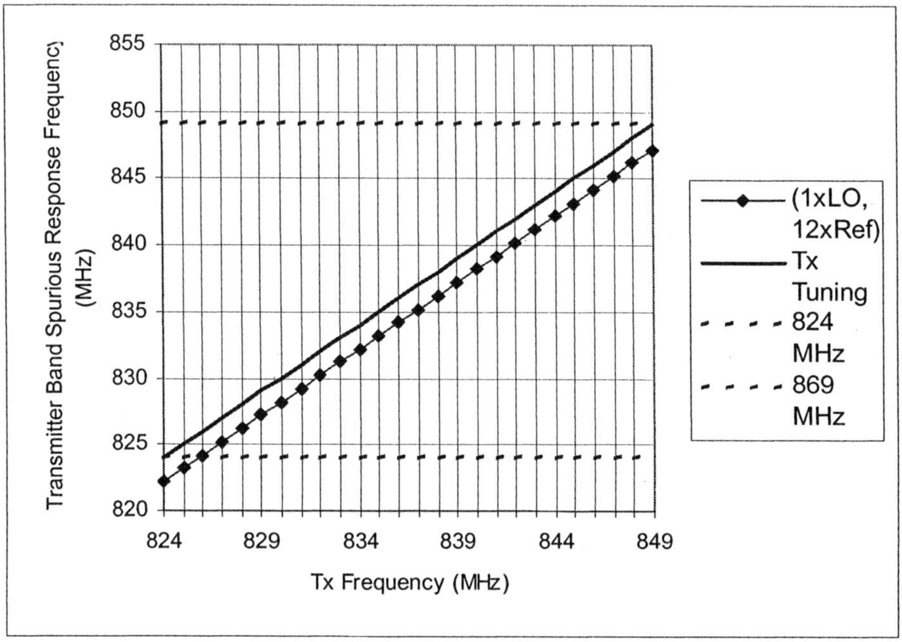

Figure 3.7. Spurious response in transmitter band

Among the spurs, it is necessary to avoid harmonics of the following fundamental signals falling into the center frequency tunable RF channel bandwidth and the IF channel bandwidth of the receiver:

- Reference signal harmonics
- Transmitter IF signal harmonics
- Transmitter VHF VCO signal harmonics
- Receiver VHF VCO signal harmonics

It will desensitize the receiver sensitivity at all the receiver channels if a harmonic of the reference signal is in the receiver IF channel bandwidth. The harmonics of the transmitter IF signal and VHF VCO signals will be low order if they fall into receiver RF channels. The low-order spurious may severely impact the receiver performance, and it is hard to be coped with if it has sneaked into the receiver channels.

The frequency planning for a half-duplex transceiver is simpler than that of the full-duplex one as described above. For example, the spurs generated by the transmitter will have no direct impact on the receiver performance of the transceiver. However, the method of analyzing spurious response and developing frequency plan for the full-duplex transceiver can be definitely used for a half-duplex system with less restriction. Other approaches of the spurious analysis and frequency planning can also be found in [1].

3.1.3. Design Consideration of Superheterodyne Transceiver

The detailed receiver and transmitter system designs are discussed in Chapter 4 and 5, respectively. Here, we are going to present the general design considerations of the superheterodyne RF transceiver, which may differ from other architecture, such as direct conversion and low IF transceivers. The transceiver design for a wireless mobile system mainlyconcerns electrical performance, *automatic gain control* (AGC) system, power consumption, and overall cost.

3.1.3.1. Receiver Sensitivity, Linearity, and Selectivity

Receiver sensitivity is defined as the *minimum detectable desired signal strength* (*MDS*) to obtain a certain *bit error rate* (*BER*) or *frame error rate* (*FER*) or *package error rate* (*PER*). The sensitivity is one of most important specifications of a receiver, and it is determined by the overall noise figure of the receiver and processing gain/loss. In a superheterodyne receiver, *linearity* is mainly measured based on the third-order distortion level of the receiver, which is represented by the *third-order intercept point* (IP_3). The third-order intercept point IP_3 is defined as the intersection point of the linear extension of the fundamental signal output versus input power characteristic and the third order intermodulation (IM) product line as shown in Fig. 3.8. The horizontal value of the third-order intercept point is called *input third-order intercept point* (IIP_3) and the vertical value is the *output third-order intercept point* (OIP_3) (see Appendix 3A for more details). In the receiver design the IIP_3 is often employed, and the OIP_3 is more naturally used in the transmitter design. Selectivity of a receiver is the characteristic of the receiver that allows it to identify the desired signal at one frequency apart from those at all other frequencies. It is mainly determined by RF, IF, and BB filters used in the receiver. These filter responses must be sharp enough to sufficiently suppress the interference from adjacent channels and other sources, but they also need to be broad enough to pass the desired signal with acceptable amplitude and phase distortions.

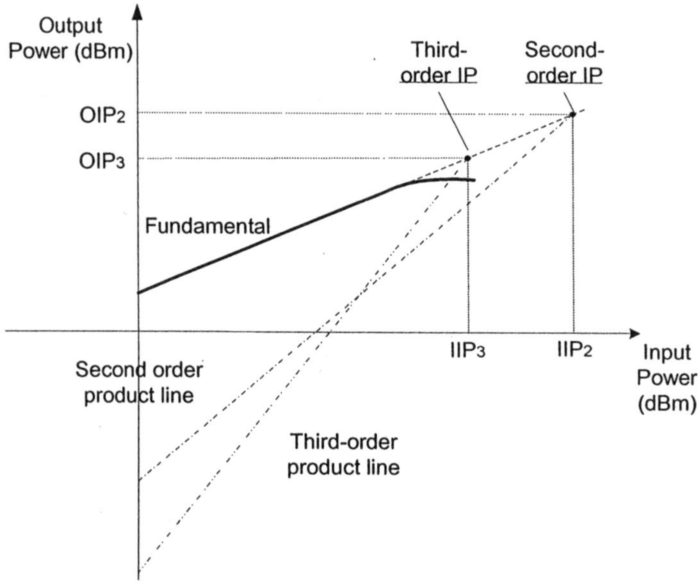

Figure 3.8. Third and second intercept points and output to input power relationships of a system or device with nonlinearity

The sensitivity and linearity or the noise figure (*NF*) and IIP_3, of a receiver are tightly dependent on the gain distribution over the receiver chain. To obtain a lower noise figure or a higher sensitivity receiver, it is preferred to make the front-end block from the antenna port through the LNA to the RF down-converter input port (see Fig. 3.1) with a high gain. In this case, the receiver noise figure may be mainly determined by the front-end block noise figure, and the back section of the receiver chain from the down-converter input to the ADC output has minor impact on the overall noise figure. However, the high front-end gain will degrade the receiver linearity since the overall IIP_3 of the receiver decreases with the front-end block gain increase. In receiver design, it is necessary to have proper gain distribution, which is able to take care of the tradeoff between sensitivity and linearity, or noise figure and IIP_3, of the receiver for an appropriate performance.

In general, it is desirable for a device to have a low noise figure and a high IIP_3. However, in reality it is difficult for an active device, an amplifier or a mixer, under a certain current consumption to obtain ideal performance on both the noise figure and the IIP_3. A compromise in the circuit design of the active device may be necessary to trade a higher noise figure for a better IIP_3 or vice versa depending on overall receiver performance. In the receiver design, it might be helpful to use the ratio of the IIP_3 to the noise figure as a measure of the quality of a receiver or a device with certain current consumption. The measure of the receiver or the device quality is expressed as

$$Q = IIP_3 - NF \, , \tag{3.1.10}$$

where IIP_3 is the third-order intercept point in dBm, and NF is the noise figure in dB. This quality measure or quality factor Q can be utilized to optimize the gain distribution of the receiver chain, which means providing a reasonably low overall noise figure and a sufficiently high IIP_3 with an acceptably low current consumption.

As a rule of thumb, for a good receiver design the Q factor of the first down-converter is higher than the overall Q factor of the superheterodyne receiver around 10 dB for active mixers and 15 dB or more for passive mixers since the down-converter is usually the last stage before the IF channel filter and it may suffer from the attack of very strong interferers amplified by the receiver front-end. The Q factor of a GSM mobile station receiver, which meets the minimum performance specification (GSM 05.05) issued by ETSI [2], is approximately –30.5 dBm (IIP_{3_min},≅ -19.5 dBm and NF_{max} ≅ 11 dB), and the Q factor of a CDMA mobile station receiver meeting IS-98D [3] is around –24 dBm (IIP_{3_min},≅ -14 dBm and NF_{max} ≅ 10 dB). Therefore, the Q factors of the RF down-converter used in GSM and CDMA mobile station receivers should be higher than –16 dBm and –9 dBm, respectively. It usually requires the spending of more current for achieving a higher Q factor for a device or a receiver. The CDMA mobile station receiver thus consumes more current than the GSM receiver does. As discussed in the previous section, most of the gain in a superheterodyne receiver is developed in the IF block. There are two reasons for this. The first one is the demanded Q factors of the stages after the IF channel filter becoming quite low (< -20 or even < -30 dBm) and their current consumption low. The second one is that the gain in the IF block is more easily controlled in a continuous way than the gain of amplifiers in the RF and analog BB blocks. It should be noticed that the preselector, an RF BPF, connected between the antenna and the LNA will not affect the overall quality measure of the receiver since the preselector is a passive device having a noise figure equal to its insertion loss and thus it makes the overall receiver noise figure and the IIP_3 both increasing the same amount of their values and keeping their difference — i.e., quality measure no change.

Blocking characteristics of a receiver are mainly determined by the selectivity of the receiver and the phase noise and spurs of synthesizers used as local oscillators of the receiver. The RF BPF performed as a pre-selector or a part of duplexer copes with out of receiver operating band blockers or interferers. The operation frequency bands of the wireless mobile communication systems are usually shared by different protocol systems. The strong in-band (within the receiver band) interference must be sufficiently suppressed by the IF channel filter and the BB LPF. These channel filters will further suppress the out-of-band interferers as well. In addition, the phase noise and the spurs of the synthesizer LOs, particularly the RF LO, should be designed to be low enough to minimize the mixing product level of the interference with the phase noise and/or the spurs of the LOs since these mixing products fall in the channel bandwidth and the IF and BB channel filter are not able to suppress them.

3.1.3.2. Transmitter Output Power, Spectrum, and Modulation Accuracy

The up-link frequency bands presented in Table 3.1 are the transmitter frequency bands of mobile stations for different protocol wireless communication systems. The frequency accuracy of the transmission carrier is usually in between $\pm 20\times 10^{-6}$ ($\pm$20 ppm) and $\pm 5\times 10^{-8}$ ($\pm$0.05 ppm) depending on systems, and it is assured by the reference oscillator, which is usually a *temperature compensated crystal oscillator* (TCXO) or a *voltage control temperature compensated crystal oscillator* (VCTCXO). There is usually an automatic frequency control (AFC) loop in mobile stations to control the VCTCXO frequency and to make it frequency tracking the received carrier frequency.

The transmission power level of a mobile station directly impacts the overall power consumption of a transceiver and also affects the talk time of a mobile station using a finite-capacity battery. The most popularly used nominal maximum transmission power levels for the mobile stations operating in different wireless communication systems are presented in Table 3.2. The maximum transmission power range is from 21 dBm to 33 dBm, and the power is either measured based on the *effective radiated power* (ERP), or the *effective Isotropic radiated power* (EIPR) with an antenna gain equal to or unequal to 0dBi. The ERP is defined as the product of the power supplied to the antenna and the antenna gain relative to a half-wave dipole in a given direction [3]. The gain of a half-wave dipole antenna is 2.15 dBi [4]. If the antenna gain of a mobile station is 1.5 dBi or 0.65 dB less than that of the half-wave dipole antenna, in order to meet the minimum 23 dBm ERP requirement, the transmission power measured at the antenna port of a cellular band CDMA mobile station must be equal to or greater than 23.65 dBm. The EIRP is the product of the power supplied to the antenna and the antenna gain in a direction relative to an isotropic antenna, or 0 dBi gain. As an example, the transmission power of a PCS band CDMA mobile station is measured based on the EIRP, and it needs only 21.5 dBm output power at the antenna port to meet the 23 dBm nominal output power requirement if the antenna gain of the mobile station is 1.5 dBi.

Table 3.2. Maximum output power of different system mobile stations

Systems	Power Class	Nominal Power (dBm)	Tolerance (dB)	Note
AMPS	III	28	-4, +2	ERP
CDMA Cell	III	23	+7	ERP
CDMA PCS	II	23	+7	EIRP
GSM 900	IV	33	-2, +2	Ant. Port
GSM 1800	I	30	-2, +2	Ant. Port
TDMA	III	28	-4, +2	ERP
WCDMA	IV	21	-2, +2	Ant. Port

At the maximum output power, the power amplifier (PA) in a mobile station transmitter usually spends the most of the power consumption. It is important for a mobile station to utilize a PA with high efficiency to reduce current consumption and to significantly increase the talk time. At present, the efficiency of a class AB PA providing 25 to 35 dBm medium power is around 35 to 40 % and the efficiency of a class C PA is approximately 45 to 55 %. What class of PA should be employed in a mobile station is entirely determined by the modulation scheme used for its information transmission. In the AMPS and GSM (GPRS) systems, FM and GMSK modulation schemes are used, respectively. In both mobile stations, the class C PA can be adopted since the envelop of the FM and the GMSK modulation waveforms is constant. In other systems, such as CDMA and TDMA systems, class AB PAs must be used because the phase modulated (PM) transmission signal in these systems is always accompanied by an amplitude modulation (AM), which demands certain linearity in the transmitter chain, especially the linearity of the PA since it is the last active stage of the transmitter chain. It is best to have a PA with a *1 dB compression point* being 2 to 3 dB higher than its possible maximum output power if the power consumption is acceptable.

To prevent spurious emissions of a mobile station transmitter from interfering with other mobile stations or systems, the transmission spectrum of the mobile stations is well defined in wireless communication systems. The emission level in the adjacent/alternate channels and up to frequency offset 2 to 4 MHz on each side of the transmission carrier is mainly determined by the modulation scheme and the base-band pulse shape filtering. Further filtering including an antialiasing low-pass filter in the analog base-band of the transmitter is necessary to sufficiently suppress the close-in emission and the emission out of the transmitter band. Certain predistortion may be needed to compensate the extra group delay distortion introduced by the analog BB filters. For transmission signals with amplitude modulation like CDMA and TDMA signals, the emission power in the adjacent channel and the alternate adjacent channel will be raised by the nonlinearity of the transmitter. The PA usually dominates the nonlinearity of the transmitter at the maximum output power. Thus linearity of the PA need be good enough to keep the *adjacent channel power* (ACP) within the specified mask. The emission out of the transmission band can be effectively suppressed by using RF band-pass filters (RF SAW filter and/or duplexer).

For frequency modulated transmission signals, such as a GMSK signal, an offset phase locked loop (OPLL) may be used to perform the RF BPF and the frequency up-converter both functions [5–6]. A block diagram of the OPLL is shown in Fig. 3.9. Although its configuration may be more complicated than an RF SAW filter plus an RF up-converter, the OPLL can be integrated in the transmitter chip to save an RF SAW, and it provides a tunable RF BPF with a bandwidth (500 KHz to 2 MHz) much narrower than that of an RF SAW filter.

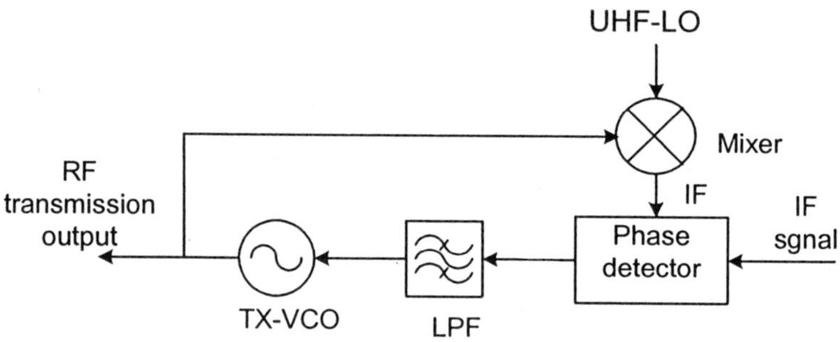

Figure 3.9. Block diagram of offset phase locked loop

Another important specification of the transmitter is the *modulation accuracy*. The mathematical expressions of various modulation accuracy definitions is presented in Chapter 5. Here some general considerations of transmitter system design on the modulation accuracy are discussed. The most popularly used measure of the modulation accuracy is the *error vector magnitude* (*EVM*), which results from the difference between the ideal waveform and the actual transmission waveform. This difference is referred to as *error vector*. The EVM is defined as the square root of the ratio of the mean error vector power to the mean reference power expressed as a %. The close-in phase noise of the UHF and VHF synthesizers in the transmitter may degrade the modulation accuracy. To minimize this influence, the close-in phase noise within the synthesizer loop bandwidth shall be lower than –75 dBc/Hz or the power ratio of the close-in phase noise integrated over the channel bandwidth shall be less than –30 dB. The group delay distortion of a narrow-band filter in the transmitter chain may also increase the EVM of the transmission. To obtain a low EVM, it is better to keep the ISI caused by the narrow-band filter group delay distortion at a relative level below –30 dB as well. The third factor that impacts the modulation accuracy is the carrier leakage. This factor may dominate the degradation of the modulation accuracy when the transmission power is low. It is desirable to make the carrier leakage 25 dB lower than the transmission signal even at very low output power level. The specification of the TDMA and the WCDMA mobile stations on the EVM is 12.5 and 17.5 % or less, respectively.

In the CDMA system, a waveform quality factor ρ is used instead of the EVM to measure the modulation accuracy. It is defined as the normalized correlated power between the actual waveform and the ideal transmission waveform. The minimum waveform quality factor for a CDMA mobile station is 0.944, but in the practical CDMA mobile station, ρ is usually designed for greater than 0.98. All the factors degrading EVM mentioned in the previous paragraph also impact the waveform quality factor in a way very similar to the EVM. The modulation accuracy in the GSM system is measured by using phase error, which is obtained by computing the difference between the phase of the transmitted waveform and the phase of the expected one. In the GSM mobile station the RMS phase error shall not be greater than 5° with a maximum peak deviation less than 20°. The phase error of the GMSK may be caused by the nonlinearity of the I/Q modulator, the loop bandwidth of the OPLL, and the in-channel phase noise of the UHF VCO in the transmitter of the GSM mobile station.

3.1.3.3. Dynamic Range and AGC systems

Depending on wireless communication systems, the minimum highest received signal level under which the mobile stations shall be able to work properly with a message error rate below the specification is in the range of –25 dBm to –20 dBm, and the dynamic range of the mobile stations is around 80 to 85 dBc. The maximum output power of mobile stations is described in Table 3.2. Only the mobile station transmitter used in the CDMA system has a high dynamic range of the transmission power, from 23 dBm or higher to –50 dBm or lower since the CDMA system has the near and far field effect. The WCDMA has a similar dynamic range, from +21 dBm to –44 dBm. But the dynamic range of the GSM mobile transmitter is only 30 dB, and it is even lower than this in the other wireless systems.

When designing the receiver AGC, the following points need be considered:

- In the mobile station receiver design, it is normal to achieve receiver sensitivity 3 to 5 dB better than the minimum performance requirement and to have a 5 dB margin of the maximum acceptable reception signal level. In addition, the overall receiver gain variation over temperature, frequency, and the inaccuracy of gain curve calibration may typicaly be 10 to 15 dB. Considering all these factors, the receiver AGC control range should be 20 to 25 dB wider than the dynamic range specified by the minimum performance of various protocol standards.
- The maximum gain of the receiver chain is determined by the ADC input set level. This level should be kept constant under the automatic gain control. To have a high signal-to-noise ratio the input level of the ADC should be set as high as possible. However, it is necessary to leave head room for the possible *peak-to-average ratio* (PAR) of the received signal, maximum DC offset, and the slow constructive fading.
- The control accuracy of the receiver AGC of mobile stations operating in most of wireless communication systems is not high — around ±4 dB to ±8 dB. In the CDMA and WCDMA systems, the receiver AGC of the mobile stations needs better control accuracy since the received signal level of the CDMA mobile station will determine its transmission power level and the power control accuracy will affect the system user capacity. An achievable gain control accuracy of the CDMA mobile station is approximately ±2.0 to ±2.5 dB. The main limitations are the gain control errors caused by imperfect calibration and gain curve fitting, temperature variation and frequency variation, and the measurement error of the received signal strength, which is generally measured by means of a *receiver signal strength indicator* (RSSI).
- The time constant of the receiver AGC system is around a couple of milliseconds for the CDMA and WCDMA mobile stations. The AGC control cycle of the GSM and TDMA mobile station receivers is approximately 4.62 msec and 20 msec, respectively.

Similar to the receiver AGC, the transmitter AGC should also have a 15 to 20 dB design margin to cover possible gain variations with temperature, frequency and other influence factors. In the CDMA and WCDMA systems, the open-loop power control accuracies are ±9.5 dB, and the time response for a 20 dB gain change should within 24 msec. The range of the closed-loop power control needs to be ±24 dB, and its relative accuracy depends on the control step size — e.g., 1 ± 0.5 dB and 0.5 ± 0.3 dB. The gain control curve of the transmitter chain is usually not linear. Depending on the order of the nonlinearity of the gain control curve, finite points will be measured, and then a curve-fitting approach is used to obtain entire control curve within an allowable tolerance. The transmitter output power of the other wireless systems is stepped instead of continuous. The tolerance of the controlled output power is ±2 dB to ±5 dB, depending on the output level.

The input BB signal of the RF transmitter chain comes from the DAC. The signal level from the DAC output is usually quite high, and it may be close to the maximum voltage swing of the DAC to make the signal with a high signal-to-noise ratio. This signal needs to be filtered to further reduce possible out-of-channel emission levels and to significantly reduce aliasing products before modulating the carrier and amplification. In the circuit design, a conversion from the voltage to current is needed in the beginning of the transmitter chain to handle the high input signal voltage.

To reduce the power consumption, most of the gain control in the transmitter should be implemented in its IF block. The RF gain control is probably 1/5 to 1/3 of overall transmitter gain control range.

3.1.3.4. Other Considerations

Low power consumption is essential to a mobile station operating based on a battery power supply. This is accomplished by choosing low current consumption circuits and devices, but it is also necessary to have efficient power management. For instance, the circuit bias current can be drawn differently at different signal levels to save current consumption, or all the circuits should be turned off when they are not being used as often as possible. The general rule of power management is to take all possible advantages and to achieve the maximum power saving.

To minimize the cost and size of a mobile station, it is important to utilize highly integrated RF circuits based on the modern GaAs, SiGe, and CMOS semiconductor technologies. A single *RF integrated circuit* (RFIC) may contain a whole transceiver, including UHF and VHF PLLs, and some matching networks for external devices, such as filters and UHF VCO.

However, at present only the half-duplex RF transceiver of the mobile stations for GSM, TDMA, etc., wireless communication systems is possible on single RFIC. The integration solution for the full-duplex transceivers is still two or multiple IC chips since the isolation of the present semiconductor technologies is not good enough for the full-duplex system. It is apparent that using the highly integrated circuits can reduce the part counts, size, and cost of the mobile stations.

3.2. Direct-Conversion (Zero IF) Architecture

The *direct conversion* means that an RF signal is directly down-converted to a BB signal or vice versa without any intermediate frequency stages, and therefore it is also referred to as *zero IF* architecture. The direct-conversion receiver is also referred to as a *homodyne* when the LO is phase-locked with the carrier of the received signal. The direct-conversion architecture has many attractive features. The direct-conversion receiver has no IF, and thus the expensive IF passive filter (SAW filter) can be eliminated, and then the cost and size of the overall transceiver are reduced. The channel filtering of the direct-conversion receiver is implemented in the analog base-band by means of active low-pass filter. The bandwidth of the active filter can be designed as adjustable. Since the bandwidth of the channel LPF is adjustable, it is easy to design the direct-conversion receiver for multimode operation with a common analog base-band circuitry and even a common RF front-end from the preselector to the RF down-converter if all the modes running in the same frequency band. This architecture does not need a frequency plan, which is usually very time-consuming work and hard to validate. The direct conversion has no image.

The configuration of the direct-conversion radio looks simpler than that of the superheterodyne radio, but its implementation is much more difficult since there are a number of technical challenges in the direct conversion receiver. Compared with the receiver, the direct-conversion transmitter has fewer issues, and it is relatively easy to be implemented. However, the direction-conversion transmitter may not provide as much profit as the direct conversion receiver does. For example, no IF channel filter saving is obtained by using the direct-conversion transmitter since the superheterodyne transmitter does not use the IF channel filter either. The outstanding feature of the direct-conversion transmitter is that its transmission contains much less spurious products than the superheterodyne transmitter. In this section, the technical difficulties and the corresponding solutions of designing and developing the direct-conversion receiver system are mainly addressed.

3.2.1. Configuration of Direct-Conversion Radio

All the direct-conversion transceivers of mobile stations operating in various protocol wireless communication systems have certain technical difficulties, but the hardest ones in their design and implementation are those of full duplex transceivers. The main focus of this section is on the full-duplex direct conversion transceiver. A block diagram of a direct conversion full-duplex transceiver is presented in Fig. 3.10. This is one possible configuration of the direct conversion full-duplex transceiver, but it is not a unique one.

The RF portion of the direct conversion transceiver is similar to that of the superheterodyne transceiver. The receiver and the transmitter now have separate UHF synthesizers and VCOs since their operating frequency bands are different, as presented in Table3.1, where the down-link frequency band and the up-link frequency band are for the receiver and the transmitter of mobile stations, respectively. In Fig. 3.10, the UHF VCO in the receiver is running at twice of the receiver operating frequency. The VCO signal is applied to a frequency by-two-divider. The two outputs of the divider with a 90° phaseshift each other are used to drive the receiver I and Q down-converter, respectively. In the transmitter, LO of the I/Q modulator is generated by a LO generator, which mainly consists of a UHF VCO, a frequency divider, and a single side band mixer (see Section 3.2.3.6). The UHF VCO in the LO generator is intentionally designed to make it running at a frequency offset from the LO generator output frequency. If the LO generator output frequency is equal to the transmission frequency, a $\pi/2$ phase shifter is used in the I/Q modulator as shown in Fig. 3.10, or otherwise a by-two-divider is employed to generate the 90° phase shift as used in the receiver I/Q quadrature down-converter. It is important that in the direct conversion transmitter, the UHF VCO should operate at the transmission frequency and also at its harmonic frequency. In this way, we can effectively get rid of the reverse-modulation problem of the VCO, which is caused by the feedback injection of the strong and modulated transmission signal or its harmonics generated by the power amplifier.

The function of the receiver BPF in the duplexer is to suppress the leakage power of the transmission and other out of receiver band interference. There is no image problem for the direct-conversion receiver since it does not have an image band. The received signal after preselecting of the duplexer is amplified by an LNA, and it is further filtered by an RF

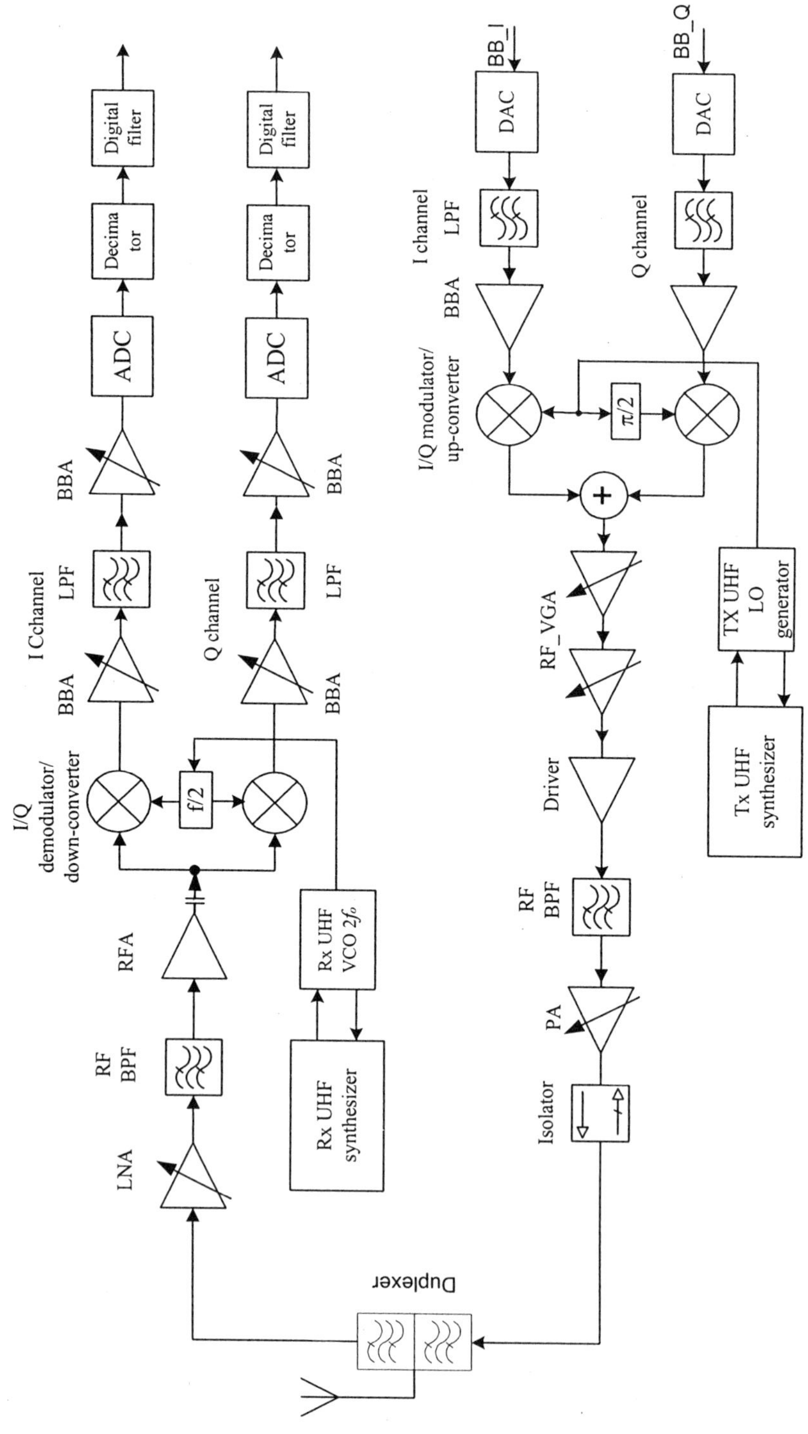

Figure 3.10. Block diagram of direct-conversion transceiver

filter. Now this filter rejection to the transmission leakage need be higher than that required in the superheterodyne receiver to control the transmission leakage self-mixing problem and to relax the second-order distortion requirement of the down-converter, which is discussed in Section 3.2.3. The filtered RF signal is then directly down-converted into I and Q channel BB signals by an I /Q down-converter also called as *quadrature demodulator*. The BB signals in the I and Q channels are synchronously amplified, but their 90° phase offset will be kept unchanged as possible. In the direct-conversion receiver, approximately 75% of the overall receiver gain is obtained from the analog base-band block when the receiver operates at its high gain mode. The gain in the RF block is usually the power gain, and the gain in the BB block is the voltage gain. There exists a low-pass channel filter in each of I and Q channels. Unlike the superheterodyne receiver, the channel selectivity now mainly depends on the stop-band rejection of the low-pass filter without any passive band-pass filter assistance. The amplified and filtered BB analog signals in the I and Q channels are converted into digital signals by analog-to-digital converters (ADC), and the digital signals then pass through digital filters to further suppress nearby interferers and enhance the channel selectivity.

The I and Q BB signals coming from the transmitter digital-to-analog converters first pass through low-pass filters to make the adjacent-channel and alternate-channel emission levels further suppressed and to eliminate aliasing products. The gain distribution in the direct-conversion transmitter is opposite to that in the direct-conversion receiver. The I and Q BB blocks provide very low gain or even negative voltage gain. The filtered and amplitude attenuated I and Q BB signals both are directly up-converted to RF signals, and they are then added together by an I/Q modulator. The composite RF signal is amplified all the way up to the RF power amplifier (PA). Almost over 90% of the direct conversion transmitter gain is in the RF block from the I/Q modulator to the PA. An RF band-pass filter is inserted in between the driver amplifier and the PA to suppress the out-of-band, especially in the receiver band, noise and spurs emissions. Comparing with the superheterodyne transmitter as depicted in Fig. 3.1, the direct-conversion transmitter does not save any passive filter but the VHF synthesizer LO.

Since the main gain block of the direct-conversion receiver is the analog base-band block, the most of the automatic gain control (AGC) is executed in this block. However, there are two parallel channels, I channel and Q channel, in the analog BB block. To synchronize the gain control in both channels, it is necessary to utilize precise step gain control instead of continuous gain control as usually used in the superheterodyne receiver. The AGC of the direct-conversion transmitter is still continuous control because the main gain stages are in the single-channel RF block. The PA gain is also adjustable within 10 dB or less to save power consumption.

An alternative configuration of the direct-conversion receiver is based on high-dynamic ADCs as shown in Fig. 3.11. This direct-conversion receiver configuration has the following features. The ADC for the CDMA application is about 10 to 12 bits determined by the AGC range and the BB LPF suppression to interferers. All the gain controls are in the RF block, two steps at LNA and one step at RFA, and overall gain-control range is around 50 to 60 dB depending on the dynamic range of the ADC. Third to fifth order based-band LPF and fixed-gain BBA are used, and therefore this configuration direct conversion receiver may have less I and Q channel mismatching and DC offset issues. The detail discussion on this direct conversion receiver can be found in Chapter 6.

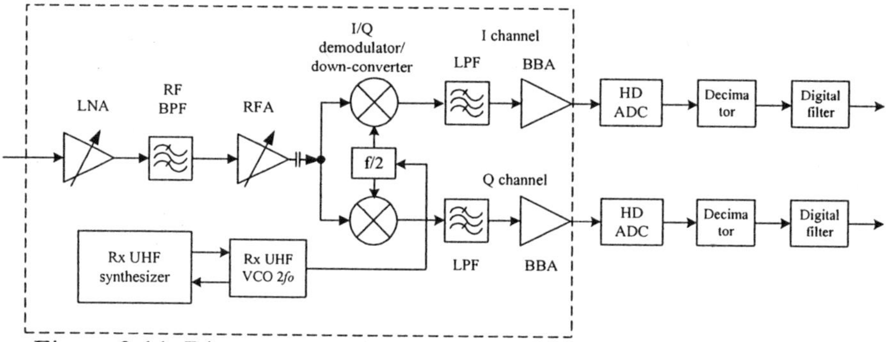

Figure 3.11. Direct-conversion receiver based on high dynamic ADC

3.2.2. Technical Challenges

The idea of the direct-conversion architecture was introduced as early as 1924. It was not practically applied to any communication transceiver for over half a century since the implementation of the direct-conversion architecture based on discrete circuits was extremely difficult. The state of the art of the semiconductor technology and the sophisticated RF IC design tools make it possible to develop direct-conversion transceivers for wireless communication mobile stations. In early 1980s, the direct-conversion receiver was just used in radio-pagers and some satellite communications. The direct-conversion architecture was broadly adopted in the GSM mobile stations in late 1990s.

In this section, the technical issues of the direct-conversion architecture are reviewed, and the possible solutions for resolving the issues are discussed.

3.2.2.1. DC Offsets

The DC offset is actually not only a direct-conversion architecture problem but also a problem for the superheterodyne architecture. However, the DC offset issue in the direct-conversion architecture is much more severe than in the superheterodyne architecture since most of signal gain in the direct-conversion receiver is developed in the BB block, and the DC offsets in this architecture can be caused by various reasons as described below.

Assuming that the direct conversion receiver is implemented on integrated circuits, the BB block circuit build-in DC offset due to the IC processing is in common for different architectures. This DC offset in the superheterodyne receiver is easy to calibrate out because the BB block gain is relative low and usually fixed there. In the direct-conversion receiver, the very high and widely varied BB block gain makes the calibration of this DC offset difficult since the DC offset voltage may be too large to be calibrated and it varies with gain change. On the other hand, the gain control of the BB block in the direct-onversion receiver is stepped, and therefore the DC offset changes also show step transient. This may cause trouble even if high-pass design is used in the BB block to eliminate DC component.

The isolation between the LO port and the RF port of the RF down-converter is finite, and thus a certain amount of LO signal leaks to its RF port and further up to the antenna port through the RF SAW filter and the LNA. The LO leakage is reflected at the interface between stages due to mismatch back into the RF down-converter, and the reflected LO leakage signal mixes with the LO signal at the down-converter producing a DC component as depicted in Fig. 3.12 This phenomenon is referred to as LO *self-mixing*.

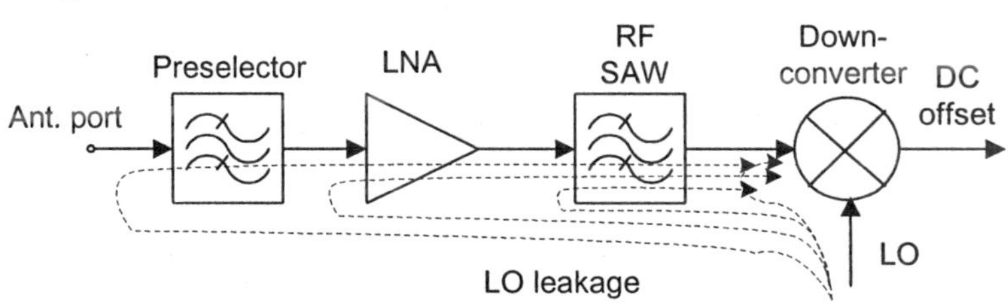

Figure 3.12. DC offset caused by LO self-mixing

In the full-duplex direct-conversion transceiver, there exists another potential self-mixing — i.e., transmission leakage self-mixing. The transmission self-mixing occurs through two paths as shown in Fig. 3.13. The first transmission leakage path is from the duplexer through the LNA and RF SAW to RF port of the down-converter. The transmission leakage passing through this path mixes with its feedthrough from the RF port to the LO port of the down-converter producing DC component at the output of the down-converter. The second transmission leakage path is from the power amplifier of the transmitter through the substrate/printed circuit board, and/or the common power supply circuitry, and the receiver LO path to the LO port of the down-converter. The transmission leakage through this path generates self-mixing DC offset in a similar way as the first path transmission leakage. In addition, the transmission leakages of the first and the second paths also mix together in the down-converter and this transmission leakage self-mixing produces the third DC offset. The transmission leakage self-mixing may generate low frequency products too other than the DC if the transmission signal is amplitude modulated (see Section 3.2.2.2).

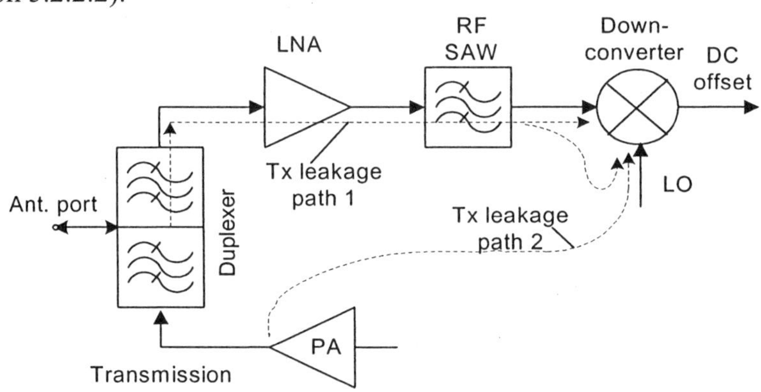

Figure 3.13. DC offset caused by transmission leakage self-mixing

When a mobile station is operating, it possibly suffers from some strong interference attack. The strong interferers can also cause self-mixing in the down-converter of the receiver and yield DC components and low-frequency components as well if any interferer is amplitude modulated. This can happen either in the full-duplex or in the half-duplex direct-conversion receiver, and the DC offset and the low-frequency products can propagate in their BB block.

The self-mixing DC offsets including the LO self-mixing may vary with time especially when the transceiver is moving during operation. In this case, the DC offset problem is much more difficult to deal with than the time-invariant DC offset.

It is apparent that the DC offset in the direct-conversion receiver must be removed or cancelled, or otherwise the receiver may not work. The gain in the BB block following the down-converter is approximately 70 to 80 dB. 200 to 250 μV DC offset appearing at the output of the down-converter will saturate the last stage of the BB amplifier or the VGA.

Utilizing AC coupling or high-pass filtering in the BB block design is one of the efficient approaches to remove the time invariant DC offset. As a rule of thumb, the corner frequency of the high-pass filter (HPF) should be approximately 0.1% of the symbol/chip rate or less to avoid any significant signal-to-noise ratio degradation [7–9]. To a low data rate or a narrow channel bandwidth system, such as IS-54 TDMA, a very narrow notch (< 50 Hz corner frequency) around the DC need be used to remove the DC offset. The coupling capacitors, in this case, will be unreasonably large for integration. Off-chip passive components may be a choice, but active DC blocks [10] can be an alternative. For the TDMA systems including the GSM system, the idle slot preceding the burst slot is possibly utilized for precharging the capacitors in the HPF with a small time constant and then the corner frequency of the HPF is switched back to its original value after the charging. For the *direct-sequence spread-spectrum* (DS-SS) systems, the receiver channel bandwidth is actually determined by the spreading chip rate instead of the data rate. The data bit in the DS-SS is spread over a pseudo random sequence, and its spectrum bandwidth is much wider than that of information data. The energy of 0.1 up to 1% of the spectrum around the DC in a DS-SS direct-conversion receiver can be removed without a significant signal-to-noise ratio loss [11].

The DC offset cancellation can be used for removing time-invariant and time-variant DC offsets [12]. The time-invariant DC offsets may be calibrated in different gain modes, and they are stored as a look-up table in a memory, or the DC offset is estimated by terminating the LNA input into a dummy load in the idle slot of the TDMA receiver and is stored in a memory. In the operation mode or in the burst slot, the stored DC offset value is fed to a subtractor in the analog BB block through a DAC to compensate the inherent DC offset based on operation gain or the estimated offset. The time-variant DC offset is measured by averaging the digitized signal containing the DC offset if the modulation scheme has a zero mean value, such as QPSK, etc. The measured DC value can be held in a latch circuit, and then it is subtracted from the input signal in the analog BB through a DAC or from the signal in the digital BB as shown in Fig. 3.14. The cancellation up-dating time can be milliseconds to hundred milliseconds, and it will rely on the averaging time of the DC offset measurement for a reasonable accuracy.

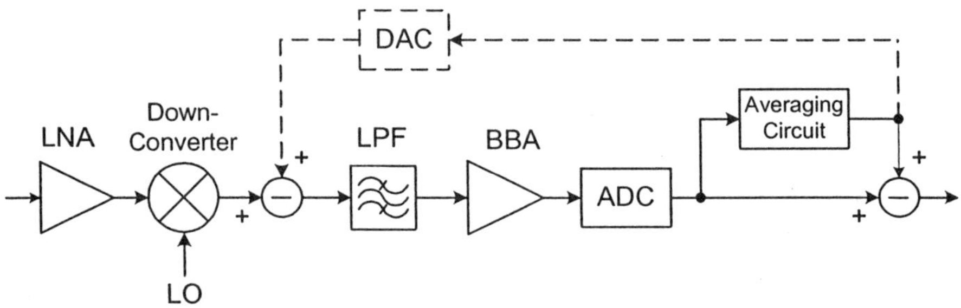

Figure 3.14. A possible configuration of DC offset cancellation

3.2.2.2. Second-Order Distortion

Second-order distortion is another severe threat to the direct-conversion transceiver if the distortion is not low enough. Frequency-closed strong interferers and/or an interferer with an amplitude modulation (AM) can be turned into low-frequency in-channel interference products including the DC component by the second-order distortion. The in-channel bandwidth interferers may propagate in the BB block and then possibly deteriorate the direct-conversion receiver performance or even block the receiver.

A device with weak linearity can be represented by the following expression:

$$y(t) = a_1 x(t) + a_2 x^2(t) + \cdots \,, \tag{3.2.1}$$

where a_1 is the gain or loss of the device, and a_2 is the second-order distortion coefficient. The second term on the right side of (3.2.1) is the second-order distortion of the device. Two strong narrow-band interferers $A\cos 2\pi f_a t + B\cos 2\pi f_b t$ passing through the device with weak second-order nonlinearity generate low-frequency products. Utilizing trigonometric function identity the low-frequency beat and the DC component caused by the second-order distortion can be expressed as

$$a_2\left(A\cos 2\pi f_a t + B\cos 2\pi f_b t\right)^2 = a_2\frac{A^2+B^2}{2} + AB\cos 2\pi\left(f_a - f_b\right) \quad \text{(3.2.2)}$$
$$+ \text{High frequency components.}$$

To an AM modulated interferer, the second-order nonlinearity demodulates the AM and the amplitude modulation is usually a low-frequency signal. Assuming the interferer has a form $\left[A + m(t)\cos 2\pi f_m t\right]\cos\left(2\pi f_c t + \varphi\right)$, the second-order distortion produces the following low-frequency products and the DC offset,

$$a_2\left\{A\left[1 + m(t)\cos 2\pi f_m t\right]\cos\left(2\pi f_c t + \varphi\right)\right\}^2$$
$$= a_2\frac{A^2}{2}\left[1 + \frac{m^2(t)}{2} + 2m(t)\cos\left(2\pi f_m t\right) + \frac{m^2(t)}{2}\cos\left(4\pi f_m t\right)\right] \quad \text{(3.2.3)}$$
$$+ \text{High frequency component.}$$

Another possible influence to the direct-conversion receiver performance due to the second-order distortion is that the second harmonic of the desired signal created by the second-order nonlinearity mixes with the second harmonic of the LO producing a BB signal of the desired signal second harmonic with a bandwidth twice wide as that of the fundamental desired signal. This second harmonic BB signal overlaps with the desired BB signal and becomes in-channel interference. However, this influence will be negligible if a differential down-converter circuit design is employed since the common mode rejection of the differential circuits can significantly suppress the even-order distortion.

The main contribution of the second-order distortion interference in the most cases results from the second-order nonlinearity of the RF I/Q down-converter in the direct conversion receiver. The low frequency and DC products caused by the second-order distortion of the LNA and the RFA in the front-end are blocked by the RF BPF and the AC coupling capacitor as depicted in Fig. 3.10.

The level of the second-order distortion products is proportional to the nonlinear coefficient a_2. The second order linearity of a device is usually defined by the second-order intercept point IP_2, as the third-order intercept point for the intermodulation products as depicted in Fig. 3.8 of Section 3.1.3.1. The IP_2 in the natural scale P_{IP2} is inversely proportional to a_2 — i.e., $P_{IP2} \propto \left|a_1/a_2\right|$, where a_1 is the fundamental signal gain of the device. To minimize the impact of the second-order distortion to the direct conversion receiver, it is necessary to use a very high IP_2 RF down-converter — for instance, greater than +55 dBm IIP_2.

On the other hand, the outcomes resulting the self-mixing of interferers and the transmission leakage as discussed in the previous section are the same as those coming from the second-order distortion. The mathematical model of the signal mixing is multiplication. Thus the multiplication of two equal signals, $x(t) \times x(t)$ (self-mixing), is equal to the square of the signal, $x^2(t)$ (second order distortion term). Their results are the same, but their mechanisms are different. The level of the self-mixing low-frequency products and the DC offset depends on the isolation between the RF and the LO ports of the RF down-converter and/or the isolations between the transmitter PA and the RF port /LO port of the down-converter. But the level of the second-order distortion products is determined by the nonlinear coefficient a_2. If the interferer or the transmission has a form $[A + m(t)\cos 2\pi f_m t]\cos(2\pi f_c t + \varphi)$, then the expression (3.2.3) can still be used except that the multiplier a_2 need be changed into an isolation related coefficient. To minimize the self-mixing low-frequency products and DC component, it is necessary to make all the isolations as high as possible.

3.2.2.3. I and Q Channel Mismatch

In the direct-conversion receiver, the received RF signal after amplifying in the RF front-end is directly down-converted into two quadrature BB signals — i.e., I and Q BB signals. The I and Q BB signals propagate and are amplified in separate I and Q paths or referred to as I and Q channels. The gain in the both analog BB paths may vary more than 80 dB. In addition, the I and Q BB signals pass through the low-pass channel filter in their own channel separately as shown in Fig. 3.10. Generally speaking, it is difficult to keep the I and Q BB signals having perfect balance in their magnitude and phase when they pass through two completely separate paths even if using the-state-of-the-art highly integrated RF circuits. A wide range of the gain control in both BB channels makes it even more difficult to maintain the balance between these two signals.

The imbalance requirements in the magnitude and the phase of I and Q signal depend on the modulation scheme and the system protocol. For example, in a CDMA mobile station receiver it will not cause significant performance degradation if the magnitude and the phase imbalances of the I and Q signals are equal to or less than 1 dB and 10 degrees, respectively. For a direct-conversion receiver, after proper calibration and tuning it is possible to obtain a magnitude imbalance ≤ 0.5 dB and a phase imbalance ≤ 5 degree.

A lower imbalance can also be achieved by using compensation in the digital domain.

To obtain a better synchronization of the gain controls in the I and Q channels and thus to be able to minimize the imbalance caused by the gain control, stepped gain controls in the analog BB block of the direct-conversion receiver are preferred. This is due to that the stepped gain control can offer much better gain accuracy than the continuous gain control. The control step size is usually a few dB, — for instance, 3, 6, or 9 dB — and in general the gain step is even over the gain control range. However, the stepped gain control may possibly cause a DC offset surge under certain gain step combinations, to achieve a gain increment.

3.2.2.4. LO Leakage Emission

Each wireless mobile system has emission restrictions based FCC regulations. In the direct-conversion receiver, the LO frequency is equal to the reception carrier frequency, and this emission is in the receiver band. The allowable emission level within the receiver band for wireless mobile stations is in the range of –60 to –80 dBm. The LO level input to the quadrature down-converter is around –5 to 0 dBm. To make the LO emission level at the antenna port of the mobile stations less than –80dBm, the reverse isolation from the down-converter LO port through LNA backward to the antenna port as shown in Fig. 3.15 is better to be greater than 85 dB.

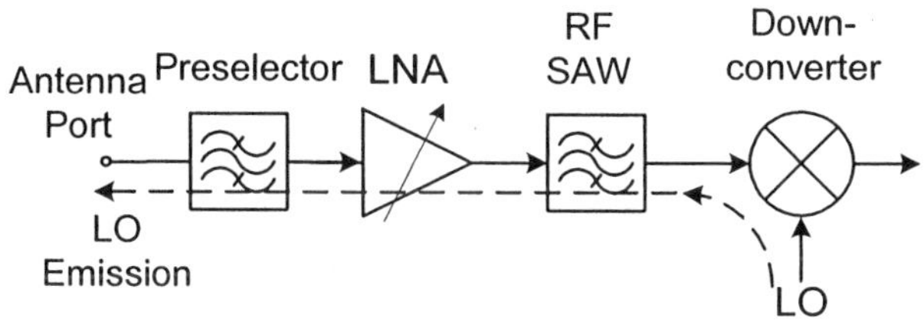

Figure 3.15. LO leakage path

Based on modern RF IC technology it is not difficult to obtain the isolation between the LO port and the RF port of UHF differential down-converters greater than 65 dB. An UHF LNA, even a single-ended LNA, at all gain modes easily achieves its reverse isolation greater than 20 dB. Thus, the conductive reverse isolation is usually high enough to reduce the LO leakage level to below –80 dBm. In addition, the insertion loss of the RF SAW and the preselector can further enhance the reverse isolation.

However, the real issue of the LO leakage emission is due to the finite semiconductor substrate isolation. The direct-conversion receiver is integrated on a very small semiconductor chip. The isolation of the semiconductor substrate depends on the technology used for the IC implementation and the circuit configuration. In general, the substrate isolation from the down-converter LO port to the LNA input port is around only 60 dB to 70 dB at 2.0 GHz if a single-ended LNA is used. An effective approach to resolving this issue is either using differential LNA or physically separating the LNA block from the rest of the direct conversion receiver — i.e., using a stand along LNA block IC.

3.2.2.5. Flicker Noise

Flicker noise is also referred to as $1/f$ noise since it is inversely proportional to the frequency f. The flicker noise gets higher when the frequency goes lower. For a direct-conversion receiver, the gain for the desired signal before it is converted to a BB signal is only around 25 dB. Flicker noise contributed by the converter, BB amplifiers, and BB filter of the direct-conversion receiver may have a visibly deteriorative effect on the desired signal.

For typical submicron MOS technologies, minimum-channel MOSFETs with a width of a few hundred microns and bias current of a few hundred microamperes exhibit a flicker noise corner frequency in the vicinity of 1 MHz [9]. The relative increase in the noise power due to the flicker noise within frequency band (f_1, f_2) can be approximately expressed as

$$\Delta N_{flck} = 10\log\left(\frac{P_{N_{flck}}}{P_{N_{thml}}}\right) \cong 10\log\left(\frac{10^6 \ln(f_2/f_1)}{f_2 - f_1}\right). \tag{3.2.4}$$

For example, assuming that $f_1 = 10$ Hz, the relative noise increase ΔN_{flck} for $f_2 = 25$ kHz and 200 kHz are 24.96 dB and 16.95 dB, respectively. It is apparent that the desired signal with narrower bandwidth will be deteriorated more. The CMOS technology is not suitable for the direct-conversion receiver requesting high sensitivity especially to narrow-band receivers.

The flicker noise is much lower for the integrated circuits in terms of SiGe and BiCOMS technologies. The direct-conversion architecture based on these semiconductor technologies makes it possible to achieve high receiver sensitivity. The flicker noise may not be a main issue if the direct-conversion receiver is developed by means of SiGe or BiCMOS technology or if the systems, such as IEEE 802.11a and 802.11g, are of wide-band.

3.2.3. Design Consideration of a Direct-Conversion Transceiver

In this section, some design considerations particularly for the direct-conversion transceiver are discussed. For a direct-conversion transceiver, a frequency plan as always used in the superheterodyne transceivers is not needed since the IF of the direct-conversion transceiver is zero Hz well defined. This can save a lot of engineering time on frequency planning, which is very time consuming. In addition, the direct-conversion has no image and IF/2 issues. However, to make the direct-conversion transceiver practically working, it is necessary to overcome all the technical challenges described in Section 3.2.2. In the following discussion, it is assumed that DC offset issue in the direct-conversion receiver is resolved by utilizing AC coupling and/or high-pass filters with proper a corner frequency.

3.2.3.1. DC Notch to Performance Influence

Using AC coupling or HPF to notch out the DC offset actually also cuts off the signal energy around the DC. This means the signal-to-noise ratio is reduced, and thus it will affect the receiver sensitivity. Actually, the waveform, the modulation, and therefore the power spectrum density of the desired signal used in a system determine the DC notch influence on the receiver performance. If a signal has a power spectrum density evenly distributed over its bandwidth, the energy loss of the signal due to the DC notch with a corner frequency 0.1% of the signal bandwidth is less than 0.01 dB. The AC coupling or the HP filtering introduces group delay distortion near the DC, and the group delay will also degrade the receiver sensitivity. In reality, the performance degradation is due not only to the signal energy loss caused by the DC notch but also to other reasons, including the group delay distortion.

For a CDMA mobile station receiver, the receiver sensitivity degradation resulting from an AC coupling/ HP filtering is less than 0.15 dB as shown in Fig. 3.16, which presents simulation results from a data rate of 9.6 kbps and a HPF corner frequency of 1.5 kHz. However, the high-pass frequency corner must be pushed down to less than 500 Hz for a CDMA mobile direct-conversion receiver to obtain good fading performance and to have less loss on detection performance of high-order modulation, such as 16 QAM, signals. There are at least two benefits that can be gained from the

DC notch although it causes approximate 0.12 dB sensitivity degradation. First, if the DC is not notched, the DC offset will desensitize the receiver even if it is small. From a relevant system simulation we know that a 0.5 LSB DC offset at the 4 bit ADC output of the CDMA receiver will cause 0.2 dB degradation of the receiver sensitivity. Removing the DC and then the DC offset can minimize this performance impairment. Second, when notching out the DC, the flicker noise near the DC is suppressed as well. Suppressing the flicker noise near the DC may improve the noise figure and the sensitivity of the receiver.

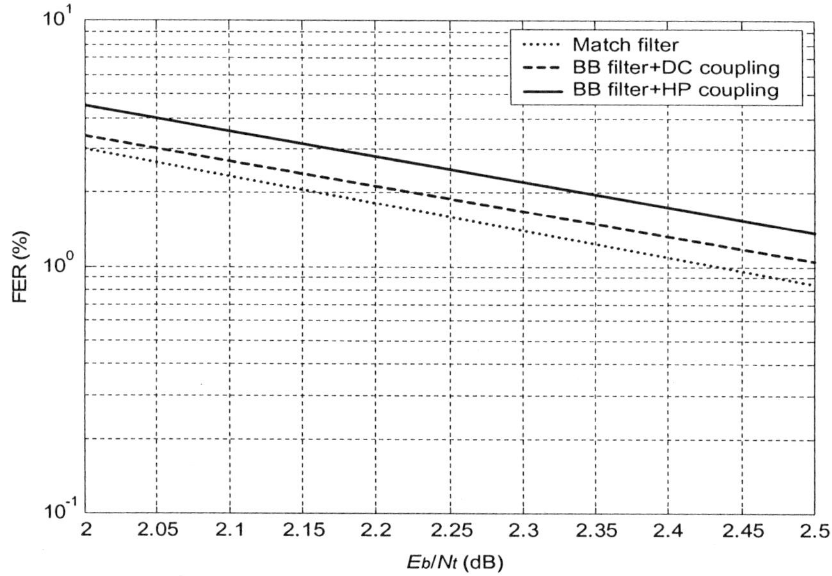

Figure 3.16. FER versus E_b/E_t of DC and AC coupling CDMA receivers

In the GSM direct conversion receiver, it will have an ignored impact on the receiver performance if the corner frequency of the DC notch is less than 100 Hz. The time constant of the HPF is quite large, and it will not be suitable for a TDMA system, like the GSM, to employ such a narrow notch since its transient will take a significant portion of the signal burst duration. The most popular approach to resolving this problem is to utilize the idle slot preceding the burst slot for precharging the capacitors in the HPF with a small time constant and then to switch the corner frequency of the HPF back to its original value after the charging.

An AMPS receiver in mobile stations has only a bandwidth approximate 25 kHz. Simulations on the AMPS receiver show that the *SINAD* reduction is negligible when using an AC coupling with a corner frequency 300 Hz or less to replace the DC coupling between the RF analog BB output port and the ADC input port. The simulation results of the SINAD versus the AC coupling capacitance of the AMPS receiver are presented in Table 3.3. The DC coupling is equivalent to having an infinite coupling capacitance. From the table we can see that the SINAD at the AMPS receiver sensitivity level has a negligible difference from that of the DC coupling when a 20 nF capacitor is used, and it is equivalent to the DC notch having a 300 Hz corner frequency. The SINAD reduces with the DC offset increasing in the DC coupling case, but the DC offset has no influence to the SINAD in the AC coupling cases as expected. Based on simulations, the SINAD of AC coupling AMPS receiver is approximately 0.5 dB lower than that of the DC coupling receiver when the mobile station is running in a Rayleigh channel with a high speed such as 100 km/hour.

Table 3.3. SINAD versus AC coupling capacitance value of an AMPS receiver

ADC Input Impedance R = 500 kΩ Coupling Capacitor	CNR (dB)	I/Q Signal RMS Level at ADC Input (mV)	DC Offset (mV)				
			0	5	10	20	40
			SINAD (dB)				
0.47 nF	3.3	250	11.06	11.06	11.06	11.06	11.06
4.7 nF	3.3	250	12.51	12.51	12.51	12.51	12.51
10 nF	3.3	250	12.69	12.69	12.69	12.69	12.69
20 nF	3.3	250	12.74	12.74	12.74	12.74	12.74
DC coupling	3.25	250	12.71	12.69	12.64	12.42	11.62

In the AMPS direct-conversion receiver using AC coupling/high-pass design, the LO frequency must be offset from the receiver signal carrier frequency. The offset frequency between the LO and receiver signal carrier should be greater than the high-pass corner frequency of the AC coupling but less than the frequency error tolerance.

3.2.3.2. Evaluation of Second-Order Input Intercept Point (IIP_2)

To avoid the problems caused by the second-order distortion, the direct-conversion receiver usually needs a very high second-order input intercept point (IIP_2). The key device, which may dominate the contribution on the second-order distortion, is the RF quadrature (I/Q) down-converter in the direct conversion receiver. The low-frequency products of the second-order distortion generated by the LNA and/or the RFA in the front-end of the receiver are easy to block with the RF BPF and the small AC coupling capacitor, respectively, as shown in Fig. 3.10. After the I/Q down-converter, all the circuits are BB low-frequency circuits, and they are not capable of mixing RF interference tones and demodulating AM RF signals. Only the second-order distortion products created by the quadrature down-converter need be really considered in the direct conversion receiver with Fig. 3.10 or a similar radio configuration.

Assuming that the allowed desired signal carrier-to-noise ratio (*CNR*) decrease resulted from the second-order distortion products is Δ dB, the decreased *CNR* can be expressed as

$$CNR - \Delta = 10\log\frac{P_S}{P_N + \Delta P_N}, \tag{3.2.5}$$

where P_S is the carrier signal power, P_N is the noise and interference power level within the receiver bandwidth, and ΔP_N is the second order distortion products within the receiver bandwidth. Considering the initial $CNR = 10\log(P_S/P_N)$, from (3.2.5) we can derive that the Δ can be represented by

$$\Delta = 10\log\left(1 + \frac{\Delta P_N}{P_N}\right). \tag{3.2.6}$$

After rearrange (3.2.6), we can obtain the relative noise or interference level increment $R_{\Delta N}$ from its original

$$R_{\Delta N} = 10\log\left(\frac{\Delta P_N}{P_N}\right) = 10\log\left(10^{\frac{\Delta}{10}} - 1\right). \tag{3.2.7}$$

A few examples of $R_{\Delta N}$ calculated from given CNR degradation Δ are presented in Table 3.4.

Table 3.4 Relative noise/interference level change vs. CNR degradation

CNR Degradation, Δ (dB)	Relative Noise/Interference Increment, $R_{\Delta N}$ (dB)
0.1	-16.3
0.5	-9.14
1.0	-5.87
2.0	-2.33
3.0	-0.02

In the following, we explain how to determine the IIP_2 of the I/Q down-converter from the allowable CNR degradation Δ. To obtain the noise/interference level at the down-converter input, we should calculate the sensitivity of the receiver excluding its front-end block as depicted in Fig. 3.17. Assuming that the overall receiver noise figure is NF_{Rx} in dB, and the noise figure and the gain of the front-end are NF_{FE} and G_{FE} both in dB, respectively, the noise figure NF_{Mxr+BB} of the receiver excluding its front-end can be calculated by means of

$$NF_{Mxr+BB} = 10\log\left(1 + 10^{\frac{G_{FE}+NF_{Rx}}{10}} - 10^{\frac{G_{FE}+NF_{FE}}{10}}\right) \tag{3.2.8}$$

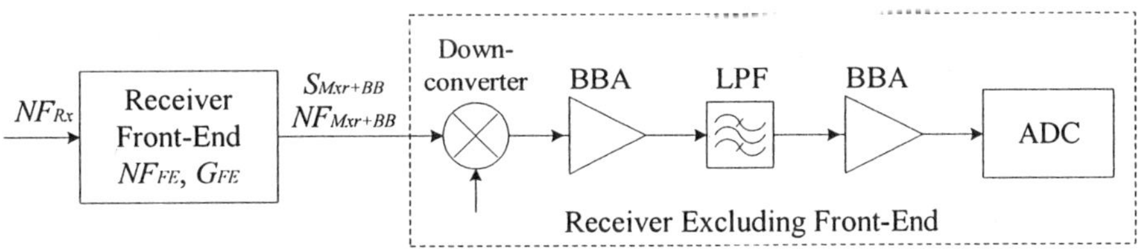

Figure 3.17. Simplified direct conversion receiver configuration

From this noise figure NF_{FE} and G_{FE} we can calculate the input noise level at the down-converter input N_{Mxr_input} by using the following formula

$$N_{Mxr\_input} = 10\log\left\{\left[10^{\frac{NF_{FE}+G_{FE}}{10}} + \left(10^{\frac{NF_{Mxr+BB}}{10}} - 1\right)\right]\cdot 10^{\frac{-174+10\log BW_{Rx}}{10}}\right\}, \tag{3.2.9}$$

where BW_{Rx} is receiver bandwidth.

The maximum allowed second order distortion product level $IMD_{2\_\max}$ generated by the down-converter can be determined by the N_{Mxr_input} and the allowed CNR degradation Δ as follows:

$$IMD_{2\_\max} = N_{Mxr\_input} + R_{\Delta N}\,. \tag{3.2.10}$$

Based on $IMD_{2\_\max}$ and the level I_{mxr_input} of two interference tones with a frequency spacing within the receiver channel bandwidth at the input of the I or the Q down-converter, we can calculate the minimum requirement of the down-converter IIP_2 as (see Appendix 3A)

$$IIP_{2\_Mxr} = 2I_{Mxr\_input} - IMD_{2\_\max}, \qquad (3.2.11)$$

where

$$I_{Mrx\_input} = I_{received} + G_{FE}. \qquad (3.2.12)$$

In (3.2.13), $I_{received}$ represents the interference tone level at the antenna port of the receiver. In fact, the formulas (3.2.6) to (3.2.11) can also be used for the overall receiver IIP_2 calculation. In this case, we have NF_{FE} and G_{FE} equal to zero.

Let us look at a calculation example based on a direct-conversion receiver for the CDMA mobile station, which is capable of handling the attack of two –30 dBm tones with frequencies in the receiver frequency band and their frequency spacing within the receiver channel bandwidth. In this case, we have $BW_{Rx} = 1.23\times10^6$ Hz. Assuming that the CDMA receiver has $NF_{Rx} = 8$ dB, $NF_{FE} = 6.3$ dB and $G_{FE} = 8$ dB, from (3.2.8) we obtain

$$NF_{Mxr+BB} = 10\log(1 + 39.81 - 26.92) = 11.43 \ \text{dB},$$

and utilizing (3.2.9) we further calculate the noise at the input of the down-converter to be

$$N_{Mxr\_input} = 10\log\left\{\left[10^{\frac{6.3+8}{10}} + \left(10^{\frac{11.43}{10}} - 1\right)\right]\cdot 10^{\frac{-174+60.9}{10}}\right\} = \text{-97.0 dBm}. \qquad (3.2.13)$$

Assuming that the allowed CNR degradation due to the second-order distortion products of the down-converter under the attack of two –30dBm tones in receiver frequency band is 3 dB, from (3.2.10) and (3.2.7) we obtain the maximum permissible second-order distortion product level to be

$$IMD_{2\_\max} = -97.0 - 0.02 = -97.02 \ \text{dBm}.$$

The required IIP_2 for the quadrature down-converter is calculated from (3.2.11) and (3.2.12) to be

$$IIP_{2\_Mxr} = -2\times(30-8) + 97.02 \cong 53 \ \text{dBm}.$$

Therefore, we need the quadrature down-converter to have an IIP_2 equal to 53 dBm or greater. The minimum IIP_2 is 45 dBm or greater if it is measured at the antenna port.

A more general formula to determine the down-converter IIP_2 is presented in (3.2.14),

$$IIP_{2\_Mxr} = 2I_{block} - 10\log\left[10^{\frac{N_{Mxr\_input}+R_{\Delta N}}{10}} - 2\cdot 10^{\frac{I_{block}-ISL_{RF\_LO}-\Delta G_{Mix}}{10}} - \sum_{k=1}^{2}\left(10^{\frac{N_{pase,k}+10\log BW_{Rx}}{10}} + 10^{\frac{N_{spu,k}}{10}}\right)10^{\frac{I_{block}}{10}}\right], \tag{3.2.14}$$

where I_{block} is the block interferer at the mixer input, ISL_{RF_LO} is the isolation between the RF and LO ports of the quadrature down-converter, ΔG_{Mix} is the mixing gain difference between the normal mixing and the self-mixing of the interference tones and their leakages at the LO Port of the quadrature down-converter, and $N_{phase,k}$ and $N_{spu,k}$ (k = 1 or 2) are the LO phase noise and spurious near these two interference tones. Strictly speaking, the calculation of the LO phase noise and spurious contributions in the square brackets of (3.2.14) should also consider the influence of the mixing gain difference from the normal mixing gain as done in the second term in the square brackets of (3.2.14). In practical calculations, these mixing gain differences are not considered since they are usually small and do not have much impact on the results.

From (3.2.14) we can estimate the needed ISL_{RF_LO} that will not significantly raise the requirement of the down-converter IIP_2 — say, the increment equal to Δ_{IIP2} or less. Assuming that the noise and spurious terms on the right side of (3.2.14) are small and can be ignored, the required minimum ISL_{RF_LO} can be expressed as

$$ISL_{RF\_LO} = I_{block} - N_{Mxr\_input} - R_{\Delta N} - \Delta G_{Mix} + 3 - 10\log\left(1 - 10^{\frac{-\Delta_{IIP2}}{10}}\right). \tag{3.2.15}$$

Substituting $N_{Mxr\_input} + R_{\Delta N} = -97\,\text{dBm}$ and I_{block} = -22 dBm used in previous IIP_2 calculation into (3.2.15), and further assuming the allowed IIP_2 increment Δ_{IIP2} =0.5 dB and ΔG_{Mix} = 0 (the worst case), we obtain the ISL_{RF_LO} from (3.2.15) to be

$$ISL_{RF\_LO} = -22 + 97 + 3 - 10\log\left(1 - 10^{\frac{-0.5}{10}}\right) = 87.6 \text{ dB}.$$

In addition to the two-tone interference, we should also consider the transmission leakage issue in the CDMA receiver. The CDMA transmission is AM modulated and the level T_x in dBm at antenna port is approximate 25 dBm. The transmission will sneak into the receiver through the duplexer and other leaking paths. The duplexer suppression R_{j_Dplxr} to the transmission is around 48 dB. The transmission leakage power into the receiver, I_{Tx}, is equal to

$$\begin{aligned} I_{Tx} &= T_x - R_{j\_Dplx} \\ &= -23 \quad \text{dBm}. \end{aligned} \tag{3.2.16}$$

There is usually an RF BPF in the receiver front-end (see Fig. 3.10) to further reject the transmission leakage power down, $R_{j\_BPF} = 20$ dB, or more. The transmission leakage power at the input of the I/Q down-converter, $I_{Tx@Mxr\_input}$, is

$$\begin{aligned} I_{Tx@Mxr\_input} &= I_{Tx} + G_{FE} - R_{j\_BPF} \\ &= -23 + 8 -- 20 = -35 \quad \text{dBm}. \end{aligned} \tag{3.2.17}$$

The second-order nonlinearity of the I/Q down-converter will convert the AM transmission leakage $I_{Tx@Mxr\_input}$ into a DC component and low frequency interference products. These second-order distortion products will also deteriorate the *CNR* of the CDMA signal. The transmission leakage always exists while the CDMA receiver is running. To minimize the influence of these second-order distortion products to the receiver performance, the allowed *CNR* degradation in this case is defined as 0.1 dB. From (3.2.13) and Table 3.4, we can calculate the allowed second-order distortion product level, $IMD_{2\_allowed}$, at the input of the down-converter

$$IMD_{2\_allowed} = -97 - 16.3 = -113.3 \text{ dBm}.$$

Based on $IMD_{2\_allowed}$ and $I_{Tx@Mxr\_input}$, we determine the required minimum IIP_2 of the I and Q each down-converter to be

$$IIP_{2\_Mxr} = 2I_{Tx@Mxr\_input} - IMD_{2\_allowed}$$
$$= 2(-35) + 113.3 = 43.3 \text{ dBm}. \tag{3.2.18}$$

The required IIP_2 of the I and Q each down-converter resulting from the transmission leakage is 10 dB lower than that determined by two external interference tones.

In fact, the second-order distortion products of the AM transmission leakage are not all in the desired signal bandwidth as analyzed in Appendix 3B. The power ratio of the effective interference portion of the second-order distortion to the overall IMD_{2_Tx} consisting of the DC and low-frequency products depends on the magnitude PDF of the AM transmission waveform. The typical magnitude PDF of CDMA mobile station transmission waveform (IS-98C voice data) is presented in Fig.3.18. Using (3B.4), (3B.5), and the PDF data given in Fig. 3.18, we obtain

$$\overline{IMD_{2\_DC}} = 1.06 \quad \text{dB}$$

$$\overline{IMD_{2\_LF}} = -5.59 \quad \text{dB},$$

and from (3B.6) the normalized IMD_{2_Tx} is calculated as

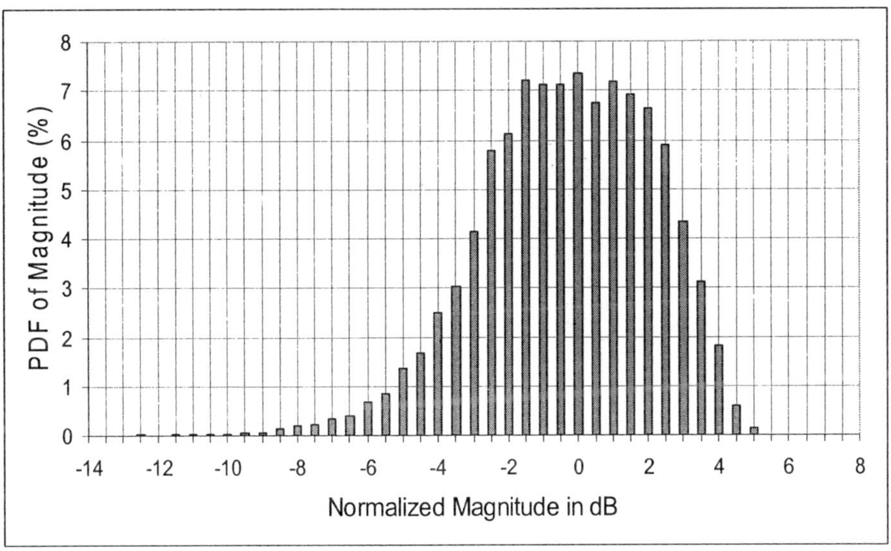

Figure 3.18. PDF of a CDMA mobile station transmission waveform

$$\overline{IMD_{2\_Tx}} = 10\log(0.28 + 1.28) = 1.93 \text{ dB}.$$

Therefore, the power ratio $PR_{IMD_{2\_LF}/IMD_{2\_Tx}}$ for the CDMA mobile station transmission is

$$PR_{IMD_{2\_LF}/IMD_{2\_Tx}} = \overline{IMD_{2\_LF}} - \overline{IMD_{2\_Tx}}$$
$$= -5.59 - 1.93 \cong -7.52 \text{ dB}.$$

Thus, the low-frequency product of the second order occupies approximately 24% of the overall IMD_{2_Tx} power.

As described in Appendix 3B, only a portion of the second order distortion low-frequency product spectrum is within the desired signal BB bandwidth. The effective interference portion of the second-order distortion, IMD_{2_effect}, is approximately half of the IMD_{2_LF} power. The power ratio of the effective interference portion to IMD_{2_Tx}, $PR_{IMD_{2\_effect}/IMD_{2\_Tx}}$, is now equal to

$$PR_{IMD_{2\_effect}/IMD_{2\_Tx}} \cong -7.52 - 3 = -10.52 \text{ dB}.$$

This means that $IMD_{2\_\max\_allowed}$ is allowed to increase 10.52 dB, or

$$IMD_{2\_allowed} = -113.3 + 10.52 = -102.78 \text{ dB}.$$

The minimum IIP_2 requirement of the down-converter to cope with AM of the transmission leakage is obtained by substituting this new $IMD_{2\_allowed}$ value into (3.2.18)

$$IIP_{2\_Mxr} = 2(-35) + 102.78 = 32.78 \text{ dBm}.$$

However, the IIP_2 of the I and Q down-converter is actually determined by the two-tone interference requirement instead of the transmission leakage. The former requests $IIP_{2\_Mxr} = 53\text{dBm}$, which is much higher that derived from the transmission leakage.

3.2.3.3. Transmission Leakage Self-Mixing and Isolation Requirements

The transmission lcakage self-mixing is another source of degrading *CNR*. The transmission leakage through the duplexer via LNA and RF BPF to the RF input of the down-converter (i.e., through Path 1 as defined in Fig. 3.13), and the leakage from the output of the PA to the LO input of the down-converter (i.e., through Path 2 defined in the same figure mix in the I/Q down-converter and generate a DC offset and a low frequency interference product very similar to the products resulted from the second-order distortion of the down-converter). The self-mixing interference product I_{SM} in dB can be estimated as follows

$$I_{SM} = I_{Tx\_path1} + I_{Tx\_path2} - LO + \Delta G_{Mxr}, \tag{3.2.19}$$

where I_{Tx_path1} and I_{Tx_path2} in dBm are the transmission leakages through Path1 1 and Path 2 to the down-converter, respectively, LO is local oscillator driving power in dBm, ΔG_{Mxr} in dB is the conversion gain change from the normal down-converter mixing to the transmission leakage self-mixing.

The isolation between the output of the PA to the LO input of the I/Q down-converter, ISL_{PA_out/LO_in}, is defined as

$$ISL_{PA\_out/LO\_in} = TX_{PA\_out} - I_{Tx\_path2}, \tag{3.2.20 }$$

where TX_{PA_out} is the transmission power at the output of the PA, and I_{Tx_path2} is the transmission leakage level at the LO input of the I/Q down-converter. If the allowed I_{Tx_path2} is allocated, the ISL_{PA_out/LO_in} is the minimum isolation requirement from the PA output leaking through the circuit board and/or common bias circuitry to the LO input of the down-converter.

We continue the CDMA direct-conversion transceiver analysis and calculation. Assuming that the allowed I_{SM} is the same as $IMD_{2\_\max\_allowed}$ = –106.3 dBm, the transmission leakage through Path 1 to the RF port of the I/Q down-converter I_{Tx_path1} is –35 dBm, and the LO level is –3 dBm, we can obtain the allowed transmission leakage level through Path 2 from (3.2.18) to be

$$I_{Tx\_path2} = -102.78 + 35 - 3 - \Delta G_{Mxr} = -70.78 - \Delta G_{Mxr} \text{ dBm}.$$

Further assuming that the transmission power at the output of the PA, TX_{PA_out}, is 28 dBm, the minimum isolation between the PA output and the LO input of the down-converter ISL_{PA_out/LO_in} shall be

$$ISL_{PA\_out/LO\_in} = 28 + 70.78 + \Delta G_{Mxr} \cong 98.8 + \Delta G_{Mxr} \text{ dB}.$$

In reality, 98.8 dB isolation between the PA output and the LO input of the down-converter is difficult to implement since the conversion gain difference ΔG_{Mxr} between the normal mixing and transmission leakage self-mixing is only a few dB less.

A feasible approach to resolving this issue is utilizing an UHF synthesizer running at the multiple of the receiver carrier frequency, and using frequency divider to achieve the right LO frequency. Another is running the synthesizer at subharmonic frequency of the receiver carrier frequency and using multiplier to obtain the right LO frequency. In these cases, the transmission leakage from the PA is also divided or multiplied before reaching the LO input of the down-converter. This means that the frequency of the transmission leakage product has been greatly shifted when it appears at the LO input. In this case, it is equivalent to making the conversion gain change ΔG_{Mxr} in (3.2.18) becoming very lower — say, –60 dB or lower. Thus the isolation requirement reduces to

$$ISL_{PA\_out/LO\_in} = 98.8 - 60 = 38.8 \text{ dB}.$$

This level of isolation is easy to be implemented from the circuit board design or the decoupling of the common bias circuitry.

Another transmission leakage self-mixing is the mixing between the leakage through Path1 to the RF port of the down-converter I_{Tx_path1} and part of I_{Tx_path1} leaking to the LO port of the down-converter due to the finite isolation between the RF and the LO ports of the converter ISL_{RF_LO}. Based on the allowed maximum self-mixing product level, I_{SM_max}, we can determine the minimum isolation requirement between the RF and the LO ports of the I/Q down-converter ISL_{RF_LO} as follows

$$ISL_{RF\_LO} = I_{Tx\_path1} - I_{SM\_Max} - \Delta G_{Mix}. \quad (3.2.21)$$

Using the same CDMA receiver example, we have $I_{Tx\_path1} = -35$ dBm, and $I_{SM\_max} = IMD_{2\_allowed} = -102.78$ dBm. In the worst case, ΔG_{Mix} is 0 dB. From (3.2.21), we thus obtain the required minimum isolation:

$$ISL_{RF\_LO} = -35 + 102.78 \cong 68.8 \text{ dB}.$$

To minimize the impact on the effective IIP_2 of the quadrature down-converter, which can be calculated by using (3.2.15), the required ISL_{RF_LO} needs to be 87.6 dB. This isolation requirement is much higher than that resulting from the transmission leakage self-mixing.

3.2.3.4. Receiver RF Chain Reverse Isolation

The LO frequency of the direct-conversion receiver is exactly same as the carrier of the received desired signal. The LO leakage through the I/Q down-converter via RF BPF, LNA, and duplexer to the antenna becomes an receiver in-band emission. The allowed maximum LO emission level of mobile station receivers operating in different wireless communication systems are listed in Table 3.5.

Table 3.5. Allowed maximum emissions in receiver band of different system receivers

Systems	Maximum Emission in Receiver Band
AMPS	-81 dBm
CDMA	-76 dBm
GSM (900)	-79 dBm
GSM (1800)	-71 dBm
TDMA	-80 dBm
WCDMA	-60 dBm

To control the LO emission level at the antenna port, we need the I/Q down-converter to have a high isolation between the LO port and the RF port, and the LNA to possess a high reverse isolation. We refer to the isolation from the LO input port of the I/Q down-converter via the LNA to the antenna port as the reverse isolation. The UHF LNA reverse isolation is around 20 to 25 dB. The total insertion loss of the duplexer and the RF BPF, $L_{filters}$, is approximate 5 dB. Based on Table 3.5, we estimate the isolation between the RF port and the LO port shall be 55 to 60 dB system dependent. Using the CDMA receiver as an example, the allowed maximum emission is –76 dBm, and considering 4 dB margin we employ –80 dBm emission in our calculation. Assuming that the LO level may be as high as 0 dBm and the LNA reverse isolation ISL_{LNA} is 20dB, we estimate the minimum isolation between the RF port and the LO port ISL_{LO_RF} to be

$$\begin{aligned} ISL_{LO\_RF} &= LO - ISL_{LNA} - L_{filters} - I_{emission} \\ &= 0 - 20 - 5 + 80 = 55 \text{ dB}. \end{aligned} \tag{3.2.22}$$

If at low gain mode the LNA is bypassed, in this case the LNA has no or only a few dB reverse isolation, and the ISL_{LO_RF} must be greater than 75 dB.

In general, the ISL_{LO_RF} is equal to ISL_{RF_LO}. Thus the isolation value between the RF and LO ports of the I/Q down-converter should be designed to achieve the highest one among the results from the calculations of the LO leakage emission level, the transmission leakage self-mixing, and the allowed IIP_2 increment. In our case, the ISL_{RF_LO} = 87.6 dB or higher resulted from $IIP2$ increment restriction should be used.

When the direct-conversion receiver is implemented on a silicon chip of highly integrated circuits, the substrate isolation may dominate the LO emission level instead of the conductive emission. In this case, the circuit configuration will significantly affect the LO emission level at the antenna port. The LO leakage level at the input of the LNA through the substrate may reduce more than 20 dB if a differential LNA is used instead of a single end LNA, and a differential I/Q down-converter is employed in both cases.

3.2.3.5. AGC System for Direct-Conversion Receiver

The AGC system in the direct-conversion receiver is different from the AGC in the superheterodyne receiver mainly in two aspects. First, in the direct conversion receiver, To minimize the mismatch between the I and the Q channels, step gain controls are used in the BB block of both channels. On the other hand, the gain control in the superheterodyne receiver is almost continuous, and most of gain control takes place in the IF block. Second, the channel filtering in the direct-conversion receiver mainly relies on the BB LPF in the I and Q channels with no passive filter assistance. The BB LPF has a finite rejection (around 65 dB) to interferers in receiver band. The residual interferer level after the BB LPF filtering is still comparable with or even higher than the desired signal level at the input of the ADC. Thus the output of the ADC cannot be directly used for the RSSI to estimate in-channel-bandwidth signal power since it includes a significant portion of the residual interferer power. Further digital filtering is necessary to suppress the residual interferer down to negligible level and to assure the RSSI providing accurate power estimation. However, in the superheterodyne receiver, a high selective IF SAW filter is usually employed as the first channel filter, and then the BB LPF in the I and Q channels further performs channel filtering. In this case, the total rejection to the in-receiver-band interferers is easily over 85 dB, and at the input of the ADC, therefore, the residual interferer level comparing with the desired signal is negligible small.

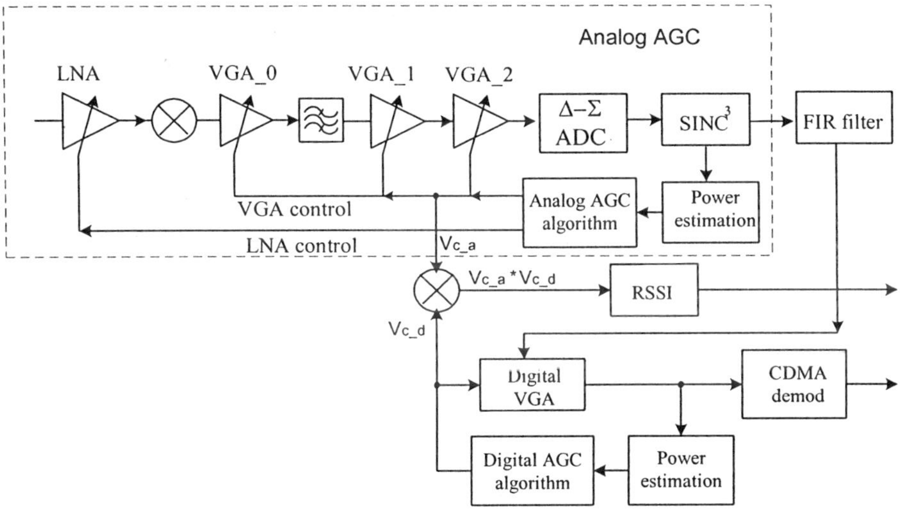

Figure 3.19. Simplified direct conversion CDMA receiver AGC system

The direct conversion receiver AGC system may contain two AGC loops. A simplified block diagram of a CDMA direct-conversion receiver AGC system is presented in Fig. 3.19. The AGC loop similarly used in the superheterodyne receiver is referred to as analog AGC loop, which operates based the total received power possibly including residual interferer power. The VGA_i ($i = 0$, 1, and 2) gain and the LNA gain in the analog AGC loop are sequentially adjusted with received signal and residual interferer power to maintain the input level to the ADC being a constant. The second loop is called a digital AGC loop, which is completely in the digital domain, and its function is to measure and track the in-channel-bandwidth power only. The digital VGA gain is adjusted with the in-channel-bandwidth signal resulted from further filtered received signal to keep the output of the digital VGA being a constant. It is easy to prove the received in-channel-bandwidth power is inversely proportional to the multiplication of the two control voltages, V_{ca} and V_{cd}. Since the overall analog gain g_a in natural scale is uniquely determined by the control voltage V_{ca} and the digital gain, g_d, in natural scale is proportional to the control 'voltage' V_{cd}, respectively. The outputs from the analog VGA_2 and the digital VGA should be constants due to the AGC loop function. Thus we have the following equations:

$$g_a(P_d + P_I) = \text{Constant\_1} \quad (3.2.23)$$

and

$$g_d(g_a \cdot P_d) = \text{Constant\_2}, \quad (3.2.24)$$

where P_d and P_I are the in-channel-bandwidth signal power and the residual interference power both in mW. From (3.2.24), we obtain

$$P_d = \frac{\text{Constant\_2}}{g_d \cdot g_a} \propto \frac{\text{K}}{V_{ca} \cdot V_{cd}}. \quad (3.2.25)$$

Therefore, the RSSI can be derived from the product of V_{ca} and V_{cd} as presented in (3.2.25). We should notice that the output of the analog VGA_2 or the input of the ADC in the direct-conversion receiver is not really a constant, and it fluctuates within ± ($\Delta G_{step}/2$) around the constant value where ΔG_{step} is the gain step size.

The advantages of this AGC approach is that the ADC dynamic range can be efficiently used when there is no interference accompanying with the desired signal, and thus the existing ADC can be used without modification. It is apparent that this AGC system is more complicated than that used in the superheterodyne receiver.

3.2.3.6. Direct-Conversion Transmitter

A typical configuration of the direct-conversion transmitter can be found in Fig. 3.10. Usually, the direct-conversion transmitter may not have so much saving in cost as the direct-conversion receiver since the superheterodyne transmitter does not need an IF SAW filter to clean up unwanted spurious. The direct-conversion transmitter may need a simple frequency plan to avoid spurious locating in the bands where the GPS, bluetooth, and other devices incorporate with the wireless mobile transceiver if a LO generator is used in the direct-conversion transmitter design.

In a full-duplex direct-conversion transceiver, it become possible that the UHF VCO for the transmitter is able to be integrated on the same chip with the other transmitter circuits. This is because separate VCOs are used in the receiver and the transmitter of the full-duplex direct conversion transceiver and because the specification requirements for the transmitter VCO are much lower than those of the receiver VCO. To avoid the LO pulling issue by loading variation, the VCO running frequency shall be different from the transmitter operating frequency. The VCO may operate at a frequency twice of the transmitter carrier frequency, and the LO frequency is then obtained in terms of a by 2 frequency divider as used in the direct-conversion receiver. This approach potentially has the VCO reverse modulation problem — i.e., the second-order harmonic of the modulated transmitter signal, which is mainly generated by the PA due to its considerable nonlinearity, feedbacks to the VCO running at twice of the transmission frequency and modulates the VCO. The reverse modulation may degrade the modulation accuracy or the waveform quality of the transmission signal if the reverse modulation of the VCO is severe enough. A better solution to local oscillator (LO) frequency generation is to employ a referred to as LO generator that possesses a VCO running at a frequency offset from the LO frequency.

One of the LO generator configuration including the synthesizer is presented in Fig. 3.20. The relationship between the LO frequency f_{LO} and the VCO frequency f_{VCO} is given in (3.2.26):

$$f_{LO} = \frac{m \pm 1}{m} f_{VCO}, \qquad (3.2.26)$$

where m is a dividing integer number of the frequency divider. When using an integer N synthesizer, the integer m can only be chosen from those that meet the equation (3.2.27):

$$\frac{m \pm 1}{m} \cdot \frac{f_r}{R} = k \cdot \Delta f_{CH\_Space}. \qquad (3.2.27)$$

In (3.2.27), f_r is the reference clock frequency, R is a dividing integer number of the reference clock divider, k is an integer number, and Δf_{CH_Space} is channel spacing frequency, and it is 30 kHz and 50 kHz for cellular and PCS band mobile systems, respectively. m = 4, 6 or 8, and k = 1, 2, or 4 are popularly used in the mobile station transmitter design.

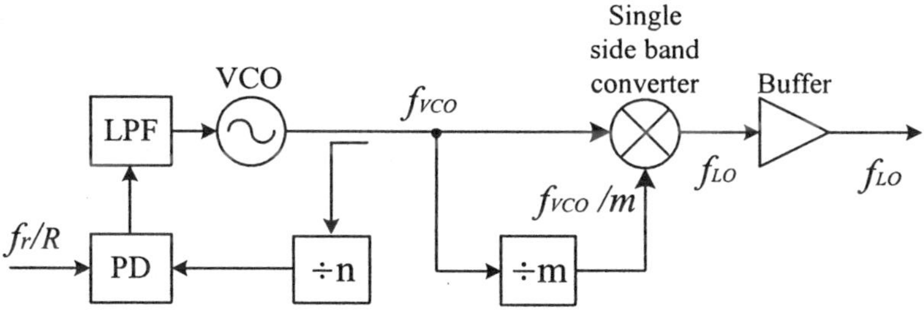

Figure 3.20. A configuration of the LO generator

Most gain controls are in the RF block of the transmitter. The gain is almost continuously controlled as in the superheterodyne transmitter. It is also possible to control the BB gain in the I and Q channels if the step gain adjustment is used. To reduce current consumption, the bias of each stage in the transmitter chain is usually dynamically controlled, incorporating the transmitter AGC. The level of the BB I and Q signals delivered by the DACs of the digital base-band can be linearly adjusted by controlling the reference voltage V_{ref} or reference current I_{ref} of the DACs. Thus the AGC range of the transmitter is extended. On the other hand, the BB signal level adjustment based on the transmission power can further reduce the overall current consumption of the transmitter.

To suppress the LO leakage emission in a single chip direct conversion transmitter, differential circuits may need for all the stages from the LPF, I/Q modulator, up to PA driver output. Usually the transmitter IC at present excludes the PA.

3.3. Low IF Architecture

The IF in the *low IF* radio architecture is system dependent, and it can be as low as, e.g., half of two times the bandwidth of the desired signal. The main advantage of the low IF architecture over the direct conversion one is that this architecture has no DC offset problem because the desired signal is off the DC by the IF. Properly choosing the low IF can remove the low-frequency interference product that resulted from the down-converter second-order nonlinearity demodulating AM interferer out of the desired signal bandwidth. In addition, the low IF architecture is also able to significantly reduce the near DC flicker noise impact on the receiver performance. This architecture, thus, is quite attractive for the highly integrated transceivers based on the CMOS technology since the flicker noise in the CMOS circuits is much higher than that in the GaAs, BiCMOS, and SiGe circuits. The main drawback of this architecture is the image-rejection problem because the IF is so low that the image band interference is close to the desired signal and it is difficult to separate the image from the desired signal by using any passive UHF BPF without degrading the receiver sensitivity.

3.3.1. Configuration of Low IF Radio

The low IF architecture is used mainly in the receiver. In the transmitter, this architecture may not have apparent advantages over either the superheterodyne or the direct conversion architecture. A possible block diagram of the low IF architecture receiver [13], [14] is presented in Fig. 3.21, where we can also find the block diagram of a superheterodyne transmitter with an *offset phase locked loop* (OPLL) performing the RF up-converter and the final transmission RF filtering both functions, which is often used in the GSM mobile stations. Using the GSM transceiver as the low IF architecture example is due to this architecture more suitable to the half-duplex system than to the full-duplex system if a very high image rejection is needed.

3.3.1.1. Low IF Architecture Receiver

The low IF receiver presented in Fig. 3.21 utilizes a digital *dual quadrature converter* to achieve a high image rejection. This configuration is definitely not unique for the low IF receiver. The same function can be achieved by using *polyphase band-pass filters* and a conventional down-converter as described in [15] and [16].

In the block diagram of Fig. 3.21, the antenna is connected to either the receiver or the transmitter through a transmitter/receiver (Tx/Rx) switch. It is quite often used in a half-duplex transceiver since the receiver and the transmitter operate at different time slots, but they usually share an antenna.

The receiver has two operation modes — i.e., normal receiving and calibration modes. In the normal receiving mode, the mode switch connects the preselector and LNA to the quadrature (I/Q) down-converter, and the calibration switch is in the open state. The preselector in the low IF architecture receiver is just a receiver band-pass filter, and it does not have any image-rejection function at all. The LNA usually dominates the overall noise figure of the receiver as the other architecture receivers. After passing through the preselector and amplified by the LNA, the received RF desired signal mixes with UHF quadrature LO signals (assuming the LO frequency lower than the received signal frequency) in the I/Q down-converter to generate low IF (100 kHz in GSM) I and Q signals. The low IF I and Q signals are amplified by low-frequency amplifiers (LFA) and then separated from high-frequency mixing products by using low-pass filters (LPF). At the input of the analog-to-digital converters (ADC), the imbalance of the I and Q signals is around 0.5 to 0.75 dB in amplitude and 3° to 5° in phase, respectively. This I and Q imbalance will cause -22 to -25 dB image crosstalk in the desired signal band. To avoid the DC offset propagation, AC coupling or high pass design can be used in between the down-converter output and the input of the ADC.

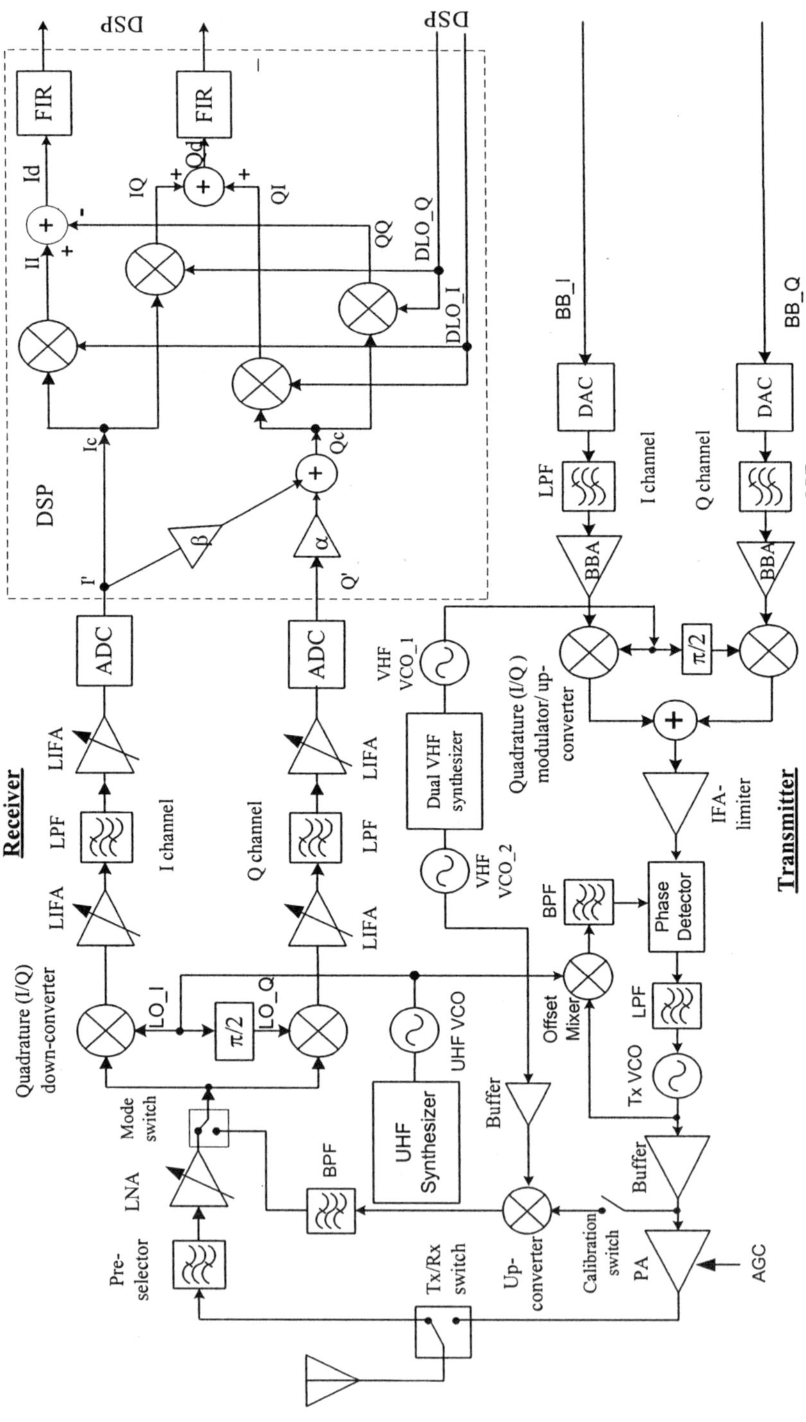

Figure 3.21. Block diagram of low IF rceiver and superheterodyne transmitter

The low-frequency analog I and Q signals are finally converted to digital signals by the ADCs. After the ADCs, all signal processes are in the digital domain. The digital I and Q signals, I' and Q', carry the same imbalance errors as that contained in the low-frequency analog I and Q signals. The imbalance errors of the I' and Q' signals are properly compensated before they are further down-converted into digital BB I and Q signals. In the low IF receiver configuration of Fig. 3.21, the compensation is executed only on the Q' signal. The compensated Q' has a form of Qc = αQ' + βI', where α = 1 and β = 0 while the I and Q channel signals are perfectly balanced. On the other hand, the I' signal has no change — i.e., Ic = I' propagating to the following stage. The Ic and Qc signals then mix with digital LO I and Q signals, respectively, in the digital dual quadrature down-converter, and four digital base-band signals (II, IQ, QI, and QQ) are generated. The final I channel output of the digital dual quadrature converter Id is the subtraction of the II and QQ signals — i.e., II – QQ, and the Q channel output Qd is the summation of the IQ and QI signals — i.e., IQ + QI. After the error compensation, the imbalance between the digital I and Q signals can drop to less than 0.4 degree in phase and less than 0.03 dB in amplitude, and thus the image rejection can achieve 50 dB or even higher [14].

When the receiver operates in the calibration mode, the mode switch is turned to connect the calibration signal generated by the transmitter, and the calibration switch is closed. The calibration signal has the same frequency as the receiver signal carrier, and it is directly fed to I/Q down-converter. The calibration signal is processed in the receiver as the desired signal except α and β are set to 1 and 0, respectively, and the digital dual quadrature down-converter provides not only the normal I and Q signals I_{cal} and Q_{cal} but also image I and Q signals I_{img} and Q_{img} by swapping the adding and subtracting operation — i.e., I_{cal} = II – QQ, Q_{cal} = IQ – QI, I_{img} = II + QQ and Q_{img} =IQ – QI. The compensation multipliers α and β can be calculated as follows (see Section 3.3.2.1.):

$$\alpha = \frac{1}{(1+\delta)\cos\varepsilon} \tag{3.3.1}$$

and

$$\beta = \tan\varepsilon\,, \tag{3.3.2}$$

where δ and ε can expressed as

$$\delta = -2\frac{I_{cal}I_{img} - Q_{cal}Q_{img}}{I_{cal}^2 + Q_{cal}^2} \tag{3.3.3}$$

and

$$\varepsilon = \tan^{-1}\frac{2(I_{cal}Q_{img} + Q_{cal}I_{img})}{I_{cal}^2 + Q_{cal}^2}. \tag{3.3.4}$$

The idle slots of the half-duplex system can be used for the I and Q imbalance error compensation. How often the compensation need be carried out will depend on the system design and circuit performance.

3.3.1.2. Superheterodyne Transmitter with OPLL

The transmitter block diagram in Fig. 3.21 is a typical GSM mobile station transmitter. In the normal transmission mode, the VHF VCO_1 is powered up and running at 44.9 MHz, but the VHF VCO_2 and all circuits related to the receiver calibration signal generation are powered down. The I and Q BB signals coming from the transmitter DACs modulate the 44.999 MHz quadrature signals in the quadrature modulator to form a GMSK IF signal. This IF signal in the offset phase locked loop (OPLL) is up-converted to a GSM RF transmission signal with a carrier within 890 to 915 MHz band. The OPLL consists of a phase-comparator, an offset mixer, a Tx-VCO, and a loop filter. The GMSK modulated IF signal inputs to the phase-comparator as a reference signal. The OPLL has a broad enough bandwidth to reproduce GMSK modulation at the output of the Tx-VCO with an acceptably low phase error, but the bandwidth must be narrow enough to suppress unwanted spurious and noise emissions down to a level, which is way below the specified requirement. The GMSK modulated RF signal from the Tx-VCO output is then amplified to about 30 dBm by the buffer and power amplifiers for final transmission from the antenna.

In the calibration mode, the BB I and Q signals contain no modulation information. The output of the quadrature modulator is an unmodulated tone with a frequency 44.9 MHz. This IF tone inputs to the OPLL to produce a RF tone signal with a frequency in the transmitter band. In the calibration mode the VHF VCO_2 is also powered up and running at 45.1 MHz. The calibration switch is closed. Thus the RF transmission tone from the output of the buffer amplifier through the calibration switch up-converted with the VCO_2 45.1 MHz signal to produce a receiver frequency tone. The output of the up-converter passes through the receiver band filter Rx BPF to obtain a clean receiver calibration tone. During the receiver calibration, the PA is turned off to save on power consumption.

3.3.2. Approaches to Achieve High Image Rejection

The main issue of the low IF receiver architecture is the image problem since the IF is too low to separate the image from the desired signal by means of an RF BPF, which is small enough to be able to be employed in the mobile stations. The imbalance between I and Q channel signals in the low IF receiver determines the possible maximum image rejection. To achieve high image rejection, it is necessary to minimize the imbalance of the I and Q signals by means of the complex quadrature down-converter or a combination of the quadrature down-converter and complex band-pass filters as mentioned in the previous subsection. In this subsection, we discuss how high image rejection can be achieved in detail.

3.3.2.1. I and Q Signal Imbalance and Image Rejection

The imbalance of the low IF I and Q channel signals causes the image crosstalk, and it degrades the image rejection. The image rejection *IR* in dB and the imbalances of the amplitude and the phase of the I and Q signals has the following relationship (refer to the derivation in Appendix 3C):

$$IR = 10\log\frac{1+2(1+\delta)\cos\varepsilon+(1+\delta)^2}{1-2(1+\delta)\cos\varepsilon+(1+\delta)^2}, \tag{3.3.5}$$

where ε is the I and Q phase imbalance from the nominal 90° offset in degree, δ is the I and Q amplitude imbalance, and it is usually represented in dB by using the formula $10\log(1+\delta)$.

A plot of image responses versus the phase imbalance ε and the amplitude imbalance $|10\log(1+\delta)|$ is presented in Fig. 3.21. In this figure, each curve represents a certain image-rejection response — i.e., each curve is characterized by the image rejection. The curves represent the image rejections from 25 dB to 60 dB in 5 dB steps. For a given amplitude imbalance, the image rejection is not so sensitive to the phase imbalance in a certain error angle range, and similarly for a given phase imbalance, the image rejection is not sensitive to the amplitude imbalance in a certain range either. For example, for a 0.3 dB amplitude imbalance, the image rejection keeps at 30 dB while the phase imbalance changes from 0 to 2 degree, and for a 4 degree phase imbalance, the same image rejection can maintain when the amplitude imbalance increases from 0 dB to 0.13 dB.

When a 100 kHz low IF is used in a GSM mobile station, the adjacent and alternate channel signals, which are 200 kHz and 400 kHz away from the desired signal carrier, respectively, may become an image interference of the desired signal. In the GSM specification [2], it defines the adjacent channel interference, and alternate channel interference can be at least 9 dB and 41 dB, respectively, higher than the desired signal. The adjacent and alternate channel interference specification applies for a desired signal input level of 20 dB above the reference sensitivity level. The CNR at the reference sensitivity level of the GSM mobile station is approximately 8 dB. The CNR at the test signal level, thus, is 28 dB. If the allowed CNR degradation under the adjacent and alternate channel interference impact, which generates in-channel interference 16.4 dB higher than the desired signal, is only 3 dB, the required image rejection is at least (16.4+28-3) = 41.4 dB. From Fig. 3.22 we can see that to achieve the 42 dB image rejection or higher, the amplitude and phase imbalances of the I and Q channel signals must be less than 0.1 dB and 1.0 degrees, respectively. However, based on GSM specification, actually a 24.4 dB image rejection is enough to cop with the adjacent and alternate channel interference (see Section 3.3.3).

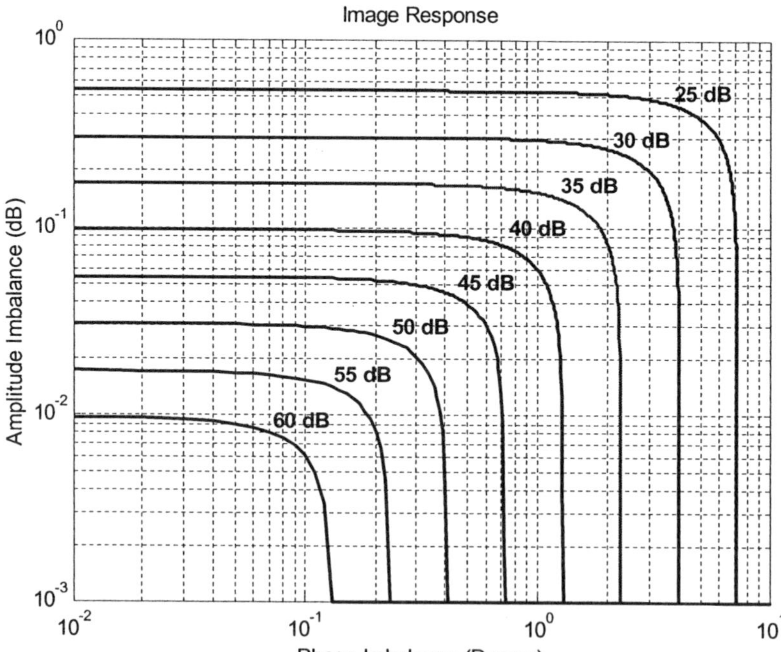

Figure 3.22. Image responses versus amplitude and phase imbalances

3.3.2.2. Digital Dual Quadrature Down-Converter Approach

To achieve higher than 40 dB image rejection, the imbalance errors of thc analog I and Q signals must be corrected. One effective approach to obtain high image rejection is using the digital dual quadrature down-converter [14]. The low IF receiver described in Fig. 3.22 is the one utilizing the digital dual quadrature down-converter. To explain how the high image rejection is able to be achieved, a simplified low IF receiver diagram consisting of only the analog and the digital quadrature down-converters is presented in Fig. 3.23.

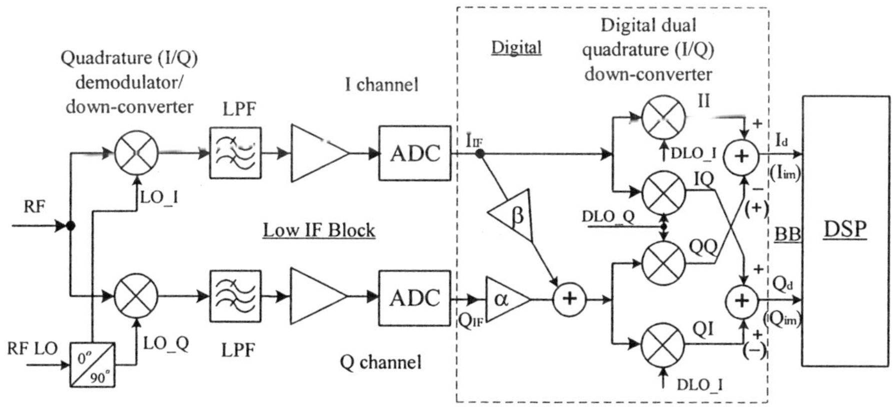

Figure 3.23. High image rejection configuration of using digital dual I/Q down-converter

The imbalanced low IF I and Q signals, I_{IF} and Q_{IF}, can be expressed as (see Appendix 3C)

$$I_{IF} = \frac{1}{2}\cos[\omega_{IF}t + \varphi(t)] \tag{3.3.6}$$

and

$$Q_{IF} = \frac{-(1+\delta)}{2}\sin[\omega_{IF}t + \varphi(t) - \varepsilon], \tag{3.3.7}$$

where δ and ε are the amplitude and phase imbalances, respectively, as defined in (3.3.5), and the amplitude of I_{IF} and Q_{IF} in (3.3.6) and (3.3.7) has been normalized to unity. In the following derivation, we also use the normalized amplitude.

If there is no imbalance compensation executed, — i.e., $\alpha = 1$ and $\beta = 0$ — the I and Q channel imbalance caused by the analog circuits will propagate to the outputs of the digital dual quadrature down-converter. Since the imbalance introduced by the digital dual quadrature converter is usually negligible, the down-concerted into base-band signals at the outputs of the dual quadrature converter are

$$II = \frac{1}{4}\cos\varphi(t) \tag{3.3.8}$$

$$IQ = -\frac{1}{4}\sin\varphi(t) \tag{3.3.9}$$

$$QI = -\frac{1+\delta}{4}\sin[\varphi(t)-\varepsilon] \tag{3.3.10}$$

and

$$QQ = -\frac{1+\delta}{4}\cos[\varphi(t)-\varepsilon]. \tag{3.3.11}$$

From (3.3.8) to (3.3.11), we can obtain the desired I and Q signals, I_d and Q_d,

$$I_d = II - QQ \cong \frac{1}{2}\cos\varepsilon\cdot\cos\varphi(t) \tag{3.3.12}$$

and

$$Q_d = IQ + QI \cong \frac{1}{2}\cos\varepsilon\cdot\sin\varphi(t). \tag{3.3.13}$$

The image I and Q products I_{im} and Q_{im}, can also be obtained from (3.3.8) to (3.3.11) by switching the adding and subtracting signs in (3.2.12) and (3.3.13), and they are expressed as

$$I_{im} = II + QQ \cong -\frac{1}{4}[\delta\cos\varepsilon\cdot\cos\varphi(t) + \sin\varepsilon\cdot\sin\varphi(t)] \tag{3.3.14}$$

and

$$Q_{im} = IQ - QI \cong \frac{1}{4}[\delta\cos\varepsilon\cdot\sin\varphi(t) - \sin\varepsilon\cdot\cos\varphi(t)]. \tag{3.3.15}$$

In the normal receiver operation mode, only the desired I and Q signals are used, but in the calibration mode the desired and the image I and Q signals will be measured by switching the adding and subtracting operations of the dual quadrature down-converter outputs from the original signs to the ones in the parentheses as shown in Fig. 3.22. A test tone is applied to the input of the RF I/Q down-converter while the receiver is running in the calibration mode. In this case, (3.3.12), (3.3.13), (3.3.14), and (3.3.15) are also held, but $\varphi(t)$ may become a constant instead of the time function if a CW test tone is used.

From the approximate equations (3.3.12) to (3.3.15), the imbalanced amplitude and phase values δ and ε can be determined as follows. The ratios of the I_{im}/I_d and R_{im}/R_d are expressed, respectively, as

$$\frac{I_{im}}{I_d} = -\frac{1}{2}(\delta + \tan\varepsilon \cdot \tan\varphi) = -\frac{1}{2}\left(\delta - \tan\varepsilon \cdot \frac{Q_d}{I_d}\right) \tag{3.3.16}$$

and

$$\frac{Q_{im}}{Q_d} = -\frac{1}{2}(\delta - \tan\varepsilon \cdot \cot\varphi) = -\frac{1}{2}\left(\delta + \tan\varepsilon \cdot \frac{I_d}{Q_d}\right), \tag{3.3.17}$$

where the following expression resulted from (3.3.12) and (3.3.13) is used

$$\frac{I_d}{Q_d} = -\cot\varphi\,. \tag{3.3.18}$$

Solving (3.3.16) and (3.3.17), we obtain the amplitude and phase imbalances represented by the measured I_d, Q_d, I_{im} and Q_{im} values

$$\delta = -\frac{2(I_d I_{im} - Q_d Q_{im})}{I_d^2 + Q_d^2} \tag{3.3.19}$$

and

$$\varepsilon = \tan^{-1}\frac{2(I_d Q_{im} + Q_d I_{im})}{I_d^2 + Q_d^2}\,. \tag{3.3.20}$$

(3.3.19) and (3.3.20) are the exactly same as (3.3.3) and (3.3.4), respectively, if letting $I_d = I_{cal}$ and $Q_d = Q_{cal}$. The above derivation clearly explains that where (3.3.3) and (3.3.4) are resulted from.

The compensation multipliers α and β should be determined by the imbalances δ and ε and they can be derived based on the consideration that the I and Q signals shall be balanced after the imbalance error compensation. The expressions of the imbalanced low IF I and Q signals have been given in (3.3.6) and (3.3.7). The compensation is executed only in the Q channel, and in the I channel we have $I_C = I_{IF}$. The compensated Q channel signal Q_C can be written as

$$Q_C = -\frac{1}{2}\{\alpha \cos\varepsilon \cdot (1+\delta)\sin[\omega_{IF}t + \varphi(t)] + [\beta - \alpha(1+\delta)\sin\varepsilon]\cdot \cos[\omega_{IF}t + \varphi(t)]\}. \tag{3.3.21}$$

In the balanced condition, we must have the following equalities:

$$\alpha(1+\delta)\cos\varepsilon = 1$$

and

$$\beta - \alpha(1+\delta)\sin\varepsilon = 0 .$$

From these two qualities, it is easy to obtain the compensation multipliers for correcting the imbalance errors:

$$\alpha = \frac{1}{(1+\delta)\cos\varepsilon} \tag{3.3.1}$$

and

$$\beta = \tan\varepsilon . \tag{3.3.2}$$

Although the same approach as [14] is used here, the form of the formulas presented in this subsection is different from that given in that reference article. This is because the imbalances used in this section are defined differently than those in [14].

3.3.2.2. Polyphase Band-Pass Filter Approach

Using *polyphase band-pass filter* is an alternative approach to achieve high image rejection. In this case, it may be not necessary to have dynamic imbalance error correction or adaptive I/Q channel mismatch cancellation if the image-rejection requirement is not very high. For a conventional GSM mobile station using a 100 kHz low IF receiver, the minimum image rejection required is around 30 dB if only the adjacent and alternate channel signals need be concerned as the image interference sources. Fig. 3.24 presents the block diagram of a simplified low IF receiver using an analog polyphase band-pass IF filter.

In this receiver configuration, to obtain high image rejection, the I/Q imbalances caused by the RF quadrature down-converter and the polyphase filter must be low enough. The I and Q channel mismatch of the RF quadrature down-converter will generate an image crosstalk in the desired

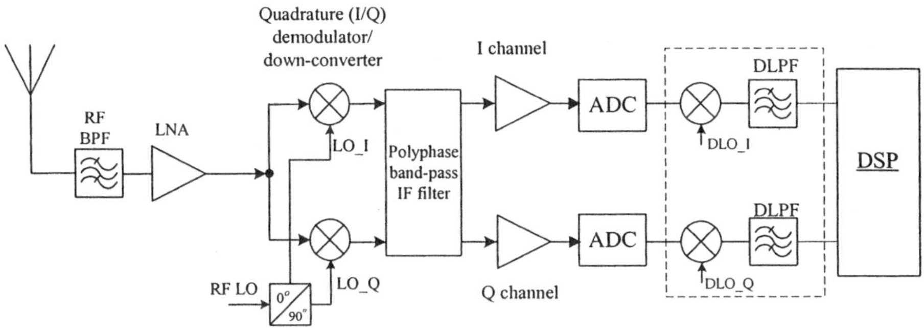

Figure 3.24. Block diagram of low IF receiver with polyphase BPF

signal band. This can in principle be explained by using complex signal expressions. Assuming that a desired signal low IF signal has an angle frequency $\omega_{IF} = 2\pi f_{IF}$ and the image signal has a angle frequency $\omega_{im} = 2\pi f_{im} = -2\pi f_{IF}$, we express the image signal as

$$S_{im} = A_{im} e^{-j[\omega_{IF}t+\vartheta(t)]} = A_{im}\cos[\omega_{IF}t + \vartheta(t)] - jA_{im}\sin[\omega_{IF}t + \vartheta(t)], \quad (3.3.22)$$

where A_{im} is the amplitude of the image, and $\vartheta(t)$ is the angle modulation. If the RF quadrature down-converter is not perfectly balanced and creates amplitude and phase imbalances 2Δ and 2ϕ, for simplicity we express the complex image signal with the I and Q imbalances at the output of the quadrature down-converter as

$$\begin{aligned} &A_{im}(1+\Delta)\cos[\omega_{IF}t + \vartheta(t) + \phi] - jA_{im}(1-\Delta)\sin[\omega_{IF}t + \vartheta(t) - \phi] \\ &\cong A_{im}\cos\phi \cdot e^{-j[\omega_{IF}t+\vartheta(t)]} + \Delta A_{im}\cos\phi \cdot e^{j[\omega_{IF}t+\vartheta(t)]} + jA_{im}\sin\phi \cdot e^{j[\omega_{IF}t+\vartheta(t)]} \end{aligned} \quad (3.3.23)$$

The second and the third terms on the right side of (3.3.23) are the image crosstalks in the desired frequency band, which resulte from the I and Q imbalance. It is apparent that the image crosstalks cannot be gotten rid of by using the polyphase band-pass filter once they are generated by the preceding RF quadrature down-converter. In this case it is necessary to restrain the phase imbalance of the RF down-converter 2ϕ to less than 1 degree to achieve a rejection greater than 40 dB since the amplitude imbalance can be usually calibrated out by adjusting the I/Q channel gain. The 90 degree phase shifter or called as quadrature generator can be implemented by means of a polyphase filter for a low phase imbalance — say, less than 0.5 degree.

An asymmetric polyphase filter possesses a different transfer function for positive and negative frequency components. This means that it is capable to amplify the desired signal and to suppress the image signal [17]. The polyphase filter has two (I_i and Q_i) inputs and two (I_o and Q_o) outputs, and therefore, it has four transfer functions from each input to each output, H_{II}, H_{IQ}, H_{QI}, and H_{QQ}, as shown in Fig. 3.25. In the ideal case, $H_{II} = H_{QQ}$ and $H_{IQ} = H_{QI}$. For a low IF receiver using a LO frequency lower than that of the desired signal, it will require the polyphase filter with a pass-band from positive to positive frequencies, with an attenuation from negative to negative frequencies, and without signal transfer from the positive to negative frequencies and vice versa. In general, the polyphase band-pass filter can be synthesized to have only a pass-band either at positive or negative frequencies. The transfer function of a polyphase band-pass filter can be obtained from the linear frequency transformation of a corresponding low-pass filter transfer function $H_{BP}(j\omega) = H_{LP}(j\omega - j\omega_{IF})$ [15], [17].

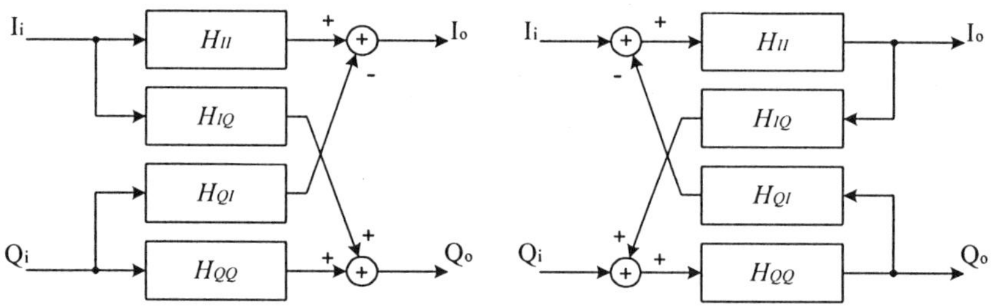

Figure 3.25. Block diagrams of polyphase filters

The maximum image rejection of an analog polyphase filter is determined mainly by the component imbalance between the I and Q paths. To achieve a 55 dB or higher image rejection the imbalance of components between two paths must be 0.2% or less. The dynamic range of a polyphase filter is dependent not only on the image rejection but also on the image interference level to the desired signal level before filtering. In the low IF GSM receiver, the adjacent and alternate channel signals may turn into in-band interference of the desired signal, and the interference can be 15 dB higher than the desired signal. If the receiver requests a higher than 30 dB image rejection, the polyphase BPF should have a greater than 45 dB dynamic range. It depends on the dynamic range requirement and applications so that a 6 to 10 bit ADC may be needed.

The overall image rejection of Fig. 3.24's low IF receiver is determined mainly by the imbalance performance of the RF quadrature down-converter if the image rejection of the polyphase BPF is much higher than that of the quadrature down-converter. For very high image rejection, such as higher than 55 dB, it can be achieved by using digital polyphase filter and imbalance error compensation in addition to a high dynamic ADC.

3.3.3. Some Design Considerations

3.3.3.1. Comparison of Low IF Receiver Configurations

In the low IF receiver the main technical challenge is the image interference suppression or rejection since the IF is so low that it is impossible to filter out the image in the RF stage. The two approaches described in the previous subsection are the most popular ones to achieve high image rejection. The first approach using a digital dual down-converter and dynamic imbalance error correction provides high-quality image rejection, which can be over 50 dB. The configuration of the first approach is much more complicated than the second approach. It needs extra circuits to generate calibration signal and more processing in the digital domain. Although the second approach based on the analog polyphase band-pass filter does not need extra circuits for calibration, the analog active polyphase filter is usually quite power hungry. The compact configuration of the second approach provides only middle range image rejection — around 40 dB.

It will depend on applications and the image rejection requirement to choose the configuration of a low IF receiver. It is apparent that the configuration of employing the dual quadrature converter with imbalance error correction is suitable for the system requesting high image rejection. In fact, the complexity of this approach is mainly in the DSP area and therefore the corresponding current consumption increase may not be significant. On the other hand, the imbalance compensation in most applications need not perform for every signal burst if the temperature or signal strength does not dramatically change. The GSM mobile station receiver with the low IF architecture is often seen based on this configuration although the image rejection for the low IF GSM receiver may need only over 30 dB. However, the imbalance compensation method described in Section 3.3.2.2 cannot be applied for the receiver in a full-duplex transceiver. The low IF receiver with an analog polyphase band-pass does not have such restriction, but its image rejection performance is not as good as the previous one.

3.3.3.2. Determination of Image Rejection

The low IF is usually in the range of half to two times bandwidth of the desired signal. In this case, the adjacent or the alternate channel signal turns into image interference of the desired signal. Using the GSM mobile station as an example, the adjacent channel frequency spacing is 200 kHz, and the alternate channel frequency spacing is 400 KHz. For a 100 kHz IF receiver, it is apparent that the adjacent channel signal is exactly in the image frequency band of the desired signal. In addition to this, the alternate channel signal becomes the adjacent channel interference after the frequency down-conversion from the RF to the IF as depicted in Fig. 3.26. In this figure, it is assumed that the desired signal carrier frequency is higher than the LO frequency 100 kHz. The lower frequency adjacent channel interference is the image interference of the desired signal and its spectrum will be mirrored and overlap with the desired signal after the frequency down-converter. After the down-conversion the image spectrum is denoted by dashed line as shown in Fig. 3.26. In a similar way, the lower frequency alternate interference becomes the adjacent channel interference as depicted in the same figure.

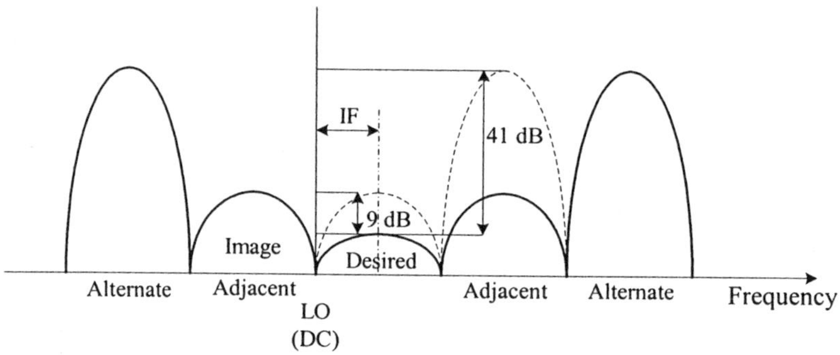

Figure 3.26. The adjacent and the alternate channel interference

The minimum image rejection requirement of a low IF receiver IR_{min} can be roughly estimated as follows:

$$IR_{\min} = CNR_{\min} + \Delta S_d + \Delta I_{inband} - \Delta CNR\,, \tag{3.3.24}$$

where CNR_{min} is the carrier-to-noise ratio at the receiver sensitivity, ΔS_d is the assigned relative level of the desired signal above the sensitivity level for the adjacent and alternate channel interference tests, ΔI_{inband} is the relative interference level within the desired signal band, and ΔCNR is the allowed CNR degradation. All the variables in (3.3.24) are in dB.

The specification of GSM mobile station defines that the adjacent channel interference and the alternate channel interference are 9 dB and 41 dB above the desired signal, respectively, as presented in Fig. 3.26. The desired signal level is defined as ΔS_d = 20 dB higher than the sensitivity level for the adjacent channel interference test, and the CNR_{min} of a GSM mobile receiver is around 8 dB. For the same BER as that at the receiver sensitivity, the CNR can be degraded 20 dB. Thus, using (3.3.24) we obtain the minimum image rejection to be

$$IR_{\min} = 8 + 20 + 9 - 20 = 17 \text{ dB}.$$

In fact, the image-rejection requirement of a low IF GSM receiver is determined not only by this image interference but also by the alternate channel interference, which is converted to the adjacent channel after the RF down-conversion. The main reasons are that first, its level is 41 dB higher than the desired, and second, the real spectra of GSM desired, adjacent, and alternate signals are not really separated as shown in Fig. 3.26, but they actually overlap each other. The power spectrum of the adjacent channel GSM signal has approximate 0.28% or -25.5 dB of the total power to be in the desired signal band. Now the adjacent channel interference converted from the lower-frequency alternate channel is 41 dB above the desired signal, and 0.28% of the overall alternate channel signal power is about 15.5 dB higher than the desired signal. Considering this interference and the image interference both, the corresponding image-rejection minimum requirement of the low IF GSM receiver is

$$IR_{\min} = 8 + 20 + 15.5 - 20 = 23.5 \text{ dB}.$$

In general, a 30 dB image rejection for the low IF GSM mobile receiver will be sufficient to handle the adjacent channel interference and the alternate channel interference. The 30 dB image rejection is not easy but also not very difficult to achieve, even if no imbalance error correction is used. From Fig. 3.22, we know that it requests the amplitude and the phase imbalances of the I and Q channel signals equal to or less than 0.25 dB and 4 degree, respectively.

3.3.3.3. Low IF Receiver AGC

This is similar to the case of the direct conversion receiver. With the gain control in the I and Q channels, it is better to use discrete step control for high accuracy and excellent balance. This is particularly important for the gain control stages before the BPF of the low IF receiver based on the polyphase filter since any gain imbalance (also the phase imbalance in the control range there) will contribute to the restraint on the possible maximum image rejection. In the low IF receiver with the imbalance compensation, the gain control imbalance in the I and Q two channels may not be so critical since the imbalance can be calibrated and then compensated based on the stored calibration information.

A digital AGC may be needed as in the direct conversion receiver when the LPF or the BPF does not have enough suppression to the blocking and other interferers. More details on the digital AGC can be found in 3.2.3.5.

3.3.3.4. Transmitter with OPLL

The transmitter given in Fig. 3.21 is a superheterodyne GSM transmitter by using offset phase locked loop (OPLL) as a tracking bandpass filter turned to the desired transmission signal. The OPLL used in the GSM transmitter should have the following characteristics: a sufficient suppression level of the transmission noise, a small phase error, and a fast settling. The band-pass characteristic of the OPLL makes it able to replace the transmission SAW and the duplexer. The fast settling can lower the current consumption.

The bandwidth of the OPLL must be narrow enough to suppress noise emission in the receiver band to below –79 dBm, and it must also be broad enough to reproduce the input at the output of the OPLL with a rms phase error less than 5 degree. The optimize bandwidth of the OPLL is in between 0.6 and 2.6 MHz [18].

3.4. Band-pass Sampling Radio Architecture

At present, it is still not practical for mobile stations to use the so-called *software radio* architecture, which should ideally have the ADC placed in the RF front-end near the antenna as possible and operating at an RF sampling frequency slightly higher than twice the greatest carrier frequency of interest, and the resulting samples are processed on a programmable signal processor. For a 1.9 GHz PCS band signal, the sampling rate of the ADC in the ideal software radio should be greater than 4.0 GHz. The main issue here is that the current technology is not mature enough to provide a device of processing samples at such a high rate and with acceptable power consumption for mobile stations. An alternate solution is to use the *band-pass sampling* architecture, which possibly possesses some features of the ideal software radio.

The band-pass sampling also referred to as *harmonic sampling* is the techniques of sampling at rates lower than the highest frequency of interest to achieve frequency conversion from RF to low IF or base-band through *intentional aliasing* and to be able to exactly reconstruct the information content of the sampled analog signal if it is a band-pass signal [19], [20] and [21]. The sampling rate requirement is no longer based on the RF carrier, but rather on the information bandwidth of the signal. Thus the resulting processing rate can be significantly reduced.

The radio architecture will be much simpler than the other architectures presented in the previous sections if the band-pass sampling is directly employed at RF. In this case, the analog RF block contains only band-pass filters and low-noise amplifiers before a high-performance ADC carrying out sampling and digitizing. It should be noticed that the ratio of the RF carrier to the *undersampling* rate used in the band-pass sampling architecture for RF transceivers is usually not high. The main reason for this is due to the noise density of present ADCs operating at RF being high and increasing with the harmonic order of the sampling rate. On the other hand, it is apparent that the band-pass sampling can also be applied in the super-heterodyne receiver to replace the I/Q down-converter and then the BB I and Q channels are created in digital domain. In this section, the RF band-pass sampling will be mainly discussed.

3.4.1. Basics of Band-pass Sampling

The sampling theory shows that, to avoid aliasing and to completely reconstruct the signal, the sampling rate must be at least twice of the highest frequency component in the signal as described in Section 2.4.1. It is implicit that the useful information of signal covers the entire band from zero frequency to cutoff frequency. However, the RF signals used in the wireless communications are usually narrow-band in nature since the signal bandwidths are only 0.003 to 0.2% of their carrier frequencies. In these cases, the minimum uniform sampling rate to avoid aliasing depends on the signal bandwidth instead of the highest frequency of interest. The minimum sampling rate for aliasing-free can be as low as twice of the signal bandwidth if the carrier frequency of the signal is properly chosen. However, the minimum sampling rate $f_{s,\min} = 2 \cdot BW$ (where BW is the signal bandwidth) just has its theoretical value in the sense that not only any imperfections in the implementation based on this sampling rate will cause aliasing, but also the band-pass sampling unlike the BB sampling case may still have aliasing problem even if the sampling rate is greater than this $f_{s,min}$.

Assume that a band-pass analog signal has its lowest frequency of interest f_L and the highest frequency of interest f_H or the bandwidth of the signal equals $BW = (f_H - f_L)$. The band-pass analog signal can be exactly reconstructed after sampling and digitizing if the sampling rate f_s meets the following two inequalities [20]:

$$\frac{(n-1)f_s}{2} < f_L \tag{3.4.1}$$

and

$$f_H < \frac{nf_s}{2}, \tag{3.4.2}$$

where n is an integer given by $1 \le n \le \lfloor f_H / BW \rfloor$ ($\lfloor \circ \rfloor$ denotes the largest integer). The sampling rate f_s meeting (3.4.1) and (3.4.2) also means the resulting spectra of the sampled signal have no overlapping or aliasing as clearly shown in Fig. 3.27.

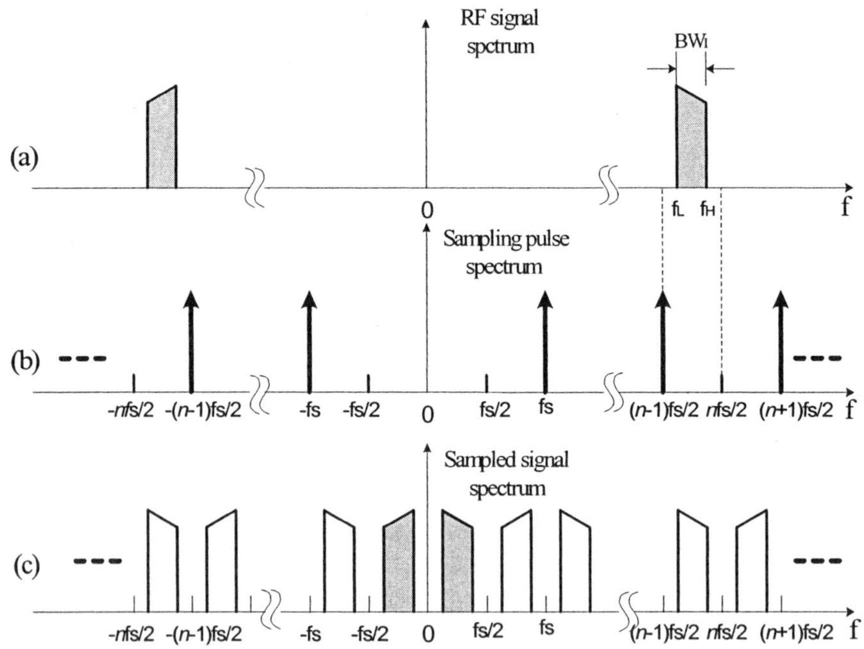

Figure 3.27. Spectra of band-pass sampling: (a) RF signal spectrum, (b) sampling pulse spectrum, and (c) sampled signal spectrum

From inequalities (3.4.1) and (3.4.2) we can determine the acceptable uniform sampling rates for *aliasing-free* to be

$$\frac{2f_H}{n} \le f_s \le \frac{2f_L}{n-1}. \tag{3.4.3}$$

The maximum allowable value of n for the band-pass signal with the lowest and highest frequencies f_L and f_H , n_{max}, is

$$n_{\max} = \left\lfloor \frac{f_H}{f_H - f_L} \right\rfloor. \tag{3.4.4}$$

Equation (3.4.3) can be described graphically as shown in Fig. 3.28 when n = 1, 2, ..., 5. Where the normalized sampling frequency f_s/BW versus the normalized highest frequency f_H/BW is plotted as presented in [19]. The areas inside the wedges are the permissible zones for sampling without aliasing. The shadowed area represents the sampling rates that result in aliasing. It is apparent the aliasing-free ranges of the sampling rate and the highest signal frequency of interest, Δ_S and Δ_{fH}, increase with normalized sampling rate and the highest signal frequency. The smaller the

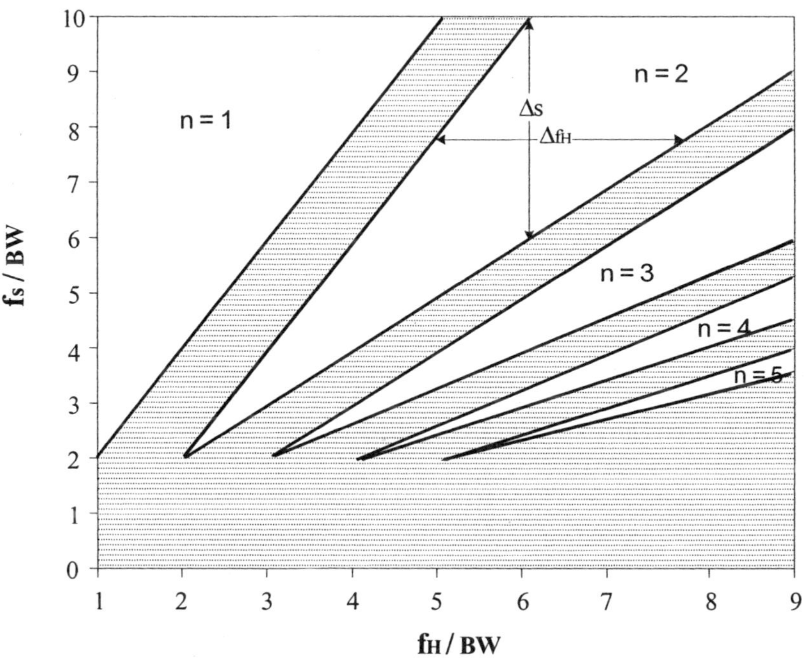

Figure 3.28. Permissible zones for uniform sampling without aliasing

integer number n is, the broader the permissible area for sampling without aliasing will be. The n is usually low less than 10 when the band-pass sampling technique is used for converting an RF signal to a low IF or base-band signal.

In the mobile station applications, it is more practical to consider the signal bandwidth of interest, $BW = (f_H - f_L)$, as the stop bandwidth of the channel filter BW_S and the signal information bandwidth as the pass-band of the channel filter $BW_P = BW_I$, as depicted in Fig. 3.29a. In addition, the center frequency of the pass-band f_c usually represents the carrier frequency of the desired signal. From Fig. 3.29, the frequency difference between the edge of the stop-band bandwidth and the sampling harmonic frequency mf_s nearest to the center frequency f_c, Δ_1 is

$$\Delta_1 = mf_s - (f_c - BW_s/2). \tag{3.4.5}$$

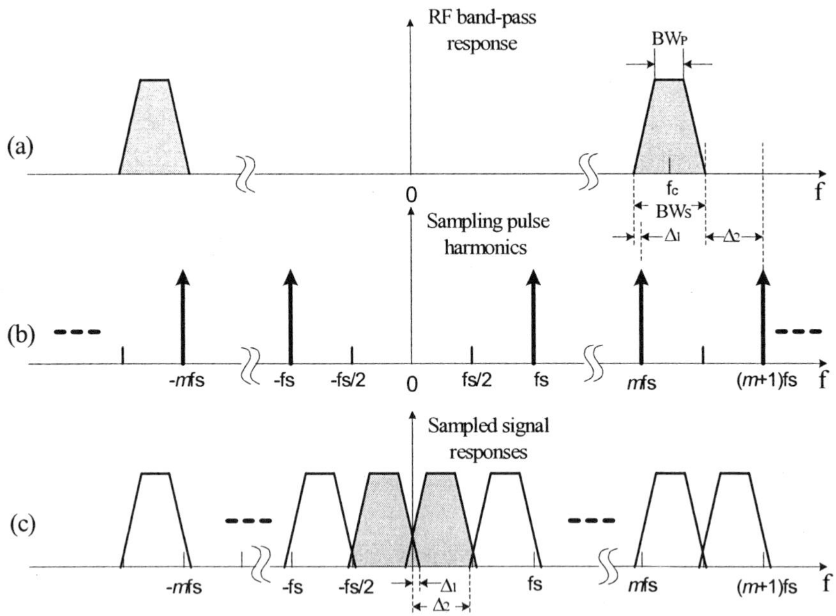

Figure 3.29. RF band-pass response, sampling pulse fundamental and harmonics, and sampled signal responses

The frequency difference between the sampling harmonic $(m+1)f_s$ and another edge of the stop-band bandwidth, Δ_2 is

$$\Delta_2 = (m+1)f_s - (f_c + BW_s/2). \tag{3.4.6}$$

It is apparent from Fig. 3.29(c) that for sampling without aliasing — i.e., no spectrum overlapping into pass-band the Δ_1 and Δ_2 must meet the following inequalities, respectively,

$$\Delta_1 < (f_c - mf_s) - \frac{BW_P}{2} \tag{3.4.7}$$

and

$$\Delta_2 > (f_c - mf_s) + \frac{BW_P}{2}. \tag{3.4.8}$$

Substituting (3.4.5) and (3.4.6) into (3.4.7) and (3.4.8), respectively, the sampling rate without aliasing can be derived as [21]

$$\frac{2f_c + \frac{1}{2}(BW_P + BW_S)}{n} < f_s < \frac{2f_c - \frac{1}{2}(BW_P + BW_S)}{n-1}, \tag{3.4.9}$$

where $n = 2m + 1$. In fact, n in (3.4.9) can be also an even integer or $n = 2m$, and in this case the sampled signal spectrum is inverted [21]. (3.4.9) will turn into (3.4.3) when letting

$$\frac{1}{2}(BW_P + BW_S) = f_H - f_L \quad \text{and} \quad f_c = \frac{f_H + f_L}{2}.$$

The sampling rate precision can be estimated in terms of the difference of the maximum and minimum allowed sampling rates, Δf_s, as follows [19]:

$$\Delta f_s = \frac{2(f_H - BW)}{n-1} - \frac{2f_H}{n}.$$

The relative precision required of the sampling rate is

$$\frac{\Delta f_s}{BW} = \frac{2}{n(n-1)}\left(\frac{f_H}{BW} - n\right) \approx 0\left(\frac{1}{n^2}\right). \tag{3.4.10}$$

The band-pass sampling relocates the RF band-pass signal to a low-pass position. The resulting signal-to-noise ratio is poorer than that from an analog quadrature down-converter. The signal-to-noise ratio for the sampled signal SNR_s is degraded by at least the noise aliased from the bands between DC and the RF pass-band, and it becomes

$$SNR_s = \frac{P_S}{P_{N_{in}} + (n-1)P_{N_{out}}}, \tag{3.4.11}$$

where P_S is the spectral power density of the band-pass signal, $P_{N_{in}}$ and $P_{N_{out}}$ are in-band and out-of-band noise power densities, respectively, and n is a positive integer number less than or equal to n_{max} given in (3.4.4). Since there is often a lot of gain in front of the ADC, the in-band noise power density is usually greater than the out-of-band noise power density — i.e., $P_{N_{in}} >> P_{N_{out}}$ — and thus the SNR_s is mainly determined by $P_S / P_{N_{in}}$. However, if $P_{N_{in}} \cong P_{N_{out}}$ and $n >> 1$, the degradation of the signal-to-noise ratio in dB can be approximately expressed as

$$D_{SNR} \cong 10\log(n). \tag{3.4.12}$$

The degradation of the signal-to-noise ratio in the band-pas sampling is caused by the sampling jitter and the quantization noise of the ADC as in the case of BB sampling. The estimation of the SNR_s degradation due to the sampling jitter and the quantization noise can be found in Section 2.4.2.

3.4.2. Configuration of Band-pass Sampling Radio Architecture

A block diagram of the band-pass sampling architecture radio for a full-duplex transceiver, such as a CDMA transceiver, is presented in Fig. 3.30. In this block diagram the receiver and the transmitter both use the harmonic sampling to translate the carrier frequency from an RF to a low IF or vice versa. A proper low IF is needed to avoid the aliasing, and the IF should be low enough for the DSP to handle it. Comparing the band-pass sampling radio with the others, it is apparent that the RF analog block configuration of this radio architecture is much simpler than any one of the architectures presented in the previous sections. The radio architecture based on directly sampling the RF signal is also referred to as digital direct conversion.

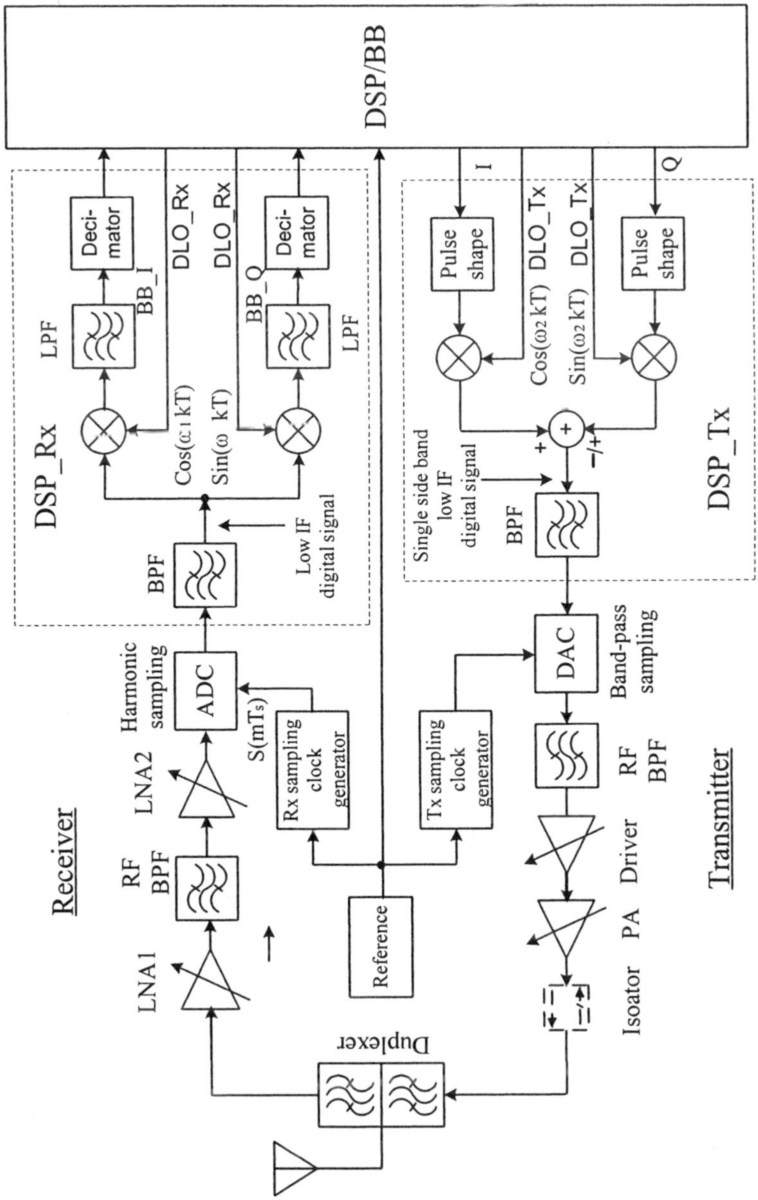

Figure 3.30. Block diagram of band-pass sampling architecture radio

The duplexer is necessary for a full-duplex transceiver, and it also plays the role of a preselector in the receiver. The RF front-end of this receiver now consists of two stages of gain adjustable LNA for achieving a low noise figure and certain gain control range and an RF band-pass SAW filter for blocking out-of-band interference and decreasing the aliased noise. The RF SAW is placed right after the first LNA in a CDMA mobile station receiver to further suppress the transmission leakage and to minimize cross-modulation of the single-tone interferer near the carrier resulting from the third-order nonlinearity of the sequential stages. From the point view of reducing the aliased noise at the output of the undersampling ADC, it may be better to locate the RF SAW in front of the ADC. However, since the order of the harmonic for undersampling a modulated RF signal is usually not high (say, second- to fifth-order), and the LNA2 gain is just around 15 dB, witching the position of the RF SAW and the LNA2 will not improve the signal-to-noise ratio of the ADC output much. But this arrangement — i.e., tow LNAs placed successively before the RF SAW will severely degrade the single tone desensitization performance of a CDMA receiver.

The filtered and amplified RF signal is then sampled at, for example, approximate one third or one fourth of the received signal-carrier frequency at the ADC, and the analog RF signal is converted into a few MHz low IF digital signal in the CDMA receiver. The low IF digital signal is equally split to two branches, and the two branch digital signals mix with a pair of quadrature digital LO signals generating base-band I and Q signals, respectively. The base-band I and Q signal are low-pass filtered for further processing. The quadrature down-conversion and the low-pass filtering can be performed in terms of DSP. In this architecture, there is no channel filter in the analog domain, and the digital low-pass filters play the channel-filtering role. This demands that the ADC must possess a high dynamic range to simultaneously deal with very strong interferers and the weak desired signal. In the CDMA case, the ADC needs at least 16 bits.

The band-pass sampling transmitter is depicted in the lower portion of Fig. 3.30. The pulse shaping of the I and Q BB channel waveforms and then filtering of the shaped waveforms to suppress the spectral level in adjacent channels are executed in the DSP the same as in the other transmitter architectures. However, in this architecture the quadrature modulation to convert the I and Q BB signals to a single side band modulated low IF signal is also carried out in the DSP. The low IF will be around a couple of or several MHz. The digital low IF signal is oversampled at a rate of one 2nd to one 5th of the transmission RF frequency, and the oversampled digital is then converted to analog signals in the high-performance digital-to-analog converter (DAC). The following RF band-pass filter selects one of the spectral replicas of the signal centered at the transmission carrier frequency and suppresses the other spectral replicas and out-of-filter band noise. The output signal from the filter is boosted to a level that is capable driving the power amplifier (PA). The desired signal finally gains enough power for transmission at the PA stage, and it then passes through the isolator and the duplexer to the antenna. The gain control range of the PA and the RFA/driver plus adjustable voltage level of the DAC output shall cover the dynamic range of the transmission output power.

In fact, the band-pass sampling or the harmonic sampling technique is used not only for directly undersampling RF signals but more often for undersampling IF signals. The technique of undersampling IF signal can be used in the superheterodyne transceiver. The block diagram of superheterodyne receiver employing the band-pass sampling technique is shown Fig. 3.31. Comparing this block diagram with the receiver part of Fig. 3.1, we can see that only one ADC is used here and it is moved forward to the IF block output — i.e., the output of the IF SAW filter. The ADC is sampled at a rate close to the IF subharmonic — i.e., the rate $f_s = (f_{IF} - f_{LIF})/n$ and $n = \lfloor f_{IF}/f_s \rfloor$. The output of the ADC contains one replica of the digital signal spectra with a low center frequency f_{LIF}. This low carrier digital signal represents the desired signal, and it is converted to BB I and Q signals by the digital quadrature down-converter. The BB I and Q signals go through low-pass filters, which function as a channel filter to suppress interferers near the desired signal. After filtering, the BB I and Q signals rout to the digital BB for further process. The quadrature down-conversion and the low-pass filtering can be implemented by means of DSP. In this receiver, the section from the ADC input to the digital LPF outputs of the I and Q channels play the same role as the block from the analog quadrature down-converter to the I and Q ADC outputs in the receiver of Fig. 3.1.

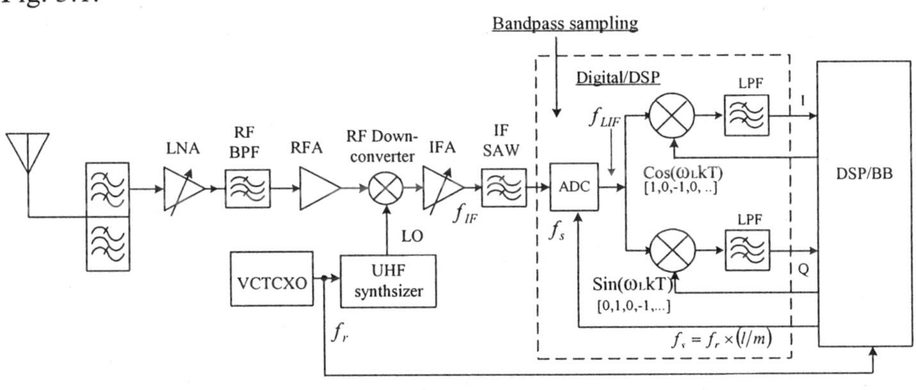

Figure 3.31. Band-pass sampling application in a superheterodyne receiver

The key components in the band-pass sampling architecture are the ADC and DAC. Based on present mixed-signal technology, the high dynamic ADC is more mature for a band-pass sampling IF signal than for a band-pass sampling RF signal. The required bandwidth for a sampling IF signal is much less than the bandwidth demanded by a digitizing RF signal. The dynamic range requirement for the ADC used in a band-pass sampling IF signal as in case of Fig. 3.31 is usually 5 to 6 bits less than that applied in the case of a band-pass sampling RF signal as in Fig. 3.30 since the IF SAW in the Fig. 3.31 receiver helps to suppress interferer level and thus to relax the dynamic requirement of the ADC.

3.4.3. Design Considerations

The performance of the band-pass sampling architecture transceiver, especially for the configuration of Fig. 3.30, depends mainly on the performance of the ADC and DAC. In other words, the RF transceiver based on band-pass sampling architecture can be applied in the wireless mobile stations only if we can obtain the high-performance ADC and DAC running at RF/IF frequency with reasonable low power consumption — less than 10 mA, for example. In the RF system design of this transceiver, we concentrate mainly on the system technical issues associated with the performance and operation conditions of the ADC/DAC. Other design considerations, which may affect the overall system performance, are also addressed.

3.4.3.1. Band-pass Sampling Rate and Residual Carrier

At present most of the wireless mobile systems operate in the frequency bands of 800 to 1000 MHz., and 1700 to 2200 MHz. The band-pass sampling rate is usually the 2nd- to the 10th-order subharmonic of the frequencies in these bands, depending on the actual system operation frequency and the ADC performance. The CDMA mobile station receiver in the cellular band, for example, operates at a channel frequency located in the frequency band between 869 MHz and 894 MHz. The band-pass sampling rate in this case will be close to 1/3 or 1/4 of the operation frequency — i.e., in the region of 289 to 294 Msamples/sec or 217 to 223 Msamples/sec. This means that the RF signal is sampled at the 3rd- or 4th-order harmonic of the sampling rate. The reason for using a low order harmonic to sample the RF signal is as follows. The theoretical maximum signal-to-noise ratio at the ADC output, SNR_{ADC}, can be estimated by using [22]

$$SNR_{ADC} = 6.02 \cdot b + 1.76 + 10\log\left(\frac{f_s}{2f_H}\right), \tag{3.4.13}$$

where b is the number of bits of the ADC, f_s is the sampling frequency or rate, and f_H is the highest frequency of the RF band-pass signal. The third term on the right side of (3.4.13) can be explained as the SNR improvement or degradation due to over or under Nyquist rate sampling. Thus the SNR degradation of the ADC operating in band-pass sampling can be estimated by means of the third term on the right side of (3.4.13). For examples, if the sampling rate is 1/3 of the highest frequency of interest, the SNR degradation is approximate 7.8 dB, and the degradation increases to 13 dB when the sampling rate drops down to 1/10 of the highest frequency of the RF signal. It is obvious that the signal-to-noise ratio of the ADC output dramatically decreases with the ratio of the highest frequency of interest to the sampling rate f_H/f_s increasing. To keep the ADC dynamic range high and the equivalent noise figure low, it is necessary to employ a low-order harmonic to sample the RF band-pass signal.

To achieve band-pass sampling without aliasing, the sampling rate f_s shall meet (3.4.9) or (3.4.3), and it is not exactly equal to the subharmonic of the RF signal carrier frequency f_c as discussed in Section 3.4.1. There is always a residual carrier with a relatively low frequency, hundreds kHz to a few MHz, in the aliasing-free frequency translation by means of the band-pass sampling. The resulting low intermediate frequency f_{LIF} is a function of the sampling rate and the RF signal carrier frequency f_c, and it can be expressed as [23]

$$f_{LIF} = \begin{cases} rem(f_c, f_s), & \text{if } \lfloor 2f_c/f_s \rfloor \text{ is even} \\ f_s - rem(f_c, f_s), & \text{if } \lfloor 2f_c/f_s \rfloor \text{ is odd,} \end{cases} \tag{3.4.14}$$

where *rem*(a, b) is the remainder after division of a by b, and $\lfloor \circ \rfloor$ denotes the largest integer. In the case of $f_{LIF} = f_s - rem(f_c, f_s)$, the signal spectrum is flipped. On the other hand, when the sampling frequency meets (3.4.9), the following two inequalities hold:

$$0 < f_{LIF} - \frac{BW_P}{2} \quad \text{and} \quad f_{LIF} + \frac{BW_P}{2} < \frac{f_s}{2}, \tag{3.4.15}$$

where BW_P is the pass-band of the channel filter, which is equal to or slightly wider than the signal information bandwidth BW_I.

When the band-pass sampling architecture is applied in the wireless mobile communication systems, it is better to have the sampling rate greater than twice of the receiver and/or the transmitter operation bandwidth B — i.e., $fs > 2B$. In the PCS band, for example, 60 MHz from 1930 MHz to 1990 MHz is allocated for mobile station receiver use, and 60 MHz from 1850 MHz to 1910 MHz for mobile station transmitter operation bandwidth. For a half-duplex band-pass sampling transceiver operating in the PCS band, the sampling rate should be greater than 120 MHz. However, for a full duplex band-pass sampling transceiver, the minimum sampling rate should be higher than twice of the separation between the receiver and the transmitter operating frequencies plus the desired signal bandwidth — i.e.,

$$f_s > 2 \cdot \left(B_a + B_s + BW_I\right), \tag{3.4.16}$$

where B_a is the receiver/transmitter operation bandwidth, and B_s is the band separation between the receiver and the transmitter operation bands as described in Section 3.1.2. For a CDMA band-pass sampling receiver operating in the PCS band, the sampling rate of the ADC should be higher than 2×(60 + 20 + 1.25) = 162.5 MSamples per second. On the other hand, if the sampling rate does not meet (3.4.16), to avoid the transmitter leakage aliasing into receiver channel bandwidth the following condition must be hold:

$$\left|f_{IF\_Tx} - f_{IF\_Rx}\right| \geq \frac{BW_{I\_Tx} + BW_{I\_Rx}}{2}. \tag{3.4.17}$$

In (3.4.17), f_{IF_Tx} and f_{IF_Rx} are the lowest spectral replica center frequencies of the sampled transmitter and receiver signals, respectively, and BW_{I_Tx} and BW_{I_Rx} are the transmission and reception signal information bandwidths, respectively.

If designing a CDMA band-pass sampling architecture receiver operating in the Cellular band (869 to 894 MHz), the sampling rate and the IF can be determined as follows. To reduce the sampling rate but not to degrade SNR_{ADC} too much, the 4th harmonic sampling is chosen — i.e., sampling rate is in the region of 217 to 224 MSample/sec. The resultant IF is determined by means of not only the CDMA signal information bandwidth BW_I = 1.23 MHz but also the specified interference tone positions at frequencies ±900 kHz and ±1.7 MHz offset from the carrier, defined in CDMA standards IS-98D. If choosing the offset frequency of the specified farthest tone, 1.7 MHz, as the resultant IF or $f_{IF\_Rx} = 1.7\,\text{MHz}$, the sampling rate for channel 400 or $f_c = 880\,\text{MHz}$ is

$$f_s = \frac{880 - 1.7}{4} = 219.575 \ \text{MSample/sec}$$

In the cellular band, B_a = 25 MHz and B_s = 20 MHz, and thus the corresponding transmitter frequency is equal to 835 MHz, and the resultant IF is $f_{IF\_Tx} = 2 \times 219.575 - 835 = 43.3\,\text{MHz}$. It is apparent that the sampling rate meets the aliasing-free criteria (3.4.9) and (3.4.16), and the resultant IFs meet (3.4.15) and (3.4.17) with plenty of margin.

Reducing the sampling rate the ADC may achieve better performance, lower current consumption, or lower cost. However, there is a practical limitation that the ADC must be still able to effectively operate on the highest frequency component of interest [22]. The ADC/DAC used in the band-pass sampling should also specify the performance at the highest frequency of the sampled signal.

3.4.3.2. ADC Noise figure and Receiver Sensitivity

Noise behavior is fundamentally poor in band-pass sampling applications because of aliasing. The aliasing noise is one of the most severe problems in the band-pass sampling architecture, especially in the case of high order harmonic sampling. In the RF, the noise behavior of a device is described by means of noise figure. The equivalent noise figure of the key device ADC in the band-pass sampling case is usually high, and it can be approximately estimated by utilizing the following expression (see Appendix 3D):

$$F_{ADC} = 1 + \left(\frac{2f_H}{f_s} - 1 \right) + \frac{1}{4kTR_S} \left(P_{Nq} + P_{Nj} \right) \left(1 + \frac{R_S}{R_L} \right)^2, \qquad (3.4.18)$$

where k = 1.38×10^{-23} J/^{O}K, T = 300 ^{O}K, R_S is the source resistance, usually 50Ω, R_L is the load impedance/ADC input impedance, P_{Nq} is quantization noise density, and P_{Nj} is the jitter noise density. For example, an ADC with a dynamic range of 90 dB or an effective 15 bits is employed in a band-pass sampling architecture CDMA receiver, and the noise figure of the ADC is estimated. The highest frequency of interest for this calculation is 894 MHz, and the sampling rate is approximately one fourth of the receiver operating frequency — i.e., in the range of 220 ± 3 MS/sec. The maximum voltage peak-to-peak swing of ADCs used in a wireless mobile station is around 1 $V_{p\text{-}p}$. The quantization noise density in this case is

$$P_{Nq} = \frac{V_{p-p}^2}{L_q^2 6 f_s}\left(\frac{2f_H}{f_s}\right) = \frac{8}{32768^2 \times 6 \times 220 \times 10^6} = 5.6 \times 10^{-18}.$$

The maximum noise density resulted from the sample time jitter P_{Nj} is assumed to be half of LSB or $P_{Nj} = (1/2^{16})^2 / (220\times10^6) = 1.1\times10^{-18}$. If R_S =50 Ω and $R_L = R_S$, from (3.4.18) the noise figure in dB can be calculated as

$$NF_{ADC} = 10\log\left(8 + \frac{4}{4\times1.38\times10^{-23}\times300}\left(5.6\times10^{-18} + 1.1\times10^{-18}\right)\right) \cong 16 \text{ dB}.$$

However, the quantization noise is no longer dominating the ADC noise behavior when the ADC operates at RF, and the noise of analog circuits in front of the quantizer of the ADC has significant contribution to the overall noise figure. In this case, the overall noise figure of the ADC operating at band-pass sampling condition can be calculated by the following formula

$$F_{ADC\_O} = F_{ADC\_a}\left(\frac{2B_{n\_a}}{f_s}\right) + \frac{F_{ADC\_q}}{g_a^2}, \tag{3.4.19}$$

where F_{ADC_a} and g_a are the noise figure and voltage gain of the analog circuits of the ADC, respectively, B_{n_a} is the noise bandwidth of the analog circuits, f_s is the sampling rate, and F_{ADC_q} is the quantizer noise figure as given in (3.4.18). If $F_{ADC\_a} = 25$, $g_a = 1$, $B_{n\_a} \cong 1000$ MHz, $f_s = 220$ MS/s, and F_{ADC_q} is the same as given in the previous example or 39.8, the overall noise figure of this ADC is

$$NF_{ADC\_O} = 10\log\left[25\left(\frac{2\times1000}{220}\right) + \frac{39.8}{1}\right] = 24.3 \text{ dB}.$$

At present, the equivalent noise figure of a high resolution ADC band-pass sampling at a rate of hundreds MS/s is possibly in the range of 20 to 30 dB, depending on the ratio of $(2f_H/f_s)$ if available.

Since the equivalent noise of the ADC band-pass sampling at hundreds MS/s is high, it becomes necessary to have an RF front-end with low noise figure and high enough gain before the ADC to achieve good receiver sensitivity. Referring to Fig. 3.30, the receiver front-end consists of the duplexer, LNA1, RF SAW, and LNA2. The corresponding blocks and the ADC are redrawn in Fig. 3.32, and the receiver sensitivity and the dynamic range can be analyzed based on this block diagram. If the noise bandwidth of the LNA2 is less than half of the sampling rate, the overall noise figure of the front-end and the ADC, F_{Rx}, can be calculated in terms of the following cascaded noise figure formula (see Chapter 4):

$$F_{Rx} = F_{FE} + \frac{F_{ADC}}{g_{FE}}, \tag{3.4.20}$$

where F_{FE} is the noise figure of the front-end, g_{FE} is the power gain of the front-end, and F_{ADC} is the equivalent noise figure of the ADC. From (3.4.20), we can see that to achieve high receiver sensitivity — i.e., low receiver noise figure — a high enough front-end power gain and a low front-end noise figure are definitely needed when the ADC noise figure is high.

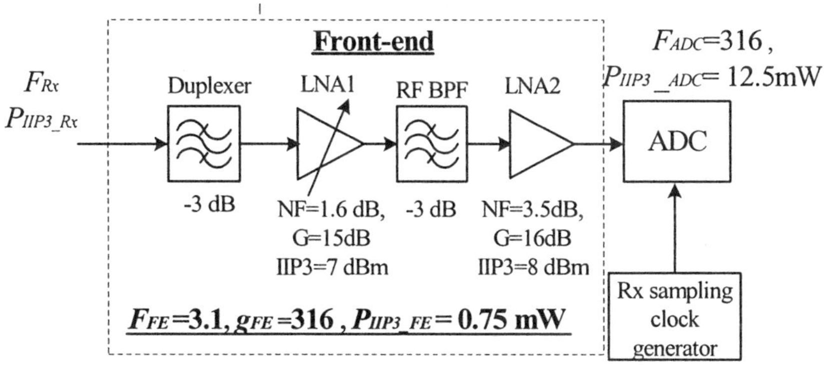

Figure 3.32. RF front-end and ADC block diagram of band-pass sampling receiver

Using the noise figure and the gain data presented in Fig. 3.32, we then obtain

$$F_{Rcvr} = 3.1 + \frac{316-1}{316} \cong 4.1 \text{ or } 6.1\,\text{dB}.$$

In the CDMA mobile station case, this noise figure approximately corresponds to a receiver sensitivity –108 dBm since the receiver noise bandwidth is 1.25 MHz, and the requested CNR for 0.5 % FER is around –1.5 dB. For an ADC with a noise figure 25 dB, the front-end should have a noise figure 5 dB and a power gain 25 dB to obtain a receiver sensitivity having 4 dB margin in the CDMA mobile station receiver.

In general, the gain of the RF front-end of the band-pass sampling architecture receiver is higher than those of the other architecture receivers for controlling the overall noise figure and developing a high enough signal level to the input of the ADC. Depending on possible interference level, system linearity requirement, power supply voltage, and allowed current consumption, the RF front-end overall gain may be up to 30 dB or so.

It should be noted that we need to use (3.4.19) to calculate LNA2 aliasing noise to the ADC output noise contribution if the LNA2 noise bandwidth is broader than half of the sampling rate. In this case we should calculate the cascaded noise figure of the LNA2 and the ADC first by means of (3.4.19) and then calculate the overall noise figure of the rest blocks of the front-end and the LNA2+ADC block.

3.4.3.3. Dynamic Range and Linearity

The dynamic range of the ADC used in the band-pass sampling receiver depends on the interference level that the mobile station copes with and the sensitivity of the mobile station receiver. Assuming that the receiver works properly under the attack of an interferer with strength I in dBm when the received desired signal level is S_d in dBm, the minimum dynamic range DR_{ADC_min} shall be equal to or greater than

$$DR_{ADC\_\min} = I - S_d + CNR + \Delta G_{LNA} + PAR_r + \Delta D_F \, , \qquad (3.4.21)$$

where CNR is the carrier-to-noise ratio at the given desired signal S_d, ΔG_{LNA} is the possible LNA gain variation, PAR_r is the peak-to-average ratio of the received signal, and ΔD_F is the possible magnitude variation caused by constructive fading.

In the minimum performance specification of CDMA mobile stations, for example, the receiver under the attack of a –30 dBm interference tone attack must still maintain a frame error rate (FER) of less than 1% when the desired CDMA signal is at –101 dBm. Considering 3 dB of margin for the interference attack, we have $I = -27$ dBm instead of –30 dBm. The CNR for the CDMA receiver is around 5 dB at S_d = -101 dBm.

For a noise-like signal with a normal distribution, the PAR_r is approximately 6 dB if considering two times standard deviation. Assuming that ΔG_{LNA} = 3dB and ΔD_F = 4 dB, from (3.4.20) the minimum effective dynamic range of the ADC should be

$$DR_{ADC\_\min} = -27 + 101 + 5 + 3 + 6 + 4 = 92 \text{ dB}.$$

A 15 bit ADC is usually to have $6.02 \times 15 + 1.76 = 92.1$ dB dynamic range when the quantization noise dominates the ADC output noise. It may need 16 bits or more to obtain an effective 92 dB dynamic range when operating at RF the analog circuitry noise of the Δ-Σ ADC is usually much high than the quantization noise and dominates output noise of the ADC.

The linearity of the band-pass sampling receiver similar to the other architecture receivers can be characterized based on the third-order input intercept point (IIP_3). Either in the superheterodyne receiver or in the direct conversion receiver, there are channel filters present in front of the ADC, but now in the band-pass receiver the ADC is directly connected to the RF front-end without any channel filter before it. The linearity or the IIP_3 requirement of the ADC used in the RF band-pass sampling architecture must be much higher that applied in the other architectures. Referring to Fig. 3.32, the overall third-order input intercept point IIP_{3_Rx} is determined by (see Chapter 4)

$$P_{IIP3\_Rx} = \left(\frac{1}{P_{IIP3\_FE}} + \frac{g_{FE}}{P_{IIP3\_ADC}} \right)^{-1} \tag{3.4.22}$$

In this formula, IIP_{3_FE} and IIP_{3_ADC} are the third-order input intercept points of the front-end and the ADC, respectively.

Still using the data given in Fig. 3.32 for a CDMA receiver as an example, to meet the intermodulation spurious response attenuation specification — i.e., the FER of the cellular band CDMA receiver shall be less than 1% when two equal power interference tones with a power at least –43 dBm and frequency offset 0.9 MHz and 1.7 MHz, respectively, attack the receiver, and the desired signal is at –101 dBm. In this case, if IIP_{3_FE} = 0.75 mW or –1.25 dBm and g_{FE} = 316, the minimum IIP_{3_ADC} for just meeting the specification –43 dBm is approximately 12.5 mW or 11 dBm, and from (3.4.22) the overall input intercept point IIP_{3_Rx} is

$$P_{IIP3\_Rx} = \left(\frac{1}{0.75} + \frac{316}{12.5} \right)^{-1} = 0.038 \text{ mW or -14.3 dBm}.$$

However, the IIP_3 of the ADC even operating at RF or high IF is probably not difficult to 14 to 15 dBm. Thus the overall IIP_3 or IIP_{3_Rx} will become −11.4 to −10.5 dBm, and the corresponding margin of the intermodulation spurious response attenuation specification will be approximately 2 to 2.5 dB.

Finally, in the CDMA mobile station receiver there is a specification referred to as single-tone desensitization, which is caused by cross-modulation of the AM transmission leakage to an interference tone close to the desired CDMA signal, and it mainly depends on the linearity of the front-end section before the RF BPF if this filter has high enough rejection — say, more than 15 dB, to the transmission leakage. This is why the RF BPF is placed in between LNA1 and LNA2 as depicted in Fig. 3.30 and Fig. 3.32. From the point of view of reducing overall noise figure, in the band-pass sampling architecture receiver the RF BPF is better to be directly allocated just before the ADC especially when the noise bandwidth of the LNA2 is much broader than the half of the sampling frequency. In this later placement, the cascaded IIP_3 of the two LNAs measured with one tone in the receiver, and the second tone in the transmitter band will be too low to cope with the single-tone desensitization issue.

3.4.3.4. AGC in a Band-pass Sampling Receiver

The AGC system in the band-pass sampling architecture receiver is simpler than the one in the other architecture receivers. The receiver dynamic range can be covered through combining the LNA automatic gain control and the high dynamic ADC. The LNA gain control range depends on the ADC dynamic range. In the above CDMA receiver example, the ADC has a 90 dB effective dynamic range. The maximum reception signal power in the CDMA mobile station may go up to –20 dBm and the sensitivity level is around –108 dBm. Thus the CDMA desired signal variation range is 88 dB. The 90 dB ADC is enough to cover the desired signal dynamic range, but the LNA gain still needs to change to meet a certain linearity requirement when the reception signal power is getting high.

For example, according to the CDMA mobile station minimum performance requirements IS-98D [3], the intermodulation interference tone level is raised to –32 dBm and –21 dBm from –43 dBm, respectively, when the received signal increases to –90 dBm and –79 dBm from the level –101 dBm. To handle higher interference tones, the linearity or the IIP_3 of the receiver needs to be higher. This can be implemented through reducing the LNA1 gain and increasing its IIP_3. In the above example, the LNA1 gain reduces in two steps 12.5 dB each from 15 dB. The corresponding IIP_3 and noise figure in each gain step are 10/17 dBm and 8/15dB as presented in Table 3.6. We may have more than 3 dB margins on the intermodulation spurious response attenuation in the middle and low LNA1 gain modes. On the other hand, if the LNA1 gain IIP_3 and noise figure are fixed, the receiver will fail to meet the intermodulation spurious response attenuation specification when the intermodulation interference tones are raised with the desired reception signal. The LNA1 gain is controlled based on the received in-channel bandwidth signal. One control scheme for this example can be as shown in Fig. 3.33. The LNA1 gain is switched from 15 dB to 2.5 dB and from 2.5 dB to –10 dB at received signal level –93 dBm and –82 dBm, respectively, when the received signal increases. The LNA1 gain is switched back from low to mid gain and from mid to high gain at –85 dBm and –96 dBm, respectively, while the received signal strength decreases. The gain control scheme will work properly only if the receiver sensitivities are higher than –96 dBm and –85 dBm in the mid and low gain modes.

Table 3.6 LNA1 gain control and corresponding parameter set-up

LNA1	High Gain	Mid Gain	Low Gain
Gain (dB)	15	2.5	-10
IIP_3 (dBm)	7	10	17
Noise Figure (dB)	1.6	8	15

The overall dynamic range of the receiver DR_{Rx} in dB is the sum of the AGC range of the LNA ΔG_{LNA} and the ADC effective dynamic range D_{ADC}, and this can be expressed as

$$DR_{Rx} = DR_{ADC} + \Delta G_{LNA} \tag{3.4.23}$$

The total dynamic range for the receiver described in the above example is then equal to 25 + 90 = 115 dB.

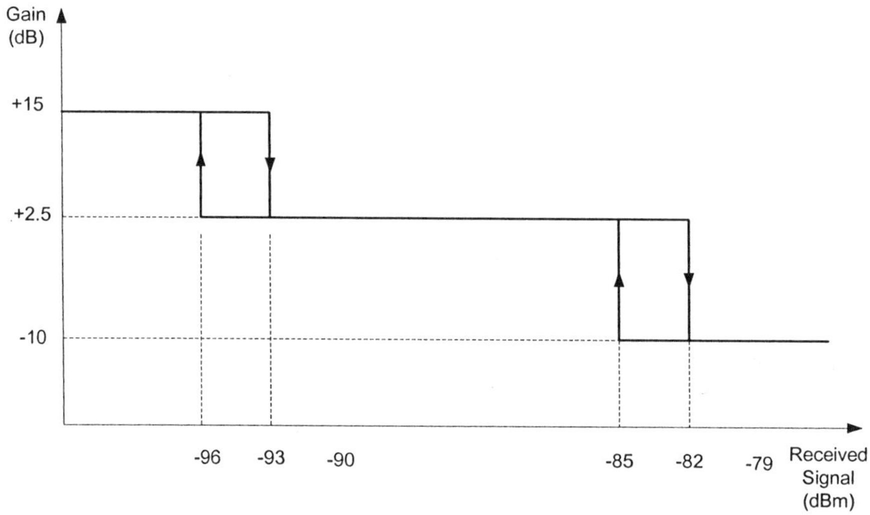

Figure 3.33. Front-end LNA gain control scheme example

3.4.3.5. Band-pass Sampling Transmitter

In the band-pass sampling transmitter, a signal process, which is reversed from that in the receiver, is employed. After pulse shaping, the digital BB I and Q signals are first up-converted, and then they are summed into a single side-band low IF digital signal as depicted in Fig. 3.34. The up-converted Q channel digital signal in normal case will subtract from the up-converted I channel signal to form the single side-band low IF digital signal. It is also possible that the low IF digital signal is constructed from the addition of the I and Q channel signals. After band-pass filtering, the low IF digital signal in a high performance DAC is converted into an analog signal at a sampling rate of

$$f_{s\_Tx} = \frac{f_{c\_Tx} \mp f_{LIF\_Tx}}{n}. \tag{3.4.24}$$

In (3.4.24) f_{s_Tx}, f_{c_Tx}, and f_{LIF_Tx} are the transmitter sampling rate, transmission signal center frequency or carrier frequency, and the low IF used in transmitter, respectively, and n is an integer number, which is the harmonic order of the band-pass sampling rate, and it is usually the same as the harmonic number used in the receiver but not necessary. The two signs in the numerator of (3.4.24) right side are used for the case of the low IF signal with normal spectrum, and for the case of the low IF signal having a flipped spectrum, which is formed from the addition of the I and Q channel signal, respectively.

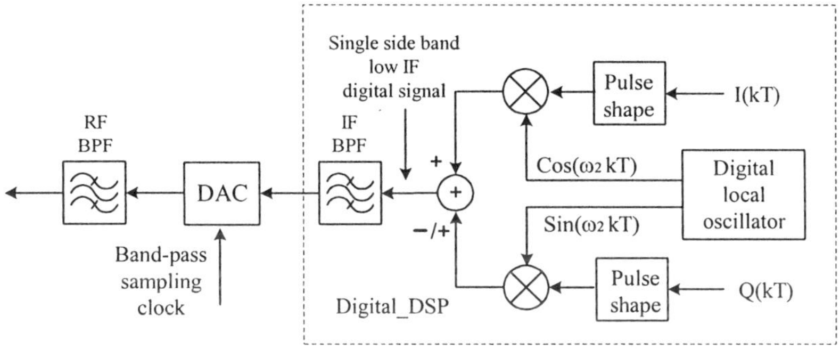

Figure 3.34. Digital quadrature modulator and band-pass sampling up-converter blocks in the band-pass sampling transmitter

The DAC in this transmitter shall be able to operate at sampling rate of hundreds MS/s, and with a full power analog input bandwidth up to the highest frequency of the transmitter frequency band. The basic concepts of the band-pass sampling and the derived results from the discussions with respect to ADC as described in Section 3.4.1 can also be used here for the band-pass sampling DAC. To obtain high signal-to-noise ratio output, the DCA in this application needs 14 to 16 bits or even higher depending on the performance of the DAC.

The analog output of the DAC contains the desired band-pass signal spectra periodically. The RF band-pass filter following the DAC selects the spectrum with the right carrier frequency and deeply suppresses all the unwanted spectra to make out of transmission band emissions within defined specifications. The bandwidth of this RF BPF covers the whole mobile station transmission band, such as 25 MHz for mobile transmitters operating in the cellular band and 60 MHz for those transmitters operating in the PCS band. However, the close-in emission level is controlled mainly by the frequency response characteristics of the digital BPF centered at the low IF. The bandwidth of this filter is approximately equal to the channel bandwidth of the mobile transmitters.

The rest of this transmitter including the driver amplifier and the PA are similar to that in the other architecture transmitters. It will not further discuss here.

3.4.3.6. Band-pass Sampling for the Superheterodyne Receiver

The band-pass sampling architecture can also be applied in the superheterodyne receiver as depicted in Fig. 3.31. In this application, a modulated IF signal in the ADC is undersampled and converted to a digital signal with a low frequency carrier. The sampling rate is a fraction of the IF. However, the following facts need to be carefully considered when determining the sampling rate of the ADC used in this architecture receiver. The wireless mobile station receivers are supposed to operate over a frequency band instead of a few fixed frequencies. It is not desirable that the selected sampling rate will make another channel signal in-receiver-band to potentially become an in-channel-bandwidth interference since the channel filter, a SAW filter, has a finite rejection, 35 to 40 dB, to the other channel signals. Considering this, it is better to choose the sampling rate greater than the span of the mobile receiver operation band, such as 25 MHz for the Cellular band systems and 60 MHz for the PCS band systems. For full-duplex mobile transceivers, the transmission leakage to the receiver side is quite high, around –30 to –25 dBm, although the duplexer has suppressed most of the transmission leakage power, and it is necessary to avoid that the transmission leakage signal through the band-pass sampling is converted into an in-channel-bandwidth interferer of the desired signal. The most safe way to avoid this happening is to choose a band-pass sampling rate greater than the difference between the receiver and the transmitter operating frequencies, such as 45 MHz for the cellular band transceivers and 80 MHz for the PCS band transceivers, but this may not be necessary.

The dynamic range of the ADC in this architecture receiver is much lower than that used in RF band-pass sampling receiver since the ADC is preceded by a channel filter with a 35 to 40 dB rejection to close-in interferers. In this case an 8 to 10 bit ADC is probably enough for the most applications. However, a higher dynamic range — say, 12 to 14 bits — ADC is definitely helpful for reducing the AGC range.

The advantages of using this architecture receiver are as follows. The analog circuits reduce to the RF LNA, RF down-converter, and possiblly one-stage IF amplifier, and these circuits can be built in a single chip integrated circuit. The I and Q channel magnitude and phase balances will be much better than those in the conventional superheterodyne receiver since the I and Q channels are formed in the digital domain. The digital LPF usually has much less group delay distortion than the corresponding analog LPF with the same out-of-band rejection performance. In addition, the AGC range can decrease with the ADC dynamic range increasing.

Appendix 3A. Intermodulation Distortion Formulas

In the weak and memoryless nonlinear case, we can use power series to model the nonlinearity of a device such as an amplifier or a mixer. If the input signal and output signal power of the device is P_i and P_o, respectively, then the output of the nonlinear device can be represented by its input as follows:

$$P_o = \sum_{m=1}^{n} P_m = \sum_{m=1}^{n} g_m P_i^m , \tag{3A.1}$$

where g_m ($m = 1, 2, \ldots, n$) is the power gain when m =1, and the nonlinear gain coefficients when $m \neq 1$, and P_m is the m-th order distortion power of the total output power P_o, and it is related to the input signal power P_i as

$$P_m = g_m P_i^m \qquad (m = 1,2,3,\cdots n) \tag{3A.2}$$

The power P_i, P_o, and P_m (m = 1, 2, …, n) all are in natural scale. In the output terms of (3A.2), let us to look at two special cases — i.e., the fundamental term and the mth order term as follows:

$$P_1 = g_1 P_i \tag{3A.3}$$

and

$$P_m = g_m P_i^m . \tag{3A.4}$$

In fact, (3A.3) describes the linear relationship between the output and input, and (3A.4) is a generic high-order distortion representative. Converting (3A.3) and (3A.4) into dB scale, we obtain

$$S_1 = G_1 + S_i \tag{3A.5}$$

and

$$S_m = G_m + mS_i , \tag{3A.6}$$

where

$$S_1 = 10\log P_1 , \quad G_1 = 10\log(g_1) , \quad S_i = 10\log P_i$$

$$S_m = 10\log P_m , \quad G_m = 10\log(g_m) , \text{ and } \quad mS_i = 10\log P_i^m .$$

Equation (3A.5) and (3A.6) represent two straight lines in the output power vs. input power plan as shown in Fig. 3A.1.

From Fig. 3A.1 and using (3A.5) and (3A.6), we are able to obtain two expressions of the output power at the intercept point, OIP_m, when the input power equals IIP_m. From (3A.5) or the fundamental line in Fig. 3A.1, we have

$$OIP_m = G_1 + IIP_m \tag{3A.7}$$

from (3A.6) or the mth order line, we obtain

$$OIP_m = G_m + mIIP_m \tag{3A.8}$$

Using (3A. 7), (3A.8), (3A.3), and (3A.4), we derive the following equation:

$$(m-1)(IIP_m - S_i) = S_1 - S_m . \tag{3A.9}$$

Now we use some notations that we are more familiar with in our RF system design. Assuming that the input is an interference, i.e. — $S_i = I$ — and the mth order distortion S_m is represented by the equivalent level at the device input IMD_m and plus the linear gain of the device G_1 or $S_m = IMDm + G_1$, then after considering $S_1 = I + G_1$ we finally obtain the generic IMD formula (3A.10) from (3A.9):

$$(m-1)(IIP_m - I) = I - IMD_m$$

or

$$IMD_m = mI - (m-1)IIP_m \tag{3A.10a}$$

or

$$IIP_m = \frac{mI - IMD_m}{m-1} . \tag{3A.10b}$$

For examples, for $m = 2$ and 3, we have

$$IIP_2 = 2I - IMD_2 \tag{3A.11}$$

and

$$IIP_3 = \frac{1}{2}\left(3I - IMD_3\right). \tag{3A.12}$$

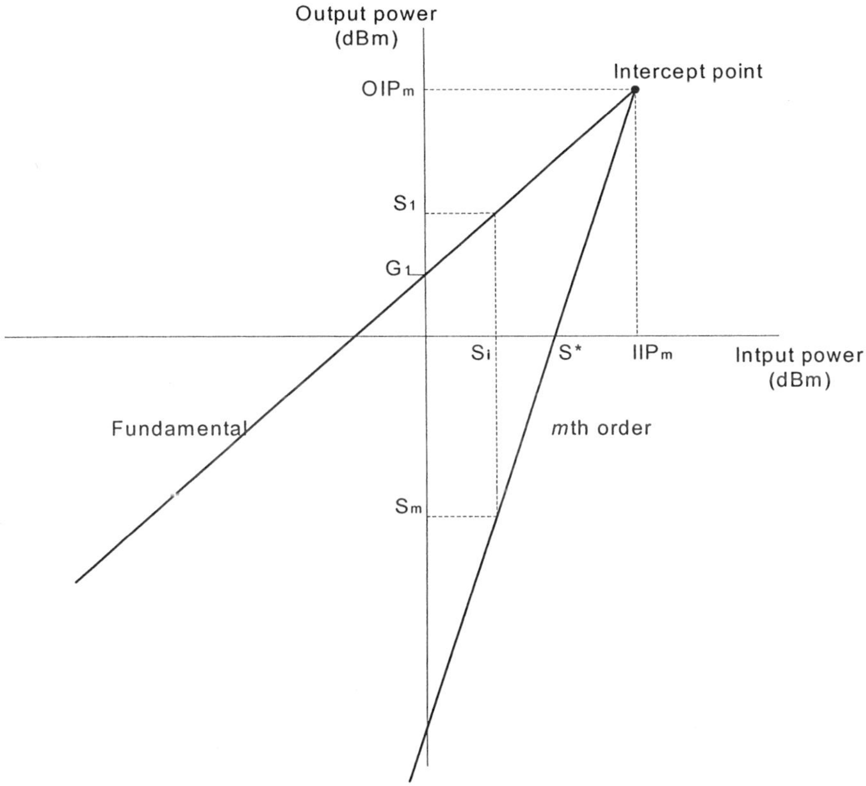

Figure 3A.1. Fundamental and *m*th order intermodulation distortion response

In general, (3A.10) is not waveform dependent but it is power level dependent since it is derived from signal average power expression instead of signal voltage under weak nonlinear and memoryless assumption. It is an approximate result, but it is accurate enough to handle the receiver interference level issues.

Appendix 3B. Effective Interference Evaluation of Second-Order Distortion Products

The second-order distortion products of the AM transmission leakage are not all in the desired signal bandwidth as depicted in Fig. 3B.1. The second-order distortion IMD_{2_Tx} that we concern mainly consists of a

DC product IMD_{2_DC} and a low frequency product IMD_{2_LF} excluding all high-frequency distortion products. IMD_{2_Tx} is expressed by IMD_{2_DC} and IMD_{2_LF} as

$$IMD_{2\_Tx} = 10\log\left(10^{\frac{IMD_{2\_DC}}{10}} + 10^{\frac{IMD_{2\_LF}}{}}\right). \tag{3B.1}$$

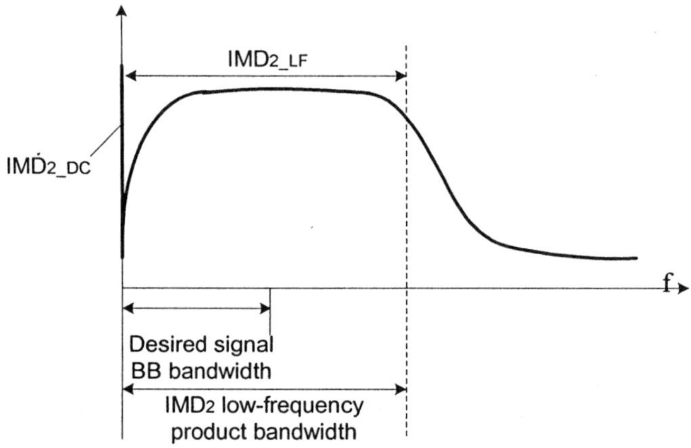

Figure 3B.1. Second-order distortion spectra of CDMA transmission leakage

To evaluate how much the second-order distortion product will really impact the CNR. The power ratios of the low-frequency distortion product to the overall IMD_{2_Tx} need be estimated. This estimation begins with the *probability density function* (PDF), $p(x)$ of the AM transmission waveform. The average magnitude square x_a^2 of the transmission waveform is

$$x_a^2 = \int_0^\infty x^2 p(x)dx. \tag{3B.2}$$

Normalizing the waveform magnitude square by the average magnitude x_a^2, we obtain $\overline{x^2} = x^2 / x_a^2$ or express it in dB

$$\overline{X} = 10\log\overline{x^2} = 10\log\frac{x^2}{x_a^2}. \tag{3B.3}$$

The average of the normalized magnitude square is corresponding to the normalized DC product $\overline{IMD_{2\_DC}}$:

$$\overline{IMD_{2\_DC}} = 10\log\left[\int_{-\infty}^{\infty}\left(10^{\frac{\overline{X}}{10}}\right)^2 p\left(\overline{X}\right)d\overline{X}\right]. \tag{3B.4}$$

The normalized low-frequency product of the second-order distortion $\overline{IMD_{2\_LF}}$ is

$$\overline{IMD_{2\_LF}} = 10\log\left\{\int_{-\infty}^{\infty}\left[\left(10^{\frac{\overline{X}}{10}}\right)^2 - 10^{\frac{\overline{IMD_{2\_DC}}}{10}}\right]^2 p\left(\overline{X}\right)d\overline{X}\right\}. \tag{3B.5}$$

Therefore, the normalized $\overline{IMD_{2\_Tx}}$ is expressed in

$$\overline{IMD_{2\_Tx}} = 10\log\left(10^{\frac{\overline{IMD_{2\_DC}}}{10}} + 10^{\frac{\overline{IMD_{2\_LF}}}{10}}\right). \tag{3B.6}$$

The power ratio of the low-frequency product IMD_{2_LF} to the second order product IMD_{2_Tx} can be expressed as

$$PR_{IMD_{2\_LF}/IMD_{2\_Tx}} = 10\log\left(\frac{10^{\frac{\overline{IMD_{2\_LF}}}{10}}}{10^{\frac{\overline{IMD_{2\_DC}}}{10}} + 10^{\frac{\overline{IMD_{2\_LF}}}{10}}}\right). \tag{3B.7}$$

As we have seen in Fig. 3B.1, only a portion of the second-order distortion low-frequency product spectrum is within the desired signal BB bandwidth. This portion is approximately half of the overall IMD_{2_LF} spectrum. Thus the power ratio of the effective interference portion of the second-order distortion, IMD_{2_effect}, to the IMD_{2_Tx}, $PR_{IMD_{2\_effect}/IMD_{2\_Tx}}$, can be easily calculated as

$$PR_{IMD_{2\_effect} / IMD_{2\_Tx}} \cong PR_{IMD_{2\_LF} / IMD_{2\_Tx}} - 3. \tag{3B.8}$$

Finally, the effective interference portion of the second-order distortion product IMD_{2_effect} is expressed as

$$IMD_{2\_effect} = IMD_{2\_Tx} + PR_{IMD_{2\_effect} / IMD_{2\_Tx}}. \tag{3B.9}$$

This expression means that in the case of using DC notch the effective interference IMD_{2_effect} is $|PR_{IMD_{2\_effect} / IMD_{2\_Tx}}|$ dB lower than the overall second order distortion IMD_{2_Tx}. (3B.9) can also be translated into that for a given *CNR* degradation the allowed maximum $IMD_{2\_\max\_allowed}$ in (3.2.17) is able to increase $|PR_{IMD_{2\_effect} / IMD_{2\_Tx}}|$ dB, and this is equivalent to relaxing the IIP_2 requirement of the down-converter $|PR_{IMD_{2\_effect} / IMD_{2\_Tx}}|$ dB.

Appendix 3C. I and Q Imbalance and Image-Rejection Formula

The I and Q signal imbalances limit the image rejection and the dynamic range of a receiver or a transmitter. The image-rejection formula (3.3.4) can derived based a simplified quadrature conversion system model as shown in Fig. 3C.1. The received RF desired signal has a frequency ω_{RF} an angle modulation $\varphi(t)$ and an amplitude of normalized to one. Assuming that all the imbalances in the I and Q channels are concentrated on the imbalances of the quadrature LO signals, the amplitude normalized I and Q LO signals are expressed as $\cos(\omega_{LO} t)$ and $(1+\delta)\sin(\omega_{RF} t + \varepsilon)$, respectively, where δ is the amplitude imbalance and ε is the phase imbalance.

If the gain of the converter is unity, the *I* and *Q* signals at the down-converter outputs have expressions as

$$\begin{aligned} I &= \cos[\omega_{RF} t + \varphi(t)] \cdot \cos(\omega_{LO} t) \\ &= \frac{1}{4}\left(e^{j[(\omega_{RF}+\omega_{LO})t+\varphi(t)]} + e^{-j[(\omega_{RF}+\omega_{LO})t+\varphi(t)]} + e^{j[\omega_{IF} t+\varphi(t)]} + e^{-j[\omega_{IF} t+\varphi(t)]}\right) \end{aligned}$$

and

$$Q = \cos[\omega_{RF}t + \varphi(t)] \cdot (1+\delta)\sin(\omega_{LO}t + \varepsilon)$$
$$= \frac{(1+\delta)}{4 \cdot j}\left\{e^{j[(\omega_{RF}+\omega_{LO})t+\varphi(t)+\varepsilon]} - e^{-j[(\omega_{RF}+\omega_{LO})t+\varphi(t)+\varepsilon]} - e^{j[\omega_{IF}t+\varphi(t)-\varepsilon]} + e^{-j[\omega_{IF}t+\varphi(t)-\varepsilon]}\right\},$$

where ω_{IF} is the low IF frequency.

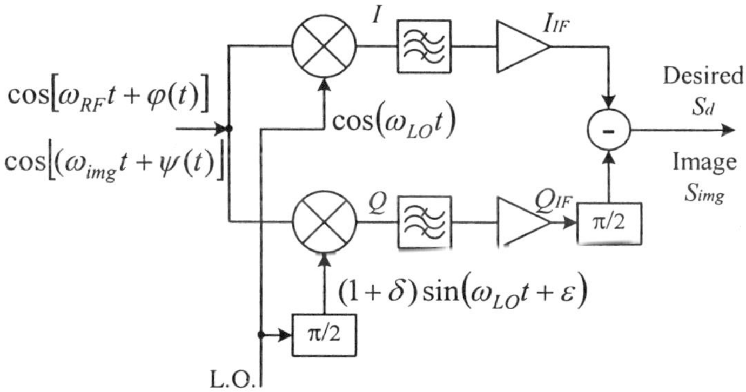

Figure 3C.1. Simplified quadrature conversion system model

After low-pass filtering, if not considering the channel gain or loss, we obtain two low IF signals having forms,

$$I_{IF} = \frac{1}{4}\left(e^{j[\omega_{IF}t+\varphi(t)]} + e^{-j[\omega_{IF}t+\varphi(t)]}\right) = \frac{1}{2}\cos[\omega_{IF}t + \varphi(t)] \tag{3C.1}$$

and

$$Q_{IF} = \frac{-(1+\delta)}{4 \cdot j}\left(e^{j[\omega_{IF}t+\varphi(t)-\varepsilon]} - e^{-j[\omega_{IF}t+\varphi(t)-\varepsilon]}\right) = \frac{-(1+\delta)}{2}\sin[\omega_{IF}t + \varphi(t) - \varepsilon]. \tag{3C.2}$$

The desired low IF signal S_d can be obtained from the I_{IF} signal subtracting 90° phase shifted Q_{IF} signal. It is expressed as

$$\begin{aligned} S_d &= \frac{1}{2}\{\cos[\omega_{IF}t + \varphi(t)] + (1+\delta)\cos[\omega_{IF}t + \varphi(t) - \varepsilon]\} \\ &= \frac{1}{2}\sqrt{1 + 2(1+\delta)\cos\varepsilon + (1+\delta)^2}\,\cos[\omega_{IF}t + \varphi(t) - \theta], \end{aligned} \tag{3C.3}$$

where

$$\theta = \tan^{-1}\frac{(1+\delta)\sin\varepsilon}{1+(1+\delta)\cos\varepsilon}. \tag{3C.4}$$

For the image signal it possesses a frequency $\omega_{img} = \omega_{RF} - 2\omega_{IF}$, an angle modulation $\psi(t)$, and an amplitude of normalized to unity. In this case the low IF I and Q signals become

$$I_{IF} = \frac{1}{4}\left(e^{j[\omega_{IF}t-\psi(t)]} + e^{-j[\omega_{IF}t-\psi(t)]}\right) = \frac{1}{2}\cos[\omega_{IF}t - \psi(t)] \tag{3C.5}$$

and

$$Q_{IF} = \frac{(1+\delta)}{4\cdot j}\left(e^{j[\omega_{IF}t-\psi(t)+\varepsilon]} - e^{-j[\omega_{IF}t-\psi(t)+\varepsilon]}\right) = \frac{(1+\delta)}{2}\sin[\omega_{IF}t - \psi(t)+\varepsilon]. \tag{3C.6}$$

In a similar way as the desired signal, the low IF image S_{im} can result from the I_{IF} signal subtracting 90° phase shifted Q_{IF} signal. It has the form

$$\begin{aligned} S_{im} &= \frac{1}{2}\{\cos[\omega_{IF}t - \psi(t)] - (1+\delta)\cos[\omega_{IF}t - \psi(t)+\varepsilon]\} \\ &= \frac{1}{2}\sqrt{1-2(1+\delta)\cos\varepsilon + (1+\delta)^2}\cos[\omega_{IF}t - \psi(t)+\varepsilon - \theta_{im}], \end{aligned} \tag{3C.7}$$

where

$$\theta_{im} = \tan^{-1}\frac{(1+\delta)\sin\varepsilon}{1-(1+\delta)\cos\varepsilon}. \tag{3C.8}$$

The image rejection IR is defined as the ratio of the desired signal amplitude to the image amplitude, and the ratio is expressed in dB as

$$IR = 10\log\frac{1+2(1+\delta)\cos\varepsilon + (1+\delta)^2}{1-2(1+\delta)\cos\varepsilon + (1+\delta)^2}. \tag{3C.9}$$

Appendix 3D. Estimation of ADC Equivalent Noise Figure

For an ADC operating in BB oversampling at a rate f_s, the output noise is dominated by its quantization noise. As discussed in Chapter 2, the quantization noise is represented by

$$\sigma_q^2 = \frac{\Delta q^2}{12} = \frac{V_{p-p}^2}{12L_q^2},$$

where Δq is quantization step or quantile interval (see Section 2.4.2), $V_{p\text{-}p}$ is the peak-to-peak voltage swing in volt, L_q is the number of quantization levels defined in (2.4.24). Assuming that the quantization noise uniformly distributes over frequency 0 to f_s /2, the quantization noise density P_{Nq} is

$$P_{Nq} = \frac{V_{p-p}^2}{6L_q^2 f_s}. \tag{3D.1}$$

However, in the case of band-pass sampling a signal with the highest frequency of interest f_H, the resultant quantization noise density increases approximately $(2f_H / f_s)$ times due to aliasing , and it turns into

$$P_{Nq} = \frac{V_{p-p}^2}{6L_q^2 f_s}\left(\frac{2f_H}{f_s}\right). \tag{3D.2}$$

When the sampling frequency is in hundreds MS/s, the noise resulted from the sampling clock jitter has significant contribution to the overall ADC output noise. The jitter noise density P_{Nj} can be expressed as

$$P_{Nj} = P_S \cdot \sigma_a^2. \tag{3D.3}$$

In this equation, P_S is the sampled signal power, and it is equal to $A^2/2$ for a sinusoidal signal with an amplitude A volts across an unit resistor, σ_a^2 is the variance of angular aperture error, and the sinusoidal signal with a frequency f_c has a form

$$\sigma_a^2 = (2\pi f_c \sigma_j)^2 \tag{3D.4}$$

where σ_j is the standard deviation of the sampling clock time jitter in second.

The equivalent noise figure is defined as the ratio of the maximum available signal-to-noise ratio from the source to the maximum available signal-to-noise ratio from the output of the ADC [21]. If the source voltage is e_S and the source resistance is R_S, the maximum available signal-to-noise ratio from the source in a 1 Hz is

$$SNR_{in} = \frac{e_S^2}{4kTR_S}, \tag{3D.5}$$

where k = Boltzman constant, 1.38×10^{-23} $J/^{o}K$, and T is absolute temperature, 300°K. The ADC output signal-to-noise ratio in a 1 Hz is given by

$$SNR_{out} = \frac{e_S^2 (R_L/(R_S + R_L))^2}{4kTR_s (R_L/(R_S + R_L))^2 + (P_{Nq} + P_{Nj})}, \tag{3D.6}$$

where R_L is load resistance, P_{Nq} and P_{Nj} are given by (3D.1) and (3D.3), respectively.

From (3D.5) and (3D.6), the ADC noise figure can be expressed as

$$F_{ADC} = \frac{SNR_{in}}{SNR_{out}} = 1 + \frac{1}{4KTR_s}(P_{Nq} + P_{Nj})\left(1 + \frac{R_s}{R_L}\right)^2. \tag{3D.7}$$

For the case of $R_L = R_S$, the ADC noise figure expression turns into

$$F_{ADC} = 1 + \frac{1}{KTR_s}(P_{Nq} + P_{Nj}). \tag{3D.8}$$

In the band-pass sampling case, the ADC output thermal noise can be $(2 f_H / f_s)$ times higher than that in the oversampling case, and the noise figure has a form

$$F_{ADC} = \frac{2f_H}{f_s} + \frac{1}{4KTR_s}(P_{Nq} + P_{Nj})\left(1 + \frac{R_s}{R_L}\right)^2. \tag{3D.9}$$

In (3D.9), P_{Nq} is represented by (3D.2).

When the ADC operates at high IF or RF, the noise of the analog circuits in the ADC may dominate the overall noise behavior of the ADC. Assuming that the ADC is imaginarily split into analog circuitry and digitization two portions as shown in Fig. 3D.1, the resultant noise figure is expressed as

$$F_{ADC\_o} = F_{ADC\_a}\left(\frac{2B_{n\_a}}{f_s}\right) + \frac{F_{ADC\_q}}{g_a^2}, \tag{3D.10}$$

where F_{ADC_a} and g_a are the noise figure and voltage gain of the analog circuitry of the ADC, respectively, B_{n_a} is the noise bandwidth of the analog circuitry, f_s is the sampling rate, and F_{ADC_q} is the noise figure of the digitization portion of the ADC, which is given in (3D.9).

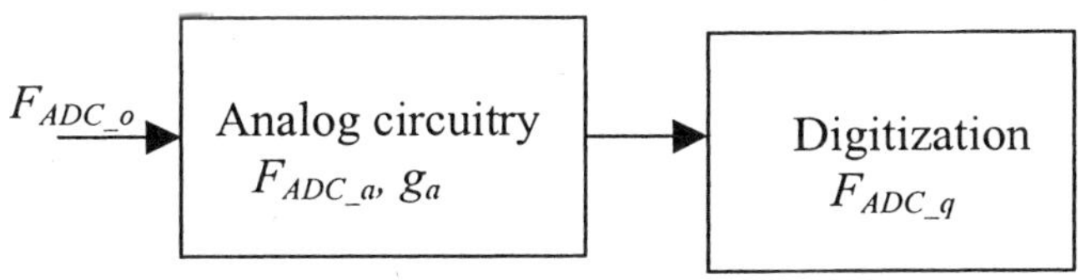

Figure 3D.1. ADC block diagram for its noise figure calculation

References

[1]. U. L. Rohde, J. Whitaker, and T.T.N. Bucher, *Communications Receivers*, 2nd ed., McGraw Hill, 1996.

[2]. ETSI, *Digital Cellular Telecommunications System (Phase 2+): Radio Transmission and Reception* (GSM 05.05 version 7.3.0), 1998.

[3]. TIA/EIA-98-D, *Recommended Minimum Performance Standards for cdma2000 Spread spectrum Mobile Stations*, Release A, March, 2001.

[4]. M. E. Van Valkenburg et al., *Reference Data for Engineers: Radio, Electronics, Computer, and Communications*, 8th ed., SAMS, Prince Hall Computer Publishing, 1993.

[5]. K. Murota and K. Hirade, "GMSK Modulation for Digital Mobile Radio Telephone," *IEEE Trans. Communications*, vol. COM-29, no.7, pp. 1044–1050, July 1981.

[6]. T. Yamawaki et al., "A 2.7-V GSM RF Transceiver IC," *IEEE J. Solid-State Circuits*, vol. 32, no. 12, pp. 2089–2096, Dec. 1997.

[7]. A. Bateman and D. M. Haine, "Direct Conversion Transceiver Design for Compact Low-cost Portable Radio Terminals," *Proceedings of IEEE, Vehicular Technology Conf.*, pp. 57–62 , May 1989.

[8]. A. A. Abidi, "Direct-conversion Radio Transceiver for Digital Communications," *IEEE J. Solid-State Circuits*, vol. 30, pp. 1399–1410, Dec. 1995.

[9]. B. Razavi, "Design Consideration for Direct-Conversion Receiver," *IEEE Trans. Circuits and Systems — II: Analog and Digital Signal Processing*, vol. 44, no. 6, pp. 428–435, June 1997.

[10]. J. F. Wilson et al, "A Single-Chip VHF and UHF Receiver for Radio Paging," *IEEE J. Solid-State Circuits*, vol. 26, pp. 1944–1950, Dec. 1991.

[11]. C. D. Hull, J. L. Tham, and R. R. Chu, "A Direct Conversion Receiver for 900 MHz (IMS Band) Spread-Spectrum Digital Cordless Telephone," *IEEE J. Solid-State Circuits*, vol. 31, pp. 1955 – 1963, Dec. 1996.

[12]. H. Yoshida, H. Tsurumi, and Y. Suzuki, "DC Offset Canceller in a Direct Conversion Receiver for QPSK Signal Reception," *Proc. of IEEE Int. Symposium. on Personal, Indoor and Mobile Radio Communications*, vol.3, pp. 1314-1318, Sept. 1998.

[13]. J. Crols and M. S. J. Steyaert, "A Single-Chip 900 MHz CMOS Receiver Front-End with a High Performance Low-IF Topology," *IEEE J. Solid-State Circuits*, vol. 30, No. 12, pp. 1483–1492, Dec. 1995.

[14]. J. P. F. Glas, "Digital I/Q Imbalance Compensation in a Low-IF Receiver," *Proceedings of IEEE Global Telecommunications Conference*, vol. 3, pp. 1461–1466, Nov. 1998.

[15]. J. Crols and M.S.J. Steyaert, "Low-IF Topologies for High-Performance Analog Front Ends of Fully Integrated Receivers," *IEEE Trans. Circuits and Systems — II: Analog and Digital Signal Processing*, vol. 45, no. 3, pp. 269–282, Mar. 1998.

[16]. L. Yu and W. M. Snelgerove, "A Novel Adaptive Mismatch Cancellation System for Quadrature IF Radio Receiver," *IEEE Trans. Circuits and Systems —II: Analog and Digital Signal Processing*, vol. 46, no. 6, pp. 789 – 801, June 1999.

[17]. J. Crols and M. Steyaert, "An Analog Polyphase filter for a High Performance Low-IF Receiver," *1955 Symposium on VLSI Circuits Digest of Technical Papers*, pp. 87–88, 1955

[18]. T. Yamawaki et al., "A 2.7 V GSM RF Transceiver IC," *IEEE Journal of Solid-State Circuits*, vol. 32, no. 12, pp. 2089 – 2096, Dec. 1997.

[19]. R. G. Vaughan, N. L. Scott, and R. D. White, "The Theory of Bandpass Sampling," *IEEE Trans. Signal Processing*, vol. 39, no. 9, pp. 1973 – 1984, Sept. 1991.

[20]. G. Hill, "The Benefits of Undersampling," *Electron. Design*, pp. 69 – 79, July 1994.

[21]. M. E. Frerking, "Digital Signal Processing in Communication Systems," Chapman & Hall, 1994.

[22]. J.A. Wepman, "Analog-to Digital Converters and Their Applications in Radio Receivers," *IEEE Communications Magazine*, PP. 39-45, May, 1995.

[23]. D. A. Akos, M. Stockmaster, J. B. Y. Tsui, and J. Caschera, "Direct Bandpass Sampling of Multiple Distinct RF Signals," IEEE Trans. Communications, vol. 47, no. 7, pp. 983–988, July, 1999.

Associated References

[1]. B. Razavi, "Challenges in Portable RF Transceiver Design," *IEEE Circuits and Devices Magazine*, vol. 12, no. 5, pp. 12–25, Sept. 1996.

[2]. J.C. Clifton et al., "RF Transceiver Architectures for Wireless Local Loop Systems," *IEE Colloquium on RF & Microwave Circuits for Commercial Wireless Applications*, pp. 11/1–11/8, Feb. 1997.

[3]. J. L. Mehta, "Transceiver Architectures for Wireless ICs," *RF Design*, pp. 76–96, Feb. 2001.

[4]. S. Mattison, "Architecture and Technology for Multistandard Transceivers," *Proceedings of the 2001 Bipolar/BiCMOS Circuits and Technology Meeting*, pp. 82–85, Oct. 2001.

[5]. M .S. W. Chen and R. W. Brodersen, "A Subsampling Radio Architecture by Analytic Signaling," *Proceedings of the 2004 IEEE International Conference on Acoustics, Speech, and Signal Processing*, vol. 4, pp. iv-533–iv-536, May 2004.

[6]. F.J. Harris, C, Dick, and M. Rice, "Digital Receivers and Transmitters Using Polyphase Filter Banks for wireless Communications," *IEEE Trans. On Microwave Theory and Techniques*, vol. 51, no. 4, pp. 1395–1412, April 2003.

[7]. P. Fines, "Radio Architectures Employing DSP Techniques," *IEE Workshop on Microwave and Millimeter-Wave Communications – Wireless Revolution*, pp. 10/1–10/5, Nov. 1995.

[8]. T. H. Lee, H. Samavati, and H. R. Rategh, "5-GHz CMOS Wireless LANs," *IEEE Trans. Microwave Theory and Techniques*, vol. 50, no.1, pp. 268–280, Jan. 2002.

[9]. S. W. Chung, S. Y. Lee, and K. H. Park, "An Energy-Efficient OFDM Ultra-Wideband Digital Radio Architecture," *2004 IEEE Workshop on Signal Processing Systems*, pp. 211–216, Oct. 2004.

[10]. J. P. K. Gilb, "Bluetooth Radio Architectures," *2000 IEEE Radio Frequency Integrated Circuits Symposium*, pp. 3–6, 2000

[11]. Z. Yuanjin and C. B. Terry, "Self Tuned Fully Integrated High Image Rejection Low IF Receiver Architecture and Performance," *Proceedings of the 2003 International Symposium on Circuits and Systems*, vol. 2, pp. II165 - II168, May 2003.

[12]. C. C. Chen and C. C Huang, "On the Architecture and Performance of a Hybrid Image Rejection Receiver," *IEEE Journal Selected Area in Communications*, vol. 19, no. 6, pp. 1029–1040, June 2001.

[13]. B. Lindoff, "Using a Direct Conversion Receiver in EDGE Terminals – A New DC Offset Compensation Algorithm," *2000 IEEE International Symposium on Personal, Indoor and Mobile Radio Communications*, vol. 2, pp. 959–963, Sept. 2000.

[14]. M. Valkama, M. Renfors and V. Koivunen, "Advanced Methods for I/Q Imbalance Compensation in Communication Receivers," *IEEE Trans. On Signal Processing*, vol. 49, no. 10, pp. 2335–2344, Oct. 2001.

[15]. J. K. Cavers and M. W. Liao, "Adaptive Compensation for Imbalance and Offset Losses in Direct Conversion Transceiver," *IEEE Trans. On Vehicular Technology*, vol. 42, no. 4, pp. 581–588, Nov. 1993.

[16]. E. Cetin, I. Kale and R.C.S. Morling, "Adaptive Compensation of Analog Front-End I/Q Mismatches in Digital Receivers," *2001 IEEE*

International Symposium on Circuits and Systems, vol. 4, pp. 370–373 , May 2001.

[17]. E. Grayver and B. Daneshrad, "A Low Power FSK Receiver for Space Applications," *2000 IEEE Wireless Communications and Networking Conference*, vol. 2, pp. 713 – 718, Sept. 2000.

[18]. S. M. Rodrigure et al, "Next Generation Broadband Digital Receiver Technology," *15th Annual AESS/IEEE Symposium*, pp. 13–20, May 1998.

[19]. K. Kalbasi, "Simulating Trade-Offs in W-CDMA/EDGE Receiver Front Ends," *Communication Systems Design*, vol. 8, no. 1, Jan. 2002.

[20]. P. Delvy, "A Receiver Comparison for GSM/GPRS Applications," *Wireless Design and Development*, pp. 22–26, Nov. 2004.

[21]. J. A. Wepman, "ADCs and Their Applications in Radio Receivers," IEEE Communications Magazine, vol. 33, pp. 39–45, May 1995.

[22]. H. M. Seo et al., "Relationship Between ADC Performance and Requirements of Digital-IF Receiver for WCDMA Base-Station," *IEEE Trans. Vehicular Technology*, vol. 52, no. 5, pp. 1398–1480, Sept. 2003.

[23]. K.A. Stewart, "Effect of Sample Clock Jitter on IF-Sampling IS-95 Receiver," *8th IEEE International Symposium on Personal, Indoor and Mobile Radio Communications*, vol. 2, pp. 366–370, Sept. 1997.

[24]. R. Schiphorst et al., " A Bluetooth-Enabled HiperLAN/2 Receiver," *2003 IEEE 58th Vehicular Technology Conference*, vol. 5, pp. 3443–3447, Oct. 2003.

[25]. M.S. Braasch and A.J. Van Dierendonck, "GPS Receiver Architectures and Measurements," *Proceedings of the IEEE*, vol. 87, no. 1, pp. 48–64, Jan. 1999.

[26]. S. Yoshizumi et al, "All Digital Transmitter Scheme and Transceiver Design for Pulse-Based Ultra-Wideband Radio," *2003 IEEE Conference on Ultra Wideband Systems and Technologies*, pp. 438–432, Nov. 2003.

[27]. W. Namgoong, "A Channelized Digital Ultrawideband Receiver," *IEEE Trans. on Wireless Communications*, vol. 2, no. 3, pp. 502–509, May 2003.

[28]. K. C. Zangi, "Impact of Wideband ADC's on the Performance of Multi-Carrier Radio Receiver," *48th IEEE Vehicular Technology Conference*, vol. 3, pp. 2155–2159, May 1998.

[29]. M. Verhelst et al., "Architectures for Low Power Ultra-Wideband Radio Receivers in the 3.1-5GHz Band for Data Rates < 10 Mbps," *Proceedings of the 2004 International Symposium on Low Power Electronics and Design*, pp. 280–285, Aug. 2001.

[30]. X. Xu et al., "Analysis of FDSS Ultra-Wideband Six-Port Receiver," *2002 IEEE Radio and Wireless Conference*, pp. 87–90, Aug. 2002.
[31]. G. Ordu et al., "A Novel Approach for IF Selection of Bluetooth Low-IF Receiver Based on System Simulations," Proceedings of IEEE International System-on-Chip Conference, pp. 43 – 46, Sept. 2003.
[32]. C. H. Tseng and S. C. Chou, "Direct Downconversion of Multiple RF Signals Using Band Pass Sampling," *2003 IEEE International Conference on Communications*, pp. 2003–2004, May 2003.
[33]. M. H. Norris, "Transmitter Architectures (GSM Handsets)," *IEE Colloquium on the Design of Digital Cellular Handsets*, pp. 4/1–4/6, March 1998.
[34]. L. Robinson, P. Aggarwal, and R. R. Surendran, "Direct Modulation Multi-Mode Transmitter," *2002 3rd International Conference 3G Mobile Communication Technologies*, pp. 206–210, May 2002.
[35]. F. Kristensen, P. Nilsson, and A. Olsson, " A Generic Transmitter for Wireless OFDM Systems," *14th IEEE 2003 International Symposium on Personal, Indoor and Mobile Radio Communication Proceedings*, pp. 2234–2238,
[36]. J. Ketola et al., "Transmitter Utilising Bandpass Delta-Sigma Modulator and Switching Mode Power Amplifier," Proceedings of the *2004 International Symposium on Circuits and Systems*, vol. 1, pp. 1-633–1-636, May 2004.
[37]. T.E. Stichelbout, "Delta-Sigma Modulator in Radio Transmitter Architectures," *1999 Emerging Technologies Symposium on Wireless Communications and Systems*, pp. 6.1–6.4, April 1999.
[38]. S. Mann et al., " A Flexible Test-Bed for Developing Hybrid Linear Transmitter Architectures," *2001 Spring IEEE Vehicular Technology Conference*, vol. 3, pp. 1983–1986, May 2001.
[39]. B. Razavi, "RF Transmitter Architectures and Circuits," *IEEE 1999 Custom Integrated Circuits Conference*, pp. 197–204, May 1999.
[40]. S. Mann et al., "Increasing Talk-Time of Mobile Radios with Efficient Linear Transmitter Architectures," *Electronics and Communication Engineering Journal*, vol. 13, issue 2, pp. 65–76, April 2001.
[41]. S. Mann et al., "Increasing Talk-Time with Efficient Linear PA's," *IEE Seminar on Tetra Market and Technology Development*, pp. 6/1 – 6/7, Feb. 2000.
[42]. D. Efstathiou, L. Fridman and Z. Zvonar, "Recent Developments in Enabling Technologies for Software Defined Radio," *IEEE Communications Magazine*, vol. 37, no. 8, pp. 112–17, Aug. 1999.
[43]. J. R. Macleod et al, "Enabling Technologies for Software Defined Radio Transceivers," *Proceedings of MILCOM*, vol. 1, pp.354 –358, Oct. 2002.

[44]. A. S. Margulies and J. Mitola III, "Software Defined Radios: A Technical Challenge and a Migration Strategy," *Proceedings of 1998 IEEE 5th International Symposium on Spread Spectrum Techniques*, pp. 551–556, 1998.

[45]. M. Rami et al., "Broadband Digital Direct Down Conversion Receiver Suitable for Software Defined Radio," *2002 IEEE International Symposium on Personal, Indoor and Mobile Radio Communications*, vol. 1, pp. 100–104, Sept. 2002.

[46]. J. Mitola, "The Software Radio Architecture," *IEEE Communications Magazine*, pp. 26 – 38, May 1995.

[47]. S. Sirkanteswara et al., "A Soft Radio Architecture for Reconfigurable Platforms," *IEEE Communications Magazine, pp. 140–147, Feb. 2000.*

[48]. J. Mitola III, "Software Radio Architecture: A Mathematical Perspective," *IEEE Journal on Selected Area in Communications*, vol. 17, no. 4, pp. 514–538, April 1999.

[49]. Z. Salcic and C. F. Mecklenbrauker, "Software Radio — Architecture Requirements, Research and Development Challenges," *2002 ICCS*, pp. 711–716, 2002.

[50]. M. Choe, H. Kang, and K. Kim, "Tolerable Range of Uniform Bandpass Sampling for Software Defined Radio," *2002 The 5th International Symposium on Wireless Personal Multimedia Communications*, vol. 2, pp. 840–842, Oct. 2002.

[51]. J. E. Junn, K. S. Barron, and W. Ruczczyk, "A Low-Power DSP Core-Based Software Radio Architecture," *IEEE Journal on Selected Areas in Communications*, vol. 14, no. 4, pp. 574–590, April 1999.

[52]. H. Erben and K. Sabatakakis, "Advanced Software Radio Architecture for 3rd Generation Mobile Systems," *48th IEEE Vehicular Technology Conference*, pp. 825–829, May 1998.

Chapter 4

Receiver System Analysis and Design

4.1. Introduction

A receiver can be implemented in terms of different architectures, such as superheterodyne, direct conversion, low IF, or band-pass sampling, as discussed in the previous chapter. No matter which architecture is employed, the receiver should possess a well-defined function and performance. This means that the mobile station receivers operating in a wireless system may be different in their architectures, but they must have many common characteristics to achieve the unique performance specified in the wireless communication system standards.

As we know, there are two kinds of duplex systems, full-duplex and half-duplex systems, used in the different wireless communication systems. The CDMA, WCDMA, and AMPS systems are full-duplex systems. In these systems, the receiver and the transmitter of either the mobile stations or the base stations are operating simultaneously, but they operate in different frequency bands. Other wireless communication systems, such as the GSM, GPRS, TDMA, and PHS systems, are the half-duplex system. In the half-duplex system, the receiver is running in time slots different from those that the transmitter is operating in, but they may use the same operation frequency. It is apparent that in the full-duplex system the transmission of the transmitter is a strong interference source to the receiver since the transmission power at the receiver antenna port can be 120 dB or even 130 dB stronger than the desired reception signal, but the transmission frequency is only tens of MHz away from that of the reception signal. The corresponding receiver must be able to work properly under the constant attack of the strong transmission interference. This makes the design of the receiver operating in a full-duplex system much harder than the one working in the half-duplex system.

The key parameters characterizing wireless mobile station receivers are the *reception sensitivity*, *intermodulation characteristics*, *adjacent channel and alternate channel selectivities*, *single-tone desensitization*, *interference blocking*, *dynamic range*, and AGC. The sensitivity of a receiver is directly related to the overall noise figure of the receiver. The linearity of a receiver, especially the third-order distortion, is the main factor to determine the *intermodulation distortion* (*IMD*) performance and the single tone desensitization as well. The channel filtering characteristics and the phase noise of the local oscillator dominate the receiver adjacent/alternate channel selectivity and the interference blocking performance of a receiver. The dynamic range of a receiver is usually achieved by using an automatic gain control and a proper ADC. In this chapter, the analyses of the important receiver parameters and the derivations of corresponding design formulas are based on the full-duplex system receiver. However, all consequences and relevant formulas resulting from the full-duplex system receiver can also be applied to the half-duplex system receiver after slight modification (what the transmission interference and related impacts need not be considered).

4.2. Sensitivity and Noise Figure of Receiver

The sensitivity of a wireless mobile receiver is defined as the weakest RF signal power that can be processed to develop a minimum signal-to-noise ratio for achieving a required error rate (BER or FER) by the system. The sensitivity signal level also varies depending on specific signal modulation and characteristics, the signal propagation channel, and external noise level.

4.2.1. Sensitivity Calculation

In the *additive white Gaussian noise channel* (*AWGN*), the noise appearing at the input of a receiver is thermal noise, and the sensitivity of the receiver can be derived from the noise figure (NF) of the RF receiver as follows. As presented in Section 2.4, the signal-to-noise ratio of the RF and the analog BB signal before digital signal processing is referred to as carrier-to-noise ratio (CNR) to distinguish it from the signal-to-noise ratio of the digital BB signal. The noise factor of the receiver from the antenna port to the output of the analog-to-digital converter (ADC) is expressed as the ratio of the input carrier-to- noise ratio $(C/N)_i$ in numeric value and the output carrier-to-noise ratio $(C/N)_o$ in numeric value

$$F_{Rx} = \frac{(C/N)_i}{(C/N)_o} = \frac{P_S / P_{Ni}}{(C/N)_o}, \qquad (4.2.1)$$

where P_S is the desired signal power at the receiver input and P_{Ni} is the integrated thermal noise power within the receiver noise bandwidth BW, which in the case of conjugate matching at the receiver input has the form of

$$P_{Ni} = kT_o \cdot BW \ , \tag{4.2.2}$$

where k = 1.38×10^{-20} mW·sec/°K is Boltzman constant and T_o = 290°K. From (4.2.1), the receiver input signal can be expressed as

$$P_S = kT_o \cdot BW \cdot F_{Rx} \cdot (C/N)_o , \tag{4.2.3}$$

Assuming that the minimum $(C/N)_o$ required for obtaining the defined error rate corresponding to the sensitivity level is equal to $(C/N)_{min}$, from (4.2.3) the sensitivity power of the receiver, S_{min} in dBm*, has the form of

$$S_{\min} = 10\log(P_{S,\min}) = -174 + 10\log(BW) + NF_{Rx} + CNR_{\min} , \tag{4.2.4}$$

where $10\cdot\log(kT_o)$ = -174 dBm/Hz is used, receiver noise bandwidth (loosely speaking, receiver bandwidth) BW is in Hz, NF_{Rx} is the overall noise figure of the receiver in dB — i.e.,

$$NF_{Rx} = 10\log(F_{Rx}) \ \text{dB}, \tag{4.2.5}$$

and

$$CNR_{\min} = 10\log(C/N)_{\min} \ \text{dB}. \tag{4.2.6}$$

The maximum CNR_{min} for various wireless system mobile receivers are presented in Table 2.4 (in Section 2.4.5.3). All CNR_{min} values given in this table are the required carrier-to-noise ratio of the receiver sensitivities in the AWGN channel or referred to as static sensitivities. From Table 2.4, we know that the maximum CNR_{min} for the CDMA mobile receiver is equal to -1 dB for the 0.5% FER, and the receiver bandwidth is approximately 1.25 MHz. Assuming that the overall noise figure NF_{Rx} is 10 dB, from (4.2.4) the receiver sensitivity is thus

---

* dBμV/m will be used when the receiver sensitivity is represented by signal electrical field strength instead of the power. The conversion between dBm and dBμV/m, and the conversion between dBm and dBμV as well are given in Appendix 4A.

$$S_{\min\_CDMA} = -174 + 60.9 + 10 - 1 = -104.1 \text{ dBm}.$$

The required CNR of a receiver for achieving a certain data error rate is determined mainly by the demodulation, decoding, and digital signal processing in the digital base-band as described in Section 2.4.5. However, the magnitude and phase frequency responses of the channel filters in the receiver RF analog section may also have notable impact on the CNR value if the bandwidth, in-band ripple, and group delay distortion of these filters are not properly defined.

In fact, it is more convenient for the RF receiver system design to employ the receiver noise figure NF_{Rx} than using the sensitivity. Rearranging (4.2.4), we obtain

$$NF_{Rx} = 174 + S_{\min} - 10\log(BW) - CNR_{\min}. \tag{4.2.7}$$

The reference sensitivity for the GSM 900 small mobile receiver is defined as –102 dBm, and the CNR_{min} is approximately 8 dB for the GMSK modulation signal to be demodulated with a BER around 3%. If the receiver bandwidth is 250 kHz, from (4.2.7) the maximum acceptable noise figure for such receiver sensitivity is

$$NF_{Rx\_GSM} = 174 - 102 - 54 - 8 = 10 \text{ dB}.$$

In practical designs, it is necessary for the receiver sensitivity to have an enough margin — say, 3 dB or more. For the above example, to achieve a –105 dBm or even better sensitivity, the overall receiver noise figure should be 7 dB or less.

4.2.2. Cascaded Noise Figure

The RF receiver usually consists of multiple stages as shown in Fig. 4.1 where g_k and F_k (k = 1, 2, …, n) are the available power gain and noise factor, respectively. A similar way of the cascaded noise factor derivation as presented in [1] will be used here. Assuming that the input impedance of each stage in the receiver is conjugate-matched with the output impedance of preceding stage, the available output noise power from the nth stage P_{N_Rx} is

$$P_{N\_Rx} = F_{Rx} \cdot kT_o \cdot BW \cdot \prod_{j=1}^{n} g_j, \tag{4.2.8}$$

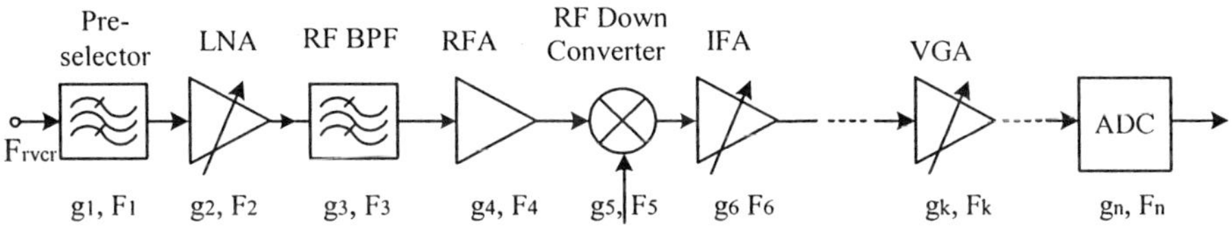

Figure 4.1. Receiver chain of multiple stages

where F_{Rx} is the overall noise factor of the receiver. The available noise introduced by the input and the first stage measured at the output of the nth stage is

$$P_{N1} = F_1 \cdot kT_o \cdot BW \cdot \prod_{j=1}^{n} g_j . \tag{4.2.9}$$

The available noise of rest stages added to the output of the nth stage is

$$P_{Ni} = (F_i - 1) \cdot kT_o \cdot BW \cdot \prod_{j=i}^{n} g_j , \tag{4.2.10}$$

where $i = 2, 3, \ldots, n$. The total available noise power P_{N_Rx} is also equal to the sum of (4.2.9) and (4.2.10):

$$P_{N\_Rcvr} = F_1 \cdot kT_o \cdot BW + \sum_{i=2}^{n} (F_i - 1) \cdot kT_o \cdot BW \cdot \prod_{j=i}^{n} g_j . \tag{4.2.11}$$

Substituting (4.2.8) into the left side of (4.2.11), after canceling the common factors $kT_o \cdot BW$ on both sides we obtain the cascaded noise factor of the receiver F_{Rx} to be

$$F_{Rx} = F_1 + \sum_{i=2}^{n} \frac{F_i - 1}{\prod_{j=1}^{i-1} g_j} . \tag{4.2.12a}$$

The noise factor F_{Rx} in dB is referred to as noise figure of the receiver NF_{Rx} here, which is expressed as

$$NF_{Rx} = 10\log\left(F_1 + \sum_{i=2}^{n} \frac{F_i - 1}{\prod_{j=1}^{i-1} g_j}\right). \tag{4.2.12b}$$

In (4.2.12b), g_k and F_k (k = 1, 2, …, n) both are in numeric value. If using noise figure of each stage, NF_k, (k = 1, 2, …, n) instead of noise factor and the power gain in dB, G_k, (k = 1, 2, …, n-1), (4.2.12b) then has the form

$$NF_{Rx} = 10\log\left(10^{\frac{NF_1}{10}} + \sum_{i=2}^{n} \frac{10^{\frac{NF_i}{10}} - 1}{\prod_{j=1}^{i-1} 10^{\frac{G_j}{10}}}\right). \tag{4.2.12c}$$

Equation (4.2.12) is also called a *Friis equation* [2]. The gains in this equation are *available power gain* based on the assumption that there is conjugate matching between stages. The assumption of conjugate match may be in reality true for the RF blocks of the receiver. However, in the IF and the analog BB blocks of the receiver the input impedance of the mth stage $Z_{i,,m}$ may not match with the output impedance of its preceding stage $Z_{o,,m-1}$ and usually $Z_{i,,m} >> Z_{o,,m-1}$. The voltage gain g_v instead of the power gain is usually used for the stages in the IF and the analog BB blocks.

Fig. 4.2 can be utilized to calculate the noise factor of each stage in the un-matched stage blocks of the receiver chain, and to determine the corresponding cascaded noise factor. Here the series voltage source $V_{N,m}$ and the parallel current source $I_{N,m}$ represent the noise of the mth stage (m =1,2, …, n_1), $R_{i,m}$ and $R_{o,m}$ are the input and the output impedances of the mth stage, respectively (since practically R_m (resistance) >> X_m (reactance) for

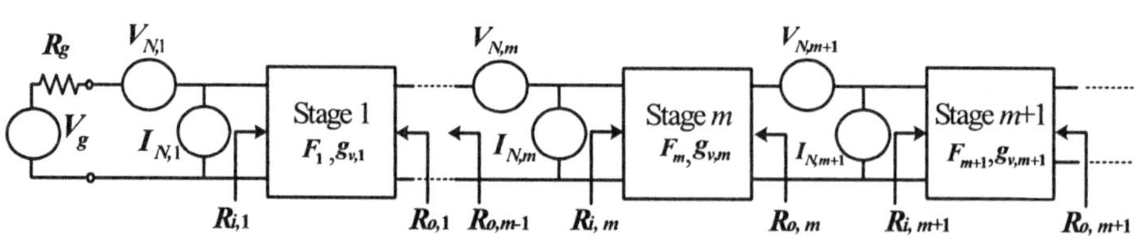

Figure 4.2. Equivalent representation of noisy receiver chain consisting of unmatched stages

both impedances in all the stages (m =1,2, …, n_I)), and R_g is the source resistance. Denote noise factor and voltage gain of each stage in Fig. 4.2 as F_m and $g_{v,m}$ (m = 1,2, …, n_1), respectively. The noise factor of the mth stage can be derived as follows [1], [3]. The mean square noise voltage generated by the output resistance $R_{o,m-1}$ is $4kT\cdot R_{o,m-1}\cdot BW$, which is the source noise of the mth stage. The mean square noise voltage at the input of the mth stage $\overline{V_{N,in\_m}^2}$ is

$$\overline{V_{N,in\_m}^2} = \left[\overline{\left(I_{N,m}R_{o,m-1} + V_{N,m}\right)^2} + 4kTR_{o,m-1}BW\right]\frac{R_{i,m}^2}{\left(R_{o,m-1} + R_{i,m}\right)^2}. \quad (4.2.13)$$

The output noise of the mth stage $\overline{V_{N,out\_m}^2}$ equals

$$\overline{V_{N,out\_m}^2} = g_{v,m}^2\overline{V_{N,in\_m}^2}\frac{R_{i,m+1}^2}{\left(R_{o,m} + R_{i,m+1}\right)^2}. \quad (4.2.14)$$

From the noise source $4kT\cdot R_{o,m-1}\cdot BW$ to output noise voltage $\overline{V_{N,out\_m}^2}$, the total voltage gain, g_{v,t_m}, is equal to

$$g_{v,t\_m} = \frac{R_{i,m}}{R_{o,m-1} + R_{i,m}} g_{v,m} \frac{R_{i,m+1}}{R_{o,m} + R_{i,m+1}}. \quad (4.2.15)$$

From (4.2.14) and (4.2.15), we obtain the noise factor of the mth stage (with a source resistance of $R_{o,m-1}$) to be

$$F_m = \frac{\overline{V_{N,out\_m}^2}}{4kTR_{o,m-1}\cdot BW\cdot g_{v,t\_m}^2} = 1 + \frac{\overline{\left(I_{N,m}R_{o,m-1} + V_{N,m}\right)^2}}{4kT\cdot R_{o,m-1}\cdot BW}. \quad (4.2.16)$$

The cascaded noise factor of two adjacent stages, the mth and the (m-1)th stages, $F_{m,\ m+1}$, can be obtained from (4.2.13) and the following equations [3]:

$$\overline{V_{N,in\_m+1}^2} = \left[\overline{V_{N,in\_m}^2}\cdot g_{v,m}^2 + \overline{\left(I_{N,m+1}R_{o,m} + V_{N,m+1}\right)^2}\right]\left(\frac{R_{i,m+1}}{R_{o,m} + R_{i,m+1}}\right)^2 \quad (4.2.17)$$

$$\overline{V_{N,out\_m+1}^2} = g_{v,m+1}^2 \overline{V_{N,in\_m+1}^2} \frac{R_{i,m+2}^2}{\left(R_{o,m+1} + R_{i,m+2}\right)^2}, \tag{4.2.18}$$

and

$$g_{v,t\_m,m+1} = \frac{R_{i,m}}{R_{o,m-1} + R_{i,m}} g_{v,m} \frac{R_{i,m+1}}{R_{o,m} + R_{i,m+1}} g_{v,m+1} \frac{R_{i,m+2}}{R_{o,m+1} + R_{i,m+2}}. \tag{4.2.19}$$

The cascaded noise factor of the two adjacent stages $F_{m,\,m+1}$ is expressed as

$$\begin{aligned} F_{m,m+1} &= \frac{\overline{V_{N,out\_m+1}^2}}{g_{v,t\_m.m+1} 4kT \cdot BW \cdot R_{o,m-1}} \\ &= F_m + \frac{F_{m+1} - 1}{g_{v,m}^2} \frac{1}{\left(\frac{R_{i,m}}{R_{o,m-1} + R_{i,m}}\right)^2} \frac{1}{\frac{R_{o,m-1}}{R_{o,m}}}, \end{aligned} \tag{4.2.20}$$

where F_{m+1} has the same expression as (4.2.16) but the subscript m of all the variables in this formula is replaced by $m + 1$.

In the general case of n_1 stages in cascade with a noise source $4kT \cdot BW \cdot R_g$ as shown in Fig. 4.2, the overall cascaded noise factor, $F_{t\_cascade}$, has a form of

$$F_{t\_cascade} = F_1 + \sum_{m=2}^{n_1} \frac{F_m - 1}{\prod_{l=1}^{m-1} g_{v,l}^2 \left(\frac{R_{i,l}}{R_{o,l-1} + R_{i,l}}\right)^2 \frac{R_{o,l-1}}{R_{o,l}}}. \tag{4.2.21}$$

In (4.2.21), F_m is given by (4.2.16), and $R_{o,\,0} = R_g$. This equation turns into the same form as the equation (4.2.12a) if letting the voltage gain multipliers of each stage, $g_{v,l}^2 \left(\frac{R_{i,l}}{R_{o,l-1} + R_{i,l}}\right)^2 \frac{R_{o,l-1}}{R_{o,l}}$, in the denominator of the fraction terms on the right side of (4.2.21) equal the available power gain of the corresponding stage g_l — i.e.,

$$g_l = g_{v,l}^2 \left(\frac{R_{i,l}}{R_{o,l-1} + R_{i,l}} \right)^2 \frac{R_{o,l-1}}{R_{o,l}} . \tag{4.2.22}$$

The input impedance of the stages in the IF and the analog BB blocks is usually much higher than the output impedance of preceding stage — i.e., $R_{i,,l} >> R_{o,,l-1}$. Thus (4.2.21) can be simplified, particularly under the assumption of $R_{o,l} \cong R_{o,l-1}$, as

$$F_{t\_cascade} \cong F_1 + \sum_{m=2}^{n_1} \frac{F_m - 1}{\prod_{l=1}^{m-1} g_{v,l}^2 \frac{R_{o,l-1}}{R_{o,l}}} \cong F_1 + \sum_{m=2}^{n_1} \frac{F_m - 1}{\prod_{l=1}^{m-1} g_{v,l}^2} . \tag{4.2.23}$$

(4.2.23) is quite convenient and useful in estimating the noise factors or noise figures of the IF and the analog BB blocks in the RF receiver engineering design calculation.

4.2.3. Receiver Desensitization Evaluation Due to Transmitter Noise Emission in the Receiver Band

There are two approaches to calculate the receiver desensitization caused by the transmitter noise emission in the receiver band in a full-duplex system. Assuming that the emission noise mainly passes through the duplexer, which connects the receiver and the transmitter to a common antenna. One is based on finding an equivalent noise figure of the duplexer, and the other one is using the concept of equivalent antenna temperature.

4.2.3.1. Equivalent Duplexer Noise Figure Method

In this analysis, the duplexer is imaged as a two-port instead of three-port device, which has one port for connecting the antenna and the second one for connecting the receiver as shown in Fig. 4.3. The third port of the duplexer is terminated with the transmitter and this port is the entrance of the interfering noise source to the receiver. The noise figure of this equivalent two-port device can be determined as follows.

Assuming that the input signal level at the antenna port is P_S, and the noise is the thermal noise with a power density of kT_o, the signal-to-noise ratio at the antenna port $(S/N)_{\text{input}}$ is simply

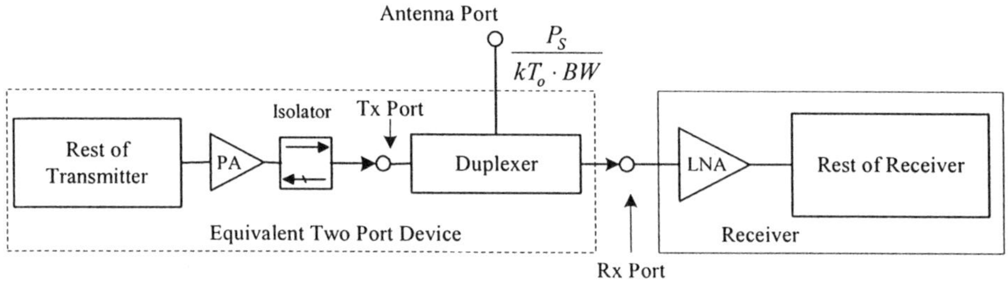

Figure 4.3. Simplified configuration of full-duplex transceiver

$$(S/N)_{input} = \frac{P_S}{kT_o \cdot BW}, \tag{4.2.24}$$

where BW is receiver noise bandwidth. If the insertion gain between the antenna port and the receiver port of the duplexer is $g_{ant\_Rx} < 1$, then the output signal power from the receiver port is equal to $g_{ant\_Rx} \cdot P_S$. Letting $P_{N,Tx\_ant,RxBand}$ mW/Hz be defined as the excess transmitter noise emission density over the thermal noise density kT_o in the receiver band and measured at the antenna port of the duplexer, the output noise from the receiver port of the duplexer is then $\left(kT_o + P_{N,Tx\_ant,RxBand} \cdot g_{ant\_Rx}\right) \cdot BW$. Here it is assumed that the excess emission noise density $P_{N,Tx\_ant,,RxBand}$ is flat in the receiver band, and BW is the noise bandwidth of the receiver. The output signal-to-noise ratio from the duplex receiver port $(S/N)_{output}$ can be written in the following form:

$$(S/N)_{output} = \frac{P_S \cdot g_{ant\_Rx}}{\left(kT_o + P_{N,Tx\_ant,RxBand} \cdot g_{ant\_Rx}\right) \cdot BW}. \tag{4.2.25}$$

The equivalent noise factor of the duplexer from the antenna port to the receiver port F_{e_dplx} is determined by the ratio of (4.2.24) to (4.2.25) — i.e.,

$$F_{e\_dplx} = \frac{(S/N)_{input}}{(S/N)_{output}} = \frac{1}{g_{ant\_Rx}} + \frac{P_{N,Tx\_ant,RxBand}}{kT_o} \tag{4.2.26a}$$

and the corresponding noise figure NF_{e_dplx} is the noise factor F_{e_dplx} expressed in dB scale — i.e.,

$$NF_{e\_dplx} = 10\log\left(\frac{1}{g_{ant\_Rx}} + \frac{P_{N,Tx\_ant,RxBand}}{kT_o}\right) \text{ dB.} \qquad (4.2.26b)$$

As we knew, the original noise figure of the duplexer's receiver filter with an insertion loss g_{ant_Rx} is $NF_{o\_dplx} = 10\log(1/g_{ant\_Rx})$. Comparing this noise figure with that given in (4.2.26), it is apparent that the excess emission noise $P_{N,Tx\_ant,,RxBand}$ results in the degradation of the duplexer noise figure. However, the $P_{N,Tx\_ant,,RxBand}$ level is usually very low and cannot be directly measured. But it can be obtained through measuring the noise emission density in the receiver band at the transmitter power amplifier (PA)/isolator output (see Fig. 4.3). If the receiver band noise emission density measured at the PA/isolator output is denoted as $N_{Tx\_inRxBand}$ in dBm/Hz, and the attenuation of the duplexer's transmitter filter to the signal/noise in the receiver band is A_{dplx_Tx} in dB, the $P_{N,Tx\_ant,,RxBand}$ level can be calculated by using the following expression:

$$P_{N,Tx\_ant,RxBand} = 10^{\frac{N_{Tx\_inRxBand} - A_{dplx\_Tx}}{10}} - kT_o, \qquad (4.2.27)$$

where kT_o is the thermal noise density in mW/Hz, and it is equal to 4×10^{-18} mW/Hz when $T_o = 290$ °K.

Let us look at an example. Assume that emission noise density in the receiver band at the PA/isolator output is $N_{Tx\_inRxBand}$ = -127.5 dBm/Hz, the attenuation of the duplexer's transmitter filter to the emission noise is A_{dplx_Tx} = 44.5 dB, and the insertion loss of the duplexer's receiver filter is g_{ant_Rx} =0.56, or $10\cdot\log(g_{ant\_Rx})$ = -2.5 dB. From (4.2.26b) and (4.2.27) we obtain the equivalent noise figure of the duplexer to be

$$NF_{e\_dplx} = 10\log\left(1.78 + \frac{10^{(-127.5-44.5)/10} - 10^{-174/10}}{10^{-174/10}}\right) \cong 3.74 \text{ dB.}$$

The noise figure of the duplexer without the transmitter emission noise NF_{dplx} is equal to $10\cdot\log(1/g_{ant\_Rx})$ = 2.5 dB. The noise figure degradation of the duplexer due to the transmitter emission noise is approximately 1.24 dB.

The noise figure increase of the duplexer from the antenna port to the receiver port directly raises the overall receiver noise figure and thus degrades the receiver sensitivity. If the noise figure of the receiver excluding the duplexer is $NF_{rx,o}$ = 3.5 dB, the overall noise figure of the receiver under the impact of the transmitter emission noise NF_{Rx} can be calculated by using (4.2.28) — i.e.,

$$NF_{Rx} = 10\log\left(F_{e\_dplx} + \frac{10^{NF_{rx,o}/10} - 1}{g_{ant\_Rx}}\right) \tag{4.2.28}$$

$$\cong 10\log\left(2.37 + \frac{2.24 - 1}{0.56}\right) \cong 6.6 \text{ dB}.$$

Comparing this receiver noise figure with the one without transmitter emission noise impact, $NF_{Rx,o} = NF_{rx,o} + NF_{o,dplx} = 3.5 + 2.5 = 6.0$ dB, the degradation of the overall receiver noise figure or the sensitivity is approximately 0.6 dB.

4.2.3.2. Equivalent Antenna Temperature Method

In the normal case without the transmitter noise-emission influence the antenna temperature is assumed to be equal to ambient temperature T_o = 290 °K, and the noise density at the input of the receiver is then N_o = 10 log(kT_o) = -174 dBm/Hz. However, the receiver in a full-duplex system is always impacted by the noise emitted from the transmitter. In this case it may be imaged as that the antenna temperature rises to a temperature T_e, and the equivalent noise density at the input of the receiver now becomes

$$kT_e = kT_o + P_{N,Tx\_Rx}. \tag{4.2.29}$$

As we know from Section 2.3, the equivalent temperature of the receiver system internal noise $T_{N,Rx}$ is determined by the receiver original noise figure $F_{Rx,o}$, and it is

$$T_{N,Rx} = (F_{Rx,o} - 1)T_o. \tag{4.2.30}$$

The overall receiver equivalent noise temperature $T_{e,Rx}$ is the sum of the temperatures, T_e and $T_{N,Rx}$ — i.e.,

$$T_{e,Rx} = T_e + (F_{Rx,o} - 1)T_o. \tag{4.2.31}$$

The alternative definition of the noise factor is the ratio of the overall receiver equivalent noise temperature $T_{e,Rx}$ to the thermal noise temperature

T_o. Thus the receiver noise factor F_{Rx} derived from the temperature ratio has the following form:

$$F_{Rx} = \frac{T_{e,Rx}}{T_o} = \frac{P_{N,Tx\_ant,RxBand}}{kT_o} + F_{Rx,o}, \qquad (4.2.32a)$$

and the corresponding noise figure NF_{Rx} is

$$NF_{Rx} = 10\log\left[\frac{P_{N,Tx\_ant,RxBand}}{kT_o} + F_{Rx,o}\right]. \qquad (4.2.32b)$$

Utilizing the same data that are used in the example in Section 4.2.3.1, from (4.2.32) we calculate the noise figure of the receiver under the transmitter emission noise influence to be

$$NF_{Rx} = 10\log\left(\frac{2.33\times10^{-18}}{4\times10^{-18}} + 3.98\right) \cong 6.6 \text{ dB}.$$

It gives us the same result as that produced by using (4.2.28).

These two approaches result in the same consequence. Either one can be used to calculate the desensitization of the receiver under the disturbance of transmitter emission noise. The acceptable desensitization of a mobile station receiver under such kind of interference is approximately a couple of tenths of dB.

4.2.4. Influence of Antenna VSWR to Receiver Noise Figure

The *voltage standing wave ratio* (*VSWR*) or *return loss* is usually used to represent the matching/mismatch condition between the antenna and the receiver input. In the case of mobile stations, the VSWR of the antenna, especially the internal antenna, may widely vary with the user hand holding gesture and position, and the distance between the antenna and the human head/body. The VSWR magnitude of a mobile station antenna can vary in value between 1.5 and 6 under the influence of different application conditions. In the following analysis, we can see that the large antenna VSWR significantly impacts receiver noise and remarkably degrades receiver sensitivity. It is definitely worth considering the antenna VSWR influence to the receiver sensitivity and taking this possible performance degradation into the sensitivity budget during the RF receiver system design.

A simplified model for the receiver noise figure analysis instead of individual amplifier stage noise analysis as in Section 4.2.2 is shown in Fig. 4.4. Where V_N and I_N are equivalent noise voltage and current of the receiver, respectively, V_g is the source voltage, and R_g is the source resistance (no source reactance shown in this figure because of conjugate matching at the receiver input assumed). In the case of no correlation between V_N and I_N, the receiver noise factor F_{Rx} can be written in the following form:

$$F_{Rx} = 1 + \frac{\overline{V_N^2}}{4kT_o \cdot BW \cdot R_g} + \frac{\overline{I_N^2 R_g^2}}{4kT_o \cdot BW \cdot R_g}, \tag{4.2.33}$$

where $\overline{V_N^2}$ and $\overline{I_N^2}$ are mean noise voltage and current square, respectively, and $kT_o \cdot BW$ is the thermal noise power in the receiver noise bandwidth BW.

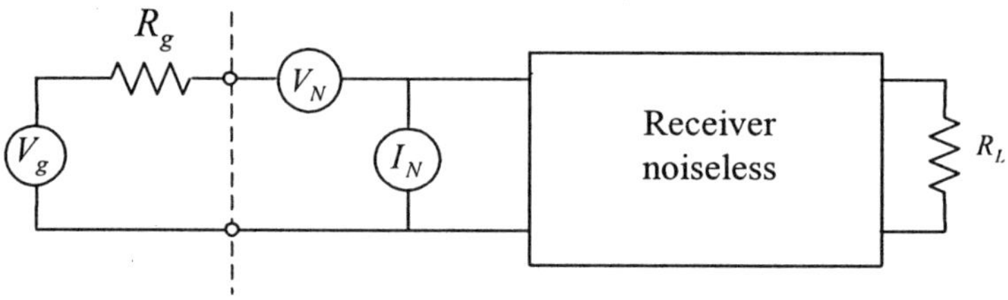

Figure 4.4. Simplified model for receiver noise figure analysis

From Section 2.3.4, we know that the $\overline{V_N^2}$ and $\overline{I_N^2}$ can be represented by an equivalent serial noise resistance R_n and an equivalent shunt conductance G_n as follows, respectively:

$$\overline{V_N^2} = 4kT_o \cdot BW \cdot R_n \quad \text{and} \quad \overline{I_N^2} = 4kT_o \cdot BW \cdot G_n. \tag{4.2.34}$$

Thus (4.2.33) turns into

$$F_{Rx} = 1 + \frac{R_n}{R_g} + G_n R_g. \tag{4.2.35}$$

The receiver noise figure F_{Rx} will be minimized if the following relationship is held:

$$R_g = \sqrt{\frac{\overline{V_N^2}}{\overline{I_N^2}}} = \sqrt{\frac{R_n}{G_n}} = R_{go} \,. \tag{4.2.36}$$

Plunging (4.2.36) into (4.2.35), the noise figure expression of the receiver is simplified as [1]

$$F_{Rx,o} = 1 + 2\sqrt{R_n G_n} \,. \tag{4.2.37}$$

The input port of the receiver is now connected to an antenna, and the corresponding equivalent circuit for the noise analysis of the receiver is shown in Fig. 4.5 where F_a is an equivalent antenna noise figure, which is equal to 1 when the antenna background noise equals the thermal noise, and R_a +j X_a is the antenna impedance presented at the receiver input port. From this figure we can obtain the receiver system noise figure F_{Rx} to be [1]

$$F_{Rx} = F_a + \frac{\overline{V_N^2}}{4kT_o \cdot BW \cdot R_a} + \frac{\overline{I_N^2}\left(R_a^2 + X_a^2\right)}{4kT_o \cdot BW \cdot R_a} \,. \tag{4.2.38}$$

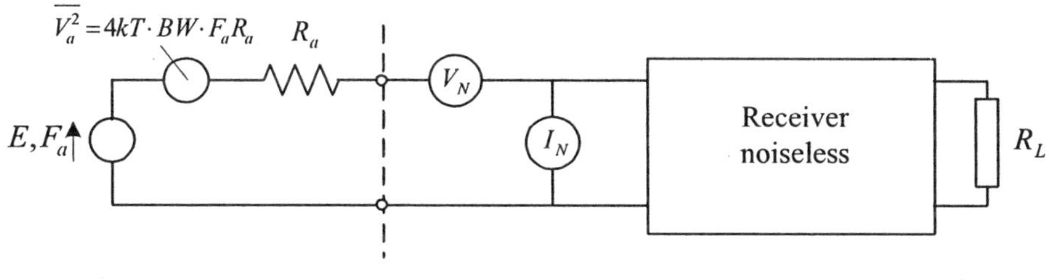

Figure 4.5. Equivalent circuit of antenna and receiver input

Using (4.2.34) and (4.2.36)), the above equation turns into

$$F_{Rx} = F_a + \sqrt{R_n G_n}\left[\frac{R_{go}}{R_a} + \frac{R_a}{R_{go}}\left(1 + \frac{X_a^2}{R_a^2}\right)\right]. \tag{4.2.39}$$

Considering that the real part of the antenna impedance, R_a, is much greater than its imaginary part X_a — i.e., $R_a >> X_a$ — and the ratio of R_a/R_{go} is the magnitude of the antenna and denoted as ρ (=R_a/R_{go}), and utilizing (4.2.37), we can approximately express the receiver noise factor formula (4.2.39) with the antenna |VSWR| = ρ as (4.2.40):

$$F_{Rx} \cong 1 + \frac{F_{Rx,o} - 1}{2}\left[\frac{1}{\rho} + \rho\right], \tag{4.2.40a}$$

where $F_{Rx,o}$ is the receiver original noise factor when the receiver noise impedance is matched with the source impedance (say 50 Ω), and $F_a = 1$ is used. The noise figure expression is

$$NF_{Rx} = 10\log\left\{1 + \frac{F_{Rx,o} - 1}{2}\left[\frac{1}{\rho} + \rho\right]\right\} \text{ dB} \tag{4.2.40b}$$

A comparison between results from the ADS simulation and (4.2.40) calculation is presented in Table 4.1. In this verification only a receiver front end instead of whole receiver chain was used, and the antenna was not matched with antenna very well on purpose. The receiver front end has noise figure approximate 1.4 dB and the antenna VSWR magnitude varies with frequency from 1.8 to 3.8. From the results given in this table, we can see that the differences between the simulation and the calculation results are within 0.2 dB. Equation (4.2.40), thus, provides us a good estimation on the influence of the antenna VSWR to receiver noise figure.

Table 4.1. Comparison of ADS simulation and (4.2.40) calculation

	Original	Antenna	Simulation	Calculation	Difference
Freq. (MHz)	NFRx,o (dB)	ρ	NFRx (dB)	NFRx (dB)	ΔNF (dB)
870	1.45	1.8	1.66	1.74	+0.05
877	1.39	2	1.74	1.59	-0.15
880	1.38	2.2	1.81	1.75	-0.06
883	1.37	2.6	1.89	1.91	+0.02
890	1.38	3.8	2.38	2.45	+0.07

The plots of the receiver noise figure versus the antenna VSWR are presented in Fig. 4.6, where the original receiver noise figures from 4 dB to 10 dB in 1 dB increment are used, respectively, as a parameter of each curve. In a similar way, the degradation of the receiver noise figure versus the antenna VSWR are depicted in Fig. 4.7. From these plots we can see that the noise figure can degrade 2 to 3 dB depending on the original noise figure when the antenna VSWR varies from 1 to 4.

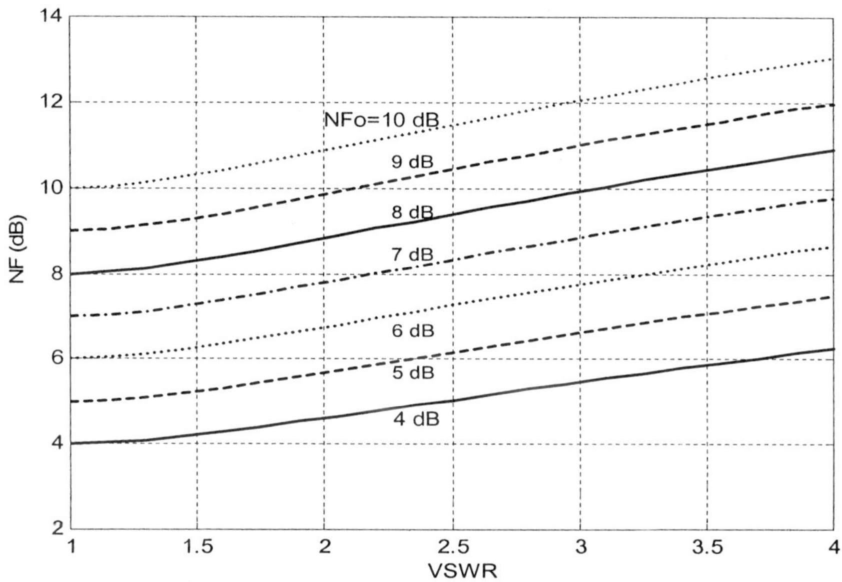

Figure 4.6 Receiver noise figure versus antenna VSWR with a parameter NF_o = 4 dB to 10 dB in 1 dB step

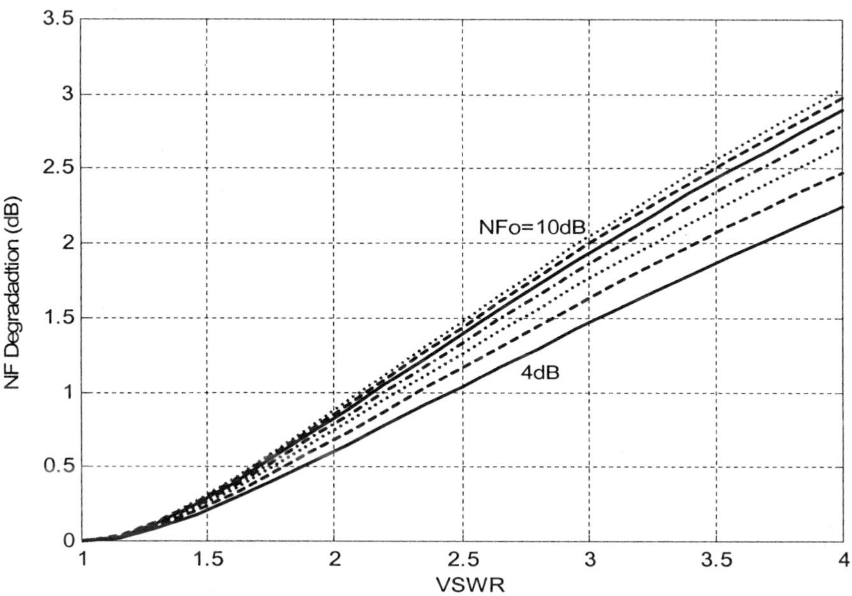

Figure 4.7. Receiver noise figure degradation versus VSWR with a parameter NF_o = 4 to 10 dB in 1 dB step

4.3. Intermodulation Characteristics

The intermodulation characteristics or referred to as intermodulation spurious response attenuation is another important measure of the receiver electrical performance. It characterizes the linearity of the receiver. In the wireless mobile systems, a two-tone test is commonly used to determine intermodulation characteristics of mobile station receivers. Corruption of signals due to intermodulation distortion (IMD) product created by two nearby strong interferers is defined to characterize receiver linearity.

The intermodulation is a nonlinear phenomenon, and it can be analyzed by means a memoryless nonlinear model given in (4.3.1):

$$y_o = a_0 + a_1 x_i + a_2 x_i^2 + a_3 x_i^3 + \cdots = \sum_{k=0}^{\infty} a_k x_i^k , \qquad (4.3.1)$$

where y_o and x_i are the output and input signals of a nonlinear system or device, respectively, and a_k (k = 1,2,3, ...) are the kth-order nonlinear coefficients. The magnitude of the output signal with respect to the magnitude of the input signal usually exhibits an odd function characteristic. This means that the odd-order terms on the right side of (4.3.1) dominate over the even-order terms. The intermodulation troublesome, in most cases, is caused by the odd-order nonlinearities, particularly the third-order nonlinearity. The even-order intermodulation distortions are usually very low, and they will not cause noticeable problems except in the direct conversion receivers.

4.3.1. Intermodulation Products and Intercept Points

The ratio of the fundamental signal, S_1 in dB, to the mth-order intermodulation product, S_m in dB, at the output of a nonlinear system or a nonlinear device is presented in (4.3.2), which is derived in Appendix 3A:

$$S_1 - S_m = (m-1)(IIP_m - S_i) , \qquad (4.3.2)$$

where S_i is the input desired signal of the system or device, and $IIP_{m,}$ is the mth-order input intercept point as depicted in Fig. 3A.1.

Now, we analyze the intermodulation distortion of the kth-stage device in the receiver chain. Assume that the inputs of the device contain a desired signal $S_{di,k}$ but also interferers with a strength $I_{i,k}$, and the mth-order intermodulation product of interferers is in the receiver channel band (although the interferers themselves are not in the receiver channel band). From Fig. 4.8, we can obtain the output signal $S_{do,k}$, and the mth-order intermodulation product denoted by $IM_{,mo,k}$ instead of S_m used in (4.3.2) as follows:

$$S_{do,k} = S_{di,k} + G_k \text{ dBm} \tag{4.3.3}$$

and

$$IM_{mo,k} = OIP_{m,k} - m(IIP_{m,k} - I_{i,k}) \text{ dBm}, \tag{4.3.4}$$

where G_k is the kth device gain, and $IIP_{m,k}$ and $OIP_{m,k}$ are input and output intercept points of the kth-stage device, respectively. Converting the intermodulation product $IM_{mo,k}$ to an equivalent product $IM_{mi,k}$ at the input of the kth device, $IM_{mi,k} = IM_{mo,k} - G_k$, we obtain

$$IM_{mi,k} = mI_{i,k} - (m-1)IIP_{m,k} \text{ dBm}. \tag{4.3.5}$$

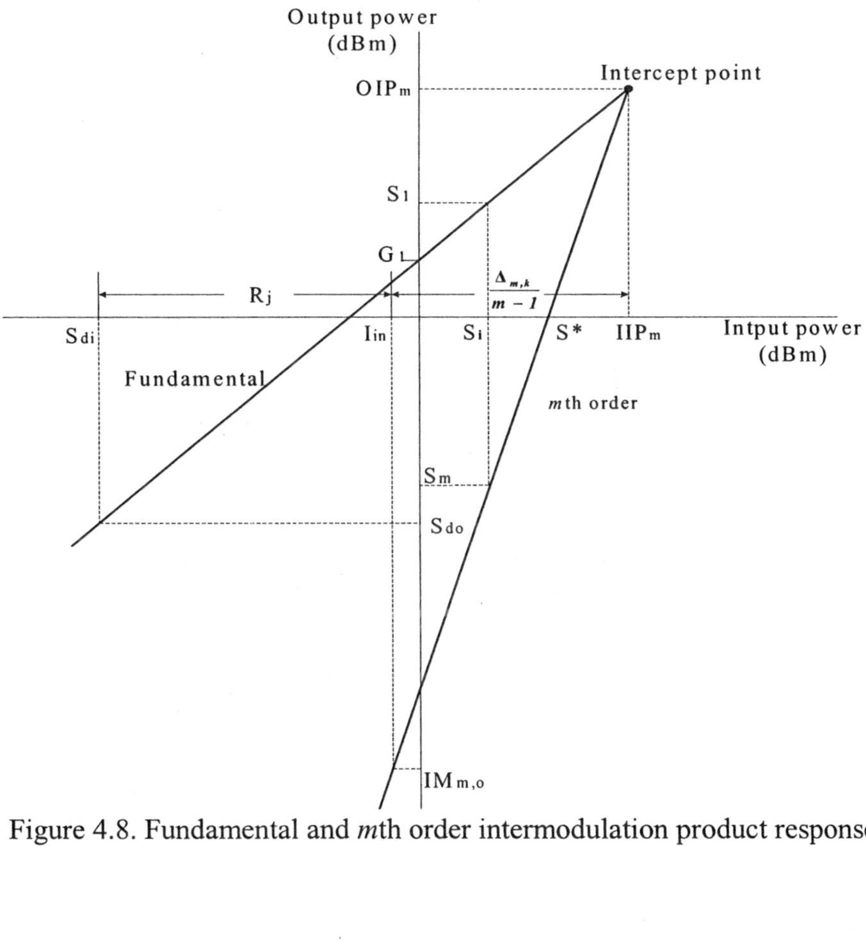

Figure 4.8. Fundamental and mth order intermodulation product response

Actually, (4.3.2) will have the same form as (4.3.5) if we replace S_i, S_I,and S_m by $I_{in,k}$, $I_{in,k} + G_k$, and $IM_{m\_in,k} + G_k$, respectively. (4.3.5) in its natural scale has the form of

$$P_{IMm\_in,k} = \frac{P_{Iin,k}^{m}}{P_{IIPm,k}^{m-1}}, \tag{4.3.6}$$

where $P_{IMm\_in,k}$, $P_{Iin,k}$, and $P_{IIPm,k}$, are the *m*th-order intermodulation product, the input interferer, and the *m*th-order input intercept point powers in mW, respectively.

From (4.3.5), the input intercept point of the *k*th-stage device is

$$IIP_{m,k} = \frac{1}{m-1}\left(mI_{in,k} - IM_{m\_in,k}\right) = I_{in,k} + \frac{\Delta_{m.k}}{m-1}, \tag{4.3.7}$$

where

$$\Delta_{m,k} = I_{in,k} - IM_{m\_in,k}. \tag{4.3.8}$$

In general, the odd-order intermodulation distortions of an analog device, such as an amplifier or a mixer, are usually much higher than the adjacent even order ones since the output signal power of these devices versus its input signal power is an odd-like function. The lowest odd order intermodulation distortion — i.e., the third-order intermodultion distortion is the most troublesome one in the RF receivers. On the other hand, the even-order intermodulation distortion, especially the second-order intermodulation distortion problem, is one of technical key issues that must be resolved to implement a direct conversion receiver. The second- and the third-order intercept point expressions can be easily derived from (4.3.7). If omitting the subscript *k*, they are as follows, respectively:

$$IIP_2 = 2I_{in} - IM_2 \tag{4.3.9}$$

and

$$IIP_3 = \frac{1}{2}\left(3I_{in} - IM_3\right) = I_{in} + \frac{\Delta_3}{2}, \tag{4.3.10}$$

where IIP_2 and IIP_3 are the second- and third-order intercept points, respectively, and IM_2 and IM_3 are the corresponding intermodulation products. (4.3.10) is often used to determine the third-order intercept point in terms of two equal test tones as depicted in Fig 4.9.

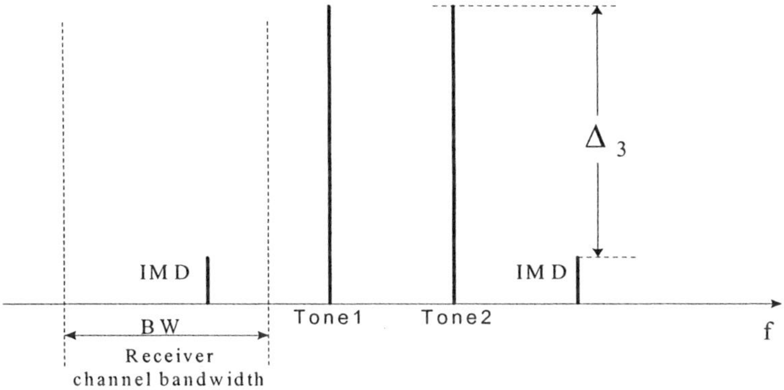

Figure 4.9. Third-order intermodulation products

In the receiver IF and BB blocks, signal voltage and voltage gain instead of signal power and power gain may be used to represent input, output, and characteristics of each stage. In this case the intermodulation products are also represented by voltage in mV or dBmV. However, the relationship between the mth-order intermodulation product voltage at the stage input, $\overline{V}_{IMm\_in}$, the mth-order input intercept point in voltage $\overline{V}_{IIPm}$, and the input interference voltage $V_{I,,i}$ has the exact same form as (4.3.6), and if omitting the subscript k, it is

$$\overline{V}_{IMm\_in} = \frac{\overline{V}^{m}_{I,in}}{\overline{V}^{m-1}_{IIPm}}. \tag{4.3.11}$$

In (4.3.11) all the voltages are rms values in mV. The corresponding expression in dBmV is similar to (4.3.7):

$$IIP_{Vm} = \frac{1}{m-1}\left(mI_{Vin} - IM_{Vm\_in}\right), \tag{4.3.12}$$

where

$$IIP_{Vm} = 20\log\left(\overline{V}_{IIPm}\right),\ I_{Vin} = 20\log\left(V_{Iin}\right),\ \text{and}\ IM_{Vm\_in} = 20\log\left(V_{IMm\_in}\right).$$

4.3.2. Cascaded Input Intercept Point

The intermodulation characteristics of an RF receiver are mainly determined by the linearity or in other words the overall intercept point of the receiver. Later on we shall see the phase noise of the synthesized local oscillators used in the receiver has certain impact on this performance as well. In the following analysis, it is assumed that the receiver consists of n stages of devices connected in cascade, for example, as shown in Fig. 4.10. The cascaded input intercept point of two different configurations — i.e., nonfrequency selective stages in cascade and frequency selective stages in cascade — will be derived from the input intercept point and the gain of individual stage in the cascade stage chain.

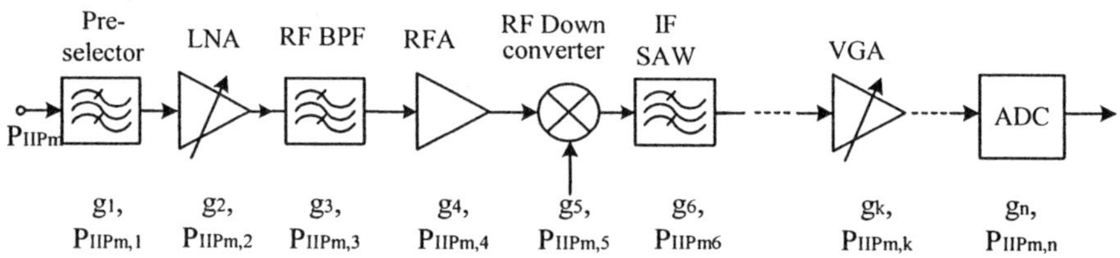

Figure 4.10. Receiver consisting of n stages of devices

4.3.2.1. Nonfrequency Selective System

A nonfrequency selective receiver system means that the interference tones, which cause the intermodulation distortion, are within the pass-band of all filters in the receiver. The interferers passing through the receiver obtain the gain same as that acquired by the desired signal and result in high intermodulation products. In reality, no receiver is practically designed in this way. However, for simplifying the analysis of the cascaded input intercept point we begin with this nonselective receiver system.

If the kth-stage gain equals g_k ($k = 1,2,3, \ldots n$), the interference level at the receiver input is P_{Iin}, and thus $P_{Iin,k} = g_1 g_2 \cdots g_{k-1} \cdot P_{Iin}$, from (4.3.6) the mth-order intermodulation product $P_{IMm\_out,k} = g_k P_{IMm\_in,k}$ generated by the kth stage is formed

$$P_{IMm\_out,k} = \frac{g_k (g_1 g_2 \cdots g_{k-1} P_{Iin})^m}{P_{IIPm,k}^{m-1}} . \tag{4.3.13}$$

This intermodulation product is amplified by a gain of $g_{k+1}\, g_{k+2} \cdots g_n$, at the output of the nth stage. Assuming the output load of the nth stage is R_L, the voltage $V_{IMm,k}$ across the load R_L resulting from the intermodulation product power $g_{k+1}\, g_{k+2} \cdots g_n \cdot P_{IMm\_out,k}$ is

$$V_{IMm,k} = \left(\frac{(g_1 g_2 \cdots g_{k-1} P_{Iin})^m}{P_{IIPm,k}^{m-1}} g_k g_{k+1} \cdots g_n \cdot R_L \right)^{\frac{1}{2}}. \tag{4.3.14}$$

The intermodulation product voltages generated by all stages in the receiver are added together at the nth-stage output load, and the corresponding total mth-order intermodulation product power P_{IMm} delivered to the load R_L in the worst case is equal to

$$\begin{aligned} P_{IMm} &= \frac{\left(\sum_{k=1}^{n} V_{IMm,k} \right)^2}{R_L} \\ &= \left[\sum_{k=1}^{n} \left(\frac{(g_1 g_2 \cdots g_{k-1} P_{Iin})^m}{P_{IIPm,k}^{m-1}} g_k g_{k+1} \cdots g_n \right)^{\frac{1}{2}} \right]^2. \end{aligned} \tag{4.3.15}$$

From (4.3.15) and referring (4.3.6), the cascaded m^{th} order input intercept point of the n stage block can be derived, and it has a form of

$$P_{IIPm} = \left(\frac{g_1 g_2 \cdots g_n P_{Iin}^m}{P_{IMm}} \right)^{\frac{1}{m-1}} = \left[\sum_{k=1}^{n} \left(\frac{\prod_{j=0}^{k-1} g_j}{P_{IIPm,k}^{m-1}} \right)^{\frac{m-1}{2}} \right]^{-\frac{2}{m-1}}. \tag{4.3.16}$$

In (4.3.16), P_{IIPm} and $P_{IIPm,k}$ (k = 1, 2, …, n) are in mW, and g_0 =1. On the other hand, if the intermodulation products created by the stages in the receiver chain are in power added together at the nth-stage output load, the mth-order intermodulation product power P_{IMm} and the cascaded mth-order input intercept point P'_{IIPm} have the following expressions, respectively:

$$P'_{IMm} = \sum_{k=1}^{n} \left(\frac{(g_1 g_2 \cdots g_{k-1} P_{Iin})^m}{P_{IIPm,k}^{m-1}} g_k g_{k+1} \cdots g_n \right) \tag{4.3.15a}$$

and

$$P'_{IIPm} = \left[\sum_{k=1}^{n} \left(\frac{\prod_{j=0}^{k-1} g_j}{P_{IIPm,k}} \right)^{m-1} \right]^{-\frac{1}{m-1}}. \tag{4.3.16a}$$

In the RF system design, (4.3.16) is generally used instead of (4.3.16a) since it will provide more design margin.

For the third-order intermodulation, the cascaded input intercept point is expressed by individual stage intercept point $P_{IIP3,k}$ ($k = 1, 2, \ldots, n$) as

$$P_{IIP3} = \frac{1}{\sum_{k=1}^{n} \frac{g_1 g_2 \cdots g_{k-1}}{P_{IIP3,k}}} = \frac{1}{\frac{1}{P_{IIP3,1}} + \frac{g_1}{P_{IIP3,2}} + \frac{g_1 g_2}{P_{IIP3,3}} + \cdots + \frac{g_1 g_2 \cdots g_{n-1}}{P_{IIP3,n}}} \text{ mW.} \tag{4.3.17}$$

The third-order cascaded input intercept point in dB scale is

$$\begin{aligned} IIP_3 &= -10\log\left(\sum_{k=1}^{n} \frac{g_1 g_2 \cdots g_{k-1}}{P_{IIP3,k}} \right) \\ &= -10\log\left(\frac{1}{P_{IIP3,1}} + \frac{g_1}{P_{IIP3,2}} + \frac{g_1 g_2}{P_{IIP3,3}} + \cdots \frac{g_1 g_2 \cdots g_{n-1}}{P_{IIP3,n}} \right) \end{aligned} \text{ dBm.} \tag{4.3.18}$$

(4.3.18) is popularly used in intermodulation calculation for the non-frequency selective receiver system. Based on (4.3.16a), the expression of the third-order cascaded input intercept point IIP'_3 is

$$IIP'_3 = -10\log\left(\frac{1}{P^2_{IIP3,1}} + \frac{g_1^2}{P^2_{IIP3,2}} + \left(\frac{g_1 g_2}{P_{IIP3,3}} \right)^2 + \cdots + \left(\frac{g_1 g_2 \cdots g_{n-1}}{P_{IIP3,n}} \right)^2 \right)^{\frac{1}{2}}. \tag{4.3.18a}$$

Again (4.3.18) is preferred for design.

4.3.2.2. Frequency Selective System

Practically all the receiver systems have frequency selectivity and suppress interferers to a great extent. It is necessary to take the receiver selectivity into account when calculating the overall input intercept point of the receiver. The selectivity of a receiver is usually implemented in the IF section and/or the analog base-band section by means of the receiver channel filters. To understand the effect of the selectivity to the input intercept point of a receiver, the analysis is in terms of two cascade stages consisting of a band-pass filter and an amplifier such as stages 6 and 7 in Fig. 4.10.

Assuming that the band-pass filter (stage 6) has a rejection ΔR_{j6} at the interferer frequency and an insertion loss IL in the pass band — i.e.,

$$G_6 = \begin{cases} IL \le 0 & \text{in - band} \\ IL - \Delta R_j & \text{at interference frequency,} \end{cases} \tag{4.3.19}$$

where ΔR_j is the filter rejection to the interference in dB. Assuming that the filter is a passive device, such as an IF SAW, with a very high input intercept point — i.e., $P_{IIPm,6} >> 1$ — the cascaded input intercept point of stages 6 and 7 $IIP'_{m,6\_7}$ can be expressed as

$$IIP'_{m,6\_7} = -10\log\left(\frac{1}{P_{IIPm,6}} + \frac{g'_6}{P_{IIPm,7}}\right) \cong -G'_6 + IIP_{m,7}\,. \tag{4.3.20}$$

In the above expression, G'_6 is equal to $G_6 = IL$ and for cascade non-selective devices

$$IIP'_{m,6\_7} = -IL + IIP_{m,7} = IIP_{m,6\_7}\,.$$

For the selective system, the interference level at the stage 7 input is now ΔR_j dB less and equals $I_{in,7}$ - ΔR_j. Thus the intermodulation product converted to the seventh-stage input $IM'_{m\_in,7}$ from (4.3.5) is now equal to

$$IM'_{m\_in,7} = m\left(I_{in,7} - \Delta R_j\right) - (m-1)IIP_{m,7}\,. \tag{4.3.21}$$

Since the intermodulation product generated by passive band-pass filter is negligible, the total intermodulation product resulting from stages 6 and 7 and converted to the sixth stage is

$$IM'_{m\_in,6\_7} = -IL + IM'_{m\_in,7} \,. \tag{4.3.22}$$

Substituting (4.3.22) into (4.3.7), we can obtain the input intercept point of cascade stages 6 and 7:

$$IIP'_{m,6\_7} = \frac{1}{m-1}\left[mI_{in,6} - IM_{m\_in,6\_7}\right] = -IL + IIP_{m,7} + \frac{m}{m-1}\Delta R_j \,. \tag{4.3.23}$$

Comparing the above equation with (4.3.20), it is apparent that the equivalent gain of the selective device (the sixth stage) G_6' is

$$G_6' = IL - \frac{m}{m-1}\Delta R_j \,.$$

It can be generalized to any frequency selective stage k_s, and the equivalent gain, G'_{m,k_S} in dB, has the following form:

$$G'_{m,k_S} = IL_{k_S} - \frac{m}{m-1}\Delta R_{j,k_S} \,, \tag{4.3.24}$$

where IL_{k_S} is the insertion loss and $\Delta R_{j,k_S}$ is rejection of the interference of the k_sth stage .

The equivalent gain of the third-order distortion for the k_Sth selective device G'_{3,k_S} is

$$G'_{3,k_S} = IL_{k_S} - \frac{3}{2}\Delta R_{j,k_S} \,, \tag{4.3.25}$$

and the equivalent gain for the second-order distortion G'_{2,k_S} is

$$G'_{2,k_S} = IL_{k_S} - 2\cdot\Delta R_{j,k_S} \,. \tag{4.3.26}$$

Let us look at an example of the filtering effect on the third-order intermodulation product. An IF SAW filter used for the receiver channel filtering has a 38 dB rejection to two interference tones. The actual rejection effect to these tones based on (4.3.25) is 1.5 times higher — i.e., 1.5 × 38 = 57 dB.

In intermodulation performance tests, two interference tones with a defined frequency separation are commonly used. These two test tones may be located on the skirt of the channel filter stop-band as depicted in Fig 4.11, and the filter has different suppression to these two interference tones since the skirt of the filter stop-band varies steeply. The value of $\Delta R_{j,k_S}$ in (4.3.25) should be the average value of the filter rejections on these two interference tones and it is calculated in the following way

$$\Delta R_{j,k_S} = \frac{2 \cdot \Delta R_{j,k_S}|_{Close\_tone} + \Delta R_{j,k_S}|_{Far\_tone}}{3}, \tag{4.3.27}$$

where $\Delta R_{j,k_S}|_{Close\_tone}$ is the rejection of the interference tone close to the carrier frequency, and $\Delta R_{j,k_S}|_{Far\_tone}$ is the rejection of the interference tone further away from the carrier.

The cascaded input intercept point of a frequency selective system can be still calculated by using (4.3.16), but in this formula the gain of all the selective stages is now given by (4.3.24) or should more accurately be the numerical value of (4.3.24)

$$g'_{m,k_S} = 10^{\frac{\left(IL_{k_S} - \frac{m}{m-1}\Delta R_{j,k_S}\right)}{10}}. \tag{4.3.28}$$

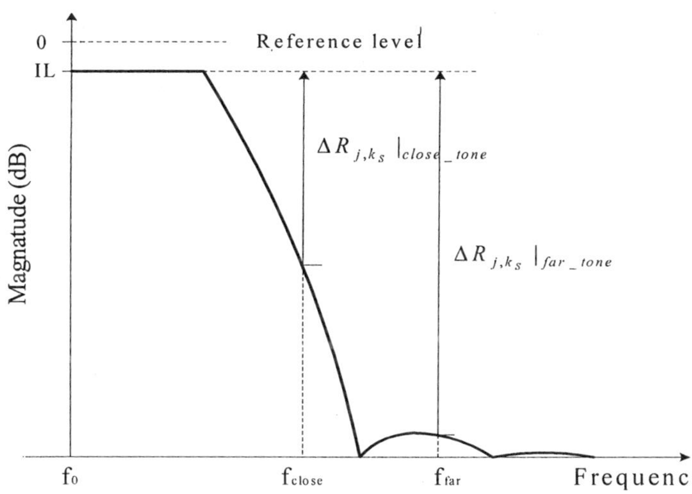

Figure 4.11. Frequency response of receiver channel band-pass filter

4.3.2.3. Cascaded IIP_m Expressions in Voltage and in Hybrid Form

In the IF and analog BB blocks, input and output signals of each stage are measured based on their voltage instead of the power, and the amplification level for individual stage is thus represented by voltage gain. The nonlinearity of the stage, in this case, is characterized in terms of the intercept point of the fundamental output voltage and the *m*th-order nonlinear output voltage versus input voltage curves. The corresponding *m*th-order cascaded input intercept point of n_v stages can be derived from (4.3.11) similar to deriving (4.3.16), and it has the form

$$\overline{V}_{IIPm} = \left(\frac{g_v \overline{V}_{Iin}^{\,m}}{\overline{V}_{IM,m}} \right)^{\frac{1}{m-1}} = \left[\sum_{k=1}^{n_v} \left(\frac{\prod_{j=0}^{k-1} g_j}{\overline{V}_{IIPm,k}} \right)^{m-1} \right]^{\frac{-1}{m-1}}, \tag{4.3.29}$$

where $g_{v,k}$ (k = 1,2, ..., n_v) is the voltage gain of the kth stage, $g_{v,0} = 1$, and $V_{IIPm,k}$ (k=1, 2, ..., n_v) is the *m*th-order input intercept point voltage of the *k*th device. For the third-order intermodulation distortion case or $m = 3$, we have the expression

$$\overline{V}_{IIP3} = \left[\sum_{k=1}^{n_v} \left(\frac{g_{v,1} g_{v,2} \cdots g_{v,k-1}}{\overline{V}_{IIP3,k}} \right)^2 \right]^{-\frac{1}{2}}, \tag{4.3.30}$$

or, written in its more commonly used form,

$$\frac{1}{\overline{V}_{IIP3}^{\,2}} = \left(\frac{1}{\overline{V}_{IIP3,1}} \right)^2 + \left(\frac{g_{v,1}}{\overline{V}_{IIP3,2}} \right)^2 + \left(\frac{g_{v,1} g_{v,2}}{\overline{V}_{IIP3,3}} \right)^2 + \cdots + \left(\frac{g_{v,1} g_{v,2} \cdots g_{v,n_v-1}}{\overline{V}_{IIP3,N}} \right)^2 . \tag{4.3.31}$$

In a true receiver chain, the front-end RF section up to the input of the RF down-converter is usually in the matching condition and the power variable and gain are employed there, but in the IF and analog base-band sections only the voltage variable is considered and the input and output ports may never be matched. The turning point from using the power variable to employing the voltage variable is usually at or just after the RF down-converter stage. Assuming the turning point is at the (n_p+1)th device in the receiver chain, it is not hard to prove that the overall *m*th-order input intercept point in power is

$$P_{IIPm} = \left[\sum_{k=1}^{n_p} \left(\frac{\prod_{j=0}^{k-1} g_j}{P_{IIPm,k}} \right)^{\frac{m-1}{2}} + \left(\frac{R_o \cdot \prod_{l=1}^{n_p} g_l}{\overline{V}^2_{IIPm|n_p+1\cdots n}} \right)^{\frac{m-1}{2}} \right]^{\frac{-2}{m-1}}, \tag{4.3.32}$$

where $g_0 = 1$, R_o is the matched input impedance of the (n_p+1)th stage, $\overline{V}^2_{IIPm|n_p+1\cdots n}$ is the cascaded input intercept point from the (n_p+1)th stage to the nth stage expressed in voltage, given by

$$\overline{V}^2_{IIPm|n_p+1\cdots n} = \left[\sum_{j=n_p+1}^{n} \left(\frac{\prod_{l=0}^{j-1} g_{v,l}}{\overline{V}_{IIPm,j}} \right)^{m-1} \right]^{\frac{-2}{m-1}}. \tag{4.3.33}$$

In (4.3.33), $\overline{V}_{IIPm,j}$ and $g_{v,j}$ are the voltage input intercept point and voltage gain of stage j ($j = n_{p+1}, n_{p+2}, \ldots n$), and $g_{v,0} = 1$.

The corresponding formula for the third-order cascaded intercept point of n stages is

$$P_{IIP3} = \frac{1}{\sum_{k=1}^{n_p} \frac{g_1 g_2 \cdots g_{k-1}}{P_{IIP3,k}} + \sum_{j=n_p+1}^{n} g_1 g_2 \cdots g_{n_p} R_o \left(\frac{g_{v,1} g_{v,2} \cdots g_{v,j-1}}{\overline{V}_{IIP3,j}} \right)^2}, \tag{4.3.34}$$

where $g_0 = 1$, R_o is the input impedance of the (n_p+1)th stage that is possibly matched with the preceding stage output impedance. From (4.3.34) we notice that voltage input intercept point $V_{IIP3,j}$ can be translated into power input intercept point $P_{IIP3,j}$ in terms of impedance R_o and by means of the following formula:

$$10 \log P_{IIP3,j} = 20 \log V_{IIP3,j} - 10 \log R_o . \tag{4.3.35}$$

If $R_o = 200\ \Omega$, $P_{IIP3,j}$ is related to $V_{IIP3,j}$ as

$$IIP_{3,j}\ (\text{dBm}) = IIP_{V,3,j}\ (\text{dBmV}_{\text{rms}}) - 53. \tag{4.3.36}$$

4.3.3. Calculation of Receiver Intermodulation Characteristics

The linearity of a receiver, which is usually represented by means of the cascaded input intercept point of the overall receiver, is the main cause of intermodulation distortion, but the intermodulation spurious response attenuation of a receiver also depends on other factors, such as the local oscillator phase noise level around the interference tones and receiver noise figure contribution. In this section, the intermodulation characteristic performance calculation takes all possible factors into account.

4.3.3.1. Allowed Degradation of the Received Desired Signal

The analysis of receiver performance may start with the *allowed maximum degradation* of the receiver input desired signal caused by noise and/or interference. A wireless communication system has a certain minimum carrier-to-noise/interference ratio, CNR_{min}, to achieve a given data error rate — i.e., a BER or a FER. The allowed maximum degradation of the receiver input desired signal $D_{max,in}$ is defined as the minimum carrier-to-noise/interference ratio CNR_{min} for a given data error rate subtracted from the receiver input desired signal, $S_{d,i}$, which is usually defined as 3 dB above the reference sensitivity level — i.e., $S_{d,i} = S_{\min\_ref} + 3$. The expression of the allowed maximum degradation of the desired signal caused by noise and/or interference at the receiver input is

$$D_{\max,in} = S_{d,i} - CNR_{\min}. \tag{4.3.37}$$

It is apparent that the allowed maximum degradation of the receiver input desired signal is actually the maximum noise and/or interference level, which deteriorates the desired signal to a carrier-to-noise/interference ratio of CNR_{min}.

In fact, a receiver has inherent noise consisting of thermal noise and the noise associated with the receiver noise figure. If the receiver noise figure is NF_{Rx} in dB, N_{nf} denotes the receiver inherent noise converted to its input port in dBm, and it has a form of

$$N_{nf} = -174 + NF_{Rx} + 10\log BW, \tag{4.3.38}$$

where BW is receiver noise bandwidth in Hz. The inherent noise of the receiver definitely reduces the maximum allowed degradation value of the input desired signal, and thus the allowed degradation of the input desired signal D_a turns into

$$D_a = 10\log\left(10^{\frac{D_{\max,in}}{10}} - 10^{\frac{N_{nf}}{10}}\right). \tag{4.3.39}$$

Assuming that S_{di} = -101 dBm, which is 3 dB above the reference sensitivity level –104 dBm of the CDMA mobile station receiver, CNR_{min} = -1 dB, which is defined in the CDMA minimum performance standard IS-98D [5], NF_{Rx} = 7 dB, and BW =1.23 MHz for a CDMA mobile receiver, from (4.3.39) we obtain

$$D_a = 10\log\left(10^{\frac{-101+1}{10}} - 10^{\frac{-174+7+60.9}{10}}\right) = -101.2 \text{ dBm}.$$

4.3.3.2. Intermodulation Distortion Resulting from Finite Receiver Linearity

The intermodulation distortion performance of a receiver depends on its linearity. The linearity of an RF receiver is characterized by using the cascaded input intercept point of different-order nonlinearity. If the mth-order cascaded input intercept point of the receiver is IIP_m, substituting S_i = I_{in} and $S_1 - S_m = I_{in} - IM_{m\_in}$ into (4.3.2) we obtain

$$I_{in} = \frac{1}{m}\left[IM_{m\_in} + (m-1)IIP_m\right]. \tag{4.3.40}$$

In (4.3.40), IM_{m_in} is the mth-order intermodulation product generated by two interference tones with the same strength I_{in} dBm.

From the previous section we know the allowed degradation of the receiver input desired signal is D_a for the data error rate corresponding to the carrier-to-noise/interference ratio, CNR_{min}. Now, assuming that this signal degradation is entirely due to the mth-order intermodulation product — i.e.,

IM_{m_in} in (4.3.40) is replaced by D_a, thus the allowed maximum interference level, $I_{in,a}$, which is also referred to as the *intermodulation spurious response attenuation*, is

$$I_{in,a} = \frac{1}{m}\left[D_a + (m-1)IIP_m\right], \tag{4.3.41a}$$

or using (4.3.39)

$$I_{in,a} = \frac{1}{m}\left[10\log\left(10^{\frac{D_{\max,in}}{10}} - 10^{\frac{N_{nf}}{10}}\right) + (m-1)IIP_m\right], \tag{4.3.41b}$$

where $D_{max.in}$ and N_{nf} are given by (4.3.37) and (4.3.38), respectively. For the third-order intermodulation, the allowable interference level at receiver input, when using (4.3.37) and (4.3.38), is

$$I_{in,a} = \frac{1}{3}\left[10\log\left(10^{\frac{S_{d,i}-CNR_{\min}}{10}} - 10^{\frac{-174+NF_{Rx}+10\log BW}{10}}\right) + 2IIP_3\right]. \tag{4.3.42}$$

However, the second term in the parentheses on the right side of (4.3.41) and (4.3.42) is usually much smaller than the first term. After neglecting the second term in this parentheses, (4.3.41b) can be simplified as

$$I_{in,a} = \frac{1}{m}\left[S_{d,i} - CNR_{\min} + (m-1)IIP_m\right]. \tag{4.3.43}$$

Using (4.3.43), we are able to approximately estimate the receiver minimum IIP_m based on the minimum requirement defined in the system standard, such as CDMA IS-98D [5] or ETSI GSM 0505 [6], on the intermodulation characteristics. The minimum performance specification usually defines the desired signal level $S_{d,i}$ in dBm and the minimum interference tone level $I_{in,min}$ in dBm or the relative level $R_{I/Sd} = I_{in,min} - S_{d,i}$ in dB. Using this power ratio $R_{I/Sd} = I_{in,min} - S_{d,i}$ and (4.3.43), we can calculate the required minimum $IIP_{m,min}$ of the receiver to be

$$IIP_{m,\min} = S_{d,i} + \frac{1}{m-1}\left(mR_{I/Sd} + CNR_{\min}\right). \tag{4.3.44}$$

In most cases, we are interested only in the third-order intermodulation product. Substituting $m = 3$ into (4.3.44), we obtain

$$\begin{aligned} IIP_{3,\min} &= S_{d,i} + \frac{1}{2}\left(3R_{I/Sd} + CNR_{\min}\right) \\ &= S_{d,i} + \frac{1}{2}\left[3\left(I_{in,\min} - S_{d,i}\right) + CNR_{\min}\right]. \end{aligned} \tag{4.3.45}$$

The minimum performance specification of the intermodulation spurious response attenuation for the CDMA mobile station defines that $S_{d,i} = -101$ dBm, $CNR_{min} = -1$ dB, and $I_{in\ min} = -43$ dBm. Using (4.3.45), we estimate the minimum third-order input intercept point requirement to be

$$IIP_{3,\min} = -101 + \frac{1}{2}(3 \times 58 - 1) = -14.5 \text{ dBm}$$

For an accurate $IIP_{3,\min}$ calculation we should use the following formula (4.3.46):

$$IIP_{3,\min} = \frac{1}{2}\left[3I_{in,\min} - 10\log\left(10^{\frac{S_{d,i} - CNR_{\min}}{10}} - 10^{\frac{-174 + NF_{Rx} + 10\log BW}{10}}\right)\right]. \tag{4.3.46}$$

Using the same data as the previous example in addition to $NF_{Rx} = 7$dB and $BW = 1.23$ MHz, plugging them into (4.3.46) we obtain the minimum IIP_3 to have a value

$$IIP_{3,\min} = \frac{1}{2}\left[3 \times (-43) - 10\log\left(10^{\frac{-101+1}{10}} - 10^{\frac{-174+7+10\log 1.23 \cdot 10^6}{10}}\right)\right] = -13.9 \text{ dBm}.$$

The difference between (4.3.45) and (4.3.46) in this example is 0.6 dB.

In practical design, performance margin is always needed. If the design margin for the intermodulation spurious response attenuation is 3 dB or more — i.e., the receiver can handle the intermodulation tone level to be equal to or greater than ($I_{in,min}$ + 3) dBm — from (4.3.46) we know the receiver input intercept point IIP_3 should be 4.5 dB greater than $IIP_{3,\min}$. Utilizing the result of the previous example, to obtain a 3 dB or more design margin for the intermodulation characteristics the IIP_3 of the CDMA mobile receiver should be –9.4 dBm or higher.

4.3.3.3. Degradation Caused by Phase Noise and Spurs of Local Oscillators

In reality, the minimum input intercept point of the receiver need be higher than that calculated from either (3.4.44) or (3.4.46). The phase noise and spurs of UHF and VHF PLL local oscillators will also contaminate the desired signal. The intermodulation interference tones mixing with the phase noise and/or spurs of the PLL LOs generate receiver in-channel bandwidth noise and spurs, which degrade the desired signal-to-noise ratio. Assuming that the average phase noise density over the receiver bandwidth BW at offset frequency equal to that of the interference tone offset from the signal carrier is N_{phase} dBc/Hz, and the spurious at a offset frequency equal to or nearby the frequency of the interference tone offset from the signal carrier is N_{spu} in dBc, their powers contributing to desired signal contamination in mW are, respectively,

$$P_{phn} = 10^{\frac{N_{phase} + 10\log BW + I_{in} - \Delta R}{10}} \tag{4.3.47}$$

and

$$P_{spu} = 10^{\frac{N_{spu} + I_{in} - \Delta R}{10}}, \tag{4.3.48}$$

where I_{in} is the intermodulation interference tone, and ΔR is the rejection to the interference tone of a filter preceding the corresponding down-converter.

Considering these contributions to signal degradation, the allowed intermodulation interference tone level at receiver input from (4.3.41) becomes

$$I_{in,a} = \frac{1}{m}\left[10\log\left(10^{\frac{D_{\max,in}}{10}} - 10^{\frac{N_{nf}}{10}} - \sum_{j=1}^{2}\sum_{k=1}^{2} P_{phn,j,k} - \sum_{j=1}^{2}\sum_{k=1}^{2} P_{spu,j,k}\right) + (m-1)IIP_m\right], \tag{4.3.49}$$

where $P_{phn,j,k}$ and $P_{spu,j,k}$ (j, k = 1, 2) are the phase noise and spurs contributions resulting from the first $j = 1$ and the second $j = 2$ LOs, and the first $k = 1$ and the second $k = 2$ interference tones, respectively. They have the same expressions as (4.3.47) and (4.3.48), but N_{phase} and N_{spu} are replaced by $N_{phase,j,k}$ and $N_{spu,j,k}$ (j, k = 1, 2), respectively. In the direct conversion receiver case, only one UHF PLL LO is needed, and therefore the double summation in the parentheses of (4.3.49) becomes a single summation.

For the third-order intermodulation, the allowed interference level is

$$I_{in,a} = \frac{1}{3}\left[10\log\left(10^{\frac{D_{\max,in}}{10}} - 10^{\frac{N_{nf}}{10}} - \sum_{j=1}^{2}\sum_{k=1}^{2}P_{phn,j,k} - \sum_{j=1}^{2}\sum_{k=1}^{2}P_{spu,j,k}\right) + 2IIP_3\right]. \tag{4.3.50}$$

For a given interference level, such as the level defined by the minimum performance requirement of system specifications $I_{in,a} = I_{in,min}$. Thus the minimum IIP_3 of the receiver can be calculated by using the following formula:

$$IIP_{3,\min} = \frac{1}{2}\left[3I_{in,\min} - 10\log\left(10^{\frac{D_{\max,in}}{10}} - 10^{\frac{N_{nf}}{10}} - \sum_{j=1}^{2}\sum_{k=1}^{2}P_{phn,j,k} - \sum_{j=1}^{2}\sum_{k=1}^{2}P_{spu,j,k}\right)\right] \tag{4.3.51}$$

Continuing the previous example and considering only UHF LO contributions, which results from the average phase noise, –136 dBc/Hz, the spurs, less than –75 dBc, around both interferers, and $\Delta R = 0$ dB since there is no channel filtering preceding the RF down-converter, the required minimum IIP_3 now becomes

$$IIP_{3,\min} = \frac{1}{2}\left[3\times(-43) - 10\log\left(10^{-10} - 10^{-\frac{106.1}{10}} - 2\times 10^{-\frac{118.1}{10}} - 2\times 10^{-\frac{118}{10}}\right)\right] = -13.7 \text{ dBm}.$$

In this example, the phase noise and spurs of the UHF LO are low enough, and they have negligible impact, only 0.2 dB, on the receiver IIP_3 requirement.

4.3.3.4. Degradation Resulting from Cross-modulation

The influence of the cross-modulation to the receiver performance is discussed in detail in Section 4.4. Here we use only the results from that section to evaluate its contribution to the degradation of the desired signal-to-noise/interference ratio. In a full-duplex transceiver the leakage of AM transmission can cross-modulate any interference tone nearby the desired signal, and partial spectrum of the cross-modulated interference tone may sneak into the receiver channel bandwidth if the interference tone is very close to the signal. The cross-modulation product within the receiver channel bandwidth may impact the receiver performance if the interference tone and the transmission leakage are high enough. In the CDMA mobile station case, the cross-modulation product, N_{CM} in dBm, is approximately expressed as

$$N_{CM} = I_{in} - 2IIP_{3,LNA} + 2(TX_{pwr} + IL_{dpx\_Rx} - R_{dplx\_Tx}) + C, \quad (4.3.52)$$

where $IIP_{3,LNA}$ is the LNA input intercept point, TX_{pwt} is the transmitter output power at the antenna port in dBm, R_{dplx} is the duplexer receiver side filter rejection to the transmission in dB, and C is a correction factor associated with waveform magnitude fluctuation and interference tone offset frequency, which is approximately –3.8 dB and –5.8 dB for the cellular and PCS band CDMA mobile station receivers, respectively. The allowed interference level at the receiver input now is reduced to

$$I_{in,a} = \frac{1}{m}\left[10\log\left(10^{\frac{D_{\max,in}}{10}} - 10^{\frac{N_{nf}}{10}} - \sum_{j=1}^{2}\sum_{k=1}^{2} P_{phn,j,k} - \sum_{j=1}^{2}\sum_{k=1}^{2} P_{spu,j,k} - 10^{\frac{N_{CM}}{10}}\right) + (m\text{-}1)IIP_m\right]. \quad (4.3.53)$$

For the third-order intermodulation, the allowed interference tone level expression is

$$I_{in,a} = \frac{1}{3}\left[10\log\left(10^{\frac{D_{\max,in}}{10}} - 10^{\frac{N_{nf}}{10}} - \sum_{j=1}^{2}\sum_{k=1}^{2} P_{phn,j,k} - \sum_{j=1}^{2}\sum_{k=1}^{2} P_{spu,j,k} - 10^{\frac{N_{CM}}{10}}\right) + 2IIP_m\right]. \quad (4.3.54)$$

The minimum third-order intercept point for a specified interference tone level $I_{in,min}$ is now

$$IIP_{3,\min} = \frac{1}{2}\left[3I_{in,\min} - 10\log\left(10^{\frac{D_{\max,in}}{10}} - 10^{\frac{N_{nf}}{10}}\right.\right.$$
$$\left.\left. - \sum_{j=1}^{2}\sum_{k=1}^{2} P_{phn,j,k} - \sum_{j=1}^{2}\sum_{k=1}^{2} P_{spu,j,k} - 10^{\frac{N_{CM}}{10}}\right)\right]. \qquad (4.3.55)$$

If $IIP_{3,LNA}$ = 8 dBm, TX_{pwt}= 25 dBm, IL_{dplx_Rr} = -2.5 dB, R_{dplx_Tx} = 48 dB, I_{in} = -43 dBm, and C = -3.8, from (4.3.52) we have N_{CM} = -113.8 dBm. Substituting this intermodulation product value and all other data used in the previous example, the required minimum IIP_3 of the receiver increases to

$$IIP_{3,\min} = \frac{1}{2}\left[3\times(-43) - 10\log\left(10^{-10} - 10^{-\frac{106.1}{10}}\right.\right.$$
$$\left.\left. - 2\times 10^{-\frac{118.1}{10}} - 2\times 10^{-\frac{118}{10}} - 10^{-\frac{113.8}{10}}\right)\right] = -13.6 \text{ dBm}.$$

Therefore, the cross-modulation interference causes only 0.1 dB increase of the IIP_3 requirement. In the normal case the cross-modulation interference and the LO phase noise and spurious do not cause much increase of the receiver IIP_3 requirement.

(4.3.53) to (4.3.55) can be used for evaluating the intermodulation spurious response performance of the overall receiver, and they are taking the main contributions to carrier-to-noise/interference into account. In the practical receiver design, it is always necessary to set design margin for key performance parameters. If the design margin for the intermodulation spurious response attenuation is 3 dB higher than the minimum requirement, in this case the LO phase noise, the LO spurs, and the cross-modulation interference products also increase 3 dB. Thus the minimum IIP_3 requirement becomes –8.7 dBm, which is 4.9dB higher than –13.6 dBm. This means the IIP_3 needs approximate 5 dB higher if we would like to improve intermodulation characteristic performance 3 dB better than the minimum requirement –43 dBm in the CDMA mobile station receiver.

4.4. Single-Tone Desensitization

Single-tone desensitization is a unique specification for CDMA mobile systems. It results from the CDMA system being incorporated with the AMPS and TDMA system in the same frequency bands, the cellular and PCS frequency bands. Furthermore, the signal bandwidth in the AMPS and TDMA systems is quite narrow, around 25 kHz, like a CW tone to the CDMA signal. The CDMA transceiver is a full-duplex system. The CDMA receiver always suffers from transmission leakage interference particularly in the LNA before a receiver RF band-pass filter, which will further suppress the transmission leakage down to an insignificant level. If a strong interference tone is present near the desired CDMA receiver signal, the amplitude modulation of the transmission leakage will cross-modulate the interference tone in the receiver LNA. The spectrum of the cross-modulated tone may partially spread into the receiver channel bandwidth while the single-tone interferer is close to the desired signal enough. The receiver will be desensitized if the cross-modulation product getting into the receiver channel band is high enough. In this section, we present a quantitative analysis of the single-tone desensitization issue based on a simplified amplitude modulation signal model instead of true CDMA reverse link signal. A more accurate analysis utilizing Volterra series can be found in reference [4].

4.4.1. Cross-Modulation Products

Assume that the single tone interferer and the transmission leakage can be approximately expressed in the following form:

$$x_i = A_I \cos\omega_I t + A_{Tx\_Leak} \cdot m_A(t)\cos\omega_{Tx} t, \tag{4.4.1}$$

where A_I is single-tone interference amplitude, ω_I is the angular frequency of the single-tone interference, A_{Tx_Leak} is the transmitter leakage signal average level, ω_{Tx} is the transmission carrier frequency, and $m_A(t)$ is the amplitude modulation of the transmission signal, which has a fundamental frequency associated with the CDMA PN sequence chip rate 1.2288 MHz, its variation depends on the pulse shaping filter and the reverse link channel configurations, and it also has the following relationships (also see Appendix 3B):

$$\lim_{T\to\infty}\frac{1}{T}\int_0^T m_A^2(t)dt = 1, \tag{4.4.2}$$

and the square of the relative power fluctuation due to AM, $\Delta P_{m_A}^2$, is

$$\Delta P_{m_A}^2 = \lim_{T\to\infty}\frac{1}{T}\int_0^T \left(m_A^2(t)-1\right)^2 dt. \tag{4.4.3}$$

The interference tone and the transmission leakage x_i is amplified in the LNA with a nonlinearity that can be approximately represented by a power series as given in (4.3.1). This amplification is mathematically equivalent to substituting (4.4.1) into (4.3.1). In the outputs of the LNA, only the terms associated with interference fundamental frequency are interesting, and they are

$$y_{o\_I}(t) = \left[a_1 A_I + \frac{3}{2}a_3 A_I A_{Tx\_Leak}^2 m_A^2(t)\right]\cos\omega_I t. \tag{4.4.4}$$

On the right side of (4.4.4), the first term is the amplified interference tone and the second term is the cross-modulation resulting from the third-order distortion products of the LNA nonlinearity. The cross-modulation term is formed by the first order of the interference tone multiplied with the second order of the transmission leakage. Thus, the spectrum of the cross-modulation has twice the bandwidth of the transmission as depicted in Fig. 4.12. In this figure, *BW* is denoted as the desired signal occupied bandwidth of the receiver and the transmitter, f_d

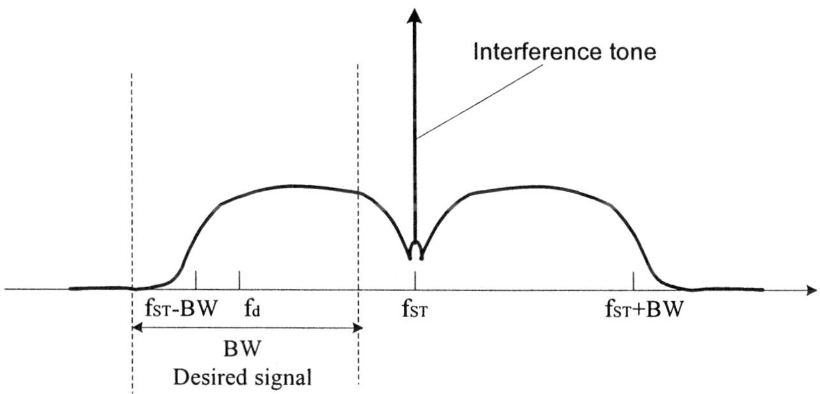

Figure 4.12. Spectrum of cross-modulated interference tone

represents the carrier frequency of the desired reception signal, and f_{ST} is the frequency of the single-tone interferer.

From (4.4.4) and (4.4.3) we can obtain that the ratio of the single tone interferer level A_{ST} and the cross-modulation product level A_{CM} is

$$\frac{A_{ST}}{A_{CM}} = \frac{|a_1|}{\frac{3}{2}|a_3| \cdot \left|\Delta P_{m_A}\right| \cdot A_{Tx\_Leak}^2}, \tag{4.4.5}$$

where A_{ST} and A_{CM} both are assumed to be the level at the LNA input, although expression (4.4.5) still holds if A_{ST} and A_{CM} both are considered as output level of the LNA. Utilizing the following relationship [3] (see Appendix 4B),

$$A_{IIP3\_LNA}^2 = \frac{4}{3}\left|\frac{a_1}{a_3}\right|, \tag{4.4.6}$$

(4.4.5) turns into

$$\frac{A_{ST}}{A_{CM}} = \frac{A_{IIP3\_LNA}^2}{2 \cdot \left|\Delta P_{m_A}\right| \cdot A_{Tx\_Leak}^2} = \frac{P_{IIP3\_LNA}}{2 \cdot \left|\Delta P_{m_A}\right| \cdot P_{Tx\_Leak}}, \tag{4.4.7}$$

where $A_{IIP3\_,LNA}$ and P_{IIP3_LNA} are the input intercept point of the LNA in voltage and in power, respectively, and P_{Tx_Leak} is the transmitter leakage power to the receiver LNA input.

Converting (4.4.7) into dB scale, the cross-modulation product can be expressed as

$$N_{CM} = 20\log A_{CM} = 2N_{Tx\_Leak} - 2\left(IIP_{3\_LNA} - IL_{dplxr\_Rx}\right) + I_{ST} + M_A + 6, \tag{4.4.8}$$

where

$$N_{Tx\_Leak} = 10\log P_{Tx\_Leak} = TX_{pwr} - \Delta R_{dplx\_Tx} \tag{4.4.9}$$

$$IIP_{3\_LNA} = 10\log P_{IIP3\_LNA} \tag{4.4.10}$$

$$I_{ST} = 20\log A_{ST} \tag{4.4.11}$$

and

$$M_A = 10\log\left(\Delta P_{m_A}^2\right). \tag{4.4.12}$$

TX_{pwr} is the transmitter output power at the antenna port in dBm, and ΔR_{dplx_Tx} is the receiver side filter rejection of the duplexer to the transmission in dB, $IL_{dplx\_Rx} < 0$ is the receiver side insertion loss of the duplexer in dB, and M_A is (4.4.3) expressed in dB and can be calculated by means of the PDF of the transmission waveform magnitude as shown in Fig. 3.18 and formula (3B.5). However, only part of the cross-modulation product spectrum extends into the receiver channel bandwidth as depicted in Fig.4.12. The cross-modulation spectrum portion in the receiver channel bandwidth depends on the offset frequency of the single-tone interferer from the receiver signal carrier.

Assuming that the spectrum of the cross-modulation is flat within $2\times BW$ around the interference tone, and the offset frequency is $\Delta f = |f_{ST} - f_d|$, the really troublesome portion of the cross-modulation product can be calculated from (4.4.13) of the modified (4.4.8)

$$N_{CM,e} = 2N_{Tx\_Leak} - 2(IIP_{3\_LNA} - IL_{dplxr\_Rx}) + I_{ST} + C, \qquad (4.4.13)$$

where C is a correction factor approximately equal to

$$C = M_A + 6 + 10\log\frac{1.5\cdot BW - \Delta f}{2\cdot BW} \qquad (4.4.14)$$

and BW is the CDMA signal occupied bandwidth, or 1.2288 MHZ.

The correction factor can be estimated as follows. M_A calculation can be carried out in the same way as $\overline{IMD_{2\_LF}}$ calculated in Section 3.2.3.2. The PDF of the AM transmission magnitude for the CDMA 9.6 kbit/sec voice data is shown in Fig. 3.18, and the corresponding M_A is approximately –5.6 dB. The interference tone offset frequency Δf for cellular and PCS bands is defined as ± 900 kHz and ± 1.25 MHz, respectively. Thus the last term on the right side of (4.4.14) is approximately –4.2 dB for the cellular band CDMA and –6.2 dB for the PCS band CDMA. Using these consequences, from (4.4.14) we obtain the correction factors to be

$$C_{cell} = -3.8\text{ dB} \qquad \text{and} \qquad C_{PCS} = -5.8\text{ dB}.$$

The correction factor may also be obtained by means of measurements. The cross-modulation level is approximately 80 dB lower than the interference tone. To accurately measure the cross-modulation, it is necessary to cancel or suppress the tone level within the dynamic range of vector signal analyzer.

4.4.2. Determination of the Allowed Single-Tone Interferer

Similarly to the analysis of intermodulation distortion performance in Section 4.4.3, letting the allowed degradation of the input desired signal at the receiver antenna port D_a as given in (4.3.39) equals the total degradation resulting from the cross-modulation, phase noise, and spurious of PLL LOs. From (4.4.13) we can derive the expression of the allowed single-tone interferer as

$$I_{ST} = 10\log\left(10^{\frac{D_{\max,in}}{10}} - 10^{\frac{N_{nf}}{10}} - P_{phn} - P_{spu}\right) + 2\left(IIP_{3\_LNA} - IL_{dplx\_Rx}\right) - 2(TX_{pwr} - R_{dplx\_Tx}) - C, \tag{4.4.15}$$

where $D_{max,in}$, N_{nf}, P_{phn}, and P_{spu} are defined in (4.3.37), (4.3.38), (4.3.47), and (4.3.48), respectively.

Utilizing (4.3.47) and (4.3.48), we rearrange (4.4.15) and obtain

$$10^{\frac{I_{ST}-[2(IIP_{3\_LNA}-IL_{dplxr\_Rx})-2(TX_{pwr}-R_{dplx\_Tx})-C]}{10}} + 10^{\frac{N_{phase}+10\log BW+I_{ST}}{10}} + 10^{\frac{N_{spu}+I_{ST}}{10}} = 10^{\frac{D_{\max,in}}{10}} - 10^{\frac{N_{nf}}{10}}.$$

Extracting $10^{I_{ST}/10}$ from all the terms on the left side of the above equation and moving the rest factor to its right side, after taking logarithm we have an explicit expression of I_{ST} as follows

$$I_{ST} = 10\log\left[\frac{10^{\frac{D_{\max,in}}{10}} - 10^{\frac{N_{nf}}{10}}}{10^{\frac{-[2(IIP_{3\_LNA}-IL_{dplxr\_Rx})-2(TX_{pwr}-R_{dplx\_Tx})-C]}{10}} + 10^{\frac{N_{phase}+10\log BW}{10}} + 10^{\frac{N_{spu}}{10}}}\right] \tag{4.4.16}$$

For an example of a Cellular band CDMA mobile station receiver, assuming that $D_{max,in}$ = -101+1 = -100 dBm, $IIP_{3,LNA}$ = 8 dB, NF_{Rcvr} = 7 dB, N_{phn} = -137 dBm/Hz, N_{spu} = -85 dBc, BW = 1.23 MHz, TX_{pwr} = 23 dBm, IL_{dplx_Rx} = 3 dB, R_{dplx} = 52 dBc, and Δf = 900 kHz, from (4.3.16) we obtain

$$I_{ST} = -25.43 \text{ dBm},$$

and in this case we have a margin 4.57 dB when compared with the IS98D specification –30 dBm.

4.5. Adjacent /Alternate Channel Selectivity and Blocking Characteristics

The *adjacent/alternate channel selectivity* measures a receiver's ability to receive a desired signal at its assigned channel frequency in the presence of adjacent/alternate channel signal at a given frequency offset from the center frequency of the assigned channel. The blocking characteristic measures a receiver ability to receive a desired signal at its assigned channel frequency in the presence of an unwanted interferer at frequencies other than those of the adjacent channels. The adjacent/alternate channel interference signal is usually modulated, but the blocking interferer is often defined as a continuous waveform (CW) tone. Either the adjacent/alternate channel selectivity or the blocking characteristics of a receiver is determined mainly by the receiver channel filter attenuation to the adjacent/alternate channel or the unwanted interferer, and the phase noise and spurs of the synthesized LOs in the adjacent/alternate channel bandwidth or around the unwanted interferer.

4.5.1. Desired Signal Level and Allowed Degradation

For most wireless mobile systems, the desired signal level at receiver input for the test of receiver blocking characteristic is defined as 3 dB above the reference sensitivity level S_{min_ref} — i.e.,

$$S_{d,i} = S_{\min\_ref} + 3 . \tag{4.5.1}$$

This definition of the desired signal level is also commonly applied in other receiver characteristics, such as intermodulation and single-tone desensitization, as described in previous sections.

However, the desired signal level for the adjacent/alternate channel selectivity test is defined differently from the other receiver characteristic tests and also differently from a mobile system to another. In the WCDMA mobile station minimum performance specification, the desired signal level for the adjacent channel selectivity test is defined as 14 dB above the reference sensitivity — i.e., $S_{d,i\_WCDMA} = -92.7$ dBm. The desired signal is 20 dB above its reference sensitivity level or S_{d,i_GSM} = –82 dBm used in the GSM mobile system. For the AMPS mobile transceiver, the desired signal level in this test is specified as 3 dB above the true sensitivity level S_{min_AMPS}, or S_{d,i_AMPS}.= S_{min_AMPS} + 3, and this means the desired signal level is not fixed in the AMPS system.

The calculation of the adjacent/alternate channel selectivity or the blocking characteristics of a receiver also begins with the allowed degradation of the desired signal as done in other performance calculations. The allowed degradation expression has the same form as (4.3.39), but the allowed degradation range for the adjacent/alternate channel selectivity test may be much wider than that for the blocking test. This is due to the desired signal level for the adjacent/alternate channel selectivity test can be 14 or even 20 dB higher than the receiver reference sensitivity:

$$D_a = 10\log\left(10^{\frac{D_{\max,in}}{10}} - 10^{\frac{N_{nf}}{10}}\right), \tag{4.3.39}$$

where

$$D_{\max,in} = S_{d,i} - CNR_{\min} \tag{4.3.37}$$

$$N_{nf} = -174 + NF_{Rcvr} + 10\log BW. \tag{4.3.38}$$

4.5.2. Formula of Adjacent/Alternate Channel Selectivity and Blocking Characteristics

The adjacent/alternate channel interference signal or blocking interferer I_i, mixing with the phase noise and spurs of the synthesized LO, generates in-receiver-channel noise and spurs, which degrade the desired signal-to-noise/interference ratio. Assuming that the phase noise density around the adjacent/alternate channel interference signal or the blocking interferer, I_i, is $N_{phase,1}$ dBm/Hz, and the spurious nearby the interference signal or interferer is $N_{spu,1}$ dB, their contributions to the degradation of the desired signal in the natural scale are, respectively,

$$P_{phn,1} = 10^{\frac{N_{phase,1} + 10\log BW + I_i}{10}} \tag{4.5.2}$$

and

$$P_{spu,1} = 10^{\frac{N_{spu,1}+I_i}{10}}. \tag{4.5.3}$$

In the superheterodyne receiver, the adjacent/alternate channel signal or the blocking interferer mixes with UHF LO signal down to an IF interferer. The interferer will be partially suppressed by the IF channel filter when it passes through this filter and routs to the second down-converter. The attenuated IF interferer mixes with the phase noise $N_{phas,2}$ and spurious $N_{spr,2}$ of the second LO and generates in-channel-band interference with a relative level from the phase noise and the spurious, respectively, as follows,

$$P_{phn,2} = 10^{\frac{N_{phas,2} - \Delta R_{IF} + 10\log BW + I_i}{10}} \tag{4.5.4}$$

and

$$P_{spu,2} = 10^{\frac{N_{spu,2} - \Delta R_{IF} + I_i}{10}}, \tag{4.5.5}$$

where ΔR_{IF} is the relative rejection of the IF channel filter to the adjacent/alternate channel signal or the blocking interferer I_i.

The total degradation of the desired signal due to the phase noises and spurs of the LOs mixing with the adjacent/alternate channel signal or the blocking interferer is

$$D_{total} = 10\log\left(\sum_{k=1}^{2} P_{phn,k} + \sum_{k=1}^{2} P_{spu,k}\right). \tag{4.5.6}$$

Letting D_{total} equal the allowed degradation presented by (4.3.39), we obtain

$$10^{\frac{D_{\max,in}}{10}} - 10^{\frac{N_{nf}}{10}} = 10^{\frac{I_i}{10}}\left(10^{\frac{N_{phn,1}+10\log BW}{10}} + 10^{\frac{N_{spu,1}}{10}} + 10^{\frac{N_{phn,2} - \Delta R_{IF} + 10\log BW}{10}} + 10^{\frac{N_{spu,2} - \Delta R_{IF}}{10}}\right). \tag{4.5.7}$$

From (4.5.7), we can derive the adjacent/alternate channel selectivity or the blocking characteristic $\Delta S_{adj/alt/block}$ to have the following form:

$$\Delta S_{adj/alt/block} = I_{adj/alt/block} - S_{d,i}$$

$$= 10\log\left(\frac{10^{\frac{S_{d,i}-CNR}{10}} - 10^{\frac{-174+10\log BW+NF}{10}}}{10^{\frac{N_{phn,1}+10\log BW}{10}} + 10^{\frac{N_{phn,2}-\Delta R_{IF}+10\log BW}{10}} + 10^{\frac{N_{spu,1}}{10}} + 10^{\frac{N_{spu,2}-\Delta R_{IF}}{10}}}\right) - S_{d,i}$$

(4.5.8)

For instance, an AMPS receiver has (1) a noise figure NF = 6.6 dB and a carrier-to-noise ratio $CNR_{,min}$ = 2.6 dB corresponding to a $SINAD$ = 12 dB at its output; (2) an UHF LO: phase noise $N_{phase,30}$ = -98dBc/Hz, and spurious $N_{spu,30}$ = -65dBc at ±30 kHz frequency offset, $N_{phase,60}$ = -116 dBc/Hz, and spurious $N_{spu,60}$ = -85 dBc at ±60 kHz frequency offset; (3) a VHF LO: phase noise $N_{phase,30}$ = -74 dBc/Hz, and spurious $N_{spu,30}$ = -60dBc at ±30 kHz frequency offset, $N_{phase,60}$ = -104 dBc/Hz, and spurious $N_{spu,60}$ = -80 dBc at ±60 kHz frequency offset; and (4) an IF channel filter: rejection $\Delta R_{IF,30}$ = 12 dB at ±30 kHz and $\Delta R_{IF,60}$ = 25 dB. From the noise figure and carrier-to-noise ratio $CNR_{,min,}$ considering receiver bandwidth BW = 30 kHz it is easy to calculate the sensitivity $S_{min} \cong -120$ dBm. Utilizing (4.5.8), we obtain the adjacent and alternate channel selectivities are, respectively,

$$\Delta S_{adj} = 41.54\text{ dB} \qquad \text{and} \qquad \Delta S_{alt} = 68.46\text{ dB}.$$

The specifications of the AMPS mobile station for the adjacent and alternate channel selectivities are ≥ 16 dB and ≥ 60 dB, respectively. Thus, the adjacent channel selectivity has a margin of 25.54 dB, and the margin of the alternate channel selectivity is approximately 8.5 dB.

The side lobe of adjacent channel interference signal may extend to the desired signal bandwidth, and it will also degrade the adjacent channel selectivity performance. The side lobe level of the adjacent spectrum depends on the modulation and the pulse-shaping filter characteristic. Usually the side lobe level is at least 35 dB blow the main lobe of the adjacent channel signal. Assuming the side lobe portion of the adjacent channel interferer spectrum in the desired signal bandwidth is Δ_{Slobe} dB lower than the adjacent channel signal power, formula (4.5.8) for the adjacent channel selectivity turns into

$$\Delta S_{adj} = I_{adj} - S_{d,i}$$
$$= 10\log\left(10^{\frac{S_{d,i}-CNR}{10}} - 10^{\frac{-174+10\log BW+NF}{10}}\right)$$
$$-10\log\left(10^{\frac{N_{phn,1}+10\log BW}{10}} + 10^{\frac{N_{phn,2}-\Delta R_{IF}+10\log BW}{10}}\right. \tag{4.5.9}$$
$$\left. + 10^{\frac{N_{spu,2}}{10}} + 10^{\frac{N_{spu,2}-\Delta R_{IF}}{10}} + 10^{\frac{-\Delta_{Slobe}}{10}}\right). - S_{d,i}$$

4.5.3. Two-Tone Blocking and AM Suppression Characteristics

The CDMA system sees the AMPS signal like a tone since its bandwidth is quite narrow comparing with the CDMA signal. Two close strong AMPS signals may become blocking interference tones for cellular-band CDMA mobile stations. It is usually not an issue for a superheterodyne receiver to cope with two strong interference tones that have a frequency spacing within the receiver channel bandwidth, but their intermodulation products are out of the channel bandwidth. There is no specification on two-tone blocking in the minimum performance requirements for CDMA mobile stations, IS-98D. However two strong in-receiver-band tones may cause a direct-conversion receiver to be completely jammed if the second-order intercept point or IIP_2 of the receiver is not high enough. An AM interferer is also able to block a direct conversion receiver in a similar way to that of the two-tone blocking.

Two strong interference tones in a direct conversion receiver may directly mix together and generate in-channel interference due to the second-order distortion of the direct conversion receiver, if the spacing of these two tones is less than the channel bandwidth. The interference product $IM_{2,in}$ can be calculated by means of (4.3.9) as

$$IM_{2,in} = 2I_{block} - IIP_{2,Rx}, \tag{4.5.10}$$

where $IIP_{2,Rx}$ is the receiver input second order distortion, and I_{block} is the level of two equal blocking interference tones.

Making the $IM_{2,in}$ equal to the allowed degradation D_a, and considering the phase noise and/or spurious contribution of the UHF synthesizer LO, we can obtain the allowed maximum blocking tone level to be

$$I_{block} = 10\log\left(\frac{10^{\frac{S_{d,i}-CNR_{\min}}{10}} - 10^{\frac{P_{nf}}{10}}}{10^{\frac{I_{block}-IIP_{2,Rx}}{10}} + \sum_{k=1}^{2} 10^{\frac{N_{phas,k}+10\log(BW)}{10}} + \sum_{k=1}^{2} 10^{\frac{N_{spur,k}}{10}}}\right). \quad (4.5.11)$$

In (4.5.11), $N_{phas,k}$ and $N_{spur,k}$ (k =1, 2) are the phase noise in dBc/Hz and spurious in dBc around the first and second blocking interferers, respectively. Actually, (4.5.11) is an equation instead of an expression since the variable I_{block} is also contained in the denominator of the fraction on the right side of the equation. In general, the contributions to signal degradation resulting from the phase noise and/or spurious of the UHF synthesizer LO are quite small and equation (4.5.11) can be simplified as

$$I_{block} = \frac{1}{2}\left[10\log\left(10^{\frac{S_{d,i}-CNR_{\min}}{10}} - 10^{\frac{P_{nf}}{10}}\right) + IIP_{2,Rx}\right]. \quad (4.5.12)$$

(4.5.12) can be used for approximate estimation of the allowed maximum blocking interference. Substituting the calculated result from (4.5.12) into the denominator of (4.5.11) right side, we can obtain a more accurate estimation of the allowed blocking interference level if we would like to solve equation (4.5.11).

For an example, a CDMA direct conversion receiver has a noise figure NF_{Rx} = 5.6 dB, $S_{d,i}$ = -101 dBm, CNR_{min} = -1.5 dB, IIP_2 = 43.5 dBm, $N_{phas,\,k}$ = -140 dBc/Hz, $N_{spur,\,k}$ < -90 dBc, and BW ≅ 1.23 MHz. Using these data and equation (4.5.11) we obtain the allowed blocking interferers equal to –28.65 dB. If employing (4.5.12), we have a result to be –28.10 dBm.

(4.5.11) and (4.5.12) can also be employed to estimate the AM suppression performance of a receiver. The AM suppression characteristics are defined in GSM specification [6]. To meet the AM suppression minimum requirement of a direct conversion receiver for the GSM mobile stations, the input second-order intercept point of the receiver needs to be equal to or greater than 45 dBm.

4.6. Receiver Dynamic Range and AGC System

4.6.1. Dynamic Range of a Receiver

The dynamic range of a mobile station receiver is the input signal power range at the antenna port of the receiver over which the data error rate (BER or FER) does not exceed a specified value. The lower end of this range depends on the receiver sensitivity level and the upper end is determined by the allowed maximum input power at which the data error does not exceed the specified value. The minimum requirements of the maximum input power at the receiver antenna port and dynamic range for the mobile stations of different wireless systems are presented in Table 4.2.

The dynamic range presented in Table 4.2 is the receiver input power range covering from the receiver reference sensitivity to the maximum input power. To be able to operate over such a wide dynamic range commonly a receiver employs an *automatic gain control* (*AGC*) system. The automatic gain control range is usually wider than the receiver dynamic range. It must also cover the receiver gain variation resulting from device production processing deviation, temperature, and voltage variations. The minimum dynamic range of a CDMA mobile station receiver, for instance, is 79 dB, but an AGC system may need a 100 dB control range to cover the gain possible variations and the dynamic range design margin.

Table 4.2 Maximum input signal and minimum dynamic range

Systems	Maximum Input Power (dBm)	Minimum Dynamic Range (dB)
AMPS	N/A	> 96
CDMA 800	-25	> 79
CDMA 1900	-25	> 79
EDGE	-26	> 72
GPRS	-26	> 73
GSM 900	-15	> 87
GSM 1800	-23	> 79
PHS	-21	> 76
TDMA	-25	> 85
WCDMA	-25	> 81.7

4.6.2. Receiver AGC System for Mobile Stations

4.6.2.1. Block Diagram of an AGC system

The main portion of a receiver AGC system used in the mobile stations is in the digital base-band and the DSP. The controlled objects of the AGC system are in the RF analog section of the receiver, and they the LNA, IF variable gain amplifier (VGA), and/or BB VGA. In this section, a typical AGC system used in a superheterodyne CDMA receiver will be discussed since the AGC systems of other mobile system receivers are very similar to this system, and they are usually simpler than this.

A block diagram of the CDMA receiver AGC system is shown in Fig. 4.13. This system consists of a gain step-controlled front-end, an IF VGA, ADCs, SINC filters, a CDMA core, an Rx AGC algorithm stored in the DSP, and a 10 bit PDM DAC. In the CDMA mobile station, the receiver AGC system function is not only to maintain the receiver chain properly operating over the dynamic range and keeping the level at the ADC input constant but also to measure the received signal strength through the receiver signal strength indicator (RSSI) and then to determine the open-loop transmission power of the transmitter. In the AGC loop the digital low-pass SINC filter is able to further suppress out-of-channel bandwidth interferers.

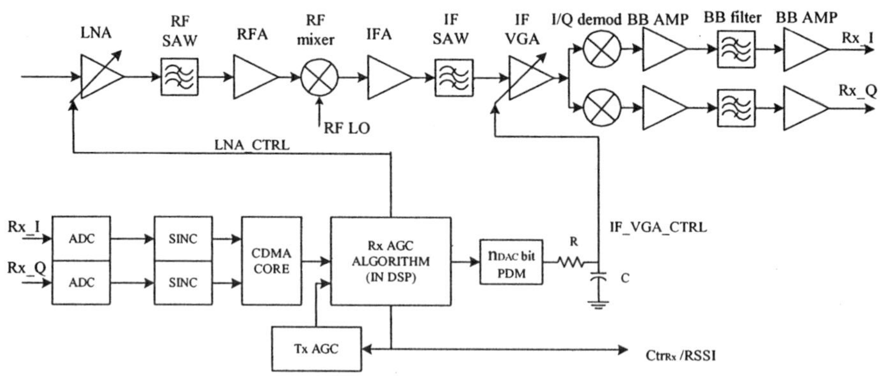

Figure 4.13. CDMA receiver AGC system block diagram

In this AGC system, the IF VGA gain is almost continuously controlled, but the LNA is step-controlled at designed received signal levels.

4.6.2.2. RF AND IF GAIN CONTROL

Assuming the LNA has three gain settings, and the nominal values of the three gain settings are G_H, G_M, and G_L dB, the LNA gain switching occurs at certain received signal strength while the signal is increasing or deceasing. The gain switching levels in a CDMA mobile station receiver are mainly chosen near to but lower than the signal levels, –90 dBm and –79 dBm, respectively, at which the intermodulation spurious attenuation characteristics of mid- and low LNA gain modes are tested.

A scheme of the LNA gain switching is shown in Fig. 4.14. To avoid gain switching back and forth resulting from received signal fluctuation and/or in-channel power change before and after switching from one gain mode to another due to the IMD product or other interference level variation, a power hysteresis is usually needed for stepped gain control. Considering the finite accuracy of the received signal strength measurement, to assure the LNA gain switch from high to mid-gain mode taking place below –90 dBm we shall set the nominal switching point a few dB, (say, 3 dB) lower than –90 dBm — i.e., -93 dBm. When switching the LNA gain from mid- to high gain mode, the nominal switching point shall be a few dB (say, 3 dB) lower than the switching point from high to mid-gain — i.e., -96 dBm. The LNA gain switching level from its mid-gain mode to low gain mode or vice versa shall be selected in a similar way as shown in Fig. 4.14.

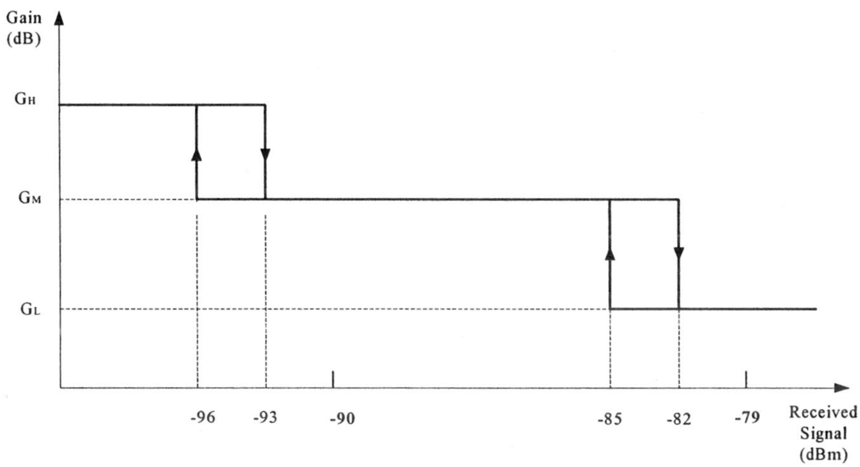

Figure 4.14. LNA gain control versus received signal strength

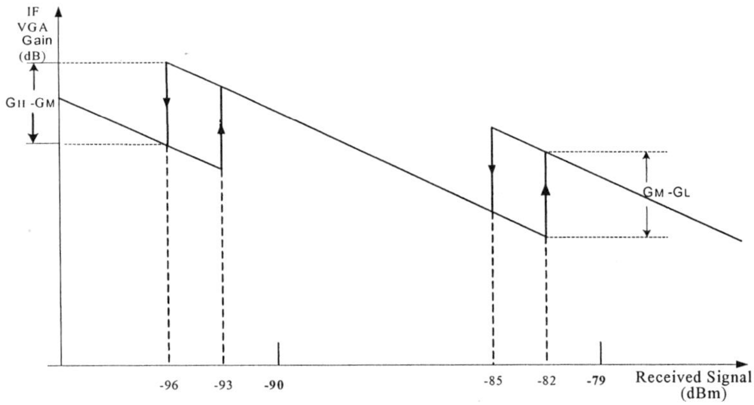

Figure 4.15. IF VGA gain control versus received signal strength

The key points in selecting LNA gain switching level are as follows. The receiver sensitivity at LNA mid- or low gain mode must be better than the level at which the LNA gain is switched from mid- to high or from low to mid-, respectively. If the RSSI accuracy is Δ_a dB, the switching level from high gain to mid-gain or from mid-gain to low gain must be lower than $-(90 + \Delta_a)$ dBm or $-(79 + \Delta_a)$, respectively. Assuming $\Delta_a = 2.5$ dB, them -93 dBm and -82 dBm may be the proper switching levels as mentioned above and depicted in Fig. 4.14 The gain-switching hysteresis should be wide enough to prevent hysteresis vanishing due to finite RSSI accuracy and gain control errors.

The IF VGA gain decreases continuously when the received signal strength increases continuously or vice versa. When the LNA gain steps up or down, the IF VGA can rapidly change its gain in the opposite direction to compensate the LNA gain variation. The IF VGA gain versus the received signal strength is depicted in Fig. 4.15. The VGA gain is adjusted to keep the voltage level constant at the ADC input.

In reality, the IF VGA gain variation with the control voltage is not a linear curve as presented in Fig.4.15. The nonlinear gain curve is often developed by means of measurement to obtain a few points, and then a curve fitting technique is used. The fitted control curve is further approximated piece-wisely by using multiple linear segments.

If the IF VGA gain curve is well behaved, it can be approximately represented by a second-order polynomial as

$$g(v) = a_o + a_1 v + a_2 v^2 \text{ dB}. \tag{4.6.1}$$

To use the second polynomial fitting the VGA gain control curve, only three gains versus three different control voltages — such as, (v_1, g_1), (v_2, g_2), and (v_3, g_3) need be measured. From these data, we can obtain the coefficients in (4.6.1), a_0, a_1, and a_2, and they are given in (4.6.2):

$$a_0 = \frac{\begin{vmatrix} g_1 & v_1 & v_1^2 \\ g_2 & v_2 & v_2^2 \\ g_3 & v_3 & v_3^2 \end{vmatrix}}{\Delta} \qquad a_1 = \frac{\begin{vmatrix} 1 & g_1 & v_1^2 \\ 1 & g_2 & v_2^2 \\ 1 & g_3 & v_3^2 \end{vmatrix}}{\Delta} \tag{4.6.2a}$$

$$a_3 = \frac{\begin{vmatrix} g_1 & v_1 & v_1^2 \\ g_2 & v_2 & v_2^2 \\ g_3 & v_3 & v_3^2 \end{vmatrix}}{\Delta} \quad \text{and} \quad \Delta = \begin{vmatrix} 1 & v_1 & v_1^2 \\ 1 & v_2 & v_2^2 \\ 1 & v_3 & v_3^2 \end{vmatrix} \tag{4.5.2b}$$

A second-order polynomial curve fitting example is presented in Fig. 4.16. High-order curve fitting may be used if the IF VGA gain control curve is quite nonlinear. From (4.6.1) and (4.6.2), we can calculate the control voltage for a given IF VGA gain g_C by utilizing

$$v_C = \frac{-a_1 + \sqrt{a_1^2 - 4a_2(a_0 - g_C)}}{2a_2} \tag{4.6.3}$$

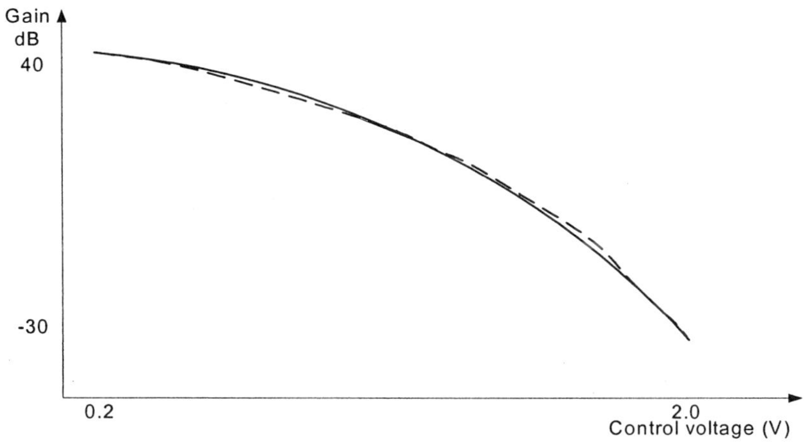

Figure 4.16. VGA gain curve (dashed line) and gain control curve (solid line) obtained from a three point fitting approach

However, we can obtain only the closed-form formula like (4.6.3) from a given gain to calculate control voltage up to the third-order polynomial equation.

The composition gain variation of the IF VGA and the LNA must cover the receiver dynamic range, the stage gain variation over temperature, processing, and frequency, and certain design margin to form a proper AGC control range. Thus the receiver AGC control range GCR_{Rx} shall be approximately expressed as

$$GCR_{Rx} \geq S_{\max} - S_{\min} + \Delta G_{R,T} + \Delta G_{R,process} + \Delta G_{R,f} + Mrgn \qquad (4.6.4)$$

where S_{max} represents the maximum input signal power, S_{min} is the receiver sensitivity, ΔG_{RT}, $\Delta G_{R,process}$, and $\Delta G_{R,f}$ are receiver gain variations due to temperature, device processing, and frequency, respectively, and $Mrgn$ represents the design margin.

4.6.2.3. Introduction to a Receiver AGC Algorithm

The AGC algorithm should keep the level at the output of the BB amplifiers in the analog base-band block or the input level to ADCs being constant, around one third of the maximum voltage swing of ΔΣ ADC depending upon the peak-to average power ratio of the signal, resolution of the ADC and other conditions by adjusting the IF VGA gain and/or the front-end gain. The received signal strength indicator (RSSI) is calculated in following way:

$$\text{Received Signal Strength (dBm)} = R_{\text{x,o}} - \frac{1}{C_R}\left(Ctr_{Rx} - Ctr_{Rx,o}\right), \qquad (4.6.5)$$

where $R_{,\text{x,o}}$ is calibrated reception power level in dBm, and its corresponding counter number is $Ctr_{Rx,o}$, $1/C_R$ is AGC control resolution in dB, Ctr_{Rx} is a countervalue proportional to received signal strength with a unit $1/C_R$ dB, and it is equal to $Ctr_{Rx,o}$ when the received signal strength is $R_{,\text{x,o}}$ dBm.

The Ctr_{Rx} value is updated based on the RF receiver gain error estimation. The gain error (ΔG_{Err}) is obtained from an integrated power of present n_{symb} symbols in logarithm scale,

$$S_{symb} = C_R \cdot 10\log\left(n_{symb} \cdot P_{symb}\right), \qquad (4.6.6)$$

comparing with a reference level S_{Ref} defined as

$$S_{ref} = C_R \cdot 10\log\left(n_{symb} \cdot P_{symb\_ref}\right). \tag{4.6.7}$$

The reference level here actually determines the setting voltage at the input of the I/Q channel ADC. The gain error is added or integrated to the Ctr_{Rx}.

Assuming the symbol rate R_S kbit, each symbol containing n_C chips and each chip having n_S samples, after adding I and Q both channel power thus the symbol power can be expressed as

$$P_{Symb} = 2 \times n_C \times n_S \times (g \cdot v_{in})^2, \tag{4.6.8}$$

where v_{in} is input level of the ADC in V, and g is a conversion gain from ADC input to AGC algorithm output, and defined as

$$g = \frac{1}{v_{\max}} \frac{g_{cdma} g_{\sin c} g_{dec}^3}{2^{n_{DAC}}}, \tag{4.6.9}$$

where v_{max} is the maximum input voltage of the ADC in mV, g_{cdma} is the CDMA core gain, g_{sinc} is the SINC filter gain, g_{dec} is the decimation ratio, and n_{DAC} is the bit number of the DAC for RSSI. The decimation ratio with a third-order power is due to that the ADC is a second-order $\Delta\Sigma$ converter and the third-order decimation filtering can optimize the noise shaping.

If we have $v_{max} = 0.8$ V, $g_{dec} = 6$, $g_{sinc} = 2800$, $g_{cdma} = 0.0415$, and $n_{DAC} = 10$, then the conversion gain is 30.64. Considering two symbol power integrated in the DSP — i.e., $n_{symb} = 2$ — and assuming $n_C = 64$, $n_S = 2$, $C_R = 256$ and $v_{in\_ref} = 180$ mVrms, from (4.6.7) the reference level in logarithm scale is

$$\begin{aligned} S_{ref} &= 256 \times 10\log(2 \cdot P_{Symb\_ref}) \\ &= 256 \times 10 \cdot \log\left(2 \cdot 256 \cdot (30.64 \times 0.18)^2\right) = 10733. \end{aligned}$$

The gain error ΔG_{Err} and the updated Ctr_{Rx} are, respectively,

$$\Delta G_{Err} = S_{ref} - S_{symb} \tag{4.6.10}$$

and

$$Ctr_{Rx} = Ctr_{Rx} + \left(1 - e^{-\frac{1}{f_{update}\tau}}\right) \cdot \Delta G_{Err} , \qquad (4.6.11)$$

where f_{update} is Ctr_{Rx} update rate equal to $R_S \cdot / n_{symb}$, and τ is the AGC loop time constant. Assuming the update rate f_{update} = 9.6 kHz, and τ = 2 msec, we have $f_s \tau$ = 19.2 and exp(-1/19.2) = 0.95.

It is apparent the AGC algorithm presented in this subsection is not unique one. Different algorithms may be used to achieve the same AGC function.

4.6.3. Dynamic Range and Other Characteristics of ADC

The dynamic range of an ADC is here defined as the maximum effective signal-to-(noise+distortion) ratio. The dynamic range requirement of an ADC used in a mobile station is basically determined by the following factors:

1. Shortage of the AGC range to the receiver dynamic range (the difference between the maximum and minimum receivable signal strengths)
2. AGC step size,
3. Required minimum carrier-to-noise ratio,
4. Ratio of in-channel band noise/interference to quantization noise,
5. Desired signal peak-to average ratio,
6. DC offset,
7. Upper and lower fading margin, and
8. Loose filtering (i.e., channel filters having no enough suppression to close-in interferers).

It is apparent that not all these factors need be considered for different receiver architectures. When using the superheterodyne architecture, for instance, we may not take factors 1, 2, and 8 into account since in this case the AGC can usually cover the full receiver dynamic range with margin, the AGC step size is quite small almost like continuous adjustment, and the IF SAW filter and the analog BB low-pass filter provides enough suppression to out of band interferers.

The ADC dynamic range for a CDMA mobile receiver can be considered by assuming that the peak-to-average voltage ratio of a noise-like CDMA signal is around 10 dB (by taking three times standard variance), the quantization noise is assumed to be 12 dB below in-channel-band noise level, the minimum CNR is –1 dB and the fading margin 3 dB to probably be enough for the CDMA mobile system. Thus the total dynamic range is approximately 24 dB. On the other hand, for a superheterodyne GSM mobile receiver, the dynamic range of its ADC should approximately cover, 16 dB for quantization noise below the in-channel-band system noise, 4 dB for DC offset, 8 dB for minimum carrier-to-noise ratio requirement, and 20 dB for constructive and destructive fading margin, and thus the ADC used in a GSM mobile station receiver has at least 48 dB dynamic range.

The dynamic range of ADC employed in the direct conversion receiver or in the low IF receiver is usually higher than the corresponding superheterodyne receiver. This is due mainly to the step gain control and loose filtering. When in the AGC system design one intentionally lets the ADC absorb part of the receiver dynamic range, in this case the ADC dynamic range may be required to be quite high.

The ADC dynamic range can also be represented in effective bits. The conversion from the ADC dynamic range to effective bits is as follows. Assuming an n_b bit ADC has an allowed peak-to-peak maximum voltage swing, $V_{p\text{-}p}$, from (2.4.21) we know the ratio of the maximum signal power $P_{S,max}$ to the quantization noise P_{qn} is

$$\left(\frac{S}{N}\right)_{q\_ADC} = \frac{P_{S,\max}}{P_{qn}} = \frac{3}{2} 2^{2n_b} . \tag{4.6.12}$$

The ADC dynamic range is defined as the signal-to-noise ratio given in (4.6.12) expressed in logarithm form as

$$DR_{ADC} = 20 \log(2^{2n_b} \sqrt{1.5}) \cong 6.02 n_b + 1.76 \text{ dB}. \tag{4.6.13}$$

From this formula and the analysis in the previous paragraph, the ADC for the CDMA mobile station receiver needs only 4 bits, while 8 bits are required for the GSM mobile station receiver. The ADC dynamic range varies from one mobile system to another and also from one receiver architecture to another. In the AMPS and TDMA mobile systems, 10 bit ADC is often used, and 12 to 13 bit ADC may be employed in a superheterodyne EDGE mobile receiver.

A typical curve of signal-to-(noise + distortion) ratio versus input voltage of a 6 effective bit ΔΣ ADC is presented in Fig. 4.17, which is obtained in terms of a sine wave signal. This curve clearly shows that the maximum effective dynamic range of a ΔΣ ADC can be actually achieved at close to half of the maximum peak-to-peak voltage swing — i.e., approximately 6 dB below the maximum peak-to-peak voltage. In fact, it is a common characteristic for the ΔΣ ADC that the maximum dynamic range is approximately at half of the allowed maximum peak-to-peak voltage.

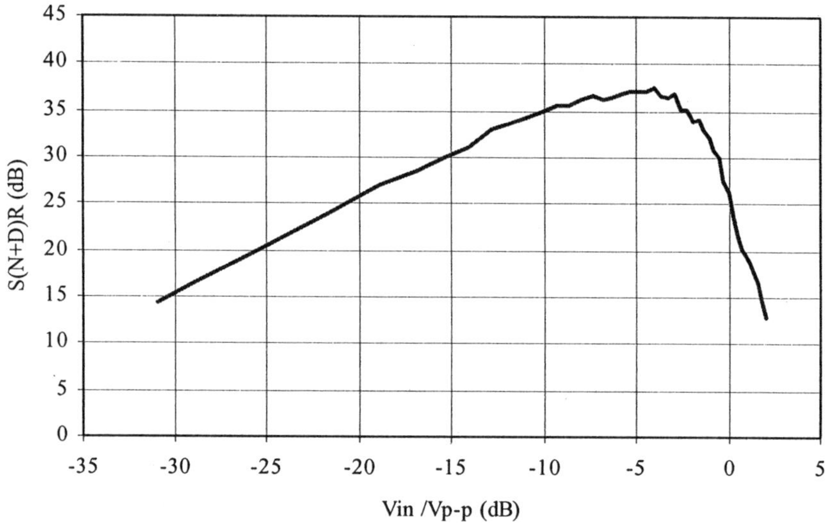

Figure 4.17. S(N+D)R versus input voltage of 6 effective bit ΔΣ ADC

In Appendix 3.D, we have already discussed the noise figure of the ADC. However, we are going to introduce an approximate noise figure expression of an ADC based on its quantization noise only. Assuming that the ADC has a noise figure F_{ADC}, the noise power generated by the ADC is

$$P_{n,ADC} = P_{qn} = kT_o(F_{ADC} - 1)\frac{f_s}{2}, \tag{4.6.14}$$

where f_s is the sampling frequency. Rearranging (4.6.14) and using (4.6.14), we obtain the noise factor, F_{ADC}, expression for the ADC to be

$$F_{ADC} = \frac{2P_{S,\max}}{kT_o f_s (S/N)_{q,ADC}} + 1 = \frac{V_{p-p}^2}{6 \cdot 2^{2n_b} \cdot Z_{in,ADC}} \cdot \frac{1}{kT_o f_s} + 1. \tag{4.6.15}$$

In (4.6.15) $Z_{in,ADC}$ is the input impedance of the ADC, k is a Boltzman constant equal to 1.38×10^{-20} mW·sec/°K and T_o = 290°K. The first term on the right side of (4.6.15) is usually much greater than 1. Neglecting 1 and taking logarithm of both sides of (4.6.15), we obtain the noise figure NF_{ADC} expression in dB (a value usually much greater than 1):

$$NF_{A/D} = 20\log V_{p-p} - 6.02 \cdot n - 10\log Z_{in,ADC} - 10\log f_s + 196.2\,. \quad (4.6.16)$$

Thus the noise figure of a 4 bit ADC with an input impedance 100 KΩ, a sampling rate 19.2 MHz, and allowed maximum peak-to-peak voltage, is approximately

$$NF_{ADC} = 20\log 1.0 - 6.02 \times 4 - 10\log 10^5$$
$$- 10\log(19.2 \cdot 10^6) + 196.2 = 49.28 \text{ dB}.$$

The more accurate noise figure calculation is given in Appendix 3D, and (4.616) just provides a rough estimation when the ADC noise figure is high.

4.7. System Design and Performance Evaluation

4.7.1. Receiver System Design Basics

When designing a receiver, we need first defined design goals based on the applications of the receiver. The goals for a receiver used in commercial mobile stations are usually the following aspects: good electrical performance, low power consumption, low cost, and small size. A tradeoff between these goals, especially the performance with the other aspects, is often necessary when setting the goals. For example, to obtain a high IIP_3 receiver for better intermodulation characteristics more current consumption is usually needed, and to lower the overall cost of a receiver the receiver performance margin may be scarified as an exchange. Among performance, power consumption, cost and size, the electrical performance is the fundamental criterion. It will not make any sense to talk about power consumption, cost, and size if a receiver cannot even have a minimum necessary performance. However, a good design is measured not only based on the electrical performance but also, sometime quite important, based on the power consumption, cost, and size. In system design we should avoid overemphasizing the performance and ignoring the other aspects.

For a mobile station receiver, its performance is represented mainly by the receiver sensitivity, intermodulation characteristics, adjacent and alternate channel selectivity, blocking characteristics, and spurious emission. Most of them, except the spurious emission that will be discussed later in this section, have been analyzed in previous sections, and the corresponding formulas given in these sections can be used for the calculation of these parameters. Design targets of these receiver parameters are normally set based on the minimum performance requirements of different wireless mobile communication systems, such as IS-98D [5] and ETSI-GSM05.05 [6], with certain margin. The design margin for different parameters may vary from one parameter to another one. In general, 3 dB is a reasonable margin for most of the receiver parameters at room temperature and the typical case, and 1.5 dB is the margin for the worst case — i.e., over temperature, frequency, and voltage. However, sensitivity is one of the most interesting parameters in the receiver performance, and the design margin for the receiver sensitivity may often be defined as 4 to 5 dB. It is also desirable to set a 4 dB margin or more for the intermodulation characteristics, which represent the receiver linearity, if the current consumption for having this margin is still acceptable. There is no unique way to define the performance design targets, and the performance target setting will always compromise with other goals (i.e., power consumption, cost, and size) setting.

The current or power consumption of a receiver will directly affect the standby time of a mobile station with a certain battery capacity and talking time too. The current consumption of the receiver varies with its operation modes. In the high-gain mode it usually consumes more current than in other gain modes since a high linearity in this case is requested to cope with high intermodulation interferers or very strong blocking interference. At present, excluding the UHF synthesizer LO, the current consumption of a superheterodyne mobile receiver operating at the high gain mode is in the region of mid- 20 to high 40 mA depending on mobile systems. For instance, the CDMA mobile receiver consumes more current than the GSM receiver. To reduce average current consumption over the different operation modes, a proper power management is necessary for controlling bias and voltage supply of devices and/or circuits in the receiver chain.

Now, the cost and the size of a receiver are more tightly associated with the receiver architectures than individual device selection. The direct conversion and the low IF receivers have lower cost and much smaller size than the superheterodyne receiver although the later usually has a better electrical performance than that of the former two. This is due to that the direct conversion and the low IF receivers do not use the IF SAW for the channel filtering, and they both can be implemented by means of highly integrated circuits. Thus the parts count of the overall receiver either the direct conversion or the low IF is much less than the superheterodyne, and the consequence is the low cost and the small size. On the other hand, the technology of the superheterodyne receiver and the parts used in it are more mutual than the other two, and thus the cost of some key parts, such as front end IC and IF-BB IC, will be relatively low. The overall cost of a superheterodyne receiver may not higher than the direct conversion or low IF one if the total volume of the receiver products is not high enough.

Based on design goals the receiver architecture should be first chosen. The advantages and drawbacks of different architectures have been discussed in great detail in Chapter 3. A fundamental receiver block diagram corresponding the chosen architecture is then developed. It can be similar to one of those presented in Fig. 3.1, Fig. 3.10, Fig. 3.21, and Fig. 3.30. No matter what receiver architecture is chosen, the most commonly used key devices in for these receiver architectures are the RF band-pass filters, RF LNA, RF down-converter, UHF synthesized LO, base-band amplifiers, base-band low-pass channel filters, and ADC. The superheterodyne receiver has additional IF channel filter, IF amplifiers, IF I/Q down-converter, and VHF LO; the low IF receiver has the low IF I/Q down-converter (demodulator) and LO in digital domain instead; and the RF down-converter is replaced by a high sampling rate and high dynamic range ADC in the band-pass sampling architecture, and its I/Q demodulator, the corresponding LO, and the channel filters are in the digital base-band similar to the low IF receiver. All these devices used in the RF and analog blocks of a receiver can be divided into passive and active categories, or according to their function they can be classified to filters, amplifiers, frequency converters, local oscillators, and analog-to-digital converters.

The receiver is formed from these devices, which are like building blocks, and it is apparent that the overall performance of the receiver depends on the characteristics of each individual building block. The objectives of the receiver system design are not only to make the designed receiver achieving the performance, current consumption, cost, and size as targeted but also to define the specifications of each individual device in the receiver and thus to ensure the targeted receiver performance and other goals achieved.

4.7.2. Basic Requirements of Key Devices in Receiver System

It is essential for a receiver system designer to understand the basic specifications of various function devices consisting of an RF receiver and to know their possible performance based on the present technology. These devices are the building blocks of the receiver, and proper selecting and specifying them are the major tasks of the receiver system design after the receiver architecture has been defined.

4.7.2.1. Filters

Filters used in the receiver include the duplexer, receiver operating band RF band-pass filter, the IF channel filter, and the base-band low-pass channel filter. The former three filters are generally passive filters, and the last one is usually an active filter. SAW filters are commonly used in the mobile receivers as the RF duplexer, the RF band-pass filter, and the IF channel filter since the SAW filters have relatively small size and low cost. High dielectric constant ceramic filters are often employed as the duplexer and the RF band-pass filter especially for the applications in frequency bands above 1.6 GHz. The ceramic filters have relatively low insertion loss, pass-band less sensitive to the temperature variation, but their size relatively large. Recently, FBAR and BAW filters* are getting mutual for practical use to replace the ceramic filters in the frequency bands above 1.6 GHz because they have much smaller size than the ceramic filter.

The main specifications of passive band-pass filters are

1. Center frequency (MHz)
2. Pass-band bandwidth (kHz or MHz)
3. Pass-band insertion Loss (dB)
4. Pass-band ripple (dB)
5. Group delay (μsec)
6. Group delay distortion (μsec) or RMS phase variation (degree)
7. Rejections or attenuations at certain defined frequencies (dB)
8. Input impedance (Ω)
9. Output impedance (Ω)
10. Input return loss (dB)
11. Output return loss (dB)

In general, the pass-band insertion loss impacts or influence the receiver noise figure and sensitivity; the bandwidth and out-of-band rejection determine the receiver selectivity and out-of-band linearity; and the bass-band ripple and group delay distortion may affect the minimum CNR for a certain BER or FER. The input and output impedances and return losses just influence the corresponding matching networks. In the specification of a passive filter the IIP_3 is usually not included since it is so high and will not degrade the overall linearity of the receiver.

---

* The full names of FBAR and BAW filters are film bulk acoustic resonator filter and bulk acoustic wave filter, respectively. They are actually the same type of filters based on utilizing the bulk acoustic wave instead of surface acoustic wave (SAW), but named differently.

Since the filters are used for different purposes and operating at various frequency bands, it is impossible to give a typical performance for all filters. They should be discussed individually. The preselector filter or the receiver side filter of the duplexer for the CDMA and AMPS has an insertion loss approximate 2 to 3 dB, a rejection of 46 to 54 dB to the transmitter band signal, 50 Ω input and output impedances, and better than –15 dB input and output return losses. The duplexer for the WCMA may have only around 2 dB insertion loss and more than 55 dB rejection to 140 MHz away transmitter signal. The RF BPF is used in the CDMA, AMPS, and the full-duplex direct conversion receiver after the LNA. Its insertion loss is approximately 2.5 dB, and the rejection to the transmitter signal is about 20 to 30 dB depending on the operation frequency band. At present, the IF SAW filter for the channel filtering posses an insertion loss 5 to 12 dB depending on the center frequency and the bandwidth, narrower bandwidth lower insertion loss, and its rejection can be 25 to 55 dB relying on the IF and frequency offset from the IF.

The low-pass channel filter is normally an active filter consisting of low-pass networks and operational amplifiers. The specifications for the active low-pass are similar to the passive one as presented above except

1. It may have voltage gain;
2. Equivalent noise voltage or noise figure is used to replace the insertion loss specification; and
3. In-band and out-of-band IIP_3 is employed to represent linearity of the filter as used in other active device.

The equivalent noise figure of an active filter is usually quite high up to 30 to 50 dB depending on the order of the filtering and current consumption or Q factor of the filter. The out of band IIP_3 of the active filter is only around 15 to 35 dBm, and it must be taken into account in the calculation of the receiver system linearity. The out-of-band rejection of an active low-pass filter depends on the order of filter and frequency offset from the pass-band, and it can be in the range of 40 to 70 dB.

The group delay distortion of the channel filters, the IF SAW filter and the active low-pass filter, must be low enough, or otherwise the receiver may have poor performance when it is operating in multipath fading channels.

4.7.2.2. LNA

The LNA in a receiver generally determines the sensitivity of the receiver. The main specifications of an LNA are as follows:

1. Operating frequency band (MHz),
2. Nominal gain (dB),
3. Noise figure (dB),
4. IIP_3 (dBm),
5. Reverse isolation (dB),
6. Input and output impedance (Ω),
7. Input and output return loss (dB).

The single stage LNA gain for the mobile station applications can be designed from 10 to 16 dB. The noise figure of an LNA depends on which technology, GaAs, SiGe, or others, is used. The noise figure of a GaAs LNA up to 2 GHz is now less than 1 dB, in the range of 0.4 to 0.8 dB, and the SiGe LNA noise figure is slight higher, about 0.8 to 1.4 dB at 1 and 2 GHz bands. The third-order input intercept point of an LNA is only requested higher than +6 dBm when it is applied in the CDMA receiver for coping with the single tone desensitization. 10 dBm IIP_3 of an LNA is achievable without increasing current consumption. In other mobile system receivers, the requirement on the LNA IIP_3 is normally less than 0 dBm. The current consumption of an LNA operating 1 or 2 GHz band is around 4 to 6 mA.

The gain of the LNA used in the mobile station receivers is adjustable, but it is commonly step control. In the superheterodyne receiver, bypassing the LNA is popularly used instead of true gain control, but in the direct conversion receiver, the LNA gain is truly step-controlled and usually it has multiple steps. It is important for a direct conversion receiver always to have certain LNA reverse isolation in all gain steps.

4.7.2.3. Down-Converter and I/Q Demodulator

The linearity or IIP_3 of the first down-converter in a receiver normally dominates the intermodulation characteristic performance of the receiver. This stage may contribute close to 50% of overall intermodulation distortion product power from the RF receiver chain. However, a reasonable balanced design between the linearity and the current consumption of the receiver will result in the first down-converter to be main contributor to the overall intermodulation distortion product of the receiver. The specification of a down-converter has the following main requirements:

1. Operating frequency band (MHz),
2. Conversion loss or gain (dB),
3. Noise figure (dB),
4. IIP_3 (dBm),
5. IIP_2 (dBm) (for direct conversion receiver use),
6. Isolation between RF/IF and LO ports (dB),
7. Isolation between LO and IF/BB ports (dB),
8. Isolation between RF/IF and IF/BB ports (dB),
9. Nominal LO power (dBm),
10. Input and output impedance (Ω),
11. Input return loss (dB).

The same specification is also applicable to the I or Q channel down-converter of an I/Q demodulator (or referred to as quadrature down-converter). For the I/Q demodulator, the following two requirements need be added into its specification in addition to those described above:

12. Magnitude imbalance (dB) between outputs of I and Q channel down-converters,
13. Phase imbalance (degree) between outputs of I and Q channel down-converters.

The RF down-converter based on the GaAs technology generally consists of diodes and driving it needs more than 5 dBm LO power. It has a conversion loss approximate –6 dB and a noise figure close to 6 dB. The linearity or IIP_3 of this kind down-converter is relatively high, 16 to 18 dBm. The isolation between RF and LO ports is about 22 to 25 dB, but the isolation between RF and IF ports can be 25 dB or higher. The input impedance is usually matched to 50 Ω, and its return loss can be better than –12 dB. On the other hand, a 1 to 2 GHz SiGe down-converter is commonly formed in terms of using Gilbert cells in differential circuit topology. The SiGe Gilbert cell down-converter usually has a voltage gain instead of loss, and its gain can be from a few dB to more than 15 dB depending on design. Its noise figure is in the range of 5 to 8 dB, and the IIP_3 is 0 to 5 dBm relying on its gain. The LO level needed for this down-converter is only around –5 to 0 dBm much less than that for the GaAs diode down-converter. The isolation between RF and LO ports can reach as high as 60 dB or even higher. The isolation between other ports is also very high, greater than 60 dB. Its input impedance is often in 50 to 200 Ω, and it output impedance in a couple of hundreds Ω.

An RF I/Q demodulator formed from two Gilbert cell down-converters may consume more than 15 mA depending on the IIP_3 requirement. The magnitude and phase imbalances between the demodulator I and Q channel outputs can be generally equal to or less 0.1 dB and 2 degree, respectively.

For an IF I/Q demodulator, its gain can be 20 to 30 dB, the noise figure is not necessary to be low, approximate 15 to 20 dB, and the IIP_3 is in the range of –20 to –30 dBm. Isolations between ports are normally greater than 40 dB for differential circuit design. The input impedance is in a few kΩ, and the output impedance in a couple of hundreds Ohm. The current consumption is 4 to 6 mA.

4.7.2.4. IF and BB Amplifiers

The contents of the specifications for the IF and BB amplifiers are the same as those of the LNA, which are listed in Section 4.7.2.1. The IF can be tens MHz to a couple of hundreds MHz. The voltage gain of a single stage IF amplifier can be designed up to more than 40 dB, and it is easy to have their noise figure 3 to 4 dB if necessary. The IIP_3 of an IF amplifier can be from higher than 10 dBm to lower than –25 dBm depending upon the IF amplifier preceding or succeeding to the channel filter or the IF SAW. The input impedance of the IF amplifier is around a few kΩ. On the other hand, the BB amplifier used in the superheterodyne receiver has even less stringent requirements than those of the IF amplifier, but the BB amplifier in the direct conversion receiver may need similar performance as the IF amplifier as described above. It may request the noise figure equal to or better than 15 dB, and the IIP_3 greater than 25 dBm if it precedes the BB LPF. However, the equivalent noise figure can be raised to over 20 dB, and the IIP3 can be relaxed to lower than –15 dBm when the BB amplifiers succeed the LPF.

The AGC of a receiver is usually executed through controlling the IF amplifier gain and sometimes the BB amplifier gain too. These gain adjustable amplifiers are referred to as variable gain amplifier (VGA). The gain of an IF VGA is continuously adjustable, but the gain of a BB VGA is generally step-controlled. The BB VGA using step control has two reasons. There are two channels, I and Q channels, in the analog base-band block of a receiver. The gain of the BB VGAs in these two channels must be simultaneously controlled, and the step gain control is relatively easy to maintain both channel gains within an acceptable tolerance. It is not difficult to design a multistage IF or BB VGA having an adjustable gain more than 70 dB.

4.7.2.5. Synthesized LO

The synthesized LO consists of a synthesizer and a VCO through a phase locked loop. In a superheterodyne receiver, two synthesized local oscillators, an UHF and a VHF synthesized LOs, are employed, but only an UHF synthesized LO is used in a direct conversion or a low IF receiver. The main specifications of a synthesized LO are as follows:

1. Operating frequency band (MHz),
2. Output power (dBm),
3. In PLL bandwidth phase noise (dBc/Hz),
4. In PLL bandwidth spurious (dBc),
5. Out of PLL bandwidth phase noise (dBc/Hz),
6. Out of PLL bandwidth spurious (dBc), and
7. Settling time (μsec or mscc).

The close-in phase noise of the LO may affect the modulation accuracy (EVM) of a received signal (see Chapter 5). To minimize this influence, the integrated phase noise over the signal bandwidth including spurs within this bandwidth need be 30 dB lower than the LO signal level. The allowed phase noise and spurious of a synthesized LO are determined by the receiver adjacent channel selectivity and the intermodulation characteristic performance. For an example, the typical phase noise of an UHF LO used in the mobile station receivers is equal to or less than –80 dBc/Hz at 10 kHz, –125 dBc/Hz at 100 kHz offset, –140 dBc/Hz at 1MHz, –145 dBc/Hz at 2 MHz, and the noise floor is approximately at –154 dBc/Hz. For a VHF synthesized LO, the phase noise requirements at different offset frequencies can be defined much looser than those of the UHF LO, and how much relaxing, 15 or 20 dB, on the phase noise requirements depends on the IF channel filter out of band suppression performance. To avoid spurious affecting the receiver adjacent channel selectivity and IMD performance, the spurious level within the signal bandwidth should be 60 dB below the LO signal at least, and the out of band spurious near the interferers should be 85 dB lower than the LO signal.

The settling time of a UHF synthesized LO is requested from less than 5 msec to less than 150 μsec depending on wireless systems. Generally, the full-duplex system, such as the CDMA system, the allowed settling time is longer than that of the TDMA systems, especially multi-burst operating systems, such as the GPRS.

The current consumption of a synthesizer is 3 to 5 mA, and the current consumption of a voltage control oscillator (VCO) is in the range of 3 to 10 mA depending on its operating frequency and output power.

4.7.3. Receiver System Performance Evaluation

In the receiver system design, it is necessary to analyze and evaluate the receiver performance after the architecture and corresponding building blocks of a receiver in design have been selected. Through proper adjusting of the performance of each building block in the receiver and then evaluating overall receiver performance, the design targets and goals will be finally achieved.

The receiver system performance evaluation can utilizing the formulas presented in Section 4.2 to Section 4.6 and can be carried out by means of Excel spreadsheet calculation or Matlab program. In Table 4.3, an example of a CDMA mobile receiver performance evaluation is presented. In this example a superheterodyne receiver is employed, and the receiver section, which is analyzed and designed, is from the receiver antenna port to the input port of the ADC.

Table 4.3 presents a receiver performance evaluation in the typical case. It is necessary to run the performance evaluations over temperature, frequency band, voltage, different gain modes, and key device processing. The formulas on the left side in the lower portion of this table are just symbols, and they are not really used for calculations of the sensitivity (MDS), IMD, and single tone desensitization. A similar performance evaluation approach for the receiver system design by using the Matlab is given in Appendix 4D. From the system performance evaluation results, we can easily define the specifications of each building block or device in the designed receiver.

Using proper power management, such as circuit bias varying with the received signal or interferer strength, can further reduce the average current consumption. Employing denser integrated circuits can decrease the bill of material and the overall receiver cost. Thus not only the electrical performance of the receiver, but also the other design considerations can finally reach our targets or goals. Some design examples of wireless mobile transceivers will be discussed in Chapter 6.

Table 4.3 Electrical performance evaluation example of a Cellular band CDMA receiver

CELLULAR CDMA RECEIVER

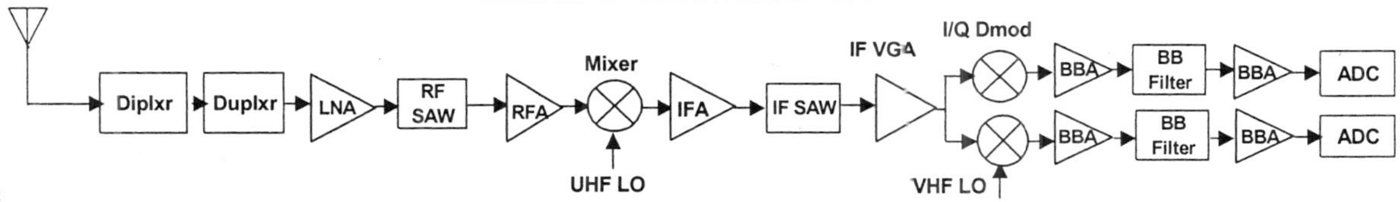

25 oC

		Diplxr	Duplxr		RF SAW		MIXER				I/QMIX	BB_AMP	BB/FILTR	BB_AMP		Total
Current (mA)								10								45
P_Gain dB		-0.4	-2.5	14	-2.5	10	-6	10	-11	18.49	30			35		95.09
V_Gain dB		-0.4	-2.5	14	-2.5	10	-6	10	-1	18.49	30			35		105.09
Fltr Rej (	InTx Band	52		In Tx Band	25			@ 900 kHz	38			@ 900 kHz	45			
(dB)	n Rx band	45						@ 1.7 MHz	50			@ 1.7 MH	55			
out dBm	-108	-108.4	-110.9	-96.9	-99.4	-89.4	-95.4	-85.4	-96.4	-77.9073	-47.9073	-47.9073	-47.9073	-12.9073	-12.9073	
out mV	0.00089	0.00085	0.00064	0.0032	0.0024	0.00758	0.0038	0.01201	0.010702	0.08997	2.84525	2.84525	2.84525	160	160	
NF dB		0.4	2.5		2.5				11	10	25		45	38	59	
Cascad NF	**5.79**	5.79	5.39	2.89	12.62	10.12	18.96	12.96	22.76	11.76	25.50	45.82	45.82	38.17		
NF Contr	100.00%	28.88%	22.47%	13.30%	1.59%	5.50%	1.08%	0.00%	1.68%	16.40%	8.13%	0.00%	0.82%	0.16%	0.01%	
IIP3 dBm	100.00	100.00	100.00	8.00	100.00	10.00	17.00	12.00	32.00	-20.00	-25.00	100.00	15.00	-21.00	40	
Cascad IIP3	**-6.26**	-6.26	-6.66	-9.16	4.92	2.42	13.26	9.64	23.43	-43.92	-25.41	15.00	15.00	-21.01		
IIP3 Cont	100.00%	0.00%	0.00%	1.92%	0.00%	17.14%	34.20%	27.17%	2.72%	0.07%	15.25%	0.00%	1.53%	0.00%	0.00%	

Rx an Tx Key Parameters									Synthesizer LO Phase Noise & Spurs					
Rx :		FER =	5.00E-03		Sd(dBm) =	-101	BW(Hz) =	1.23E+06			Sp,adj =	-85	Sp,atl =	-85
		CNR(dB)=	-2								Nphase=	-137	Nphase=	-140
Tx :		PTx(dBm)	23	23	TxPhsNois									
					in Rx band	-138	dBm/Hz							

Receiver Performance:

			Margin (dBc)	Spec
1. Sensitivity				
MDS =	-174 + 10 log(BW) + CNR + NF =	**-109.16** dBm	**5.16**	-104 dBm
2. Intermodulation Spurious Response Attenuation				
IMD =	[2(IIP3 - Sd) - CNR] /3 =	**-37.54** dBm	**5.46**	-43 dBm
3. Single Tone Desensitization				
IST =	Sd-CNR+2*IIP3-2(PTx-Rj_dplx) -40log(M_indx) =	**-24.79** dBm	**5.21**	-30 dBm

Appendix 4A. Conversion Between Power dBm and Electric Field Strength dBμV/m

The receiver sensitivity defined in Section 4.2.1 is based on the power level at the antenna connector of a receiver. However, the electrical field strength is also often employed to measure the receiver sensitivity. In this case, the antenna characteristics will definitely affect the electric field strength value. Usually it is assumed that the antenna used here is defined to have an isotropic radiation pattern and a 0 dB gain. The effective area of the isotropic antenna is

$$A_e = \frac{\lambda^2}{4\pi} = \frac{c^2}{4\pi f^2}, \tag{4A.1}$$

where λ is the wavelength of the signal carrier in meter, c is the propagation velocity of electromagnetic wave and equals 3×10^8 m/sec, and f is the carrier frequency in Hz. If the electric field strength at the aperture of the isotropic antenna is $\mathcal{E}$ V/m^2, and the impedance of free space is $120\pi\Omega$. Thus power flow density S W/m^2 is

$$P_E = \frac{\mathcal{E}^2}{120\pi}. \tag{4A.2}$$

The total power captured by the antenna is equal to

$$P_S = P_E \cdot A_e = \frac{\mathcal{E}^2}{120\pi}\frac{c^2}{4\pi f^2} = \frac{c^2\mathcal{E}^2}{480\pi^2 f^2} \text{ W}. \tag{4A.3}$$

Taking logarithm of (4A.3) on both sides and after some manipulation, we obtain

$$20\log[\mathcal{E}(\mu\text{V/m})] = 10\log[P_S(\text{mW})] + 20\log[f(\text{MHz})] + 10\log\left(5.264\times10^7\right)$$

or

$$E(\text{dB}\mu\text{V/m}) = S(\text{dBm}) + 20\log[f(\text{MHz})] + 77.2\,. \tag{4A.4}$$

The third approach to express the receiver sensitivity is by means of the source electromotive force (emf) voltage at the receiver sensitivity level when measured under the receiver input matching with the 50 Ω source resistance as shown in Fig 4A.1. The signal power delivered to the 50 Ω receiver is

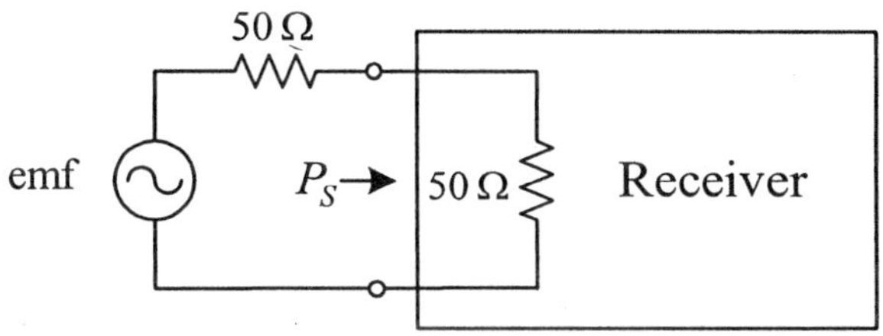

Figure 4A.1. Receiver with 50 Ω input impedance tested by means of a signal source with a 50 Ω source resistance

$$P_S = \frac{\overline{emf}^2}{4 \cdot 50} \text{ W.} \tag{4A.5}$$

This expression in dB scale turns into

$$10 \log[P_S \text{ (W)}] = 20 \log[\overline{emf} \text{ (V)}] - 23 \tag{4A.6}$$

or

$$S \text{ (dBm)} = \overline{EMF} \text{ (dB}\mu\text{V)} - 113. \tag{4A.7}$$

Appendix 4B. Proof of Relationship (4.4.6)

Substituting a two-tone input, $v_i = A\cos\omega_1 t + A\cos\omega_2 t$, to (4.3.1), we have an expansion

$$\begin{aligned} v_o &= \left(a_1 + \frac{9}{4}a_3 A^2\right) A\cos\omega_1 t + \left(a_1 + \frac{9}{4}a_3 A^2\right) A\cos\omega_2 t \\ &\quad + \frac{3}{4}a_3 A^3 \cos(2\omega_1 - \omega_2)t + \frac{3}{4}a_3 A^3 \cos(2\omega_2 - \omega_1)t + \cdots. \end{aligned} \tag{A4.1.1}$$

In the case of weak nonlinearity condition, $a_1 >> 9a_3 A^2/4$, and the input level at the intercept point where the output components at frequencies $2\omega_1$-ω_2 and $2\omega_2$-ω_1 have the same amplitude as those at frequencies ω_1 and ω_2 is noted as A_{IIP3} and it can be determined from the relationship

$$|a_1| A_{IIP3} = \frac{3}{4}|a_3| A_{IIP3}^3. \tag{A4.1.2}$$

Thus, the amplitude of input intercept point is

$$A_{IIP3}^2 = \frac{3}{4}\left|\frac{a_1}{a_3}\right|. \tag{A4.1.3}$$

This expression is the same as (4.4.6).

Appendix 4C. A Comparison of Wireless Mobile Station Minimum Performance Requirements

(See next page)

Table 4C.1. Minimum performance requirements of mobile station receivers

Systems	Referance Sensitivity(dBm)	Max. Input (dBm)	IMD (dBm)	Single Tone (dBm)	Adj. Ch. Selectivity (dBm)	Alt. Ch. Selectivity (dBm)	Blocking (dBm)	Emission (dBm)
AMPS	-116	Not Defined	> S_{min} @ 60 & 120 kHz > S_{min} +3+70 @ 330 & 660 kHz	N/A	> Smin+3+15 @ +/- 30 kHz	> S_{min}+3+60 @+/- 60 kHz		-81
CDMA 800	-104	> -25	> -43 @ 0.87 & 1.7 MHz	> -30 @ +/- 900 kHz	N/A	N/A	N/A	-76
CDMA 1900	-104	> -25	> -43 @ 1.25 & 2.05 MHz	> -30 @ +/- 1.25 MHz	N/A	N/A	N/A	-76
EDGE	-98	> -26	> -45 @ 0.8 & 1.6 MHz	N/A	> -73 Modulated @ +/- 200 kHz	> -41 Modulated @ +/- 400 kHz	< -43 CW 600 kHz < |f-fo| < 1.6 MHz	< -79
GSM 900	-102	> -15	> -49 @ 0.8 & 1.6 MHz	N/A	> -73 Modulated @ +/- 200 kHz	> -41 Modulated @ +/- 400 kHz	< -43 CW 600 kHz < |f-fo| < 1.6 MHz	< -79
GSM 1800	-102	> -23	> -49 @ 0.8 & 1.6 MHz	N/A	> -73 Modulated @ +/- 200 kHz	> -41 Modulated @ +/- 400 kHz	< -43 CW 600 kHz < |f-fo| < 1.6 MHz	< -79
GPRS	-99	> -26	> -49 @ 0.8 & 1.6 MHz	N/A	> -73 Modulated @ +/- 200 kHz	> -41 Modulated @ +/- 400 kHz	< -43 CW 600 kHz < |f-fo| < 1.6 MHz	< -79
PHS	-97	> -21	> -47 @ 0.6 & 1.2 MHz	N/A	> -37 Modulated @ +/- 600 kHz	N/A	> -47	< -54
TDMA	-110	> -25	> -45 @ 120 & 240 kHz	N/A	@ +/- 30 kHz	@+/- 60 kHz	> -45 Modulated @ +/- 90kHz > -30 Modulated @ +/- 3 MHz	< -80
WCDMA	-106.7	> -25	> -46 @ 10 & 20 MHz	N/A	> -52 Modulated @ +/- 5 MHz	> -56 Modulated @ +/- 10 MHz	< -44 Modulated @ +/- 15 MHz	< -60

Appendix 4D. An Example of Receiver Performance Evaluation by Means of Matlab

This is an example of using Matlab to evaluate the performance of a CDMA mobile receiver in Cellular band.

```
function out = cdma_rx(FileName, Cell)

% This program calculates CDMA receiver performance
% including overall receiver Gain, Noise Figure, and
% Input 3rd Order Intercept Point, Receiver Sensitivity,
% IMD and Single Tone Desensitization. Note:
% One restriction of using this program
% is: the LNA must be placed at the 3rd stage in the
% receiver chain and following up by a RF SAW, or
% otherwise this program need slightly be modified for
% Single tone Desensitization calculation.
%
%
   BWn=1.2288*10^6;          % in Hz
   k=1.3727833*10^(-20);     % Boltzman constant
   T=290;                    % Room temperature in Kelven
   No=10*log10(k*T);         % Thermal Noise in dBm/Hz
   Sd=-101;                  % Desired signal level in dBm
   CNR=-1.5;    % Requirement of carrier to noise ratio
                  for FER = 0.5%
   Dmax=Sd-CNR;            % The maximum degradation of the
                             desired signal-to-noise ratio
   Iimd_spc=-43;           % Spec of two intermodulation
                             interferers in dBm
   Ii_spc=-30;             % Spec of the single tone
                             interference in dBm
%
FileExists=exist(FileName);
if FileExists ~= 0
% ------- Input Data -----------
% Read the measurement data
%
   clear  Gain  NFdat  IIP3  Rj_Rx  Rj_Tx  Rj_IF1  Rj_IF2
N_syn;
   eval(FileName);
   if
exist('Gain')==0|exist('NFdat')==0|exist('IIP3')==0|exis
t('Rj_Rx')==0|exist('Rj_Tx')==0 ...
```

```
    |exist('Rj_IF1')==0|exist('Rj_IF2')==0|exist('N_syn')
==0, error(['There is no enough data tables in
"',Filename,'"']);
        end;
    [Mdat,Ndat]=size(Gain);
    G_dB = Gain(:,2:Ndat);
    NF_dB = NFdat(:,2:Ndat);
    IP3_dB = IIP3(:,2:Ndat);
    Fltr_Rj1_dB = Rj_IF1(:,2:Ndat);
    Fltr_Rj2_dB = Rj_IF2(:,2:Ndat);
    len=length(G_dB(1,:));
    Rx_Rj_Tx= Rj_Rx(:,3); % Duplexer Rx filter to Tx
                            transmission rejection in dBc
    Tx_Rj_Rx= Rj_Tx(:,3); % Duplexer Tx filter rejection
                            in the receiver band in dBc
    Rx_Rj_Tx1=Rj_Rx(:,5); % RF SAW rejection to the Tx
                            transmission in dBc
    IIP3_LNA= IIP3(:,4);  % LNA input intercept point
                            in dBm
    Ntx_rx= Txn_rx(:,2);  % Transmitter noise in receiver
                            band,unit in dBm/Hz
    Nphs1= N_syn(1,2);    % UHF synthesizer phase noise
                            at the first interferer
    Nphs2= N_syn(2,2);    % UHF synthesizer phase noise
                            at the second interferer
    Nspu1= N_syn(1,3);    % UHF synthesizer spurious
                            at the first interferer
    Nspu2= N_syn(2,3);    % UHF synthesizer spurious
                            at the second interferer
else
    G_dB = input('gain of all stages in dB, i.e., [10,5,-
3]=');
    NF_dB = input('noise figure of all stages in dB [0.5
1.2 6]=');
    IP3_dB = input('3rd order intercept pt of all stages
in dB [-10 0 6]=');
    Fltr_Rj1_dB = input('Rejection of each stage at first
interferer frequency offset in dB [0 0 40]=');
    Fltr_Rj2_dB = input('Rejection of each stage at
second interferer frequency offset in dB [0 0 55]=');
    len=length(G_dB(1,:));
    BWn=1.2288*10^6;          % in Hz
    k=1.3727833*10^(-20);     % Boltzman constant
    T=290;                    % Room temperature in Kelven
    No=10*log10(k*T);         % Thermal Noise in dBm/Hz
    Sd=-101;                  % Desired signal level in dBm
    Ntx_rx=-155;  % Transmitter noise in receiver band,
```

```
                        unit in dBm/Hz
   Rx_Rj_Tx=45;
   Tx_Rj_Rx=42;
   Rx_Rj_Tx1=15; % RF SAW rejection to the Tx
                   transmission in dBc
   IIP3_LNA=4;
   Ntx_rx= -155; % Transmitter noise in receiver band,
                   unit in dBm/Hz
   Nphs1= -137;  % UHF synthesizer phase noise at the
                   first interferer
   Nphs2= -140;  % UHF synthesizer phase noise at the
                   second interferer
   Nspu1= -85;   % UHF synthesizer spurious at the first
                   interferer
   Nspu2= -90;   % UHF synthesizer spurious at the
                   second interferer
end;
%
if Cell~=0
   Tx_pwr=23;    % Cell CDMA transmitter power in dBm
                   at antenna port
   Cindx=0.6;    % Cross-Modulation index for Cell CDMA
else
   Tx_pwr=15;    % PCS CDMA transmitter power in dBm at
                   antenna port
   Cindx=0.45;   % Cross-Modulation index for PCS CDMA
end;
%
%
% ------- Convert dB to power/ratio -------------------
   G = 10.^(G_dB./10);
   F = 10.^(NF_dB./10);
   IP3 = 10.^(IP3_dB./10);
   IIP3_LNA = IP3_dB(:,3);
%
% ------- calculate gain ----------------
%
   g1 = cumprod(G,2);
   g1m  =  [ones(length(G(:,1)),1)  g1(1:length(G(:,1)),
1:len-1)];
   gt = g1(1:length(G(:,1)),len);
   Gt_dB = 10*log10(gt);
   G_R_dB = G_dB-(2 .*Fltr_Rj1_dB + Fltr_Rj2_dB)./2;
   G_R = 10.^(G_R_dB./10);
   gr1 = cumprod(G_R,2);
   gr1m            =             [ones(length(G_R(:,1)),1)
gr1(1:length(G_R(:,1)), 1:len-1)];
```

```
   gr5 = cumprod(G_R(:,5:len),2);
   gr5m          =          [ones(length(gr5(:,1)),1)
gr5(1:length(gr5(:,1)), 1:length(gr5(1,:))-1)];
%
% ------- calculate noise figure ------------------
%
   F1          =          [zeros(length(G_dB(:,1)),1)
ones(length(G_dB(:,1)), len-1)];
   F2 = F-F1;
   F_div = F2./g1m;
   Ft = sum(F_div,2);
   NFt_dB = (10*log10(Ft));
%
% ------- calculate 3rd order intercept pt ---------
%
   IP3_1 = gr1m./IP3 ;
   D = sum(IP3_1,2);
   IP3t = 1./D;
   IP3t_dB = (10*log10(IP3t));
%
%------- Calculate Receiver Sensitivity -----------
%
   NFo  =  10*log10(Ft  +10.^((Ntx_rx  -Tx_Rj_Rx)  ./10)
./(k*T));
   MDS = No+NFo+10*log10(BWn)+CNR;
%
%------- Calculate Intermodulation Response Attenuation
------
%
   P_Dmax = 10 .^(Dmax/10);
   P_nf = 10.^((MDS-CNR)./10);
%  Iimd = (10*log10(P_Dmax - P_nf) + 2 .*IP3t_dB)./3;
   P_Nphs1 = 10 .^((Nphs1+10*log10(BWn))/10);
   P_Nphs2 = 10 .^((Nphs2+10*log10(BWn))/10);
   P_Nspu1 = 10 .^((Nspu1)/10);
   P_Nspu2 = 10 .^((Nspu2)/10);
   Ncm          =          -2.*IIP3_LNA+2*(Tx_pwr
Rx_Rj_Tx)+40*log10(Cindx);
   P_Ncm = 10.^(Ncm./10);
   P_sum = (P_Nphs1+P_Nphs2+P_Nspu1+P_Nspu2+P_Ncm);
   IMD = imd_cal(IP3t_dB,P_Dmax,P_nf,P_sum);
%
% --------- Calculate Single Tone Desensitization ------
-----
%
   IP3_5 = gr5m ./IP3(:,5:len);
   D_5 = sum(IP3_5,2);
```

```
   IP3t_5 = 1 ./D_5;
   IP3t_5_dB = 10*log10(IP3t_5);
   IIP3_LNA_m  =  -10*log10(1  ./(10  .^(IIP3_LNA  ./10))  +
...
               10  .^((sum(G_dB(:,3:4),2)-1.5  .*Rx_Rj_Tx1)
./10) ./10
            .^(IP3t_5_dB ./10));
   P_CMR = 10.^((-2.*IIP3_LNA_m+2*(Tx_pwr - ...
            Rx_Rj_Tx)+40*log10(Cindx))./10);
   P_NphsR=10^((Nphs1+10*log10(BWn))/10);
   P_NspuR=10^((Nspu1)/10);
   Nst=(P_Dmax-P_nf);
   Dst=(P_NphsR+P_NspuR+P_CMR);
   Ratio = Nst ./Dst;
   E = find(Ratio < 0);
   if isempty(E) == 0,
    error(['Either  Duplexer  Rejection  too  Low,  or  Tx
Power too High, or
           Synthesizer too Lousy']);
   end;
    Ist =10*log10(Ratio);
%
if FileExists ~= 0
   out = [  Gain(:,1)/1000 Gt_dB NFt_dB IP3t_dB MDS  IMD
Ist];
   disp(' ')
   disp('Freq.(GHz)      Gain(dB)      NF(dB)    IIP3(dBm)
MDS(dBm)  IMD(dBm)  Ist(dBm)');
   disp(out);
else
   out = [  Gt_dB NFt_dB IP3t_dB MDS  IMD  Ist];
   disp(' ')
   disp('    Gain(dB)      NF(dB)    IIP3(dBm)    MDS(dBm)
IMD(dBm)  Ist(dBm)');
   disp(out);
end;
%
% ------------------------- Plotting -------------------
----------------------
%
   fig=['Performance of [',DEVICE,']'];
   x=Gain(:,1);
   nx=length(Gain(:,1));
   xi=(Gain(1,1):5:Gain(nx,1));
   xmin=min(xi);
   xmax=max(xi);
   MDS_i=interp1(x,MDS,xi,'spline');
```

```
   IMD_i=interp1(x,IMD,xi,'spline');
   Ist_i=interp1(x,Ist,xi,'spline');

   figure;
   set(gcf,'unit','pixel',...
      'pos',[100 300 550 450],...
      'numbertitle','off',...
      'name',fig);
   h1=plot(xi,(MDS_i+93+1/3)*3,'r-',xi,IMD_i,'g--
',xi,Ist_i,'m-
      .',x,(MDS+93+1/3)*3,'rx',x,IMD,'g*',x,Ist,'mo');

   grid;
   legend('MDS','IMD','S.T. Desen.');
   xlabel('Frequency (MHz)');
   ylabel('IMD and Single Tone Desen. (dBm)');
   title(fig);
   xlim=get(gca,'xlim');
   ylim=get(gca,'ylim');
   set(gcf,'defaulttextfontname','Times New Roman');
   set(gcf,'defaulttextfontweig','bold');
   set(gcf,'defaulttexthorizont','left');
   set(gcf,'defaulttextfontsize',12);
   set(gcf,'defaulttextcolor','black');
   set(gca,'xlim',[xmin,xmax]);
   set(gca,'xtick',[xmin:(xmax-xmin)/5:xmax]);
   set(gca,'ylim',[-50,-20]);
   set(gca,'ytick',[-50:3:-20]);
%
   text(xmax+(xmax-xmin)/11,-50+14,'Sensitivity
   (dBm)','rot',90,'hor','center');
   for i=0:10
      text(xmax+(xmax-xmin)/16,i*3-50,num2str(i-
110),'hor','right');
   end;
   set(text(xmax, -50-
   2.5,['Name',date]),'hor','right','fontangle','italic'
,'fontsize',12);
```

References

[1]. U. Rohde, J. Whitaker, and T.T.N. Bucher, *Communications Receivers: Principles and Design*, 2nd ed., McGraw-Hill, 1997.
[2]. H.T. Friis, "Noise Figure of Radio Receivers," *Proc. IRE*, vol. 32, pp. 419-422, July 1944.
[3]. B. Razavi, *RF Microelectronics*, Prentice Hall PTR, 1998.
[4]. V. Aparin, B. Bultler, and P. Draxler, "Cross Modolation Distortion in CDMA Receivers," *2000 IEEE MTT-S Digest*, pp. 1953–1956.
[5]. TIA/EIA-98-D, *Recommended Minimum Performance Standards for cdma2000 Spread Spectrum Mobile Stations*, Release A, March, 2001.
[6]. ETSI, *Digital Cellular Telecommunications System (Phase 2+); Radio Transmission and reception (GSM 05.05 version 7.3.0)*, 1998.

Associated References

[1]. B. Stec, "Sensitivity of Broadband Microwave Receiver with Consideration Nonlinear Part," *15th International Conference on Microwave, Radar, and Wireless Communications*, vol. 3, pp. 940 – 943, May 2004.
[2]. C. S. Lee, I. S. Jung, and K.H. Tchah, "RF Receiver Sensitivity of Mobile Station in CDMA," *1999 Fifth Asia-Pacific Conference on Communications*, vol. 1, pp. 637–640, Oct. 1999.
[3]. M. C. Lowton, "Sensitivity Analysis of Radio Architectures Employing Sampling and Hold Techniques," *1995 Sixth International Conference on Radio Receiver and Associated Systems*, pp. 52–56, Sept. 1995.
[4]. P. W. East, "Microwave Intercept Receiver Sensitivity Estimation," *IEE Proc. Radar, Sonar, Navig.*, vol. 144, no. 4, pp. 186–193, Aug. 1997.
[5]. P. E. Chadwick, "Sensitivity and Range in WLAN Receiver," *IEE Colloquium on Radio LANs and MANs*, pp. 3/1–3/5, April 1995.
[6]. C. R. Iversen and T. E. Kolding, "Noise and Intercept Point Calculation for Modern Radio Receiver Planning," *IEE Proc. Commun*, vol. 148, no. 4, pp. 255–259, Aug. 2001.
[7]. K. M. Gharaibeh, K. Gard, and M. B. Steer, "Statistical Modeling of the Interaction of Multiple Signals in Nonlinear RF System," *IEEE MTT-S Digest*, pp. 143–146, June 2002.
[8]. J. Roychowdhury and A. Demir, "Estimating Noise in RF Systems," *1998 IEEE/ACM International Conference on Computer-Aided Design*, pp. 199–202, Nov. 1998.

[9]. P. Smith, "Little Known Characteristics of Phase Noise," *RF Design*, pp. 46–52, March 2004.

[10]. C.P. Chiang et al., "Mismatch Effect on Noise Figure for WLAN Receiver," *2002 45th Midwest Symposium on Circuits and Systems*, vol. 3, pp. III-587 – III-590, 2002.

[11]. J. Leonard and M. Ismail, "Link-Budget Analysis for Multistandard Receiver Architectures," *IEEE Circuits Devices Magazine*, pp. 2 – 8, Nov. 2003.

[12]. W. Sheng and E. Sanchez-Sinencio, "System Level Design of Radio Frequency Receivers for Wireless Communications," *Proceedings of 5th International Conference on ASCI*, vol. 2, pp. 930–933, Oct. 2003.

[13]. B. Sklar, "RF Design: Will the Real Eb/No Please Stand Up?" *Communication Systems Design*, pp. 23–28, April 2003.

[14]. A. P. Nash, G. Freeland, and T Bigg, "Practical W-CDMA Receiver and Transmitter System Design and Simulation," *2000 IEE First International Conference on 3G Mobile Communication Technologies*, pp. 117–121, March 2000.

[15]. Z. Yuanjin and C. B. T. Tear, "5G Wireless LAN RF Transceiver System Design: A New Optimization Approach," Proceedings of 20002 3rd International Conference on Microwave and Millimeter Wave Technology, vol. 2, pp. 1157–1161, Nov. 2002.

[16]. W. Y. A. Achmad, "RF System Issues Related to CDMA Receiver Specifications," *RF Design*, pp. Sept. 1999.

[17]. R. Cesari, "Estimate Dynamic Range for 3G A/D Converters," *Communication Systems Design*, vol. 8, no. 10, Oct. 2002.

[18]. B. Brannon, "Correlating High-Speed ADC Performance to Multicarrier 3 G Requirements," *RF Design*, pp. 22 – 28, June 2003.

[19]. N. Swanberg, J. Phelps, and M. Recouly, "WCDMA Cross Modulation Effects and Implications for Receiver Linearity Requirements," *2002 IEEE Radio and Wireless Communications*, pp. 13–18, Aug. 2002.

[20]. W. Y. Ali-Ahmad, "Effective IM2 Estimation for Two-Tone and WCDMA Modulated Blockers in Zero-IF," *RF Design*, pp. 32 – 40, April 2004.

[21]. E. Cetin, I. Kale and R. C. S. Morling, "Correction of Transmitter Gain and Phase Errors at the Receiver," *2002 IEEE International Symposium on Circuits and Systems*, vol. 4, pp. IV-109–IV 112, May 2002.

[22]. X. Huang, "On Transmitter Gain/Phase Imbalance Compensation at Receiver," *IEEE Communications on Letters*, vol. 4, no. 11, pp. 363–365, Nov. 2000.

[23]. S. Fouladifard and H. Shafiee, "A New Technique for Estimation and Compensation of IQ Imbalance in OFDM Receivers," *8th International Conference on Communication Systems*, vol. 1, pp. 224–228, Nov. 2002.

[24]. B. Lindoff, P. Main, and O. Wintzell, "Impact of RF Impairments on HSDPA Performance," *2004 IEEE International Conference on Communications*, vol. 6, pp. 3265–3269, 2004.

[25]. P. Baudin and F. Belveze, "Impact of RF Impairments on a DS-CDMA Receiver," *IEEE Trans. on Communications*, vol. 52, no. 1, pp. 31–36, Jan. 2004.

[26]. J. Feigin, "Don't Let Linearity Squeeze Your WLAN Performance," Communication Systems Design, vol. 9, no. 10, Oct. 2003.

[27]. S. Freisleben, "Semi-Analytical Computation of Error Vector Magnitude for UMTS SAW Filters," Tech. Note from EPCOS AG, Surface Acoustic Wave Devices, Munich, Germany.

[28]. F. J. O. Gonzalez et al., "A Direct Conversion Receiver Using a Novel Ultra Low Power A.G.C.," 1997 IEEE 47th Vehicular Technology Conference, vol. 2, pp. 667–670, May 1997.

[29]. I. Magrini, A. Cidronali, et al., "A Study on a Highly Linear Front-End for Low/Zero-IF Receivers in the 5.8 GHz ISM Band," IEEE MELECON 2004, pp. 155–158, May 2004.

[30]. E. Grayer and B. Daneshrad, "A Low-Power All-Digital FSK Receiver for Space Applications," *IEEE Trans. On Communications*, vol. 49, no. 5, pp. 911–921, May 2001.

[31]. B. Lindoff, "Using a Direct Conversion Receiver in Edge Terminals: a New DC Offset Compensation Algorithm," *2000 IEEE International Symposium on Personal, Indoor and Mobile Radio Communications*, vol. 2, pp. 959–963, Sept. 2000.

[32]. J. L. Mehta, "Transceiver Architectures for Wireless ICs," *RF Design*, pp. 76–96, Feb. 2001.

Chapter 5

Transmitter System Analysis and Design

5.1. Introduction

The transmitter is the companion of the receiver in wireless mobile stations. They operate simultaneously in a full-duplex system, but they may run in different time slots in a half-duplex system. Similar to a receiver, the architecture of a transmitter can be one of the following: superheterodyne, direct conversion, or band-pass sampling. The corresponding block diagrams of these architectures are presented in Fig. 3.1, Fig. 3.10, and Fig. 3.30, respectively. The low IF architecture may not be necessary for the transmitter since the DC offset is relatively easy to be tuned out or compensated, and the noise figure is not as critical in the transmitter as in the receiver. The selection of the transmitter architecture will depend on system performance requirements, size, cost, IC technology maturity, current consumption, etc. One thing that should be noticed is that the transmitter usually does not employ an IF SAW filter even in a superheterodyne transmitter. The cost reduction from using a direct conversion transmitter, thus, cannot count on the elimination of the IF SAW filter as is commonly done when employing a direct conversion receiver.

The signal processed and amplified in a transmitter unlike a received signal is deterministic since it is generated in the local digital base-band. The signal level in the transmitter is normally much higher than that in a receiver. The important parameters for a transmitter are the output power, especially the *maximum output power*, which may be seen as a counterpart of the receiver sensitivity, and the fidelity of the transmission waveform measured by *modulation accuracy*, *EVM*, or *waveform quality factor*, ρ. In CDMA, there is a new specification named as *code domain power accuracy* (*CDPA*) [1], which depends mainly on the linearity of the RF analog section of the transmitter. In addition to these parameters of the desired transmission signal, the unwanted emissions, such as *adjacent channel power* (*ACP*) and in-band and out-of-band noise/spurious emissions, are usually well defined in the mobile transmitter specifications. This is due to the fact that unwanted emissions may interfere or jam other wireless mobile stations and/or other systems.

Since the signal in the transmitter is high enough to consider the nonlinearity of an active device, such as power amplifier (PA) and even the *pre-power amplifier* (or so-called *driver amplifier*), to obtain accurate results certain nonlinear analysis approaches will be employed. Utilizing the nonlinear analysis approach can help to optimize the balance between the transmitter system linearity and power consumption efficiency when the transmission signal is amplitude modulated. The nonlinearity of the transmitter chain mostly comes from the PA and driver stages, and it results mainly in an increase of adjacent and alternate channel power emissions and a decrease of the code channel power accuracy.

It is clear that the nonlinearity is not the only issue that needs to be considered in the transmitter performance analysis. The key specifications of a mobile station transmission, such as output power, transmission spectrum, modulation accuracy, wide-band noise emission, spurious emission, etc. will be discussed, and how to make the designed transmitter be able to meet these specifications is analyzed in the following sections. On the other hand, the transmitter AGC system and power management will be also addressed.

Finally, the transmitter system design method will be discussed. For practical system design, certain knowledge regarding key devices — for instance, power amplifier, up-converter, and synthesizer — is necessary. Their basic requirements and achievable performance at present will be introduced.

5.2. Transmission Power and Spectrum

The transmission power of a mobile station is defined differently for wireless mobile systems. In the GSM and the WCDMA systems, the transmission power of the mobile station is defined as the power level at the antenna port (or connector) of the transmitter. In the cellular band CDMA and the AMPS systems, the effective radiated power (ERP) is used for measuring the transmission power of a mobile transmitter, and the effective isotropic radiated power (EIRP) is employed in the PCS band CDMA system. The definitions of the ERP and the EIRP are given in Section 3.1.3.2. They are not repeated here, but the relationships between ERP/EIRP in dBm and the transmission power at the antenna port of the mobile transmitter, TX_{pwr_ant}, in dBm are as follows:

$$ERP = TX_{pwr\_ant} + G_{ant} - 2.15 \tag{5.2.1}$$

and

$$EIRP = TX_{pwr\_ant} + G_{ant}, \tag{5.2.2}$$

where G_{ant} in dB is the antenna gain in a given direction. Generally speaking, the transmission power at the antenna port is more often used than the ERP/EIRP in the RF transmitter system analysis and design since the RF transmitter system is considered only as the section of the transmitter from the DAC to the antenna port and the antenna is excluded. However, it is easy to calculate the transmission power TX_{pwr_ant} from the requested ERP or EIRP by means of (5.2.1) or (5.2.2).

The transmission power can be measured in the frequency domain in terms of the integrated power over a bandwidth crossing the main lobe of the transmitted signal power spectrum. In the WCDMA, this integration bandwidth is clearly defined as $(1 + \alpha)\cdot f_{chip}$, where α is the roll-off factor of a root raised cosine (RRC) filter and equal to 0.22, and f_{chip} is chip rate and equal to 3.84 Mcps [1]. The main lobe of the WCDMA signal power spectral density (PSD) and the bandwidth of the transmission power measurement is shown in Fig. 5.1. The bandwidth for the transmission power measurement in the CDMA system is not well defined, but 1.23 MHz (approximately equal to the chip rate) bandwidth is often used in the power measurement of the CDMA signal. For the GSM mobile transmitter, 200

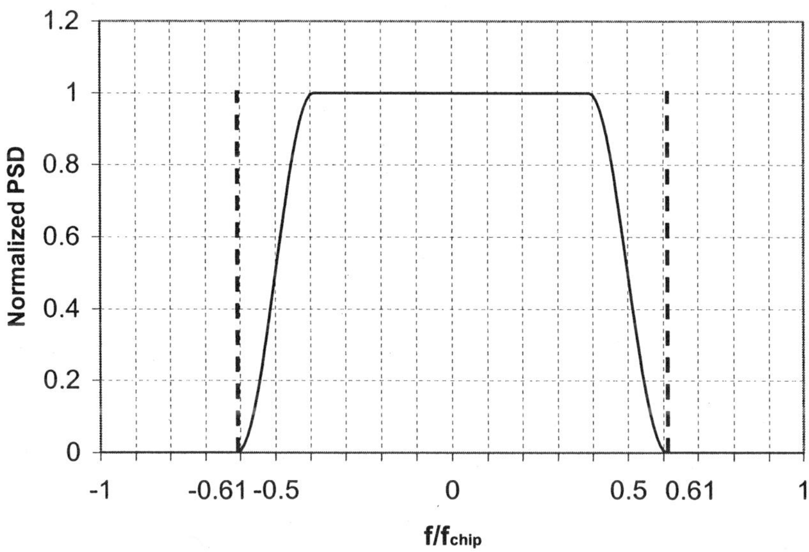

Figure 5.1. Ideal main lobe of WCDMA signal power spectrum and power measurement bandwidth, 1.22×3.84 MHz

kHz is the proper bandwidth for its transmission power measurement in the frequency domain. The main lobes of CDMA and GSM signal spectra are given in Section 2.4.4.3, Fig. 2.35, and Section 2.4.4.2, Fig. 2.33.

The maximum output power of a mobile transmitter will affect the overall wireless system performance, such as cell size and capacity, and the mobile station power consumption as well. It is well defined in all the wireless communication systems and is generally classified based on power levels and operating frequency bands. For the GSM system, in the 900 MHz band, the nominal maximum output powers of power class 2, 3, 4, and 5 are 39, 37, 33, and 29 dBm, respectively, but in the 1800 MHz band, the nominal maximum output powers of power class 1, 2, and 3, are respectively 30, 24, and 36 dBm [2]. For the WCDMA system, in the UMT (2.0 GHz) frequency band, the nominal maximum output powers of power class 1 to 4 are 33, 27, 24 and 21 dBm, respectively [1]. In the CDMA system, the maximum output power is defined slightly different from the GSM and the WCDMA systems, and there lower and upper limits of the maximum output power are used instead of the nominal value [3]. These limits for the 800 MHz Cellular band and 1900 MHz PCS band CDMA mobile transmitters are 23 and 30 dBm for their popularly used power classes 3 and 2, respectively. However, the most commonly used maximum transmission powers in the practical mobile transmitter design of various wireless communication systems are presented in Section 3.1.3.2, Table 3.2. In reality, the maximum output power of a mobile transmitter might be also limited by compromising transmission power within the tolerance range for a better transmitter linearity and/or for meeting requirements of the *specific absorption rate* (*SAR*) * [4].

5.3. Modulation Accuracy

5.3.1. Error Vector Magnitude EVM and Waveform Quality Factor ρ

Modulation accuracy is one of the key parameters for the wireless communication transmitters. The modulation accuracy is represented by error vector of modulation, which is the difference between the actual symbol location and the theoretical symbol location on the modulation vector constellation diagram. This may be better explained through an example. In the CDMA reverse link, offset QPSK (OQPSK) modulation is

* SAR is used to measure the amount of RF power that a mobile station through its antenna delivers to the human body.

used for the PN spreading code because of the consideration of power efficiency and spectrum efficiency. The RF signal with OQPSK modulation can be expressed as a composition of an in-phase and a quadrature signal as follows:

$$s(t) = a_I(t)\cos(\omega_c t) - a_Q(t)\sin(\omega_c t)\,, \tag{5.3.1}$$

where $a_I(t)$ and $a_Q(t)$ are the amplitudes of the in-phase and the quadrature signals, respectively,

$$a_I(t) = \sqrt{2}A\sum_{k=-\infty}^{k=\infty} I_k\, g(t - kT_c) \tag{5.3.2}$$

and

$$a_Q(t) = \sqrt{2}A\sum_{k=-\infty}^{k=\infty} Q_k\, g(t - kT_c - T_c/2). \tag{5.3.3}$$

In the above equations, A is the amplitude of the modulation signal, $\{I_k\}$ and $\{Q_k\}$ are I and Q PN sequences with value 1 or –1 that are actually mapped from the I and the Q spread data PN sequences by means of converting its 0 to 1, and 1 to –1. T_c is the duration of the spreading PN chips and $g\,(t)$ is the time domain response of the pulse shaping filter as given in Table 2.2 in Section 2.4.4.3, but it should be a rectangular pulse before the pulse shaping, $g\,(t) = g_r\,(t)$ defined as

$$g_r(t) = \begin{cases} 1 & 0 \le t \le T_c \\ 0 & otherwise \end{cases}. \tag{5.3.4}$$

For simplification, the rectangular pulse is used in the following analysis. Using (5.3.2) to (5.3.4), we can rewrite (5.3.1) into

$$s(t) = A\sum_{k=-\infty}^{\infty} \cos\left[\omega_c t + \phi(t - kT_c)\right], \tag{5.3.5}$$

where $\phi(t\text{-}kT_c)$ is defined as

$$\phi(t-kT_c)=\tan^{-1}\left[\frac{a_Q(t)}{a_I(t)}\right]=\begin{cases}\tan^{-1}(Q_{k-1}/I_k), & kT_c\le t<(k+1/2)T_c\\ \tan^{-1}(Q_k/I_k), & (k+1/2)T_c\le t<(k+1)T_c\end{cases}$$
$$=\begin{cases}\phi_{k-}, & kT_c\le t<(k+1/2)T_c\\ \phi_k, & (k+1/2)T_c\le t<(k+1)T_c\end{cases} \tag{5.3.6}$$

and the true value of $\phi(t-kT_c)$ = ϕ_{k-} or ϕ_k is as given in Table 5.1. The modulated signal constellation is shown as in Fig. 5.2(a).

Table 5.1. Channel I and Q sequence mapping to phase shift

Data PN Sequence		*Mapped PN Sequence*		*Phase Shift*
d_I	d_Q	a_I	a_Q	ϕ_{k-} or ϕ_k
0	0	1	1	$\pi/4$
1	0	-1	1	$3\pi/4$
1	1	-1	-1	$-3\pi/4$
0	1	1	-1	$-\pi/4$

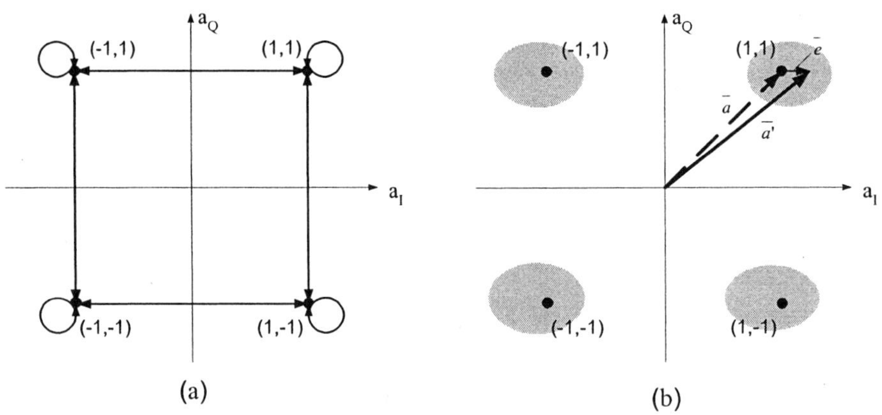

Figure 5.2. (a) Ideal OQPSK modulation constellation and (b) corresponding distorted modulation constellation

Using a discrete time variable with a time step $T_c/2$ (half chip duration) instead of the continuous time variable and considering (5.3.1), (5.3.5) and (5.3.6), the base-band modulation I/Q signal at $t = k_I \cdot T_c/2$ from (5.3.2) and (5.3.3) we obtain the following forms:

$$a_I(k_1) = A\cos\phi(k_1) \tag{5.3.7}$$

and

$$a_Q(k_1) = A\sin\phi(k_1), \tag{5.3.8}$$

where argument k_l denotes the time instant $k_l \cdot T_c/2$, and the modulation angle $\phi(k_l)$ is determined by (5.3.6) and the mapping given in Table 5.1. The modulation can also be expressed in a vector form:

$$\bar{a}(k_1) = \bar{a}_I(k_1) - j\bar{a}_Q(k_1) = A \cdot \exp[\phi(k_1)]. \tag{5.3.9}$$

The modulation is generally distorted by close-in phase noise of the modulator LO when it modulates an RF carrier. It will be further degraded when the modulated RF signal passes through a narrow-bandwidth filter due to the magnitude ripple and the group delay distortion of the filter. Thus, the distorted modulation $\bar{a}'(k_1)$ can be expressed as

$$\bar{a}'(k_1) = \bar{a}(k_1) + \bar{e}(k_1), \tag{5.3.10}$$

where $\bar{e}(k)$ represents residual error vector. The constellation diagram of the distorted modulation vector is depicted in Fig. 5.2(b).

In the wireless communications, the modulation accuracy is represented by *error vector magnitude* (*EVM*), and it is defined as the mean square error between the samples of the actual and the ideal signals, normalized by the average power of the ideal signal. The *EVM* can be mathematically represented as

$$EVM = \left[\frac{E\left\{\left|\bar{a}'(k_1) - \bar{a}(k_1)\right|^2\right\}}{E\left\{\left|\bar{a}(k_1)\right|^2\right\}}\right]^{\frac{1}{2}} = \left[\frac{E\left\{\left|\bar{e}(k_1)\right|^2\right\}}{E\left\{\left|\bar{a}(k_1)\right|^2\right\}}\right]^{\frac{1}{2}}. \tag{5.3.11}$$

In (5.3.11), $E\{\cdot\}$ denote the expectation of ensemble averages.

In the CDMA system, the waveform quality factor, ρ, is used to represent the modulation accuracy instead of *EVM*. The waveform quality factor is defined as a correlation coefficient between the actual waveform $Z(t)$ and the ideal waveform $R(t)$, and it is expressed as [3]

$$\rho = \frac{\left|\sum_{k=1}^{M} R_k Z_k^*\right|^2}{\sum_{k=1}^{M} |R_k|^2 \sum_{k=1}^{M} |Z_k|^2}, \tag{5.3.12}$$

where $R_k = R_k(t_k)$ is the kth sample of the ideal signal in the measurement interval; $Z_k = Z_k(t_k)$ is the kth sample of the actual signal in the measurement interval; and M is the measurement interval in half-chip intervals and shall be at least 1229 half-chip intervals (0.5 ms). There is an approximate relationship between ρ and the EVM when the cross-correlation between the ideal signal and the error signal is negligible [5],

$$\rho \cong \frac{1}{1 + EVM^2} \tag{5.3.13a}$$

or

$$EVM \cong \sqrt{\frac{1}{\rho} - 1} \tag{5.3.13b}$$

The proof of (5.3.13) can be found in Appendix 5A.

In the following sections we discuss the contributions to degradation of the modulation accuracy. Most of results derived from these subsections are also applicable to evaluation of the modulation accuracy in the receiver chain although they may result from the analyses of the transmitter

5.3.2. Influence of Intersymbol or Interchip Interference to EVM

The modulation accuracy of a modulated RF/IF signal can be degraded when the signal passes through a nonideal filter. The cause of this is as follows. A modulation signal usually consists of symbols or chips. The symbol or chip waveforms after filtering are distorted due to filter group delay distortion and magnitude response ripples, and one symbol or chip waveform generates interference in the adjacent and other symbols or chips. This kind of interference is referred to as *intersymbol interference* (*ISI*) or *interchip interference* (*ICI*). The ISI or ICI is the root of the degradation of the modulation accuracy when the modulated signal passes through the nonideal filter, especially if its bandwidth close to that of the modulation spectrum.

Actually, the ISI or ICI of symbols or chips may have been created at beginning to generate a transmission signal. To obtain high spectral efficiency of the transmission signal, the originally rectangular symbol or chip waveform is reshaped, and it is also called *pulse shaping* as discussed in Section 2.4.4. The pulse-shaping procedure can be mathematically represented as follows. If the rectangular symbol (or chip) is $a_{rect}(t)$, and the impulse response of the pulse-shaping filter is $h_{PS}(t)$, the shaped symbol (or chip) waveform $a_{Tx\_ideal}(t)$ can be expressed as

$$a_{Tx\_ideal}(t) = h_{PS}(t) * a_{rect}(t), \tag{5.3.14}$$

where * denotes convolution operator. Now, the symbol (or chip) waveform $a_{Tx\_ideal}(t)$ may contribute interference to adjacent and other symbol (or chip) waveforms for most pulse-shaping filters commonly used in wireless communication systems. For instance, the root raised cosine filter and modified root raised cosine filter with impulse coefficients given in Table 2.2 in Section 2.4.4.3 are employed as the pulse shaping in the WCDMA and CDMA mobile station transmitter, respectively. As we have seen in Chapter 2, these filters cause the shaped symbol or chip to have ISI or ICI. However, in the wireless systems, a complementary filter with an impulse response $h_{PS\_C}(t)$ is normally used in the corresponding receiver side to equalize the phase and magnitude distortion and thus to eliminate or to minimize the ISI or ICI caused by the pulse shaping in the transmitter. The ideal symbol or chip waveform in the corresponding receiver should be

$$a_{Rx\_ideal}(t) = h_{PS}(t) * h_{PS\_C}(t) * a_{rect}(t). \tag{5.3.15}$$

For example, a root raised cosine filter with a roll-off factor α = 0.22 is employed in the WCDMA mobile station transmitter as pulse shaping filter, and the same filter is also used in the WCDMA base station receiver as the complementary filter. These two root raised cosine filters being in cascade compose a raised cosine impulse response. The ideal chip waveform sequences in the receiver can then be expressed as

$$\sum_{k=-\infty}^{\infty} a_{Rx\_ideal}(t - kT_c) = \sum_{k=-\infty}^{\infty} h_{RC}(t - kT_c) * a_{rect}(t - kT_c). \tag{5.3.16}$$

Symbol or chip waveforms shaped by the raised cosine filter do not contribute any interference to their adjacent and other symbol or chip waveforms as shown in Fig. 5.3. Therefore, the modulation accuracy degradation resulting from pulse shaping is usually not considered when we analyze and calculate the transmission modulation accuracy. On the other hand, the modulation accuracy definition of the mobile station transmitters is generally based on the difference between the actual transmission waveform and the ideal waveform, which is formed by the symbols or chips with a waveform given in (5.3.14). It is apparent that the ISI or ICI between symbols or chips in the ideal transmission waveform has no contribution to the modulation accuracy degradation since it is a reference waveform of measuring the modulation accuracy in the transmitter.

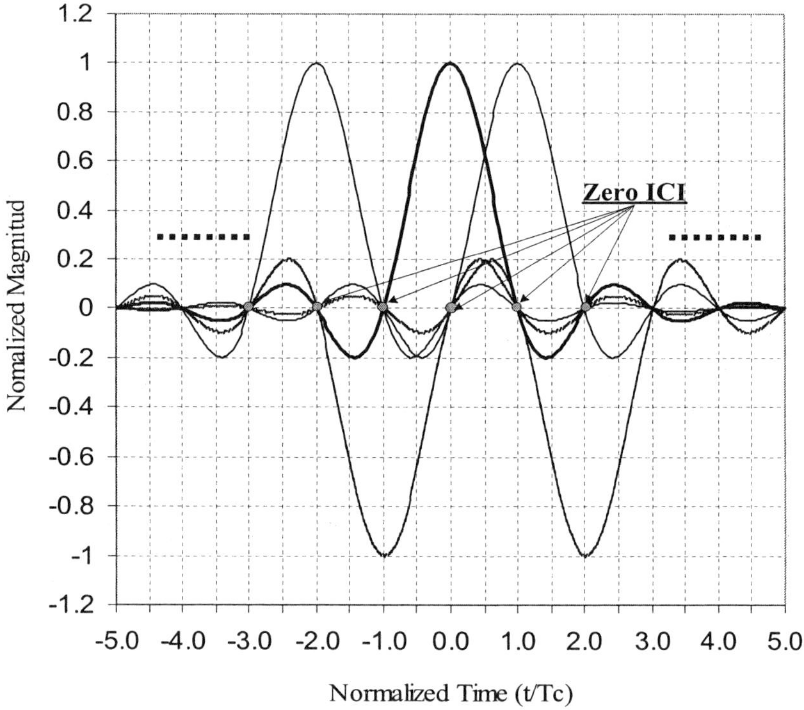

Figure 5.3. Chip or symbol waveforms shaped by a raised cosine filter having no interference from adjacent and other waveforms

The filters in the transmitter path other than the pulse-shaping filter may cause the degradation of the modulation accuracy degradation especially when the pass-band of the filter, such as a channel filter, is close to the bandwidth of the transmission signal. Assuming the impulse response of the filter is $h_{fltr}(t)$, the shaped symbol or chip waveform $a_{Tx\_ideal}(t)$ after passing through the filter turns into

$$a_{Tx}(t) = h_{fltr}(t) * a_{Tx\_ideal}(t). \tag{5.3.17}$$

The corresponding *EVM* can be expressed by the impulse response function as [6]

$$EVM_{ISI} = \sqrt{\frac{\sum_{k=-\infty,k\neq 0}^{k=\infty} \left| h_{fltr}(t_o + kT_s) \right|^2}{\left| h_{fltr}(t_o) \right|^2}}. \tag{5.3.18}$$

The terms in the numerator in the square root of (5.3.18) right side are ISIs or ICIs to the symbol or the chip at t_o, and they are contributed from adjacent and other symbols or chips. Each term in the square root of (5.3.18) can be also obtained by means of the following formula:

$$\Delta I_{ISI}(\pm k) = \frac{\left| h_{fltr}(t_o \pm kT_s) \right|}{\left| h_{fltr}(t_o) \right|} = \frac{\int_{t_o-\delta t}^{t_o+\delta t} \left| h_{fltr}(t \pm kT_s) \right| dt}{\int_{t_o-\delta t}^{t_o+\delta t} \left| h_{fltr}(t) \right| dt}, \tag{5.3.19}$$

where k is equal to 1, 2, 3, $\cdots$, and $2\delta t$ is the duration of sampling pulse. Utilizing (5.3.19), we can rewrite (5.3.18) as

$$EVM_{ISI} = \sqrt{\sum_{k=-\infty}^{\infty} \Delta I_{ICI}^2(k)}. \tag{5.3.20}$$

In recent transmitter design of wireless mobile stations, no channel filtering is really used in the IF block of the transmitters. Some low-pass filters may be employed in the I and the Q channels of analog base-band of the transmitters to suppress the emission power at adjacent channels, but the bandwidth of these low-pass filters is usually wider than that of the channel filter, and thus they have slight impact on the modulation accuracy. The EVM resulting from the low-pass filtering is approximately 5% or less. In addition to this, predistortion can also be used to minimize the impact on the modulation accuracy due to the phase and magnitude distortions of the low-pass filters if necessary.

However, channel filtering is definitely used in the receivers of mobile stations. An example presented in Section 2.4.5.2, the intersymbol interferences resulting from preceding and successive symbols for a π/4QPSK modulated signal with symbols shaped by the raised cosine filter having a roll-off factor $\alpha = 0.5$ after channel filtering are increased from zero to 0.075 and 0.066, respectively. Based on these data and using (5.3.20), we can approximately estimate the modulation accuracy of this filtered signal to be

$$EVM_{ISI} = \sqrt{0.075^2 + 0.066^2} = 9.98\,\%.$$

The modulation accuracy degradation of the received signal will cause the signal-to-noise/interference ratio of the demodulated base-band signal decrease.

5.3.3. Influence of Close-in Phase Noise of Synthesized LO to EVM

Another main contribution to the degradation of the modulation accuracy is the close-in phase noise of the VHF and UHF synthesizers applied as local oscillator of the modulator and the up-converter in the transmitter. The influence of the phase noise on the *EVM* can be estimated as follows. Assuming that the vector error in (5.3.10) is caused by the synthesizer phase noise $\phi_n(t)$, the degraded signal can be expressed as

$$\bar{a}'(t) = \bar{a}(t) + \bar{e}(t) = \bar{a}(t)\exp(j\phi_n(t)). \qquad (5.3.21)$$

The magnitude of the vector error then is

$$\begin{aligned} |\bar{e}(t)|^2 &= |\bar{a}'(t) - \bar{a}(t)|^2 = |\bar{a}(t)(e^{j\phi_n(t)} - 1)|^2 \\ &= |\bar{a}(t)[\cos\phi_n(t) - 1 + j\sin\phi_n(t)]|^2 \\ &= |\bar{a}(t)|^2\left[\overline{\cos\phi_n(t) - 1}^2 + \sin^2\phi_n(t)\right] \cong 4[\phi_n(t)/2]^2 = \phi_n^2(t), \end{aligned} \qquad (5.3.22)$$

where we use normalized magnitude $|\bar{a}(t)| = 1$. The statistical average of $\phi_n^2(t)$ is the autocorrelation function of the phase noise. The autocorrelation function and the power spectrum density $S_n(f)$ have the following relationship

$$\lim_{n\to\infty} \frac{1}{nT_S} \int_{-T_S/2}^{T_S/2} E\{\phi_n^2(t)\} dt = \int_{-\infty}^{\infty} S_n(f) df = P_{Nphase} \tag{5.3.23}$$

Based on the definition of the *EVM* (5.3.11) we obtain the *EVM* resulting from the phase noise of the synthesizers to be

$$EVM_{Nphase} = \sqrt{\lim_{nT_c\to\infty} \frac{\int_{-nT_c}^{nT_c} E\{\phi_n^2(t)\}}{nT_c}} = \sqrt{\int_{-\infty}^{\infty} S_n(f) df} = \sqrt{P_{n,phase}} \, . \tag{5.3.24}$$

Usually, the phase noise within the loop bandwidth of synthesizers used in the mobile transmitters is in the range of –60 to –80 dBc/Hz. In the case of the synthesizer loop bandwidth being reasonably wide, $P_{n,phase}$ can be estimated by means of the following approximate formula:

$$P_{Nphase} \cong 2 \cdot 10^{\frac{N_{phase}}{10}} \cdot BW_{synth\_loop} \, , \tag{5.3.25}$$

where N_{phase} is the average phase noise, in dBc/Hz, within the synthesizer loop bandwidth, and BW_{synth_loop} is the bandwidth of the synthesizer loop filter in Hz. Substituting (5.3.25) to (5.3.24), we obtain

$$EVM_{Nphase} = \sqrt{P_{Nphase}} \cong \sqrt{2 \cdot 10^{\frac{N_{phase}}{10}} BW_{synth\_loop}} \, . \tag{5.3.26}$$

When more than one synthesizers are used, the *EVM* formula becomes

$$EVM_{Nphase} = \sqrt{\sum_{k=1}^{n} P_{Nphase,k}} \, . \tag{5.3.27}$$

For example, assuming that there are two frequency up-conversions to bring a BB signal to an RF signal in a CDMA transmitter, and the close-in phase noises of the VHF and the UHF synthesizers integrated in a bandwidth of 1.23 MHz are –26 dBc and –28 dBc, respectively, the *EVM* due to the phase noise in this case is

$$EVM_{Nphase} = \sqrt{P_{Nphase\_VHF} + P_{Nphase\_UHF}} \cong \sqrt{10^{-\frac{26}{10}} + 10^{-\frac{28}{10}}} = 6.4\% .$$

Considering the *EVM* due to the low-pass filter to be 5%, the combined *EVM* is

$$EVM_T = \sqrt{EVM_{ICI}^2 + EVM_{n,phase}^2} = \sqrt{0.05^2 + 0.064^2} = 8.1\% .$$

The corresponding waveform quality factor ρ from (5.3.13a) is

$$\rho_T = \frac{1}{EVM_T{}^2 + 1} = \frac{1}{0.0066 + 1} = 0.993 .$$

This EVM can be improved since the integrated phase noise used in the above calculation is quite high, and the phase noise of each synthesizer is easy cut down 2 dB at least.

5.3.4. Carrier Leakage Degrading the Modulation Accuracy

The DC offset in the BB I and Q channels will cause carrier leakage, and it will degrade the modulation accuracy of the transmission signal. This can be explained by using the following simple example. Assuming that the DC offsets in the base-band I and Q channels are ΔI_{dc} and ΔQ_{dc}, respectively, the BB signals, $a'_I(t)$ and $a'_Q(t)$, in the I and the Q channels are represented by

$$a'_I(t) = I(t)\cos\phi(t) + \Delta I_{dc} \tag{5.3.28}$$

and

$$a'_Q(t) = Q(t)\sin\phi(t) + \Delta Q_{dc} , \tag{5.3.29}$$

where $I(t)$ and $Q(t)$are the amplitudes of the I and the Q BB signals, and $\phi(t)$ is the phase of the corresponding I and Q BB signals.

At the output of the modulator, the I and Q quadrature signals turn into a signal with an IF or RF carrier, and it can be expressed as

$$\begin{aligned} f_{Tx}(t) &= a_I'(t)\cos\omega_c t - a_Q'(t)\sin\omega_c t \\ &\cong A(t)\cos[\omega_c t + \varphi(t)] + \Delta_{dc}\cos(\omega_c t + \Delta\theta), \end{aligned} \tag{5.3.30}$$

where ω_c is the carrier angle frequency $\omega_c = 2\pi f_c$, $A(t)$ is obtained from $a_I(t)$ and $a_Q(t)$ as

$$A(t) = \sqrt{I^2(t)\cos^2\phi(t) + Q^2(t)\sin^2\phi(t)}\ , \tag{5.3.31}$$

$\varphi(t)$ represents phase modulation equal to

$$\varphi(t) = \tan^{-1}\left(\frac{Q(t)\cdot\sin\phi(t)}{I(t)\cdot\cos\phi(t)}\right), \tag{5.3.32}$$

Δ_{dc} and $\Delta\theta$ are the overall DC offset of the I and Q channels and the phase shift, respectively,

$$\Delta_{dc} = \sqrt{\Delta I_{dc}^2 + \Delta Q_{dc}^2} \tag{5.3.33}$$

and

$$\Delta\theta = \tan^{-1}(\Delta Q_{dc}/\Delta I_{dc}). \tag{5.3.34}$$

The last term on the right side of (5.3.30) is the *carrier leakage*, also called *carrier feed through* (*CFT*).

Here, we define the ratio of the carrier leakage power to the desired signal transmission power as the carrier suppression C_S in dB — i.e.,

$$C_S = 10\log\frac{P_{CFT}}{P_{Tx}} = 20\log\frac{V_{CFT\_rms}}{V_{Tx\_avg}}. \tag{5.3.35}$$

It is clear that the rms voltage of the carrier leakage is equal to

$$V_{CFT\_rms} = \Delta_{dc}/\sqrt{2}. \tag{5.3.36}$$

However, for an RF signal with AM, the rms voltage of its average signal power V_{Tx_avg} can be calculated from the peak amplitude A_{max} of the desired signal and the *peak-to-average ratio* (*PAR*) as follows:

$$V_{Tx\_avg} = 10^{\frac{20\log A_{\max} - PAR - 3}{20}}. \tag{5.3.37}$$

For example, the *PAR* of the CDMA IS-95 voice is approximately 3.9 dB, and the peak swing of the base-band I and Q signals is $V_{p-p} = 2A_{\max} = 1.0$ V. Then we have the rms voltage of

$$V_{Tx\_avg} = 10^{\frac{20\log(0..5\times\sqrt{2})-3.9-3}{20}} \cong 0.32 \text{ V}.$$

The DC offset in the BB I and Q channels causes carrier feed through, but this is not the only root cause of this problem. The modulator and the frequency up-converter may cause carrier leakage too because the isolation between the LO port and the RF/IF port of these devices is finite around –30 dBc depending on design and IC processing technology. The carrier leakage, no matter what is its cause, will degrade the modulation accuracy of the transmission signal, and the corresponding EVM_{CFT} can be calculated by using the overall carrier suppression C_S as

$$EVM_{CFT} = \sqrt{10^{\frac{C_S}{10}}} = \sqrt{\sum_{k=1}^{n} 10^{\frac{C_{S,k}}{10}}}, \tag{5.3.38}$$

where $C_{S,k}$ represents the carrier suppression of the kth individual contribution to the overall carrier leakage, and n is the number of the contributors. For instance, assuming that the overall leakage results from the I/Q DC offset, $C_{S,1}$ = -27 dBc, and the finite isolation between the LO and IF/RF of the modulator and the up-converter, $C_{S,2}$ = -28 dBc, the modulation accuracy due to the carrier leakage is

$$EVM_{CFT} = \sqrt{10^{\frac{-27}{10}} + 10^{\frac{-28}{10}}} = \sqrt{(1.995 + 1.585)\times 10^{-3}} = 5.98\%.$$

On the other hand, based on the budget of the *EVM*, we can allocate the allowed carrier feed-through levels resulting from the DC offset in the I and the Q channels and the finite isolation between the LO and the IF/RF ports of the modulator and the up-converter. From the carrier leakage levels allocated to the DC offset and the isolation, we can determine the allowed I and Q channel DC offset from (5.3.35) – (5.3.36) and (5.3.33), but the minimum requirement of the isolation between the LO and IF/RF ports is directly obtained from (55.3.33) after replacing its denominator with the LO power or rms voltage. Let us look at an example to determine the allowed maximum DC offset if demanding the $EVM_{CFT} < 4.5\%$ caused by the carrier feed through due to the DC offset and the $V_{Tx\_avg} \cong 0.32\,\text{V}$. From (5.3.38) and (5.3.35), we can directly estimate the allowed carrier leakage rms voltage from the given EVM_{CFT}

$$V_{CFT\_rms} = V_{Tx\_avg} \cdot EVM_{CFT} = 0.32 \times 0.045 = 0.014\ \text{V}.$$

Therefore, the allowed overall DC offset of the I and Q channels is obtained from (5.3.36), or

$$\Delta_{dc} = \sqrt{2} \times 0.014 \times 1000 = 19.8\ \text{mV}.$$

Actually, the allowed maximum DC offset in each of the I and Q channels is approximately 14.0 mV.

5.3.5. Modulation Accuracy Degradations Resulting from Other Factors

The other factors, which may also affect the transmission modulation accuracy, are the amplitude and phase imbalances of the I and Q channel BB signals, the nonlinearity of the power amplifier, LO re-modulated by the transmission reflection, and the in-channel-band phase noise at very low transmission output power. They are generally not the main contribution to modulation accuracy degradation, but in certain situations, such as at low transmission output power, some of them may dominate the modulation accuracy performance.

5.3.5.1. Degradation Due to I and Q Imbalance

The amplitude and phase imbalances between the I and the Q channel BB signals generate an image of the transmission signal, which has the same carrier and spectral bandwidth as the transmission signal. Assuming that the overall normalized amplitude imbalance caused by the I and Q channel and the quadrature modulator is $1{:}(1+\delta)$, and the phase imbalance of the I and Q channel signals is ε degree off from 90° and the phase error σ caused by the quadrature modulator is relatively small, the transmission signal will contain the unwanted image product, and it can be approximately expressed as (see Appendix 5B)

$$f_{Tx}(t) \cong A(t)\frac{\sqrt{1+2(1+\delta)\cos(\varepsilon+\sigma)+(1+\delta)^2}}{2}\cos[\omega_c t+\phi(t)+\theta]$$
$$+A(t)\frac{\sqrt{1-2(1+\delta)\cos(\varepsilon-\sigma)+(1+\delta)^2}}{2}\cos[\omega_c t-\phi(t)+\varphi]. \qquad (5.3.3)$$

The second term on the right side of (5.3.39) is the unwanted image product, which has an image spectrum of the desired signal and occurs in the same bandwidth as the desired one (refer to Fig. 5B.1 in Appendix 5B).

Defining the ratio of the image product power to the desired transmission power as image suppression, IMG_S, (image rejection is used in Section 3.3, which is an inverse of the image suppression defined here), and it is usually presented in logarithm form:

$$IMG_S = 10\log\frac{1-2(1+\delta)\cos(\varepsilon-\sigma)+(1+\delta)^2}{1+2(1+\delta)\cos(\varepsilon+\sigma)+(1+\delta)^2}. \qquad (5.3.40)$$

The EVM due to the image product is

$$EVM_{img} = \sqrt{10^{\frac{IMG_S}{10}}}. \qquad (5.3.41)$$

It is not difficult to obtain an image suppression of a transmission below –30 dBc after proper tuning. In the case of IMG_S = -30 dBc, the EVM_{img} is approximately

$$EVM_{img} = \sqrt{10^{\frac{-30}{10}}} \cong 3.2\%.$$

Generally, the image product does not degrade the modulation accuracy much. If IMG_S = -30 dBc, and the phase imbalance is mainly caused by the I and Q channels — i.e., $\sigma \ll \varepsilon$ – utilizing $IR = -IMG_S$ from Fig. 3.22 we can estimate the allowed amplitude and phase imbalances to be 0.3 dB and 1°.

5.3.5.2. Nonlinearity Influence on EVM

For a transmission signal with amplitude modulation such as CDMA, WCDMA, TDMA, and EDGE transmission signals, the nonlinearity in the transmitter chain may affect its modulation accuracy. The transmitter nonlinearity mainly results from the power amplifier since considering the efficiency the PA in the mobile station transmitters demanding linear operation is usually running in class AB instead of class A. The nonlinearity influence to the modulation accuracy can be estimated as follows. It is assumed that only the signal amplitude equal to and greater than the level of 1 dB compression point of the power amplifier, P_{-1}, will affect the modulation accuracy. We define the excess value of the output power level P_{Tx} over the output P_{-1} as

$$\Delta_{dB} = P_{Tx} - (P_{-1})_{out}. \tag{5.3.42}$$

The modulation error caused by the excess amplitude Δ_{dB} can be approximately estimated by $\left(10^{(\Delta_{dB}+1)/20} - 1\right)$, and the corresponding *EVM* calculation need take the amplitude PDF $P_{tg,k}$ into account. The accumulated modulation error or the EVM_{nonlin} due to the nonlinearity can be estimated by means of the following approximation (clipping at P_{-1} assumed):

$$EVM_{nonlin} = \int_0^{\infty} P_{tg}(\Delta_{dB}) \left(10^{\frac{\Delta_{dB}+1}{20}} - 1\right) \cdot d\Delta_{dB} \cong \sum_{k=1}^{n} P_{tg,k} \cdot \left(10^{\frac{\Delta_{dB,k}+1}{20}} - 1\right). \tag{5.3.43}$$

Assuming that a PA output power of a CDMA transmitter is 28 dBm and its one dB compression point $(P_{-1})_{out}$ is 31 dBm, in this case the average transmission output power is 3.0 dB below the $(P_{-1})_{out}$. The PDF of the transmission signal is presented in Fig. 5.4. Based on this PDF and (5.3.43), we can evaluate the EVM_{nonlin}:

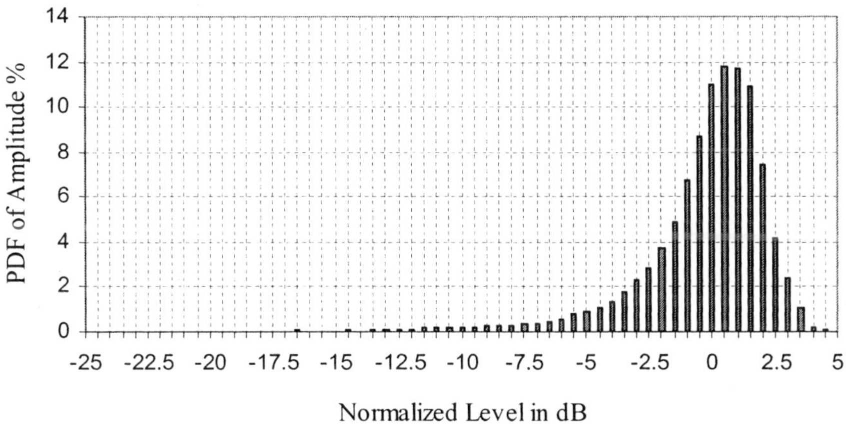

Figure 5.4. Amplitude probability distribution of CDMA IS-2000 voice transmission

$$EVM_{nonlin} = \left[4.2\times\left(10^{1/20}-1\right)+2.4\times\left(10^{1.5/20}-1\right)+1.0\times\left(10^{2/20}-1\right)\right.$$
$$\left.+1.0\times\left(10^{2/20}-1\right)+0.2\times\left(10^{2.5/20}-1\right)+0.1\times\left(10^{3/20}-1\right)\right]\% = 1.3\%.$$

In this example, we can see that the PA nonlinearity does not significantly impact the modulation accuracy. It is apparent that the impact will depend on the PA nonlinearity and the peak-to-average ratio of the transmission signal. The EVM_{nonlin} increases to 3.3% if the transmission signal is the CDMA IS-95 voice instead of the IS-2000 voice in the previous example.

5.3.5.3. Impact of In-Channel Bandwidth Noise

The emission noise within the channel bandwidth may not be an issue to the modulation accuracy when the transmission power is high. However, it may dominate the modulation accuracy of the transmission signal when the transmission power is very low — for instance, only –50 dBm is transmitted from the CDMA mobile stations while they move close to base stations.

The formula of calculating the EVM_{noise} caused by the emission noise within the channel bandwidth is

$$EVM_{noise} = \sqrt{10^{\frac{N_{in-ch-band}-Tx}{10}}}, \tag{5.3.44}$$

where $N_{in\text{-}ch\text{-}band}$ is the integrated noise over the channel bandwidth, and Tx is transmission power in dBm. Both can be measured at the antenna port of the mobile stations. For example, assuming that the transmission power of a CDMA mobile station is –50 dBm and the integrated noise over the channel bandwidth 1.23 MHz is –71.5 dBm, the EVM at this output power level is

$$EVM_{noise} = \sqrt{10^{\frac{-71.5+50}{10}}} = 8.4\%,$$

5.3.5.4. Modulation Error Resulting from Reverse Modulation of LO

The LO of the VHF or UHF quadrature modulator may be modulated by the reflected harmonic of the transmission signal from the load of the modulator if the carrier frequency of the harmonic signal is equal to LO frequency. This phenomenon is called *reverse modulation.* In general, the reverse modulation may occur when the modulator has a heavy load or needs to deliver too much output power. This is rarely seen in the superheterodyne transmitter, but it may happen in the direct conversion transmitter since the VCO in the synthesizer for the direct conversion is usually running at harmonic or subharmonic frequency of the transmission carrier. Nevertheless, the reverse modulation impact on the modulation accuracy can be reduced to a negligible extent, as long as we have a proper design and implementation of the quadrature modulator of the transmitter.

Assuming the integrated reverse modulation noise of the synthesized LO over the transmission signal bandwidth is $|N_{rm}|$ dB below the LO level, the modulation accuracy or EVM_{rm} caused by the reverse modulation is

$$EVM_{rm} = \sqrt{10^{\frac{N_{rm}}{10}}}\,, \tag{5.3.45}$$

where $N_{rm} < 0$ in dBc. The reverse modulation noise influence will be insignificant when $N_{rm} < -30$ dBc, and this is not difficult to implement for an RF transmitter based on highly integrated circuits.

5.3.6. Total EVM and Waveform Quality Factor

If all factors contributing to the degradation of the modulation accuracy are uncorrelated, the overall *EVM* of the transmission signal can be expressed as

$$\begin{aligned} EVM_{total} &= \sqrt{\sum_k EVM_k^2} \\ &= \sqrt{EVM_{ISI}^2 + \sum_{i=1}^{2} EVM_{Nphase,i}^2 + EVM_{CFT}^2 + EVM_{img}^2 + \cdots}\,. \end{aligned} \tag{5.3.46}$$

The corresponding overall transmission waveform quality factor ρ_{total} is approximately

$$\rho_{total} = \frac{1}{1 + \sum_k EVM_k^2}$$
$$= \frac{1}{1 + EVM_{ISI}^2 + \sum_{i=1}^{2} EVM_{Nphase,i}^2 + EVM_{CFT}^2 + EVM_{img}^2 + \cdots}. \quad (5.3.47)$$

Using the example results presented in Sections 5.3.3 to 5.3.5 for different modulation accuracy degradations, we estimate the overall modulation accuracy EVM_{total} of a CDMA mobile station transmitting IS-95 voice signal at high output power to be

$$EVM_{total} = \sqrt{0.064^2 + 0.060^2 + 0.032^2 + 0.033^2} = 9.9\%$$

and

$$\rho_{Total} = \frac{1}{1 + 0.099^2} = 0.990\,.$$

The degradation due to ICI is not included in the above calculation. In the case of very low output power (e.g., –50 dBm), the corresponding EVM and ρ are, respectively,

$$EVM_{total\_LowPwr} = \sqrt{0.099^2 + 0.084^2} = 13.0\%$$

and

$$\rho_{Total\_LowPwr} = \frac{1}{1 + 0.13^2} = 0.983.$$

5.4. Adjacent and Alternate Channel Power

Unwanted emissions from mobile station transmitters are usually tightly restricted so as not to interfere with other radio systems. Among emission specifications of mobile transmitters the most important ones are the emission levels in the adjacent and the alternate channels, and they are stringently controlled in wireless mobile systems. The adjacent/alternate channel emission power specification is generally defined as ratio of the power integrated over an assigned bandwidth in the adjacent/alternate channel to the total desired transmission power. Referring to Fig. 5.5, the *adjacent channel power ratio* (*ACPR*) can be expressed as

$$ACPR = \frac{\int_{f_1}^{f_1+\Delta B_{ACP}} SPD(f) \cdot df}{\int_{f_o-BW/2}^{f_o+BW/2} SPD(f) \cdot df}. \tag{5.4.1}$$

The bandwidth of measuring adjacent channel power ΔB_{ACP} varies with different mobile systems. In the GSM and CDMA mobile stations, ΔB_{ACP} is 30 kHz, AMPS uses 3 kHz, and it is 3.84 MHz in the WCDMA.

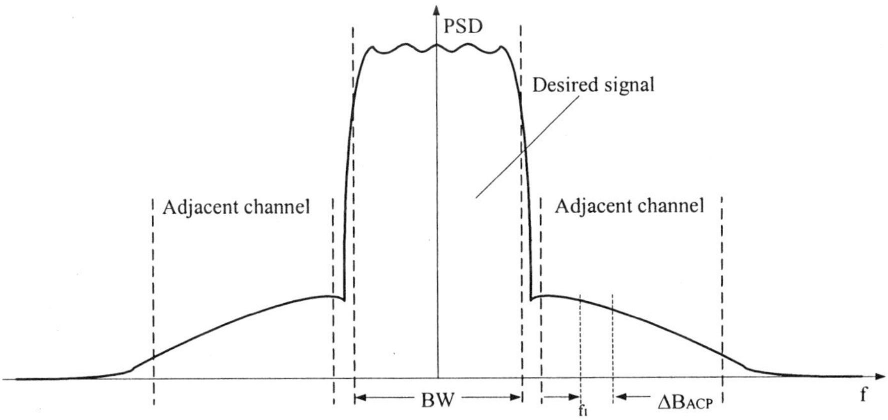

Figure 5.5. Calculation of adjacent channel power ratio

In the half-duplex or TDMA system, the transmitter is operating in burst mode condition. Improper ramp up or ramp down of the transmission signal pulse in the active time slots will regrowth spectral power in the adjacent channels. It is important for the transmitter operating these systems to properly control the rise and fall edges of the transmission signal pulses.

5.4.1. Low-Pass Equivalent Behavioral Model Approach

The adjacent/alternate channel powers mainly result from spectral regrowth caused by the nonlinearity of the transmitter chain, which mostly comes from the power amplifier (PA) and the PA driver. The *ACPR* of a digitally modulated transmission signal cannot be accurately determined from intermodulation distortion of discrete tones. In this case, a behavioral nonlinear model of amplifiers is needed. This amplifier model, a low-pass representation, is developed from AM-AM and AM-PM measurement or simulation data as described in Section 2.2.3. The AM-AM and AM-PM nonlinear characteristics of a device are also referred to as *envelope nonlinearity* since they only depend on the input signal amplitude.

The only concern of the spectral regrowth in the transmitter amplifiers is with the nonlinearity that generates distortion products within adjacent and alternate channels. For this purpose, a band-pass nonlinear model [8] can be used. If the pass-band of the transmitter is a small percentage of the carrier frequency, the nonlinearity can be only characterized by odd-order terms of a power series or by odd function series, such as Fourier sine series, etc. [9]. An equivalent low-pass representation, which characterizes the effects of the amplifier nonlinearity on the signal envelope, is desirable for the nonlinear simulation. This low-pass equivalent nonlinear model can be represented by means of a complex envelope transfer characteristic function, $\tilde{y}(A(t))$, as [10]

$$\tilde{y}(A(t)) = f(A(t))e^{jg(A(t))}, \tag{5.4.2}$$

where $f(A(t))$ is the magnitude of the transfer characteristic, and $g(A(t))$ is the phase shift determined by AM-PM measurements. The block diagram representation of the nonlinear amplifier low-pass equivalent is given in Fig. 2.5.

The magnitude of the transfer characteristic $f(A(t))$ can be expressed as an odd-order power series over the range of amplitude of the envelope signal as

$$f(A(t)) = \sum_{k=0}^{n} a_{2k+1} \cdot A(t)^{2k+1}, \tag{5.4.3}$$

where the coefficients a_{2k+1} (k = 0, 1, …, n) can be derived from the nonlinear device AM-AM characteristic measured on a network analyzer with a single tone signal. For the CDMA signal, the coefficients a_{2k+1} ($k = 0$, 1, …, n) can be determined as follows. Assume that the measured AM-AM characteristic curve $A_{out}(A_{in})$ can be expressed by an odd order power series of the test tone amplitude A_{in} as

$$A_{out}(A_{in}) = \sum_{k=0}^{n} c_{2j+1} \cdot A_{in}^{2j+1}. \tag{5.4.4}$$

The expansion coefficient a_{2k+1} in (5.4.3) then is obtained from the following relationship [11]:

$$a_{2k+1} = c_{2k+1} \frac{2^{2k} \cdot k! \cdot (k+1)!}{(2k+1)!}. \tag{5.4.5}$$

The phase shift of the complex transfer characteristic function $g(A(t))$ can be directly obtained from AM-PM measurements on the network analyzer with a single tone signal. The AM-PM nonlinear curve $\Phi_{out}(A_{in})$ is usually an even function, and it can be fitted in terms of an even order power series as

$$\Phi_{out}(A_{in}) = \sum_{k=0}^{n} b_{2k} A_{in}^{2k}. \tag{5.4.6a}$$

The corresponding phase shift of the envelope transfer characteristic function is expressed as

$$g(A(t)) = \sum_{k=0}^{n} b_{2k} A(t)^{2k}. \tag{5.4.6b}$$

Now let us look at an example of the spectral regrowth of a CDMA reverse link signal passing through a GaAs FET power amplifier with an AM-AM and an AM-PM nonlinear characteristics as shown in Fig. 5.6. Using these nonlinear characteristics and (5.4.3) to (5.4.6), we can determine the corresponding complex transfer characteristic function (5.4.2), which represents the low-pass equivalent of the nonlinear amplifier. The calculation of the output spectrum and the spectral regrowth is based on the model presented in Fig. 2.5, and the detail procedure is described in Appendix 5C. The calculated results of the spectra and *ACPR*s for a CDMA (IS-95 reverse link) signal at the input and output ports of the power amplifier are shown in Fig. 5.7(a) and (b), respectively.

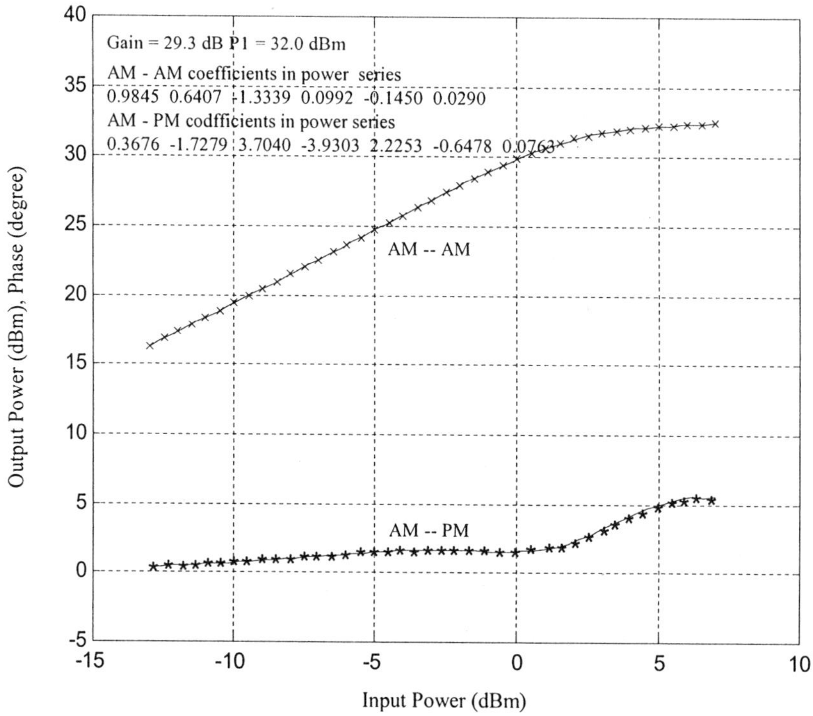

Figure 5.6. AM-AM and AM-PM nonlinear characteristics of a PA

From Fig 5.7, we can clearly see that the adjacent and the alternate channel power ratios, *ACPR*1 and *ACPR*2, of the CDMA reverse link transmission signal are degraded at the output of the power amplifier due to the spectral regrowth caused by the nonlinearity of the PA. The *ACPR*1 and the *ACPR*2 are measured at 885 kHz and 1.98 MHz offset from the carrier of the transmission signal from the Cellular band mobile stations. In the CDMA system, adjacent/alternate channel power is measured over a 30 kHz bandwidth. The *ACPR*s are not only dependent on the nonlinear characteristic of the PA but also on the peak-to-average ratio (PAR) of the amplified signal.

The simulation approach based on the low-pass equivalent behavioral model can provide quite accurate evaluation of *ACPR*s. However, it may not be convenient for the overall transmitter system analysis. Some approximate calculations are introduced in following sections.

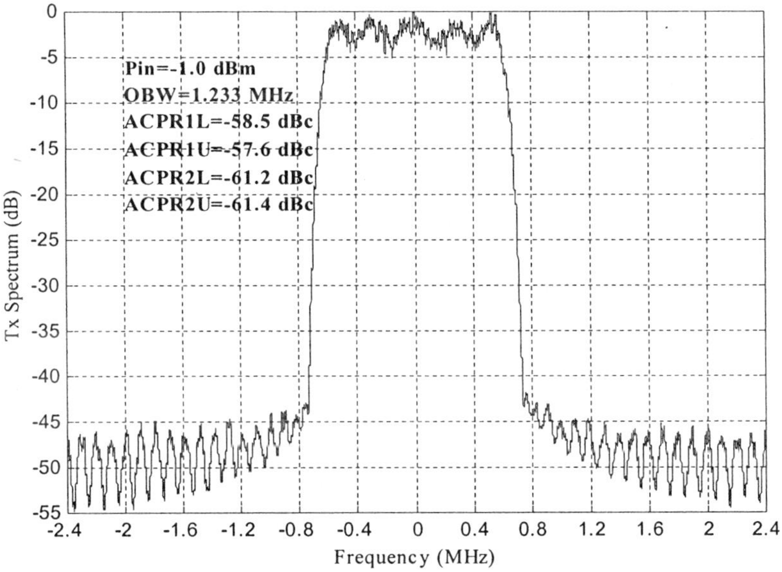

(a) CDMA signal spectrum at the input port of the PA

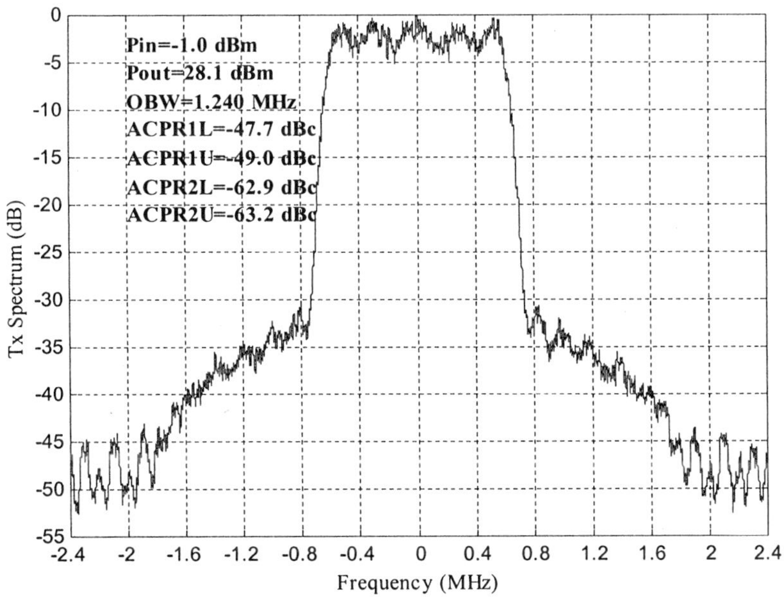

(b) CDMA signal spectrum at the output port of the PA

Figure 5.7. Spectral regrowth and ACPR evaluation based on low-pass equivalent of nonlinear amplifier

5.4.2. Multitone Techniques

The two-tone measurements are commonly used in determining the intermodulation distortion (IMD) characteristics as described in the previous chapter. In the digital mobile communications, signals are quite complicated, and their spectral regrowth caused by the nonlinearity of the transmitter chain cannot be accurately analyzed in terms of the two-tone IMD. It becomes necessary to apply the multitone signals to analytically assess the *ACPR* of transmission signals. Closed form formulas of *ACPR* derived from the multitone based third order distortion are presented in [12]–[14]. The *n*-tone $ACPR_n$ formula for a nonlinear device output has an expression

$$ACPR_n = IMR_2 - 10\log\left(\frac{n^3}{2\left(\frac{2n^3 - 3n^2 - 2n}{3} + \varepsilon\right) + \left(n^2 - \varepsilon\right)}\right), \quad (5.4.7)$$

where

$$IMR_2 = 2\left(P_{2\_Tone} - OIP_3\right) - 6 \text{ dB} \quad (5.4.8)$$

and

$$\varepsilon = \text{mod}\left(\frac{n}{2}\right) = \text{the division remainder of } n \text{ by } 2$$

In (5.4.8), OIP_3 is the output third-order intercept point of the nonlinear device, and P_{2_Tone} is the total power of the two tones at the output of the device. In the $n = 2$ case, the second term on the right side of (5.4.7) is equal to 3, and therefore the two-tone $ACPR_2$ has the form of

$$ACPR_2 = 2\left(P_{2\_Tone} - OIP_3\right) - 9 \text{ dB}. \quad (5.4.9)$$

When the signal is band-pass random noise instead of multiple tones with equal level, the corresponding $ACPR_{RN}$ can be expressed as

$$ACPR_{RN} = IMR_2 + 10\log\left(\frac{4}{3}\right) = 2\left(P_{2\_Tone} - OIP_3\right) - 4.75 \text{ dB}. \quad (5.4.10)$$

Comparing this with (5.4.9), we can see that the *ACPR* is degraded 4.25 dB while the two signal tones are replaced by the band-pass random noise with an integrated power P_{RN} same as the total power of the two signal tones, or $P_{RN} = P_{2\_Tone}$. This is due to the peak-to-average ratio (PAR) of the random noise being higher than that of the equal tones. The band-pass random noise has a PAR of approximate 8 dB or higher depending on the bandwidth, but the PAR of two equal tones is only 3 dB.

For mobile transmission signals with an amplitude modulation, the PAR, where the peak power is generally defined as 99% of amplitude CDF, may vary from less than to greater than 3 dB. The PAR of the CDMA IS-95 reverse link is approximate 3.85 dB, and the PAR of different channel configurations of CDMA 2000 reverse link varies from 3.2 dB to 5.4 dB. The general formula for the *ACPR* of a transmission signal at the output of a device, such as driver or power amplifier with an output intercept point, OIP_3, can be expressed as

$$ACPR_{Tx} \cong ACPR_2 + C \cong 2(Tx - OIP_3) - 9 + C_o\,, \qquad (5.4.11)$$

where Tx is the transmission signal power at the output of the device under analyzed, and C_o is a correction factor depending on the PAR and signal configuration, and it can be approximately obtained as

$$C_o \cong 0.85 \cdot (PAR - 3) \text{ dB}. \qquad (5.4.12)$$

In the mobile communication systems, the adjacent channel power may be measured in a bandwidth ΔB_{ACP} (see Fig. 5.5) different from the desired transmission signal bandwidth BW. In this case, the *ACPR* approximation (5.4.11) turns into

$$ACPR_{Tx} \cong 2(Tx - OIP_3) - 9 + C_o + 10\log\left(\frac{\Delta B_{ACP}}{BW}\right). \qquad (5.4.13)$$

Using (5.4.13) to estimate the *ACPR* of an IS-95 CDMA reverse link signal amplified by the same PA as given in the previous subsection as an example, the PA has an $OIP_3 \cong 39.6$ dBm, the amplified transmission signal power is 26.5 dBm, $\Delta B_{ACP} = 30$ kHz, and $BW = 1.23$ MHz. Thus the *ACPR* is

$$ACPR_{Tx} \cong 2(26.5 - 39.6) - 9 + 0.72 + 10\log\left(\frac{30\times10^3}{1.23\times10^6}\right) = -50.48 \text{ dBc}.$$

Comparing this with the simulation results presented in Fig. 5.7(b), the average error is approximate 1.0 dB. However, in the simulation of the previous section's example the input signal has a finite *ACPR*, approximate –59.7 dBc (see Fig. 5.7 (a)). If considering this input signal *ACPR*, the PA output signal *ACPR* becomes –49.98 dBc instead of –50.48 dBc. In fact, the accuracy of the calculation result from (5.4.13) also depends on the power back-off from the 1 dB compression point of the PA. The contribution to the adjacent channel power of the fifth-order and other higher-order nonlinear distortions increases with the output power close to 1 dB compression point of the PA, but (5.4.13) only considers the third-order nonlinearity.

Generally speaking, the alternate channel power regrowth is mainly caused by the fifth order nonlinear distortion of the PA or other devices. Thus (5.4.13) cannot be used for calculating the alternate channel power ratio since it is derived based on the assumption of the third order nonlinearity only.

5.4.3. ACPR of Cascaded Stages in Transmitter Chain

The *ACPR* of a transmitter consisting of multiple stages connected in cascade can be derived in terms of (5.4.13) and the cascaded OIP_3 formula (5.4.14)

$$OIP_3 = -10\log\left(\frac{1}{P_{OIP\_n}} + \sum_{k=1}^{n-1}\frac{1}{P_{OIP\_k}\cdot\prod_{i=k+1}^{n} g_i}\right), \tag{5.4.14}$$

where g_i is the ith stage gain in natural scale, and P_{OIP_k} is the OIP_3 of the kth stage in natural scale. Assuming that the *ACPR* of an input signal to the transmitter chain is denoted as $ACPR_i$ and the overall output third order intercept point is OIP_3 in dBm, the *ACPR* increment due to the transmitter chain nonlinearity can be calculated by using (5.4.13) and it is expressed as

$$\begin{aligned} 10\log\left(10^{\frac{ACPR_{Tx}}{10}} - 10^{\frac{ACPRi}{10}}\right) &= 2(Tx - OIP_3) + 10\log\frac{\Delta B_{ACP}}{BW} - 9 + C_o \\ &= 2(Tx - OIP_3) + C, \end{aligned} \tag{5.4.15}$$

where $ACPR_{Tx}$ is the $ACPR$ of the output signals, and C is a constant equal to the sum of the last three terms on the right side.

Substituting (5.4.14) to (5.4.15), we obtain

$$10\log\left(10^{\frac{ACPR_{Tx}}{10}}-10^{\frac{ACPRi}{10}}\right)=2\left(Tx+10\log\left(\frac{1}{P_{OIP\_n}}+\sum_{k=1}^{n-1}\frac{1}{P_{OIP\_k}\cdot\prod_{i=k+1}^{n}g_i}\right)\right)+C. \tag{5.4.16}$$

On the other hand, (5.4.15) after rearrangement turns into

$$OIP_3=Tx+\frac{1}{2}\left[C-10\log\left(10^{\frac{ACPR_n}{10}}-10^{\frac{ACPRi}{10}}\right)\right]. \tag{5.4.17}$$

Similar formulas can be derived for each stage in the transmitter chain. If assuming that the $ACPR$ of the input signal for measuring the kth stage $ACPR_k$ is negligible low, then the output intercept point $OIP_{3,k}$ in dBm can be expressed in $ACPR_k$ and the output power $P_{o,k}$ as

$$OIP_{3\_k}=P_{o\_k}+\frac{1}{2}\left[C-ACPR_k\right] \tag{5.4.18}$$

or in natural scale as

$$P_{OIP\_k}=10^{\frac{2P_{o\_k}-ACPR_k+C}{20}}. \tag{5.4.19}$$

Plugging (5.4.19) into (5.4.16), we obtain

$$10\log\left(10^{\frac{ACPR_{Tx}}{10}}-10^{\frac{ACPR_i}{10}}\right)=2\left[Tx+10\log\left(\frac{1}{10^{\frac{2P_{o\_n}-ACPR_n+C}{20}}}+\sum_{k=1}^{n-1}\frac{1}{10^{\frac{2P_{o\_k}-ACPR_k+C}{20}}\cdot\prod_{i=k+1}^{n}g_i}\right)\right]+C. \tag{5.4.20}$$

Utilizing the following relationships —

$$Tx = -10\log 10^{\frac{p_{o\_k}}{10}} \cdot \prod_{i=k+1}^{n} g_i$$

and

$$\frac{1}{2}C = 10\log 10^{\frac{C}{20}}$$

(5.4.20) can be turned into

$$ACPR_o = 10\log\left\{\left[\sum_{k=1}^{n}\left(10^{\frac{ACPR_k}{10}}\right)^{\frac{1}{2}}\right]^2 + 10^{\frac{ACPR_i}{10}}\right\}. \quad (5.4.21)$$

In fact, not all contributions to $ACPR$ from different stages are completely correlated, but only partially correlated. In this case (5.4.21) should be expressed as

$$ACPR_{Tx} = 10\log\left[\sum_{k=1}^{n} 10^{\frac{ACPR_k}{10}} + 10^{\frac{ACPR_i}{10}} + \sum_{k=1}^{n}\sum_{\substack{j=1\\ j\neq k}}^{n} \alpha_{k,j} \cdot \left(10^{\frac{ACPR_k}{10}}\right)^{\frac{1}{2}}\left(10^{\frac{ACPR_j}{10}}\right)^{\frac{1}{2}}\right], \quad (5.4.22)$$

where $\alpha_{k,j}$ (k, j = 1, 2, …, n; $k \neq j$) is correlation coefficients between different stage contributions, and $0 \le \alpha_{k,j} \le 1$.

For an example, a transmitter contains mainly two nonlinear devices, a transmitter IC and a PA as depicted in Fig. 5.8. Assuming that the

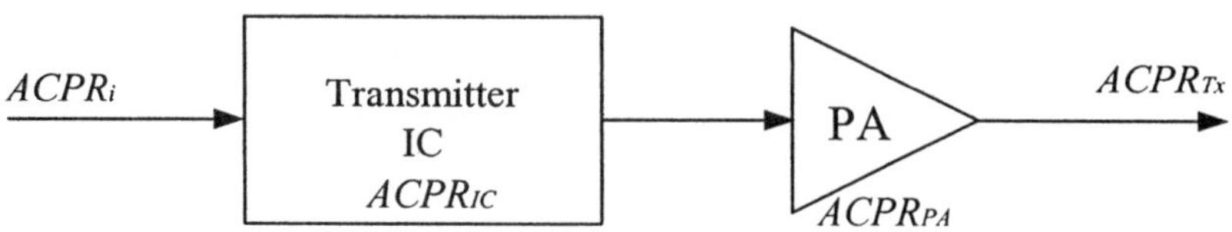

Figure 5.8. High-level block diagram of a transmitter

$ACPR_i$ of the input signal is –60dBc, and the measured $ACPR_{IC}$ and $ACPR_{PA}$ are –56 dBc and –52 dBc, respectively, and the correlation coefficient between the spectral regrowths of the IC and PA is approximately 0.45, then the $ACPR_{Tx}$ of the transmission signal at the PA output is

$$ACPR_{Tx} = 10\log\left(10^{\frac{-56}{10}} + 10^{\frac{-52}{10}} + 10^{\frac{-60}{10}} + 0.9 \times 10^{\frac{-56}{20}} \times 10^{\frac{-52}{20}}\right) = -49.3 \text{ dBc}.$$

5.5. Noise-Emission Calculation

The noise emission from mobile station transmitters is one of the important specifications of the transmitter, especially the noise emission in the receiver band of a full-duplex mobile station. The noise emissions discussed here are those located outside of alternate channels, or otherwise the emission noise is included in the *ACPR* specifications.

In the transmitter system design, we may find that a gain distribution to obtain low noise emission will be not desired for good *ACPR* performance. In general, we like to have lower gain PA and driver amplifier for achieving low noise emission, but the gain setting of the PA and the driver to obtain a better *ACPR* performance will be completely opposite.

5.5.1. Formulas for Noise-Emission Calculation

To evaluate the transmitter noise emission, it is better to begin with estimating the contribution from individual stage in the transmitter chain. Let us look at a device that has a gain (or loss) g and a noise factor F as shown in Fig. 5.9. The noise generated by the device itself can be calculated by means of its noise factor. As we know, the noise factor can also be defined as the ratio of the converted input noise from the output noise to the thermal noise at the input,

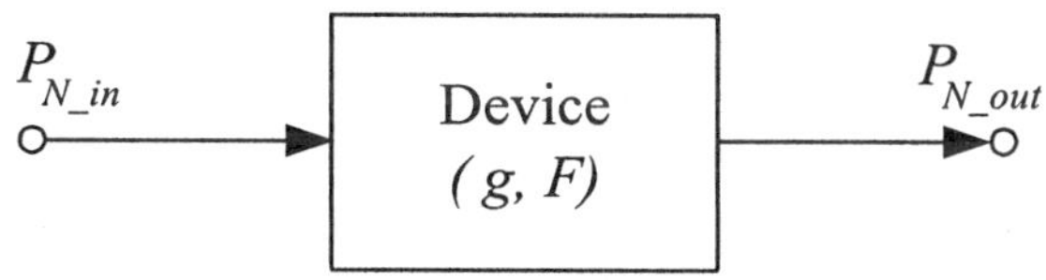

Figure 5.9. Noise emission analysis model of a device

$$F = \frac{P_{No} + P_{Nd\_in}}{P_{No}} = 1 + \frac{P_{Nd\_in}}{P_{No}}, \tag{5.5.1}$$

where P_{Nd_in} is the equivalent device noise at input port of the device in mW/Hz, and P_{No} is the thermal noise equal to $kT_o = 10^{-174/10}$ mW/Hz. The equivalent device input noise then is

$$P_{Nd\_in} = P_{No}(F-1). \tag{5.5.2}$$

Thus the noise generated by this device at its output port P_{Nd_out} is

$$P_{Nd\_out} = g \cdot P_{No}(F-1) = g \cdot kT_o \cdot (F-1). \tag{5.5.3}$$

In addition to this noise generated by the device itself, if an input noise P_{N_in} (mW/Hz) is imposed on the input of the device, the total output noise becomes

$$P_{N\_out} = g \cdot P_{N\_in} + kT_o \cdot g \cdot (F-1). \tag{5.5.4}$$

For a transmitter consisting of n stages as shown in Fig. 5.10, the noise emission from the transmitter, $P_{N\_Tx\_out}$, in mW/Hz has a similar formula as (5.5.4),

$$P_{N\_Tx\_out} = g_{Tx} \cdot P_{N\_Tx\_in} + kT_o \cdot g_{Tx} \cdot (F_{Tx} - 1) \tag{5.5.5}$$

where $P_{N\_Tx\_in}$ in mW/Hz is the noise at the transmitter input, g_{Tx} is overall transmitter gain equal to

$$g_{Tx} = \prod_{i=1}^{n} g_i, \tag{5.5.6}$$

and F_{Tx} is the overall noise factor of the transmitter calculated as

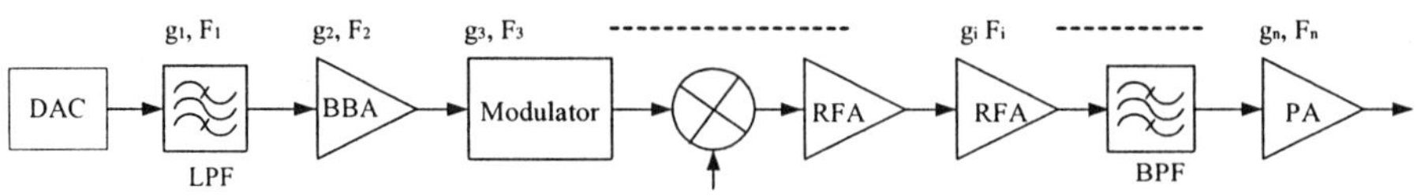

Figure 5.10. Block diagram of a transmitter chain

$$F_{Tx} = F_1 + \sum_{i=1}^{n-1} \frac{F_{i+1} - 1}{\prod_{k=1}^{i} g_k}. \tag{5.5.7}$$

When using (5.5.6) and (5.5.7), we need be careful that each stage gain g_i and noise factor F_i are not always those value of the transmitter in-band, and they must be the gain and noise factor in the emission frequency band, which we are interested in.

In the noise-emission calculation, another important concept is excess noise, which represents the amount of noise exceeding the thermal noise. The excess noise $P_{\Delta N\_Tx\_out}$ in mW/Hz is expressed as

$$P_{\Delta N\_Tx\_out} = P_{N\_Tx\_out} - kT_o = g_{Tx} \cdot (P_{N\_Tx\_in} + kT_o \cdot F_{Tx}) - kT_o \cdot (g_{Tx} + 1). \tag{5.5.8}$$

In a full-duplex transceiver, transmitter noise emission in the receiver band may desensitize the receiver sensitivity, but the desensitization is determined by the excess noise level of the transmitter noise emission instead of overall emission power as described in Chapter 4.

5.5.2. Some Important Notes in Noise-Emission Calculation

5.5.2.1. Output Noise of an Attenuator

It may cause some confusion when we calculate output noise of a loss device, such as an attenuator. While thermal noise kT_o is imposed at the input of an attenuator — say, with a 10 dB attenuation — the noise at the output of this attenuator is not $10\log(kT_o)$ - 10 = -174 -10 = -184 dBm/Hz, but it is still at the same thermal noise level, kT_o or *-174* dBm/Hz. This can be verified by using (5.5.4) as follows. Assuming the attenuator has a loss l and a noise factor $F = 1/l$, and the input noise P_{N_in} is kT_o, substituting these data into (5.5.4) we obtain

$$P_{N\_out} = l \cdot kT_o + kT_o \cdot l \cdot (1/l - 1) = kT_o. \tag{5.5.9}$$

5.5.2.2. Output Noise Floor of Device or Transmitter

The example given in the previous subsection tells us the output noise floor of a physical device is the thermal noise, kT_o. It is impossible to make a device output noise lower than the thermal noise when the device is in an environment of temperature T_o. Therefore, the following relation is always true:

$$P_{N\_out} = g \cdot P_{N\_in} + kT_o \cdot g \cdot (F-1) \geq kT_o . \tag{5.5.10}$$

5.5.2.3. Product of Noise Factor F and Gain g Greater Than One

From (5.5.10) and (5.5.4), we can come to the following conclusion:

$$g \cdot F \geq 1. \tag{5.5.11}$$

Assuming $P_{N\_in} = kT_o + P_{\Delta N}$, where $P_{\Delta N}$ represents excess noise over the thermal noise, from (5.5.10) we have

$$P_{N\_out} = g \cdot (kT_o + P_{\Delta N}) + kT_o \cdot g \cdot (F-1) \geq kT_o$$

or

$$g \cdot P_{\Delta N} + kT_o \cdot g \cdot F \geq kT_o . \tag{5.5.12}$$

P_{N_out} is minimum when $P_{\Delta N} = 0$. Thus (5.5.12) turns into

$$kT_o \cdot g \cdot F \geq kT_o \quad \text{or} \quad g \cdot F \geq 1.$$

5.5.2.4. Input Noise Floor of a Device

From sections 5.5.2.1 to 5.5.2.3, we can conclude that the minimum input noise of a physical device is the thermal noise kT_o or

$$P_{N\_in}\big|_{\min} = kT_o . \tag{5.5.13}$$

5.5.3. Noise Expressed in Voltage

It is similar to the receiver case that in analog base-band and the IF blocks voltage signal and noise are usually used instead of power ones. In the transmitter analog base-band and IF blocks, it is apparent that noise should be represented in voltage. The fundamental noise voltage formulas are presented as follows. In Fig. 5.11, a device has a voltage gain g_V and a noise factor F, the input impedance is R_i, and noise source impedance and voltage are R_s and V_n, respectively. From Section 2.3.1 in Chapter 2, the source noise in V/$\sqrt{\text{Hz}}$ has a form

$$v_n = \sqrt{4kT_o R_s} \ . \tag{5.5.14}$$

The input noise voltage of the device in V/$\sqrt{\text{Hz}}$ is

$$v_{ni} = \frac{R_i}{R_i + R_s} \sqrt{4kT_o R_s} \ . \tag{5.5.15}$$

In the match ($R_i = R_s$) case or the input impedance much greater than the source impedance ($R_i >> R_s$) case, (5.5.15) turns into (5.5.16) and (5.5.17), respectively,

$$v_{ni} = \sqrt{kT_o R_s}, \qquad \text{while } R_i = R_s \tag{5.5.16}$$

or

$$v_{ni} \cong \sqrt{4kT_o R_s}, \qquad \text{while } R_i >> R_s . \tag{5.5.17}$$

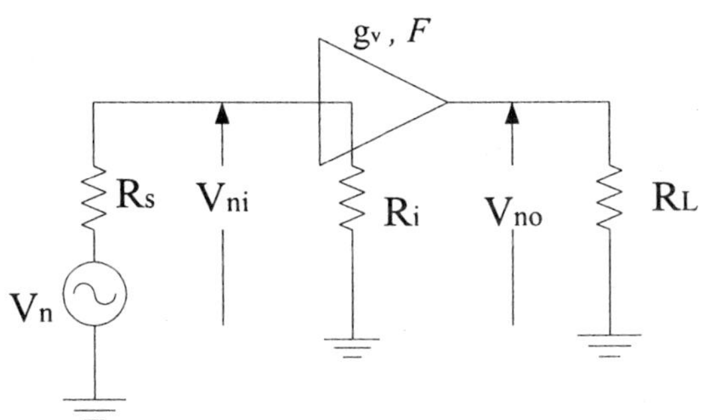

Figure 5.11. Fundamental representation of noise voltages for a device

When the noise at the input of the device is only the thermal noise, then the output noise voltage of the device in $\mathrm{V}/\sqrt{\mathrm{Hz}}$ is

$$v_{no} = \sqrt{F} \cdot g_v v_{ni} = \sqrt{F \cdot 4kT_o R_s}\,\frac{g_v R_i}{R_i + R_s}. \tag{5.5.18}$$

In the BB and the IF blocks, it is common that the output impedance of preceding stage is much smaller than the input impedance of this stage — i.e., $R_s << R_i$, (5.5.18) turns into

$$v_{no} = 2\sqrt{F \cdot kT_o R_s} \cdot g_v. \tag{5.5.19}$$

If the input noise of the device is more than the thermal noise, the output noise voltage expression of the device is

$$v_{no} = \sqrt{v_{ni}^2 + (F-1) \cdot 4kT_o R_s}\,\frac{g_v R_i}{R_i + R_s}, \tag{5.5.20}$$

or considering $R_s << R_i$, the output noise voltage is approximately expressed as

$$v_{no} = \sqrt{v_{ni}^2 + (F-1) \cdot 4kT_o R_s} \cdot g_v. \tag{5.5.21}$$

5.5.4. *Examples of Noise-Emission Calculations*

The BB block of a transmitter has a –19.5 dB gain and a 30.5 dB noise figure in the receiver frequency band. If the input of the BB block is connected to the output of a DAC, which has a source impedance 200 Ω and a noise voltage emission in the receiver frequency band of 20 $\mathrm{nV}/\sqrt{\mathrm{Hz}}$, the noise voltage at the output port of the BB block can then be calculated in terms of (5.5.21):

$$v_{no} = \sqrt{\left[\left(20 \cdot 10^{-9}\right)^2 + \left(10^{\frac{30.5}{10}} - 1\right) 3.98 \cdot 10^{-21} \cdot 4 \cdot 200\right]}\, 10^{\frac{-19.5}{20}} \cdot 10^9 \cong 6.7 \text{ nV}.$$

The RF block of a Cellular band CDMA mobile transmitter has a noise figure 45.2 dB and a gain equal to –45 dB in the receiver band.

Assuming the input noise density in the receiver frequency band is around –136.4 dBm/Hz, using (5.5.5) we can predict the emission-noise level at the transmitter output port to be

$$P_{N\_Tx\_out|Rx\_band} = 10^{\frac{-45}{10}} \cdot \left[10^{-\frac{136.4}{10}} + 10^{-\frac{174}{10}} \cdot \left(10^{\frac{45.2}{10}} - 1 \right) \right] = 10^{\frac{-173.1}{10}} \text{ mW/Hz.}$$

5.6. Some Important Considerations in System Design

The system design of a mobile transmitter should first make the transmitter performance not only meet the defined specification but also have a margin for a better reliability and a lower *field failure rate* (FFR). Lowering power consumption and reduce of the overall cost are also main tasks for the transmitter system design in addition to the electrical performance.

In a mobile station, two thirds of the average overall current consumption is spent in the RF transceiver, and almost three quarters of the RF transceiver power is consumed in the transmitter. The RF transmitter power consumption therefore will significantly impact the talking time of the mobile station since it is close to half of the total power consumption. Thus it becomes essential in the RF transmitter design to minimize power consumption in addition to providing good electrical performance.

5.6.1. Comparison of Architectures

In general, there is no IF channel filter to be used in the superheterodyne architecture transmitter for any mobile station of wireless communication systems as given in Fig. 3.1. It is not like the receiver, for which the direct conversion architecture can save an IF channel SAW filter and provides possibility for multimode operation through programming the receiver IC without using extra external components. Compared with the superheterodyne transmitter, the direct conversion architecture (Fig. 3.10) does have the advantage of generating less spurious products since it has no IF and only one frequency conversion. However, it may take more current than a superheterodyne transmitter. This is due the fact to that the gain of the transmission signal in the direct conversion transmitter is developed in the

RF block, and the RF amplifiers usually spend more current than the IF amplifiers for the same amount of gain.

In the GSM mobile station transmitter, an offset PLL is often adopted to play RF up-converter and filtering roles as shown in Fig. 3.21. In this architecture, a transmitter band RF SAW or BAW BPF can be saved although it is still a superheterodyne transmitter. Recently, a so-called *polar modulation* based on the OPLL for the phase modulation plus an AM through modulating the PA power supply is proposed for the EDGE wireless mobile transmitter and other mobile system transmitters [15]–[18]. This kind of architecture is also referred to as *envelope elimination and restoration* (EER) transmitter, and one of possible configurations of the polar modulation or EER transmitter is depicted in the block diagram of Fig. 5.12.

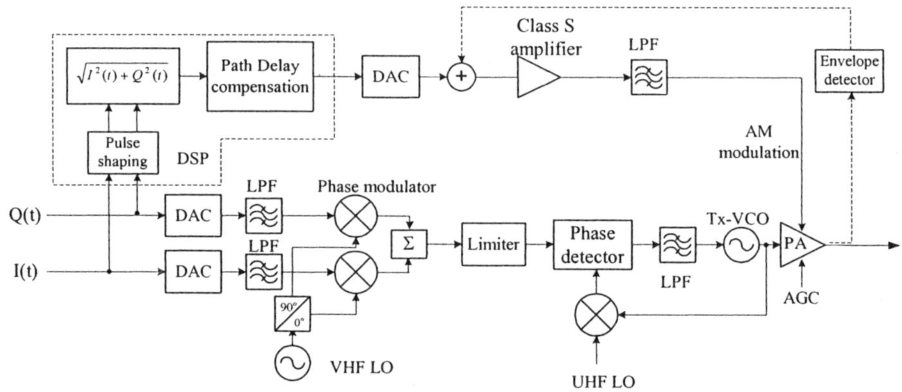

Figure 5.12. Block diagram of a mobile EER transmitter

The angle (phase) modulation of the transmission signal is executed by means of the I/Q quadrature modulator and the offset phase locked loop as shown in the bottom portion of Fig. 5.12. On the other hand, the amplitude modulation of the transmission signal is implemented through controlling the PA power supply voltage, and the simplified block diagram of the AM modulation is presented in the top portion of Fig. 5.12. The advantages of this architecture are that the power efficiency can be quite high and the talking time of the mobile station to be able to increase significantly if some relevant technical issues, such as dynamic path delay compensation between the amplitude and the phase modulation and wide range gain control, can be properly resolved. Generally speaking, this transmitter architecture is much more complicated than those with the classic modulation scheme as described in Chapter 3.

The modulated transmission signal with an IF carrier can be formed in the digital domain (usually reserved for base-band). The transmitter in term of band-pass sampling technique as depicted in Fig. 3.30 belongs to this kind of architecture. The alternate configuration of this architecture transmitter has the digital IF signal oversampled and converted into an analog signal with the same IF carrier. The corresponding block diagram of the digital IF superheterodyne transmitter is shown in Fig. 5.13. This architecture transmitter, either Fig. 3.30 or Fig. 5.13, has very low I and Q imbalance and good modulation accuracy. It uses no analog frequency up-converter in the band-pass sampling case or only one analog up-converter in the superheterodyne case. The possibility for a high level of integration and relatively low current consumption are other features of this architecture. However, it demands high-performance digital-to-analog converters and the DSP running at very high rates — say, hundreds of MSample/sec.

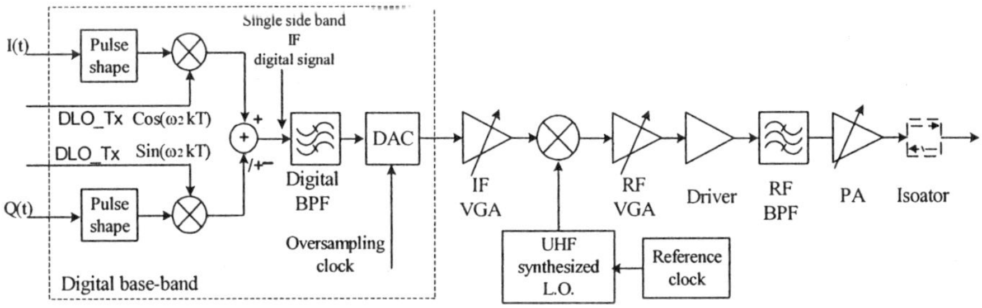

Figure 5.13. Block diagram of superheterodyne transmitter with digital IF

At present, popularly used architectures are still the classical superheterodyne and the direct conversion transmitters for the mobile stations of all kind of wireless communication systems.

5.6.2. Transmitter Chain Gain Distribution and Performance

The linearity of a transmitter will directly affect the spectral regrowth and then the adjacent/alternate channel power ratio performance. To obtain better *ACPR*, it will be desirable to have higher gain in the RF stages close to the antenna port. It is common knowledge that a higher gain in the RF stages means more power consumption. On the other hand, from a reducing-noise emission point of view, we like to have less gain in the stages close to the antenna port.

The interface of the transmission signal path in the I and Q channels between the analog base-band and the digital base-band of mobile station transmitters is commonly a DAC. The output from the DAC can be a voltage or a current signal with a large maximum swing close to its upper limit to achieve the best signal-to-noise ratio. The BB signal is usually too large for the input stages in the I and Q channels of the analog-RF transmitter to handle or otherwise these stages need spend a lot of current. Proper signal attenuation and conversion from voltage to current or vice versa is necessary before or at the quadrature modulator of the transmitter. The attenuation, in most cases, will not degrade the signal-to-noise ratio of the BB signal because the noise level from the DAC and the digital base-band is generally much higher than the thermal noise. Placing attenuation at the input of the analog-RF transmitter will increase the overall noise figure of the transmitter, but the effect on the final noise emission of the transmission is usually insignificant.

A performance comparison example of two different gain distributions of a CDMA mobile transmitter is presented in Table 5.2. The main difference in these two gain distributions is that the gain distribution (a) has an 8 dB attenuator placed at the input of the analog-RF transmitter, but the distribution (b) does not have any attenuation before the modulator. From this table, we can see that the noise emission in the receiver frequency band for the distribution (a) is –173.56 dBm/Hz comparing with –173.62 dBm/Hz for the distribution (b), or the noise emission in the receiver band is 0.06 dB higher when using an 8 dB attenuator at the input. However, the *ACPR* performance of the gain distribution (a) is approximately –49.8 dBc, which is 0.5 dB better than –49.3 dBc ACPR of the gain distribution (b). Actually, not only the linearity of (a) is greater than (b), but the current consumption of (a) is lower than (b) too. This statement can be indirectly proved by comparison of the requirements on the OIP_3 of the quadrature modulator and the VGA1 for achieving the corresponding *ACPR*. They are 6 and 11 dBm for the gain distribution (a) and 10 and 18 dBm for the gain distribution (b), respectively. This example tells us that properly allocating the gain along the transmitter chain can achieve both, to obtain better performance and to save current consumption.

Table 5.2. A performance comparison of two different gain distribution transmitter chains

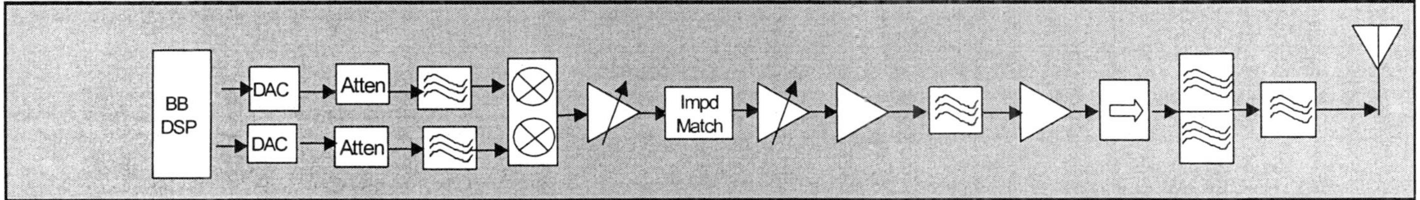

Function Block	DAC	Atten	LPF	Modltr Img-Rjc	RF VGA1	Impdnc Match	RF VGA2	Driver	Balun + SAW Fltr	PA	Directional Coupler	Duplx	Diplex	
(a) The first gain distribution with attenuation at the input														
Power Gain		**-8.00**	**0.00**	**5.00**	**4.17**	**-1.00**	**12.00**	**8.00**	**-4.00**	**27.00**	**0.40**	**-2.50**	**-0.50**	
Cascad Gain	0.00	-8.00	-8.00	-3.00	1.17	0.17	12.17	20.17	16.17	43.17	43.57	41.07	40.57	**40.57**
Pout (dBm)	-15.57	-23.57	-23.57	-18.57	-14.40	-15.40	-3.40	4.60	0.60	27.60	28.00	25.50	25.00	**25.00**
Vout (mVp-p)	2500.00	995.27	995.27	1769.86										
Rej @ Rx			20.00						30.00			42.00		
NF (dB) @ Rx Band	0.00	8.00	20.00	9.00	7.83	1.00	11.00	5.00	34.00	5.00	0.00	44.50	0.50	
Cascad NF @ Rx	0.00	8.00	28.00	37.00	37.80	37.81	38.55	38.55	39.81	41.70	41.70	51.42	51.87	**51.87**
OIP3(dBm)	1000.00	100.00	10.00	**6.00**	**11.17**	100.00	19.00	27.00	100.00	42.00	100.00	100.00	100.00	
Cascad OIP3	100.00	91.36	100.00	5.49	7.34	6.34	15.65	22.10	18.10	40.27	40.67	38.17	37.67	**37.67**
ACPR (dBc)	-59.73	-59.00	-59.00	-58.83	-58.53	-58.53	-57.57	-56.46	-56.46	-49.79	-49.79	-49.79	-49.79	**-49.79**
Noise in Rx band (dBm/Hz)	-153.37	-161.18	-173.28	-159.90	-154.95	-155.93	-143.21	-135.20	-167.96	-139.09	-138.69	-173.51	-173.56	**-173.56**
(b) The second gain distribution without attenuation at the input														
Power Gain		**0.00**	**0.00**	**5.00**	**0.17**	**-5.00**	**12.00**	**8.00**	**-4.00**	**27.00**	**0.40**	**-2.50**	**-0.50**	
Cascad Gain	0.00	0.00	0.00	5.00	5.17	0.17	12.17	20.17	16.17	43.17	43.57	41.07	40.57	**40.57**
Pout (dBm)	-15.57	-15.57	-15.57	-10.57	-10.40	-15.40	-3.40	4.60	0.60	27.60	28.00	25.50	25.00	**25.00**
Rej @ Rx		20.00							30.00			42.00		
NF (dB) @ Rx Band	0.00	20.00	2.00	9.00	9.83	5.00	16.00	5.00	34.00	5.00	0.00	44.50	0.50	
Cascad NF @ Rx	0.00	20.00	22.00	29.31	30.51	30.76	36.92	36.93	38.66	40.99	40.99	51.35	51.80	**51.80**
OIP3(dBm)	100.00	100.00	100.00	**10.00**	**18.17**	100.00	19.00	27.00	100.00	42.00	100.00	100.00	100.00	
Cascad OIP3	96.99	95.23	93.98	10.00	9.53	4.53	14.58	21.24	17.24	39.97	40.37	37.87	37.37	**37.37**
ACPR (dBc)	-59.00	-59.00	-59.00	-58.23	-58.00	-58.00	-56.86	-55.62	-55.62	-49.25	-49.25	-49.25	-49.25	**-49.25**
Noise in Rx band (dBm/Hz)	-153.37	-170.69	-169.64	-159.14	-157.89	-162.67	-144.81	-136.80	-169.10	-139.80	-139.40	-173.58	-173.62	**-173.62**

5.6.3. AGC and Power Management

The transmission power of mobile stations in different wireless communication systems is controlled either based on certain control loops or on *base station* (*BS*) commands. The maximum and the minimum output powers for most popular mobile stations are presented in Table 5.3. The CDMA systems including WCDMA need much wider power control range (over 70 dB) than other mobile systems (30 dB or less).

Table 5.3. Maximum and minimum transmission power for mobile stations

Systems	Power Class	Nominal Maximum Power (dBm)	Nominal Minimum Power (dBm)	Dynamic Range (dB)	Power Control Method
AMPS	III	28	8	≥ 20	BS commends
CDMA 800	III	23	-50	≥ 73	Open and closed loops
CDMA 1900	II	23	-50	≥ 73	Open and closed loops
GSM 900	IV	33	5	≥ 28	BS commands
DCS 1800	I	30	0	≥ 30	BS commands
TDMA	III	28	8	≥ 20	BS commands
WCDMA	IV	21	-50	≥ 71	Open and closed loops

In GSM or DCS systems, the output power of mobile stations is commanded by the base station, and changes are within specified power levels in 2 dB intervals over 30 dB. The transmission power of AMPS and TDMA mobile stations is controlled in a similar way to the GSM system but in 4 dB steps over 20 dB dynamic range. In CDMA mobile stations, the transmission power is determined based on the received signal level (RSSI), and the estimation of an open loop formula defined as [3]

$$P_{Tx\_open} = -P_{Rx} - 73 + \text{correction} \quad \text{for cellular band CDMA} \tag{5.6.1}$$

and

$$P_{Tx\_open} = -P_{Rx} - 76 + \text{correction} \quad \text{for PCS band CDMA}. \tag{5.6.2}$$

In addition to the open-loop control, the output power of the CDMA mobile transmitter is also periodically adjusted up or down in 1, 0.5, or 0.25 dB steps (set by the base station) through power control bits embedded in forward link CDMA signal frame. The latter reverse link power control is referred to as *closed-loop control*. The closed-loop power control range is ±24 dB around the open loop power estimation in the CDMA system. The transmission power control mechanism in the WCDMA mobile station is similar to that used in CDMA, but its open loop power control is used to set its output power to a specified value, and the inner (closed) loop control range is nominally from the maximum output power to less than –50 dBm with step size 1, 2, or 3 dB [1].

The output power tolerance of mobile stations is system dependent. In GSM systems, the output power tolerance of the mobile transmitters is ±3 to ±5 dB depending on the output power level [2]. In CDMA the open-loop output power tolerance of the mobile transmitters is ±9.5 dB, and the tolerance for the closed-loop control is step size dependent, ±2, ±2.5, and ±3 dB in overall 10 dB change for 1, 0.5, and 0.25 dB step, respectively. The open-loop power control tolerance in WCDMA mobile transmitter is ±9 dB, and the inner-loop power control tolerance is also step size dependent, ±2 and ±4 dB for 10 consecutive equal 1 and 2 dB steps, respectively, and ±5 dB tolerance after 7 consecutive equal 3 dB steps.

Generally speaking, the RF transmitter is under the control of certain DSP AGC algorithm to execute the transmission power adjustment. Accurately executing the transmission power control and effectively utilizing the AGC to manage the transmitter power consumption are the two key tasks for the RF transmitter AGC design.

To make accurate power control, it is desirable to design IF and RF VGAs with a gain control curve as linear possible and with least variation over temperature and IC processing. However, it is the reality that the gain control curve no matter the IF or the RF VGA exhibits certain nonlinearity, and the curve may change with temperature, operating channel frequency, and processing. Thus the gain control curves of VGAs in the RF transmitter chain need be carefully characterized, and proper curve fitting techniques and piece-wise approximation are then used to express these gain control curves for the transmitter AGC use. After power calibration, the accumulated error of the transmitter gain control must be within the specifications as described above and with a reasonable margin. The compensation of the transmitter gain control characteristic over temperature and channel frequency is usually needed. Fortunately, in most cases, a linear compensation is good enough for both temperature and frequency.

It is quite natural to incorporate the transmitter AGC to the power management of a mobile station. Reduction of the RF transmitter power consumption will significantly increase the talk time of a mobile station since the transmitter may consume half of the overall mobile station power.

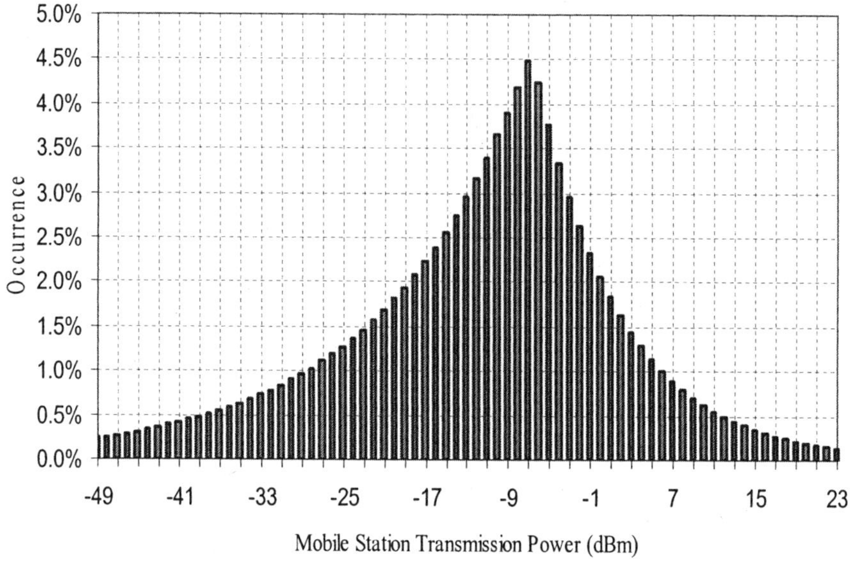

(a) Data rate 9.6 kHz in urban area

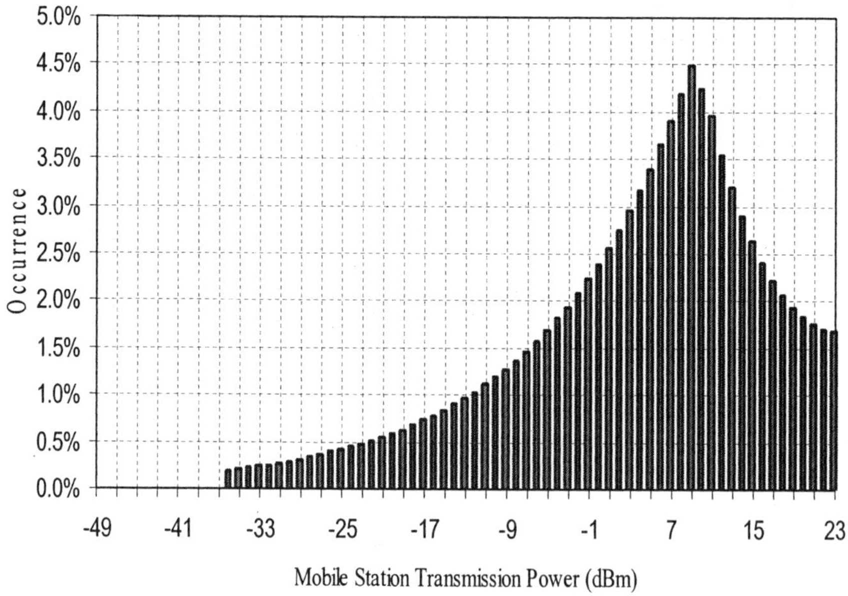

(b) Data rate 153.6 kHz in urban area

Figure 5.14. CDMA IS-2000 reverse link transmission power distributions

A general rule in the power management of the RF transmitter is to cut down the current consumption as fast as possible with transmitted power decrease but to maintain a proper linearity.

The transmission power of a mobile station in practical applications has a statistical distribution — i.e., not all the transmission power levels are equally used. For example, the reverse link transmission power statistical distributions of the CDMA mobile system are recorded in the CDG document [19]. The occurrence probabilities of the CDMA 2000 1x mobile station operating at data rate 9.6 kHz and 153.6 kHz in the urban area are given in Fig 5.14(a) and (b), respectively. The occurrence peak appears at –7 dBm mobile station transmission power in the case of data rate 9.6 kHz, and the occurrence peak in the case of data rate 153.6 kHz moves to +9 dBm transmission power.

It is expected that the transmission power statistical distribution for different mobile systems and even for the same system operating in different channel configurations will be different. To sufficiently utilize the transmission power statistical characteristic for managing the power consumption, the current of the mobile station transmitter should drop rapidly with its output power, and thus the current consumption in the statistical distribution peak region of the output power will be low as depicted in Fig. 5.15.

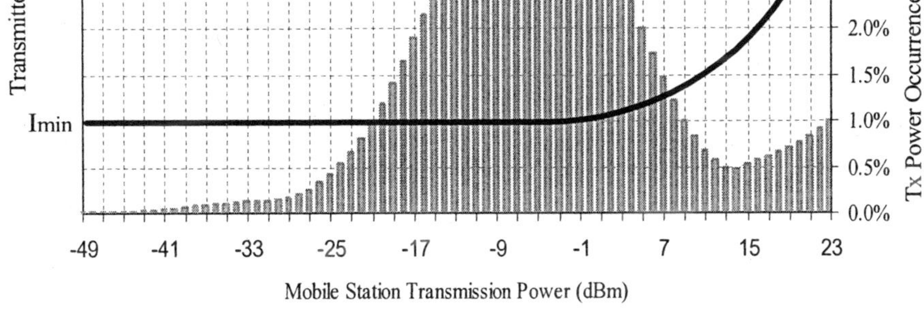

Figure 5.15. Desirable characteristic of current consumption vs. output power for mobile station RF transmitter

If we know the current consumption versus output power of a mobile station RF transmitter and the statistical distribution of mobile transmission power in practical applications as shown in Fig. 5.15, the statistical average current consumption of a mobile station RF transmitter $\bar{I}_{Tx}$ can be estimated by using the following formula:

$$\bar{I}_{Tx} = \sum_{k=1}^{n} p_{Tx\_k} \cdot I_{Tx\_k}, \tag{5.6.3}$$

where p_{Tx_k} and I_{Tx_k} are the occurrence probability and current consumption at output power P_k, respectively, and

$$\sum_{k=1}^{n} p_{Tx\_k} = 1.$$

In the case of the transmission waveform with amplitude modulation, such as CDMA, TDMA, and EDGE signal waveforms, the transmitter chain should have certain linearity to make spectral regrowth low enough. The power management thus becomes more complicated than the case of a transmission waveform with constant envelope — for example, GSM and AMPS waveforms. It is necessary to trade off between saving current and maintaining adequate linearity and to obtain significant power saving but not to sacrifice *ACPR* performance.

The AGC design for a transmitter, which demands linearity, should balance current consumption, *ACPR* performance, and noise emission. The noise emission, especially the noise in the receiver frequency band, is very important to a full-duplex system since it may desensitize the receiver sensitivity. The Vegas in BB, IF, and RF blocks of the transmitter chain and the PA gain as well need to be controlled in proper sequence and profile to optimize overall transmitter performance — i.e., to minimize the current consumption with only slight impact on *ACPR* performance and noise emission.

In the AGC optimization it is important to know the followings. To obtain a better transmitter linearity it is preferred to have the gain of the PA and the RF stages high, and on the contrary it is better to have their gain low if lower noise emission is desired. The current reduction usually impact the 1 dB compression point or OIP_3 of the transmitter and therefore the linearity of the transmitter, and in this case the transmission *ACPR* needs to be carefully monitored.

Appendix 5A. Approximate Relationship Between ρ and EVM

The CDMA signal is a noiselike signal. The power of noiselike signals is proportional to the expectation of the second power of voltage signals:

$$P_{ideal} = E\left\{\left|\bar{a}(k_1)\right|^2\right\}, \qquad P_{error} = E\left\{\left|\bar{e}(k_1)\right|^2\right\} \tag{5A.1}$$

$$\begin{aligned} P_{actual} &= E\left\{\left|\bar{a}(k_1)+\bar{e}(k_1)\right|^2\right\} = E\left\{\left|\bar{a}(k_1)\right|^2\right\} + 2E\left\{\left|\bar{a}(k_1)\right|\cdot\left|\bar{e}(k_1)\right|\right\} + E\left\{\left|\bar{e}(k_1)\right|^2\right\} \\ &= P_{ideal} + 2P_{cross} + P_{error}, \end{aligned} \tag{5A.2}$$

where P_{ineal} is the ideal CDMA signal power, P_{error} is the error signal power, P_{actual} is the actual transmission CDMA signal power, and P_{cross} is the cross power between the ideal and the error signal.

Based on the definition, the waveform quality factor ρ has the following expression

$$\rho = \frac{P_{ideal}}{P_{actual}} = \frac{P_{ideal}}{P_{ideal} + P_{cross} + P_{error}}. \tag{5A.3}$$

The cross power is equal to zero if the error signal is not correlated to the ideal signal. For instance, errors caused by white noise will be not correlated to the signal, but other distortions, such as intermodulation and group delay distortion, will have nonzero P_{cross}. If the cross power is negligibly small (5A.3) can be simplified as

$$\rho \cong \frac{P_{ideal}}{P_{ideal} + P_{error}} = \frac{1}{1 + P_{error}/P_{ideal}} = \frac{1}{1 + EVM^2}. \tag{5A.4}$$

This is (5.3. 13a), or it can be written in (5.3.13b) form

$$EVM \cong \sqrt{\frac{1}{\rho} - 1}. \tag{5A.5}$$

(5A.4) will overestimate ρ when the error signals are not correlated to the ideal signal or the cross power is not equal to zero.

Appendix 5B. Image Suppression of Transmission Signal

Assuming that the base-band I signal has an amplitude 1 and the Q signal amplitude is α, which is close to 1 but not equal to 1; and the phase between them is ε degree off 90°, they can be respectively expressed as

$$I(t) = \cos\phi(t) \tag{5B.1}$$

and

$$Q(t) = \alpha \sin[\phi(t) + \varepsilon]. \tag{5B.2}$$

The base-band I and Q quadrature signals turn into a modulated signal with an IF or RF carrier at the modulator. This signal can be expressed as

$$f_{Tx}(t) = \cos\phi(t)\cos\omega_c t - (1+\delta)\sin[\phi(t)+\varepsilon]\sin(\omega_c t + \sigma), \tag{5B.3}$$

where the amplitudes are normalized by the amplitude of cosine product term; $(1+\delta)$ is the normalized amplitude of the sine product term, which is a product of amplitude imbalances generated by the base-band I and Q channels and the quadrature modulator; and σ (<<90°} is the phase imbalance created by the modulator. The right side of (5B.3) can be expanded as

$$\begin{aligned} \cos\phi(t)\cos\omega_c t - (1+\delta)\sin[\phi(t)+\varepsilon]\sin(\omega_c t+\sigma) \\ = \frac{e^{j\phi(t)} + e^{-j\phi(t)}}{2}\frac{e^{j\omega_c t} + e^{-j\omega_c t}}{2} \\ + (1+\delta)\frac{e^{j[\phi(t)+\varepsilon]} - e^{-j[\phi(t)+\varepsilon]}}{2}\frac{e^{j[\omega_c t+\sigma]} - e^{-[j\omega_c t+\sigma]}}{2}. \end{aligned} \tag{5B.4}$$

Re-arranging the right side of (5B.4), we obtain

$$f_c(t)=\frac{1}{4}\Big[e^{j[\omega_c t+\phi(t)]}\left(1+\beta e^{j[\varepsilon+\sigma]}\right)+e^{-j[\omega_c t+\phi(t)]}\left(1+\beta e^{-j[\varepsilon+\sigma]}\right)$$
$$+e^{j[\omega_c t-\phi(t)]}\left(1-\beta e^{j[\sigma-\varepsilon]}\right)+e^{-j[\omega_c t-\phi(t)]}\left(1-\beta e^{-j[\sigma-\varepsilon]}\right)\Big]. \tag{5B.5}$$

We know that $\left[1+(1+\delta)e^{j[\varepsilon+\sigma]}\right]$ and $\left[1+(1+\delta)e^{-j[\varepsilon+\sigma]}\right]$ have the same magnitude since

$$\left|1+(1+\delta)e^{j[\varepsilon+\sigma]}\right|=\sqrt{1+2(1+\delta)\cos(\varepsilon+\sigma)+(1+\delta)^2}=\left|1+(1+\delta)e^{-j[\varepsilon+\sigma]}\right|. \tag{5B.6}$$

Similarly, we also have

$$\left|1-(1+\delta)e^{j[\delta-\varepsilon]}\right|=\sqrt{1-2(1+\delta)\cos(\varepsilon-\sigma)+(1+\delta)^2}=\left|1-(1+\delta)e^{-j[\sigma-\varepsilon]}\right|. \tag{5B.7}$$

Utilizing (5B.6) and (5B.7), the first two terms on the right side of (5B.5) can be rewritten into

$$\frac{1}{4}\left[e^{j[\omega_c t+\phi(t)]}\left(1+(1+\delta)e^{j[\varepsilon+\sigma]}\right)+e^{-j[\omega_c t+\phi(t)]}\left(1+(1+\delta)e^{-j[\varepsilon+\sigma]}\right)\right]$$
$$=\frac{\sqrt{1+2(1+\delta)\cos(\varepsilon+\sigma)+(1+\delta)^2}}{2}\cos[\omega_c t+\phi(t)+\theta], \tag{5B.8}$$

where θ is defined as

$$\theta=\tan^{-1}\frac{(1+\delta)\sin(\varepsilon+\sigma)}{1+(1+\delta)\cos(\varepsilon+\sigma)}. \tag{5B.9}$$

In a similar way, the last two terms on the right side of (5B.7) can be turned into

$$\frac{1}{4}\left\{e^{j[\omega_c t-\phi(t)]}\left[1-(1+\delta)e^{j[\sigma-\varepsilon]}\right]+e^{-j[\omega_c t-\phi(t)]}\left[1-(1+\delta)e^{-j[\delta-\varepsilon]}\right]\right\}$$
$$=\frac{\sqrt{1-2(1+\delta)\cos(\varepsilon-\sigma)+(1+\delta)^2}}{2}\cos[\omega_c t-\phi(t)+\varphi], \tag{5B.10}$$

where φ is

$$\varphi = \tan^{-1} \frac{(1+\delta)\sin(\sigma - \varepsilon)}{1-(1+\delta)\cos(\sigma - \varepsilon)}. \tag{5B.11}$$

Finally, substituting (5B.8) and (5B.10) into (5B.5), $f_c(t)$ can be expressed in the following form:

$$\begin{aligned} f_{Tx}(t) = & \frac{\sqrt{1+2(1+\delta)\cos(\varepsilon+\sigma)+(1+\delta)^2}}{2}\cos[\omega_c t + \phi(t) + \theta] \\ & + \frac{\sqrt{1-2(1+\delta)\cos(\varepsilon-\sigma)+(1+\delta)^2}}{2}\cos[\omega_c t - \phi(t) + \varphi]. \end{aligned} \tag{5B.12}$$

In (5B.12), the first term on the right side is the desired transmission signal, and the second term on the same side is the unwanted image signal. It is apparent that the image signal has a flipped spectrum of the desired signal, but it occupies the same bandwidth as the desired signal (see Fig. 5B.1). The power ratio of the image signal to desired signal in dB scale is referred to as image suppression IMG_S, and it is defined as

$$IMG_S = 10\log\frac{1-2(1+\delta)\cos(\varepsilon-\sigma)+(1+\delta)^2}{1+2(1+\delta)\cos(\varepsilon+\sigma)+(1+\delta)^2}. \tag{5B.13}$$

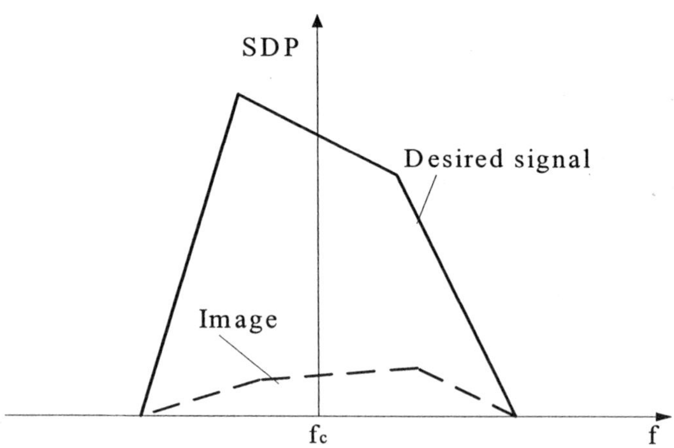

Figure 5B.1. Spectra of a desired signal and its image

Appendix 5C. Amplifier Nonlinear Simulation: ACPR Calculation

```
function nonline_sim(Files,Pin,INPP)
% nonline_sim simulates CDMA transmitter output spectrum
% and ACPRs.
%
%       Example of Application:
%          nonline_sim('padata',2,-3,[1,5,1,5]);
%
%   Input:
%      Files A string matrix that contains the name(s) of
%      the file(s) that characterize the nonlinearities.
%
%   Pin:
%      Input power level in dBm.
%
%   INPP:
%      Input parameters for the nonlinearity modeling.
%      INPP=[A_BASIS,A_TERMS,P_BASIS,P_TERMS]
%      Expansion function type:
%      1. power series: A_BASIS = 1, P_BASIS = 1.
%      Important:
%         For CDMA spectral regrowth calculation the
%         power series expansion must be used.
%      2. Fourier-sine series (good for AM-AM modeling
%         only): A_BASIS = 2
%      3. Fourier-cosine series (good for AM-PM modeling
%         only): P_BASIS = 3
%
%      Example: If INPP=[1,5,1,5], Zin=50 (Ohms), Zout=50
%         (Ohms),use 5-terms power series for amplitude
%         nonlinearity and 5-terms power series for phase
%         nonlinearity.
%
% Make Pin a row vector
%
   if nargin<2
      error('You need to specify input power (in dBm)');
   end;
   PinIsRow=1; [M_Pin,N_Pin]=size(Pin);
   if M_Pin>1
      if N_Pin~=1, error('Pin must be a column or a row
vector.'); end;
      Pin=Pin'; PinIsRow=0; N_Pin=M_Pin; M_Pin=1;
   end;
%
%
```

```
%  Read baseband I-Q signal
%
   load('i_sqn.tim');
   load('q_sqn.tim');
   i0=i_sqn(:,2);
   q0=q_sqn(:,2);
   Txtau=1/(4.9152*10^6);
%
   M_uin=length(i0);
   t0=[1:M_uin]';
   t0=t0.*Txtau;
   V0i=(i0+1j*q0);
   V0im=max(abs(V0i));
   Vi=(V0i./V0im);

Pi=10*log10(sum(abs(Vi.*conj(Vi)))*1000/length(t0)/50);
   Ratio=10^((13-Pi)/20); % 13 dBm equal to 1V voltage
                          % across a 50 Ohm load
   Vi=Vi.*Ratio*dbm2v(Pin,50);
   Pi=10*log10(sum(abs(Vi.*conj(Vi)))*1000/length(t0)/50
);
%
%
% Compress by nonlinearity
%
   DEVICES='';
   Pin0=Pin;
   [M,N]=size(Files);
   for i=1:M
      OUTP=nonline_curve(Files(i,:),INPP);
      ZI = getzi(OUTP);
      DEVICE= getdev(OUTP);
      Vo=nonline_calculation(OUTP,Vi);
      Vi = Vo;
      if i==1
      DEVICES=[DEVICES,'[',DEVICE,']'];
      else
      DEVICES=[DEVICES,'+[',DEVICE,']'];
      end;

   end;
   ZO=getzo(OUTP);
%
%  Perform Fourier transform to obtain the spectrum
%
   [UOUT,f0]=ldtft(Vo,t0);
%
```

```
%  Convert the normalized frequency to frequency in MHz
%
   f0 = f0 ./10^6;
   ind = find(abs(f0)<=0.6144);
   tmpsize = size(f0);
   Pout =10*log10((sqrt(sum(abs(UOUT).^2))).^2 ...
           .*1000./length(t0)/ZO);
   ind = find(f0 >= -2.4 & f0 <= 2.4);
   f0=f0(ind) ;
%
%  Set the RBW to be 10KHz
%
   Imax=round(0.010/diff(f0(1:2)));
   if rem(Imax,2)==1, Imax=Imax+1; end;
   UOUTSQ=0;
   for i = -Imax/2:Imax/2
       UOUTSQ=UOUTSQ + (abs(UOUT(ind+i,:)).^2);
   end;
   UOUTSQ=UOUTSQ./(ones(size(f0))*max(UOUTSQ));
%
% calculate OBW in MHz
%
   NN=length(f0);
   Nlh=floor(NN/2);
   Nrh=floor(NN/2)+1;
   lhpwr=cumsum(UOUTSQ(Nlh:-1:1,:));
   rhpwr=cumsum(UOUTSQ(Nrh:NN,:));
   LHPWR=lhpwr(Nlh,:);
   RHPWR=rhpwr(NN-Nrh+1,:);
   totpwr=LHPWR+RHPWR;
   [lhi,lhj]=find(diff([lhpwr>=0.99*ones(Nlh,...
            1)*LHPWR]));
   [rhi,rhj]=find(diff([rhpwr>=0.99*ones(NN-...
            Nrh+1,1)*RHPWR]));
   OBW = (f0(Nrh+rhi-1) - f0(Nlh-lhi+1));
   OBW = OBW';
   if ~PinIsRow, OBW=OBW'; end;
%
% calculate ACPR
%
   cind=find(f0 >= -.6144 & f0 <= 0.6144);
   CP=sum(UOUTSQ(cind,:));
%
% ACPR1 at +/- 900 kHz
%
   lacind=find(f0 >= (-0.9-0.015) & f0 <= (-0.9+0.015));
   uacind=find(f0 >= ( 0.9-0.015) & f0 <= ( 0.9+0.015));
```

```
   LACP=sum(UOUTSQ(lacind,:));
   UACP=sum(UOUTSQ(uacind,:));
   LACP=10.*log10(LACP./CP);
   UACP=10.*log10(UACP./CP);
   if ~PinIsRow, LACP=LACP'; UACP=UACP'; end;
%
% ACPR2 at +/- 1.98 MHz
%
   lacind1=find(f0 >= (-1.98-0.015) & f0 <= ...
                      (-1.98+0.015));
   uacind1=find(f0 >= ( 1.98-0.015) & f0 <= ...
                      ( 1.98+0.015));
   LACP1=sum(UOUTSQ(lacind1,:));
   UACP1=sum(UOUTSQ(uacind1,:));
   LACP1=10.*log10(LACP1./CP);
   UACP1=10.*log10(UACP1./CP);
   if ~PinIsRow, LACP1=LACP1'; UACP1=UACP1'; end;
%
   figure;

   plot(f0,10.*log10(UOUTSQ),'m-');
   grid;
   set(gca,'xlim',[-2.4,2.4],'xtick', ...
      [-2.4:0.4:2.4],'ylim',[-55,0]);
   title('[CDMA CELL Band Transmission Spectrum]');
   xlabel('Frequency (MHz)');
   ylabel('Tx Spectrum (dB)');
   text(-2.0,-3,['Pin=',sprintf('%.1f',Pin0),'dBm'],...
      'fontname','Times New Roman','fontweight','bold');
   text(-2.0,-6,['Pout=',sprintf('%.1f ',Pout),'dBm'],
...
      'fontname','Times New Roman','fontweight','bold');
   text(-2.0,-9,['OBW=',sprintf('%.3f ',OBW),'MHz'], ...
      'fontname','Times New Roman','fontweight','bold');
   text(-2.0,-12,['ACPR1L=',sprintf('%.1f,LACP), ...
      'dBc'],'fontname','Times New Roman', ...
      'fontweight','bold');
   text(-2.0,-15,['ACPR1U=',sprintf('%.1f ',UACP), ...
      'dBc'],'fontname','Times New Roman', ...
      'fontweight','bold');
   text(-2.0,-18,['ACPR2L=',sprintf('%.1f ',LACP1), ...
      'dBc'],'fontname','Times New Roman', ...
      'fontweight','bold');
   text(-2.0,-21,['ACPR2U=',sprintf('%.1f ',UACP1), ...
      'dBc'],'fontname','Times New Roman', ...
      'fontweight','bold');
```

****

```
function OUTP=nonline_curve(FileName,INPP,Plot)
% nonline_curve: Characterize amplifier nonlinearity
% based upon Pin-Pout and Pin-Phase characteristics.
%
%
% Usage: nonline_curve('padata',[2,5,3,5],1)
%
% Input:
%  FileName
%     A string, name of a matlab file that contains a
%     look-up table describing the device nonlinearity.
%     The data in the file must have a certain format.
%
%     If the file doesn't exist, then this programs
%     assumes that the characteristics of the device has
%     been known and is specified in the "INPP"
%     parameters.
%  INPP
%     If the file exists, then
%     INPP is a 4-element vector that specifies the
%     parameters of modeling
%     INPP(1)  Expansion function type for the AM-AM
%              characteristics
%     INPP(2)  Number of terms of expansion for the
%              AM-AM characteristics
%     INPP(3)  Expansion function type for the AM-PM
%              characteristics
%     INPP(4)  Number of terms of expansion for the AM-
%              PM characteristics
%
%  Expansion function type:
%     1  power series
%     2  Fourier-sine series
%        (good for AM-AM modeling only)
%     3  Fourier-cosine series
%        (good for AM-PM modeling only)
%
% Output:
%  OUTP  A row vector that characterizes the device
%  nonlinearity
%  OUTP is a 4-element vector that specifies the
%  parameters of modeling
%     OUTP(1)  Expansion function type for the AM-AM
%              characteristics
%     OUTP(2)  Number of terms of expansion for the
```

```
%                AM-AM characteristics
%      OUTP(3)   Expansion function type for the AM-PM
%                characteristics
%      OUTP(4)   Number of terms of expansion for the
%                AM-PM characteristics
%      OUTP(5)   Input impedance (Ohm)
%      OUTP(6)  Output impendance (Ohm)
%      OUTP(6+[1:INPP(4)])  Coefficients for the AM-AM
%                           characteristics
%      OUTP(6+OUTP(4)+[1:OUTP(6)])   Coefficients for the
%                           AM-PM characteristics
%      OUTP(7+OUTP(4)+OUTP(6))     Linear gain  (dB)
%      OUTP(8+OUTP(4)+OUTP(6))     P1dB (dBm)
%      OUTP(9+OUTP(4)+OUTP(6))     Maximum allowed input
%                                  level (dBm)
%      OUTP(10+OUTP(4)+OUTP(6))    Saturated output level
%                                  (dBm)
%      OUTP(11+OUTP(4)+OUTP(6))    SAP, 4-element vector,
%                                  noise coefficients
%      OUTP(12+OUTP(4)+OUTP(6):length(OUTP))  (Absolute
%                                  value) The device name
%
%---> Define defaults
%
Plot=1;
if nargin < 3
   Plot=0;
   if nargin<2
      INPP=[1,5,1,5];
   end;
end;
% DEFINE CONSTANSTS
   POLYNOMIAL= 1;
   SINE      = 2;
   COSINE    = 3;
   SERIES_TYPE=str2mat(...
   'power series:',...
   'Fourier-sine series:',...
   'Fourier-cosine series:');
%
%Get input parameters
%
A_BASIS = INPP(1);A_TERMS = INPP(2);P_BASIS =
INPP(3);P_TERMS   = INPP(4);
%
%
```

```
FileExists=exist(FileName);

if FileExists ~= 0
%
% Read the measurement data and calculate coefficients
for the assigned series type
%
   clear TABLE SAP ZI ZO DEVICE;
   eval(FileName);
   if exist('TABLE')==0, error(['There is no data table
in "',Filename,'"']); end;
   [Mtmp,Ntmp]=size(TABLE);
            if Ntmp <2, error(['Incomplete table in
"',Filename,'"']); end;
   if exist('DEVICE')==0, DEVICE='untitled'; end;  %<---
                                          %set DEVICE default
   if exist('SAP')==0, SAP=[70,50,1,4]; end; %<---set
                                                %SAP default
   if exist('ZI')==0, ZI=50; end;   %<---set ZI default
   if exist('ZO')==0, ZO=50; end;   %<---set ZO default
   Pi=TABLE(:,1);
   Po=TABLE(:,2);
   NoPhaseInput=0;
%
%  Calculates gain, P1dB, Psat
%
   NPi = length(Pi);
   PINM= Pi(NPi);
   PSAT = Po(NPi);
   gain=Po-Pi;
   GAIN = mean(gain(1:3));
   gain(4:NPi)=polyval(polyfit(Pi(4:NPi),...
               gain(4:NPi),7),Pi(4:NPi));
   ind=find(gain<=GAIN-0.7 & gain>=GAIN-1.5);
   if isempty(ind)
      P1DB = 60;
   else
      if length(ind)== 1
         P1DB=Po(ind);
         else
         P1DB = interp1(gain(ind),Pi(ind),GAIN-1, ...
                'spline')+GAIN-1;
      end;
   end;
%
%  Instroduce convenient units for input and output
%  voltage
```

```
%
   UVi = sqrt(ZI/500)*10^((P1DB+1-GAIN)/20);
   UVo = sqrt(ZO/500)*10^((P1DB+1)/20);
   vi = dbm2v_tone(Pi,ZI) ./ UVi;
   vo = dbm2v_tone(Po,ZO) ./ UVo;
   v1i= dbm2v_tone(P1DB,ZI) ./ UVi;
   max_vi=vi(length(vi));
   max_vo=vo(length(vo));
%
%  Modeling AM-AM characteristics
%
   if A_BASIS==POLYNOMIAL
%
   p = polyfit([-vi;vi],[-vo;vo],2*A_TERMS-1);
   A_COEF = p(2*A_TERMS-1:-2:1);
   n = [1:A_TERMS-1];
   fct = [1 2.^(2*n).*fctrl(n).*fctrl(n+1) ...
         ./fctrl(1+2*n)];
   if Plot == 0
      A_COEF = A_COEF.*fct;
   else
      A_COEF = A_COEF;
   end;
   else
%
   if A_BASIS==SINE
   NN = 201;
   vip = max_vi .* [0:NN-1]' ./ (NN-1);
   vop = interp1([0;vi;max_vi*11/10],[0;vo;max_vo],vip);
   n = [1:2:2*A_TERMS-1];
   KERNEL = sin((pi/2/max_vi .* vip) * n);
   A_COEF = 2/max_vi .* trapz(vip, ...
            (vop*ones(1,A_TERMS)) .* KERNEL);
   else
%
   if A_BASIS==COSINE
      error('Cannot use Fourier-cosine series for...
               expanding AM-AM characteristics.');
   else
      error('Unknown expansion function type');

   end;
   end;
   end;
%
%  Modeling AM-PM characteristics
%
```

```
   Ph=TABLE(:,3) .* (pi/180); %---> convert to radians
%------>Find the constant phase shift, and substract it.
   q = polyfit([-vi(length(vi):-1:1);vi], ...
               [Ph(length(vi):-1:1);Ph],2*6);
   Ph0 = q(2*6+1);
   Ph=Ph-Ph0;
%<------OK
   max_ph=Ph(length(Ph));
%
   if P_BASIS == POLYNOMIAL
   q = polyfit([-vi;vi],[Ph;Ph],2*P_TERMS);
   P_COEF = q(2*P_TERMS-1:-2:1);
   else
   if P_BASIS==SINE
   error('Cannot use Fourier-sine series for ...
            expanding AM-PM characteristics.');
   else
   if P_BASIS==COSINE
%
   NN = 201;
   vip = max_vi .* [0:NN-1]' ./ (NN-1);
   Php = interp1([-vi(length(vi):-
1:1);vi],[Ph(length(Ph):-1:1);Ph],vip);
   n = [0:2:2*P_TERMS-2];
   KERNEL = cos((pi/2/max_vi .* vip) * n);
   P_COEF = 2/max_vi .* trapz(vip, ...
            (Php*ones(1,P_TERMS)) .* KERNEL);
   P_COEF(1)=P_COEF(1)/2;
   else
      error('Unknown expansion function type');
   end;
   end;
   end;
   OUTP=[A_BASIS,A_TERMS,P_BASIS,P_TERMS,ZI,ZO, ...
         A_COEF,P_COEF,GAIN,P1DB,PINM,PSAT,SAP, ...
         abs(DEVICE)];
else
      error('No File Exists.');
end;
%
if ~Plot
   return;
end;


ViM = dbm2v_tone(PINM,ZI);
VoM = dbm2v_tone(PSAT,ZO);
```

```
NN=51;
if FileExists ~= 0
   Pip = Pi(1) + (PINM-Pi(1)) .* [0:NN-1]' ./ (NN-1);
else
   Pip = PINM .* [0:NN-1]' ./ (NN-1);
   Pi = []; Po=[]; Ph=[];
end;

%Pip;
fig=['Characterization of [',DEVICE,']'];
Vi = dbm2v_tone(Pip,ZI);
Vo = nonline_calculation(OUTP,Vi);
Pop= v2dbm(Vo,ZO);
Php= atan2(imag(Vo),real(Vo));
Vi';
h=findfig(fig);
if h==0;
   figure;
   set(gcf,'unit','pixel',...
      'pos',[200 200 550 450],...
      'numbertitle','off',...
      'name',fig);
else
   figure(h);
end;
rad2deg = 180/pi;

if NoPhaseInput
   plot(Pip,Pop,'r-',Pip,Php.*rad2deg,'g--',Pi,Po,'rx');
else
   plot(Pip,Pop,'r-',Pip,Php.*rad2deg,'g--
',Pi,Po,'rx',Pi,Ph.*rad2deg,'g*');
end;
grid;
xlabel('Input power (dBm)');
ylabel('Output power (dBm), Phase (deg)');
title(fig);
xlim=get(gca,'xlim');
ylim=get(gca,'ylim');
set(gcf,'defaulttextfontname','Times New Roman');
set(gcf,'defaulttexthorizont','left');
set(gcf,'defaulttextfontsize',10);
set(gcf,'defaulttextcolor','k');

text(xlim(1),ylim(2)-1*diff(ylim)/20,sprintf(...
   'Gain=%.1fdB P1=%.1fdBm ',GAIN,P1DB));
```

```
text(xlim(1),ylim(2)-2*diff(ylim)/20,sprintf('AM-AM...
   coefficients in %s',SERIES_TYPE(A_BASIS,:)));
text(xlim(1),ylim(2)-3*diff(ylim)/20,sprintf('%.4f', ...
A_COEF));
text(xlim(1),ylim(2)-4*diff(ylim)/20,sprintf('AM-PM...
   coefficients in %s',SERIES_TYPE(P_BASIS,:)));
text(xlim(1),ylim(2)-5*diff(ylim)/20,sprintf('%.4f', ...
   P_COEF));
set(gca,'xlim',xlim);
set(gca,'ylim',ylim);
```

****

```
function Vo=nonline_calculation(INPP,Vi)
% Simulates amplifier bandpass nonlinearity
%
% Usage: Vo=nonlin(INPP,Vi)
%
%  Given the device nonlinearity parameters INPP, the
%  input voltage Vi, this function calculates the output
%  voltage Vo. Note that Vi and Vo are both low-pass
%  equivalent complex voltages.
%
% Input:
%  INPP is a  vector that specifies the parameters of
%  modeling
%     INPP(1)  Expansion function type for the AM-AM
%              characteristics
%     INPP(2)  Number of terms of expansion for the
%              AM-AM characteristics
%     INPP(3)  Expansion function type for the AM-PM
%              characteristics
%     INPP(4)  Number of terms of expansion for the
%              AM-PM characteristics
%     INPP(5)  Input impedance (Ohm)
%     INPP(6)  Output impedance (Ohm)
%     INPP(6+[1:INPP(4)])  Coefficients for the AM-AM
%                          characteristics
%     INPP(6+INPP(4)+[1:INPP(6)])   Coefficients for the
%                                 AM-PM characteristics
%     INPP(7+INPP(4)+INPP(6))     Linear gain  (dB)
%     INPP(8+INPP(4)+INPP(6))     P1dB (dBm)
%     INPP(9+INPP(4)+INPP(6))     Maximum allowed input
%                                 level (dBm)
%     INPP(10+INPP(4)+INPP(6))    Saturated output level
%                                 (dBm)
```

```
%      INPP(11+INPP(4)+INPP(6))    SAP, 4-element vector,
%                                  noise coefficients
%      INPP(12+INPP(4)+INPP(6):length(INPP))  (Absolute
%                                  value) The device name
%
%      Expansion function type:
%      1  power series
%      2  Fourier-sine series
%         (good for AM-AM modeling only)
%      3  Fourier-cosine series
%         (good for AM-PM modeling only)
%
%   Vi
%      A matrix or vector that describes the input signal
%      (which in general is complex).
%      If Vi is a M by N matrix, there are N separate
%      input, each is a M by 1 vector.
%      If vi is missing, this programs plots the
%      compression curve, both measurement and fitting
%      curve.
%
% Output:
%      Vo A matrix or vector that describes the input
%      signal (which in general is complex).
%
%---> Define defaults
%

if nargin < 2
      error('No Input');
else
end;


% DEFINE CONSTANSTS
   POLYNOMIAL= 1;
   SINE       = 2;
   COSINE     = 3;
   SERIES_TYPE=str2mat(...
   'power series:',...
   'Fourier-sine series:',...
   'Fourier-cosine series:');
%
%Get input parameters
%
```

```
if length(INPP)<4
   error('The input parameters are not completely
specified.');
end;
A_BASIS = INPP(1);A_TERMS = INPP(2);P_BASIS =
INPP(3);P_TERMS    = INPP(4);
if length(INPP) < (11+A_TERMS+P_TERMS+3)
   error('The input parameters are not completely
specified.');
end;
ZI     = INPP(5);
ZO     = INPP(6);
A_COEF= INPP(7+[0:A_TERMS-1]);
P_COEF= INPP(7+A_TERMS+[0:P_TERMS-1]);
GAIN   = INPP(7+A_TERMS+P_TERMS);
P1DB   = INPP(8+A_TERMS+P_TERMS);
PINM   = INPP(9+A_TERMS+P_TERMS);
PSAT   = INPP(10+A_TERMS+P_TERMS);
SAP    = INPP(11+A_TERMS+P_TERMS+[0:3]);
DEVICE= setstr(INPP(15+A_TERMS+P_TERMS:length(INPP)));
UVi = sqrt(ZI/500)*10^((P1DB+1-GAIN)/20);
UVo = sqrt(ZO/500)*10^((P1DB+1)/20);

ViM = dbm2v_tone(PINM,ZI);
VoM = dbm2v_tone(PSAT,ZO);
ViIsRow=0;

%---------->
ViA = abs(Vi);
[tmp_M,tmp_N]=size(ViA);
if tmp_M==1
   ViIsRow=1;
   tmp_M=tmp_N;
   ViA=ViA';
end;
%
ViP = atan2(imag(Vi),real(Vi));

%<----------

max_vo=dbm2v_tone(PSAT,ZO)/dbm2v_tone(P1DB+1,ZO);

%
%Calcuate the output amplitude
%
if A_BASIS==POLYNOMIAL
   p = zeros(1,2*A_TERMS);
```

```
   p(2*A_TERMS-1:-2:1)=A_COEF;
   VoA = polyval(p,ViA./UVi).*UVo;
else
   if A_BASIS==SINE
      n = [1:2:2*A_TERMS-1];
      VoA = zeros(size(ViA));
      for i=1:tmp_N
         KERNEL = sin((pi/2/ViM .* ViA(:,i)) * n);
      VoA(:,i) = (KERNEL * A_COEF') .* UVo;
      end;
      else
         if A_BASIS==COSINE
            error('Cannot use Fourier-cosine series ...
               for expanding AM-AM characteristics.');
         else
            error('Unknown expansion function type');
         end;
   end;
end;
%
%Calcuate the phase distortion
%
if P_BASIS==POLYNOMIAL
   q = zeros(1,2*P_TERMS+1);
   q(2*P_TERMS-1:-2:1)=P_COEF;
   VoP = polyval(q,ViA./UVi);
   else
      if P_BASIS==SINE
         error('Cannot use Fourier-sine series to ...
            expand AM-AM characteristics.');
      else
         if P_BASIS==COSINE
            n = [0:2:2*P_TERMS-2];
            VoP = zeros(size(ViA));
            for i=1:tmp_N
               KERNEL = cos((pi/2/ViM .* ViA(:,i)) * n);
               VoP(:,i) = (KERNEL * P_COEF');
            end;
         else
               error('Unknown expansion function type');
         end;
      end;
end;
%
if ViIsRow
   VoA = VoA';
   VoP = VoP';
```

```
end;
VoP = VoP + ViP;      % Add the phase distortion
Vo = VoA .* exp(1j .* VoP);
```

****

```
function [H,F]=LDTFT(h,t)
% LDTFT custom Discrete-time Fouriour transform
%
% [H,W]=LDTDT(h,t)
%
%  Input:
%
%       h       input vector, whose length is N
%       t       time vector, whose length is N
%
%  Output:
%
%       H       vector of the DTFT of h
%       F       vector of the frequencies where DTFT is
computed
%
[M,N] = size(h);
ODD=0;
ROW=0;
VECTOR=0;
if M==1
   ROW=1;
   t=t';
   h=h';
   M=N;
   N=1;
end;

[M_t,N_t]=size(t);
if (M_t ~= M)
   error('LDTFT: dimensions of time and signal do not
agree');
else
   if (N_t~=1 & N_t~=N)
      error('LDTFT: dimensions of time and signal do not
agree');
   else
      if N_t==1
         VECTOR=1;
         t = t*ones(1,N);
      end;
   end;
```

```
end;
if rem(M,2)==1
   ODD=1;
   M=M-1;
   h=h(1:M,:);
   t=t(1:M,:);
end;
mid = M/2+1;
W = (2*pi/M) .* ([0:(M-1)]'*ones(1,N));
W(mid:M,:)=W(mid:M,:)-2*pi;
W = fftshift(W);
F = W .* (ones(M,1)*((1/2/pi)./(t(2,:)-t(1,:))));
n0 = M/2;
t0 = t(mid,:);
scale=1/sqrt(M);
HH = fft(h,M).*scale;
H(1:mid-1,:) = HH(mid:M,:);
H(mid:M,:) = HH(1:mid-1,:);
H = H .* exp(-1j*(W*n0+2*pi*(F.*(ones(M,1)*t0))));

if ODD
   H = [H;H(1,:)];
   F = [F;-F(1,:)];
end;

if VECTOR
        F=F(:,1);
end;

if ROW
   H=H';
   F=F';
end;
```

****

```
function [h,t]=LIDTFT(H,F)
%LIDTFT Leon's custom inverse Discrete-time Fouriour
transform
%
%
%  Last edited on 06/05/96
%
%
% [h,t]=LIDTDT(H,F) generates the inverse Discrete-time
Fouriour transform
%
%  Input:
```

```
%
%       H       input vector
%       F       frequency vector
%
%  Output:
%
%       h       vector of the IDTFT of H
%       t       vector of the tome where IDTFT is
computed
%
[M,N] = size(H);
ODD=0;
ROW=0;
VECTOR=0;
if M==1
   ROW=1;
   F=F';
   H=H';
   M=N;
   N=1;
end;

[M_F,N_F]=size(F);
if (M_F ~= M)
   error('LDTFT: dimensions of frequency and signal do
not agree');
else
   if (N_F~=1 & N_F~=N)
      error('LDTFT: dimensions of frequency and signal
do not agree');
   else
      if N_F==1
         VECTOR=1;
         F = F*ones(1,N);
      end;
   end;
end;

if rem(M,2)==1
   ODD=1;
   M=M-1;
   H=H(1:M,:);
   F=F(1:M,:);
end;
mid = M/2+1;
T = (2*pi/M) .* ([0:(M-1)]'*ones(1,N));
T(mid:M,:)=T(mid:M,:)-2*pi;
```

```
T = fftshift(T);
t = (    T ./  (ones(M,1)*(F(2,:)-F(1,:)))     )./(2*pi);
n0 = M/2;
F0 = F(mid,:);
scale=sqrt(M);
hh = ifft(H,M).*scale;
h(1:mid-1,:) = hh(mid:M,:);
h(mid:M,:) = hh(1:mid-1,:);
h = h .* exp(1j*(T*n0+2*pi*(t.*(ones(M,1)*F0))));
if ODD
   h = [h;h(1,:)];
   t = [t;-t(1,:)];
end;

if VECTOR
   t = t(:,1);
end;

if ROW
   h = h';
   t = t';
end;
```

****

```
function v = fctrl(n)
% Factorial calculation n! = n(n-1)(n-2)....1.
if n == 0
   v = 1;
else
   for k = 1:n
     if k == 1
        v = 1;
      else
        v = v*k;
      end;
    end;
end;
```

****

```
function V=dBm2V(dBm,Z)
%DBM2V convert dBm number to volt
%
%
if nargin< 2
   Z=50;
```

```
end;
%
V = 10 .^ (dBm./20) .* sqrt(Z/1000)
```

****

```
function V=dBm2V_tone(dBm,Z)
%DBM2V convert dBm number to volt
%
%
if nargin< 2
   Z=50;
end;
 V = 10 .^ (dBm./20) .* sqrt(2*Z/1000);
```

****

```
function DEVICE=getdev(INPP)
%
A_TERMS=INPP(2);
P_TERMS=INPP(4);
DEVICE= setstr(INPP(15+A_TERMS+P_TERMS:length(INPP)));
```

****

```
function [GAIN,P1DB,PINM,PSAT]=getpwr(INPP)
%
A_TERMS=INPP(2);
P_TERMS=INPP(4);
GAIN  = INPP(7+A_TERMS+P_TERMS);
P1DB  = INPP(8+A_TERMS+P_TERMS);
PINM  = INPP(9+A_TERMS+P_TERMS);
PSAT  = INPP(10+A_TERMS+P_TERMS);
```

****

```
function ZI=getZI(INPP)
%
ZI=INPP(5);
```

****

```
function ZO=getZO(INPP)
%
ZO=INPP(6);
```

References

[1]. 3GPP TS 25.101, V5.2.0, *Technical Specification Group Radio Access Networks; UE Radio Transmission and Reception (FDD)*, March 2002.

[2]. ETSI, *Digital Cellular Telecommunications System (Phase 2+); Radio Transmission and reception (GSM 05.05 version 7.3.0)*, 1998.

[3]. TIA/EIA-98-E, *Recommended Minimum Performance Standards for cdma2000 Spread Spectrum Mobile Stations*, February, 2003.

[4]. IEEE STD 1528-200X, *Recommended Practice for Determining the Peak Spatial-Average Specific Absorption Rate (SAR) in Human Body Due to Wireless Communications Devices: Experimental Techniques (Draft)*, Aug. 2001.

[5]. R. A. Birgenheier, "Overview of Code-Domain Power, Timing, and Phase Measurements," *Hewitt-Packard Journal*, Feb., 1996

[6]. S. Freisleben, "Semi-Analytical Computation of Error Vector Magnitude for UMTS SAW Filters," EPCOS AG, Surface Acoustic Wave Devices, Munich, Germany.

[7]. V. Aparin, B. Bulltler, and P. Draxler, "Cross Modulation Distortion in CDMA Receivers," *2000 IEEE MTT-S Digest*, pp. 1953–1956.

[8]. N. M. Blachman, "Detectors, Band-pass Nonlinearities and Their Optimization: Inversion of the Chebyshev Transform" *IEEE Trans. Information Theory*, vol. IT-17, no. 4, pp. 398–404, July 1971.

[9]. S. A. Maas, *Nonlinear Microwave Circuits*, Artech House, Norwood, MA, 1988

[10]. J. S. Kenney and A. Leke, "Power Amplifier Spectral Regrowth for Digital Cellular and PCS Applications," *Microwave Journal*, pp. 74–92, Oct. 1995.

[11]. K. G. Gard, H. M. Gutierrez, and M. B. Steer, "Characterization of Spectral Regrowth in Microwave Amplifiers Based on the Nonlinear Transformation of a Complex Gaussian Process," *IEEE Trans. On MTT*, vol. 47, no. 7, pp. 1059- 1060, July 1999.

[12]. J. C. Pedo and N. B. de Carvalho, "On the Use of Multitone Techniques for assessing RF Components' Intermodulation Distortion," *IEEE Trans. Microwave Theory and Techniques*, vol. 47, no. 12, pp. 2393–2402, Dec. 1999.

[13]. N. B. de Carvalho and J. C. Pedo, "Multi-Tone Intermodulation Distortion Performance of 3rd Order Distortion Microwave Circuits," *1999 IEEE MTT-S Digest*, pp. 763–766, 1999.

[14]. O. W. Ota, "Two-tone and Nine-tone Excitatins in Future Adative Predistorted Linearized Basestation Amplifiers of Cellular Radio," *Wireless ersonal Communications*, Kluware Publishers, vol. 16, no. 1, pp. 1–19, Jan. 2001.

[15]. M. Heimbach, "Polarizing RF Transmitters for Multimode Operation," *Communication System Design*, Oct. 2001.
[16]. M. Heimbach, "Polar Impact: The Digital Alternative for Multi-Mode Wireless Communications," *Applied Microwave and Wireless*, Aug. 2001.
[17]. S. Mann, M. Beach, P. Warr, and J. McGeehan, "Increasing the Talk-Time of Mobile Radio with Effective Linear Transmitter Architectures," *Electronics and Communication Engineering Journal*, Apr. 2001.
[18]. J. K. Jau and T. S. Horng, "Linear Interpolation Scheme for Compensation of Path-Delay Difference in an Envelope Elimination and Restoration Transmitter," *Proceeding of APMC2001*, pp. 1072-1075.
[19]. CDMA Development Group, *CDG System Performance Tests* (optional), Rev. 3.0 Draft, Apr. 9, 2003.

Associated References

[1]. L. Robinson, P. Aggarwal, and R. R. Surendran, "Direct Modulation Multi-Mode Transmitter," *2002 IEEE International Conference on 3G Mobile Communication Technologies*, pp. 206–210, May 2002.
[2]. G. D. Mandyam, "Quantization Issues for the Design of 3rd-Generation CDMA Wireless Handset Transmitter," pp. 1–7, 1999.
[3]. L. Angrisani and R. Colella, "Detection and Evaluation of I/Q Impairments in RF Digital Transmitters," *IEE Proc.–Sci. Meas. Technol.*, vol. 151, no. 1, pp. 39–45, Jan. 2004.
[4]. P. Naraine, "Predicting the EVM Performance of WLAN Power Amplifiers with OFDM Signals," *Microwave Journal*, vol. 47, no. 5, pp. 222–226, May 2004.
[5]. B. Andersen, "Crest Factor Analysis for Complex Signal Processing," *RF Design*, pp. 40–54, Oct. 2001.
[6]. N. Dinur and D. Wulich, "Peak-to-Average Power Ratio in High-Order OFDM," *IEEE Trans. On Communications*, vol. 49, no. 6, pp. 1063–1072, June 2001.
[7]. N. Ngajikin, N. Fisal and S. K. Yusof, "Peak to Average Power Ratio in WLAN-OFDM System," *Proceedings of 4th National Conference on Telecommunication Technology*, Shah Alam, Malaysia, pp. 123–126, 2003.

[8]. O. Gorbachov, Y. Cheng and J. S. W. Chen, "Noise and ACPR Correlation in CDMA Power Amplifiers," *RF Design*, pp. 38–44, May 2001.

[9]. F. H. Raab, "Intermodulation Distortion in Kahn-Technique Transmitter," *IEEE Trans. On Microwave Theory and Techniques*, vol. 44, no. 12, pp. 2273–2278, Dec. 1996.

[10]. S. Freisleben, "Semi-Analytical Computation of Error Vector Magnitude for UMTS SAW Filters," Tech. Note from EPCOS AG, Surface Acoustic Wave Devices, Munich, Germany.

[11]. S. J. Yi et al., "Prediction of a CDMA Output Spectrum Based on Intermodulation Products of Two-Tone Test," *IEEE Trans. On Microwave Theory and Technology*, vol. 49, no. 5, May 2001.

[12]. F. M. Ghannouchi, H. Wakana, and M. Tanaka, "A New Unequal Three-Tone Signal Method for AM-AM and AM-PM Distortion Measurements Suitable for Characterization of Satellite Communication Transmitter/transponders," *IEEE Trans. on Microwave Theory and Techniques*, vol. 48, no. 8, pp. 1404–1407, Aug. 2000.

[13]. F. M. Ghannouchi and A. Ghazel, "AM-AM and AM-PM Distortion Characterization of Satellite Transponders/Base Station Transmitters Using Spectrum Measurements," *Proceedings of 2003 International Conference on Recent Advance in Space Technologies*, pp. 141–144, Nov. 2003.

[14]. A. Laloue et al., "An Efficient Method for Nonlinear Distortion Calculation of the AM and PM Noise Spectra of FMCW Radar Transmitter," *IEEE Trans. On Microwave Theory and Techniques*, vol. 51, no. 8, pp. 1966–1976, Aug. 2003.

[15]. Application note, "On the Importance of Adaptive-Bias Techniques in the RF Section of CDMA Mobile Phones to Improve Standby and Talk-Time Performance," Wireless Semiconductor Division, Agilest Technologies.

[16]. S. Mann et al., "Increasing the Talk-time of Mobile Radios with Efficient Linear Transmitter Architectures," *Electronics and Communication Engineering Journal*, pp. 65–76, April 2001.

[17]. S. Mann et al., "Increasing Talk-Time with Efficient Linear PA's," *IEE Seminar on Tetra Market and Technology Development*, pp. 6/1–6/7, Feb. 2000.

[18]. F.H. Raab, P. Asbeck et al., "RF and Microwave Power Amplifier and Transmitter Technologies — Part 1," *High Frequency Electronics*, pp. 23–36, May 2003.

[19]. F.H. Raab, P. Asbeck et al., "RF and Microwave Power Amplifier and Transmitter Technologies — Part 2," *High Frequency Electronics*, pp. 22–36, July 2003.

[20]. F.H. Raab, P. Asbeck et al., "RF and Microwave Power Amplifier and Transmitter Technologies — Part 3," *High Frequency Electronics*, pp. 34–48, Sept. 2003.
[21]. F. H. Raab, P. Asbeck, et al., "RF and Microwave Power Amplifier and Transmitter Technologies — Part 4," *High Frequency Electronics*, pp. 38–49, Nov. 2003.
[22]. F.H. Raab, "High-Efficiency L-Band Kahn-Technique Transmitter," *1998 IEEE MTT-S Digest*, pp. 585–588, June 1998.
[23]. E.W. McCune Jr., "Multi-Mode and Multi-Band Polar Transmitter for GSM, NADC, and EDGE," *2003 IEEE Wireless Communications and Networking*, vol. 2, pp. 812–815, March 2003.
[24]. M. Heimbach, "Polarizing RF Transmitters for Multimode Operation," *Communication Systems Design*, vol.10, no. 12, Dec. 2004.
[25]. Z. Zhang and L. E. Larson, "Gain and Phase Error-Free LINC Transmitter," *IEEE Trans. On Vehicular Technology*, vol. 49, no. 5, pp.1986–1994, Sept. 2000.
[26]. W. B. Sander. S.V. Schell, and B. L. Sander, "Polar Modulator for Multi-Mode Cell Phones," *Proceedings of the IEEE 2003 Custom Integrated Circuits Conference*, pp. 439–445, Sept. 2003.
[27]. T. Sowlati et al., "Quad-Band GSM/GPRS/EDGE Polar Loop Transmitter," *IEEE Journal of Solid-State Circuits*, vol. 39, no. 12, pp. 2179–2189, Dec. 2004.
[28]. L. Sundstrom, "Spectral Sensitivity of LINC Transmitters to Quadrature Modulator Misalignments," *IEEE Trans. on Vehicular Technology*, vol. 49, no. 4, pp.1474–1486, July 2000.
[29]. S. O. Ampem-Darko and H.S. Al-Raweshidy, "A Novel Technique for Gain/Phase Error Cancellation in LINC Transmitters," *1999 IEEE 50th Vehicular Technology Conference*, vol. 4, pp. 2034–2038, Sept. 1999.
[30]. S. A. Olson and R. E. Stengel, "LINC Imbalance Correction Using Baseband Preconditioning," *1999 IEEE Radio and Wireless Conference*, pp. 179–182, Aug. 1999.
[31]. B. Shi and L. Sundstrom, "An IF CMOS Signal Component Separator Chip for LINC Transmitter," *2001 IEEE International Conference on Custom Integrated Circuits*, pp. 49–52, May 2001.
[32]. X. Zhang, L. E. Larson and P. M. Asbeck, "Calibration Scheme for LINC Transmitter," *Electronics Letters*, vol. 37, no. 5, pp. 317–318, Mar. 3001.
[33]. B. Shi and L. Sundstrom, "A Time-Continuous Optimization Method for Automatic Adjustment of Gain and Phase Imbalances in Feedforward and LINC Transmitters," pp. I-45–I-48, 2003.
[34]. R. Strandberg, P. Andreani and L. Sundstrom, "Bandwidth Considerations for a CALLUM Transmitter Architecture," *2002 IEEE*

International Symposium on Circuits and Systems, vol. 4, pp. IV-25–IV-28, May 2002.

[35]. M. Boloorian and J.P. McGeehan, "Automatic Removal of Cartesian Feedback Transmitter Imperfections," *IEE Proc.–Commun.*, vol. 144, no. 4, pp. 281–288, Aug. 1997.

[36]. S.I. Mann, M.A. Beach and K.A. Morris, "Digital Baseband Cartesian Loop Transmitter," *Electronics Letters*, vol. 37, no. 22, pp.1360–1361, Oct. 2001.

[37]. N. Sornin et al., "A Robust Cartesian Feedback Loop for a 802.11 a/b/g CMOS Transmitter," *2004 IEEE Radio Frequency Integrated Circuits Symposium*, pp. 145–148, 2004.

[38]. M. Helaoui, et al. "Low-IF 5 GHz WLAN Linearized Transmitter Using Baseband Digital Predistorter," *Proceedings of the 2003 10th IEEE International Conference on Electronics, Circuits, and Systems*, vol. 1, pp. 260 – 263, Dec. 2003.

[39]. P. B. Kenington, "Linearized Transmitters: An Enabling Technology for Software Defined Radio," *IEEE Communications Magazine*, pp. 156 – 162, Feb. 2002.

[40]. K.C. Peng, et al., "High Performance Frequency Hopping Transmitters Using Two-Point Delta-Sigma Modulation," *2004 IEEE MTT-S Digest*, pp. 2011–2014, June 2004.

Chapter 6

Applications of System Design

In this chapter, design examples of wireless mobile receiver and transmitter systems are introduced. These design examples are used to explain how to use the methods and formulas discussed in previous chapters for the receiver and transmitter designs and what the most important considerations in practical designs are. A complete design of a mobile communication transceiver is also involved. Here only the critical parameters — such as sensitivity, intermodulation spurious attenuation, and adjacent channel selectivity in the receiver side, and modulation accuracy, ACPR, and noise/spurious emissions in the transmitter side — are discussed since they basically characterize the RF performance of a mobile transceiver. The design targets of these parameters are specified, and how to achieve these targets in design are described in detail.

Two transceiver system-design examples are presented in following sections. A multimode and multiband superheterodyne transceiver design, which actually covers the designs of GSM, TDMA, AMPS and GPRS (general package radio system) mobile systems, is first discussed. Then a CDMA transceiver is designed based on the direct conversion architecture as the second example.

6.1. Multimode and Multiband Superheterodyne Transceiver

In this section, the RF system design of a mobile RF transceiver — which can operate in the GSM (GPRS), TDMA, and AMPS systems and in the 800 MHz cellular and 1900 MHz PCS dual bands — is described in detail. At present, some GSM mobile stations not only have GPRS function but also support EDGE (enhanced data rates for global evolution). In this example, the EDGE is not discussed. A superheterodyne architecture is chosen in this design, and the block diagram of the superheterodyne multimode and multiband mobile transceiver is shown in Fig. 6.1. This is a commonly used configuration of a superheterodyne mobile transceiver. The whole transceiver is built up on

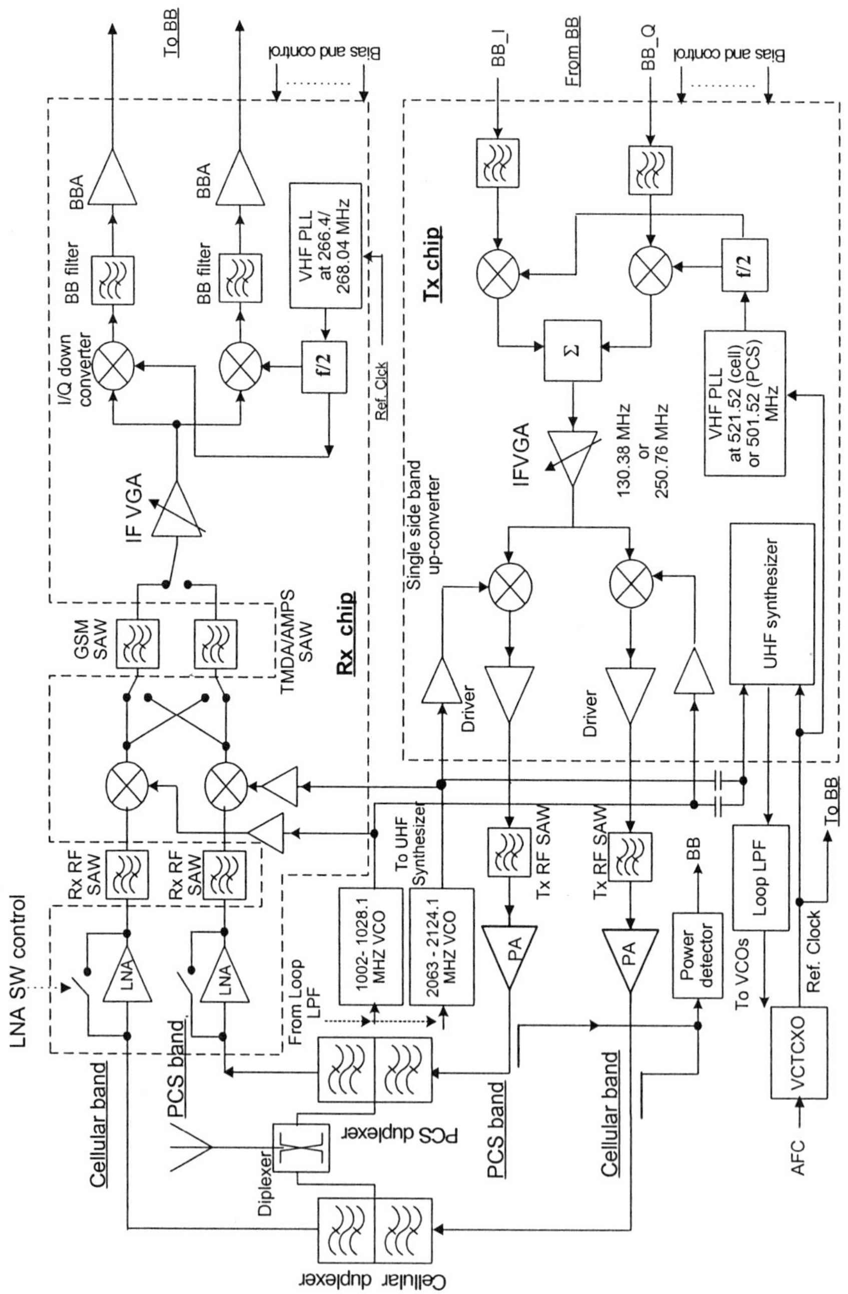

Figure 6.1. Block diagram of multimode and multiband transceiver

one receiver IC, one transmitter IC, two voltage control oscillators, two power amplifiers plus duplexers, RF and IF SAW filters.

The receiver IC consists of two sets of LNAs and down-converters for 800 MHz Cellular frequency band and 1900 MHz PCS frequency band operation, respectively, and shared blocks including an IF VGA, a quadrature demodulator (the second down-converter), and BB low-pass filters and BB amplifiers in the I and Q channels. In the transmitter IC, the constitution is similar to that of the receiver. It has shared BB I and Q channels, quadrature modulator, and IF VGA, and separated RF sections, which comprise an up-converter, an RF VGA, and a driver amplifier, for each frequency band. In addition, the frequency synthesizer running in both frequency bands is also built in the transmitter IC. The receiver and transmitter ICs can be designed and implemented by means of the SiGe/BiCMOS and CMOS technology.

In this dual band transceiver, the RF stages (including these on the IC chips and off chip devices such as PA, duplexer and RF SAW BPFs used in the receiver, and the transmitter) are separated for 800 MHz band and for 1900 MHz band. The only shared part in the RF is the diplexer, which identifies the cellular band and the PCS band signals and forwards them to the corresponding receiver from the antenna port or sends them to the antenna port from the corresponding transmitter. On the other hand, this is also a quad-mode transceiver, and it needs two different IF SAW channel filters to handle the bandwidth 200 kHz GSM and GPRS signals and the bandwidth 25 kHz TDMA and AMPS signals, respectively. The bandwidth of the low-pass filters in the BB I and Q channels of the receiver IC must be tunable according to the operation mode.

6.1.1. Selection of a Frequency Plan

This multimode transceiver operates as a half-duplex system in GSM, GPRS, and TDMA mode and as a full-duplex system only in AMPS mode. However, AMPS transceiver is running in 800 MHz cellular band. From Table 3.1, the overall frequency band span of the cellular band is 70 MHz including receiver and transmitter operation bands (25 MHz each) and a band separation (20 MHz). From the frequency planning discussion of superheterodyne transceiver in Section 3.1.2, we know that in this case the first IF should be greater than 70 MHz.

Generally speaking, designing a mobile transceiver for these commonly used wireless communication systems, we may utilize an existing and verified frequency plan except for some special reasons. In this case, commercial SAW or crystal channel filters are available, and usually there are multiple sources for designers to choose from. Thus we can save the filter development cost and obtain qualified channel filters at low cost. For example, the first IF is chosen equal to 133.2 MHz for GSM and GPRS receiver, and 134.04 MHz for the TDMA and AMPS receiver. Using slightly different IF for the GSM and for the TDMA/AMPS is due to the fact that having a common reference clock is used for synthesizing frequencies in two different channelization systems — i.e., the 200 kHz channel spacing in the GSM and the 30 kHz channel spacing for the TDMA/AMPS.

The 19.2 MHz reference clock or VCTCXO (voltage control temperature compensated crystal oscillator) frequency is popularly used in mobile stations. This multimode transceiver may employ this reference clock for all the mode operations.

If high LO injection approach — i.e., the LO frequency higher than the desired signal frequency — is used for the receiver UHF down-converter and for the transmitter UHF up-converter too, the tuning ranges of the UHF synthesizers for the 800 MHz cellular band and 1900 MHz PCS band operation are 1002 to 1029MHz and 2063 to 2125 MHz, respectively. Actually these two frequency tuning ranges can be implemented by means of a single VCO with a tuning range 2004 to 2125 MHz and a by 2 divider since the frequency range from 2004 to 2125 MHZ is about 6% of the VCO operating frequency, and this is a quite reasonable tuning range that still maintains good phase noise performance.

In a superheterodyne mobile transceiver, the UHF synthesizer LO is usually shared between its receiver and its transmitter. The transmitter IF thus is determined by the receiver IF and the spacing between the transmitter and receiver bands. In this example, the transmitter IFs are 178.2 MHz for GSM and GPRS operations and 179.04 MHz for TDMA and AMPS modes, respectively, in the 800 MHz frequency band. The transmitter intermediate frequencies in the 1900 MHz frequency band are 35 MHz higher than those in the 800 MHz frequency band transmitters — i.e., 213.2 MHz and 214.04 MHz for the GSM and TDMA transmitters, respectively.

The VHF VCOs either in the receiver or in the transmitter are running at twice the corresponding receiver IF or the transmitter IF frequency. Since we have two IFs either in the receiver or in the transmitter, the receiver VHF VCO operates at 266.4 MHz for GSM and 268.08 MHz for TDMA and AMPS, and the corresponding frequencies of the transmitter VHF VCO are 356.4 MHz and 358.08 MHz, respectively. All the frequencies discussed above are summarized in Table 6.1.

Table 6.1. Summary of possible frequencies used in this transceiver

	GSM and GPRS	TDMA and AMPS
VCTCXO (MHz)	19.2	19.2
Receiver IF (MHz)	133.2	134.04
Transmitter (MHz)	178.2	179.04
UHF VCO Tuning (MHz)	2004 – 2058	2063 – 2125
Receiver VHF VCO (MHZ)	266.4	268.08
Transmitter VHF VCO (MHz)	356.4	358.08

6.1.2. Receiver System Design

The designed transceiver in GSM, GPRS, or TDMA mode can operate in both the 800 MHz and the1900 MHz bands, but in AMPS mode it runs only in the 800 MHz band. The minimum performance requirements of the GSM, GPRS, TDMA, and AMPS mobile receivers are presented in Tables 6.2 to 6.4, respectively [1]–[3]. Our design goal is not only to make the receiver performance meeting these minimum specifications but also to have reasonable margins.

Table 6.2. Minimum performance specifications of a GSM mobile receiver

GSM & GPRS 800/1900	Specifications	Note
Frequency Band (MHz)	869–89 or 1930–1990	
Modulation	GSMK	
Symbol rate (kS/sec)	270.833	
Sensitivity (dBm)	< –102	RBER < 2%
800 MHz band GPRS sensitivity (packet data channel) (dBm)	< –100	BLER < 10%,
1900 MHz band GPRS sensitivity (packet data channel) (dBm)	< –102	BLER < 10%,
800 MHz dynamic range (dBm)	> –15	RBER < 0.1%
1900 MHz dynamic range (dBm)	> –23	RBER < 0.1%
Intermodulation spurious response Attenuation (dBm)	> –49	f1:+/-800 kHz offset (CW) f2:+/-1.6 MHz offset (Mod)
Adjacent channel selectivity (dBc)	> 9	+/-200 kHz offset, 2% BER

Table 6.2. (Cont.)

Atl. adjcent channel selectivity (dBc)	> 41	+/-400 kHz offset, 2% BER
Blocking characteristic (dBc)	> 49	+/-(600 kHz to 1.6 MHz) offset, 2% BER
Blocking characteristic (dBc)	> 66	+/-(1.6 MHz to 3 MHz) offset, 2% BER
800 MHz blocking characteristic (dBc)	> 76	> 3 MHz offset, 2% BER
1900MHz blocking characteristic (dBc)	> 73	> 3 MHz offset, 2% BER
800 MHz spurious emission (dBm/100 kHz)	< -79	In Receiver Band
1900 MHz spurious emission (dBm/100 kHz)	< -71	In Receiver Band
Spurious emission (dBm/100 kHz)	< -36	In Transmitter Band

Table 6.3. Minimum performance specifications of TDMA mobile receiver

TDMA 800/1900	Specifications	Note
Frequency band (MHz)	869–894 or 1930–1990	
Modulation	π/4 DQPSK	
Symbol rate (kS/sec)	24.3	
Sensitivity (dBm)	< -110	BER < 3%
Dynamic range (dBm)	> -25	BER < 3%
Intermodulation spurious response attenuation (dBc)	> 62	f1:+/-120 kHz offset (CW), f2:+/-240 kHz offset (CW)
Adjacent channel selectivity (dBc)	> 13	+/-30 kHz offset, 3% BER
Atl. Adj. channel selectivity (dBc)	> 42	+/-60 kHz offset, 3% BER
Spurious emission (dBm)	< -80	In receiver band

Table 6.4. Minimum performance specifications of AMPS mobile receiver

AMPS 800/1900	Specifications	Note
Frequency Band (MHz)	869 – 894	
Modulation	FM	
Noise Bandwidth (kHz)	~27	

Table 6.4. (Cont.)

Sensitivity (dBm)	< -116	SINAD = 12 dB
Dynamic range (dBm)	> -25	
Intermodulation spurious response attenuation (dBc)	> 65	f1:+/-60 kHz offset (CW) f2:+/-120 kHz offset (CW)
Intermodulation spurious response attenuation (dBc)	> 70	f1:+/-330 kHz offset (CW) f2:+/-660 kHz offset (CW)
Adjacent channel selectivity (dBc)	> 16	+/-200 kHz offset, 2% BER
Atl. Adjacent channel selectivity (dBc)	> 60	+/-400 kHz offset, 2% BER
Spurious emission (dBm)	< -80	In receiver band

6.1.2.1. Determination of Carrier-to-Noise Ratio for Receiver Performance Evaluation

In the receiver system design, we need to first determine the carrier-to-noise ratio for each operation mode at a specified BER. Starting with the GSM system, for the receiver sensitivity and other performance calculations, 2% RBER (residual bit error rate) is used as defined in the specifications (Table 6.2). It depends on the type of channels and propagation conditions that the RBER is close to BER in some channel and multi-path fading conditions, such as in propagation condition RA250, and it is better than BER, on the other hand, in other propagation conditions, such as in condition TU3 [1]. In general, it is safe to calculate the required CNR based on BER instead of RBER.

For the GSM speech channel TCH/FS class II, the *CNR* can be estimated as follows. From the BER versus E_b/N_o waterfall curve of a GMSK signal with a BT_b equal to 0.25 in Fig. 2.37 we estimate $E_b/N_o \cong 5$ dB for BER = 2% although the GSM signal has a BT_b equal to 0.3. The difference between BER versus E_b/N_o curves of the GMSK signals with BT_b = 0.3 and 0.25 is insignificant around 2% BER. In addition, the GSM signal BER performance may be degraded by the phase noise of LOs, the group delay distortion of IF and BB channel filters, and the imbalances of I and Q channel signal phase and magnitude. The required increments of the GSM signal E_b/N_o for a 2% BER due to the above impairments are presented in Table 6.5. Thus the total E_b/N_o for a 2% BER is approximately 5.6 dB. The corresponding CNR can be obtained by using (2.4.11), GSMK bit rate 271 kHz, and the receiver noise bandwidth 182 MHz as follows:

$$CNR_{GSM} = \frac{E_b}{N_o} + 10\log\frac{R_b}{BW} = 5.6 + 10\log\frac{271}{182} = 7.3 \text{ dB}.$$

In the following calculation of the GSM receiver performance, 8 dB CNR_{GSM} will be employed to reserve 0.7 dB margin. However, when the speech channel with AMR (adaptive multiple rate), the CNR for the same sensitivity of –102 dBm will be approximately 1.5 dB higher than that of the original speech channels — i.e., 9.5 dB.

Table 6.5. GSM signal E_b/N_o increment for 2% BER due to impairments

Item	Specification	E_b/N_o Degradation (dB)
Total integrated phase noise of two LOs	< -25 dBc	0.1
Group delay distortion of channel filters	< 2 μsec	0.4
I and Q imbalance in phase and magnitude	< 5° and < 0.5 dB	0.1

The requirement on the CNR in the worst case is the channel TCH/AFS5.9 in HT100 propagation condition. For a –102 dBm sensitivity 9.4 dB CNR is needed even without AMR.

In the GPRS case, the modulation scheme is still the GMSK as the GSM system. The corresponding CNR for a 10% *block error rate* (*BLER*) in packet data channels (PDCH) is in general approximately 8 dB as required by the GSM speech channels, but in the worst case, it may rise to close to 10 dB. However, the reference sensitivity of the GPRS in the worst case (PDCH/CS-4 in 1800 MHz band, for example) is relaxed to –100 dBm instead of –102 dBm.

The performance evaluation of the GSM and GPRS system RF receivers later on will be based on CNR_{min} equal to 8 dB, and the evaluation is concentrated mainly on the GSM since the results from the GSM evaluation will be also applicable to the GPRS. On the other hand, in the system design we should leave enough margin to cover the performance in the worst case including GSM speech channel with AMR.

In a similar way, we can determine the CNR for performance calculation of the TDMA receiver where the signal is π/4-DQPSK modulated. From Fig. 2.39, we can see that E_b/N_o for a 3% BER is approximately equal to 5 dB in the case of ISI free. Assume that the channel filters have a total group delay distortion 1.5 μsec, which causes 0.3 dB E_b/N_o increase to keep the 3% BER, and the other factors such as I and Q channel mismatching and the phase noise of the LOs raise E_b/N_o another 0.2 dB. Therefore, it needs total 5.5 dB to reach 3% BER. Using the same formula (2.4.11) and considering $R_b/BW = 2$ in the case of π/4-DQPSK modulation, we obtain CNR to be

$$CNR_{TDMA} \cong 5.5 + 10\log 2 = 8.5 \text{ dB}.$$

Similar to the GSM situation, we add 0.5 dB to the above CNR_{TDMA} value — i.e., $CNR_{TDMA} = 9$ dB — for performance evaluation of the TDMA mobile receiver.

The *SINAD*, which is defined by (2.4.113) in Section 2.4.5.2, is used in an analog AMPS system to measure the receiver sensitivity and other performance instead of BER. The *SINAD* value defined to measure AMPS receiver performance is 12 dB [3], and it can be converted to CNR as described in Section 2.4.5.2. Here, we are going to use $CNR_{AMPS} = 3.0$ dB — i.e., approximately 0.5 dB higher than the CNR value figured out in Fig. 2.41 — for evaluation of AMPS receiver performance.

6.1.2.2. Noise Figure

Here we discuss only the static sensitivity since unlike the other sensitivity measures, such as the sensitivity under multipath fading, it is mainly determined by the receiver noise figure. The receiver sensitivity is one of the most important specifications for the receiver. Usually we would like to have a 4 dB margin in the typical case and a 1.5 dB margin in the worst case.

The receiver static sensitivity is determined by the noise bandwidth, noise figure, and CNR. It can be calculated in terms of (4.2.4) presented in Section 4.2.1; the same formula can also be used to calculate the noise figure from a defined sensitivity level by knowing the noise bandwidth and the CNR value, which determines the sensitivity. Considering 4 dB margin, the sensitivities for the GSM, TDMA, and AMPS mobile receivers are –106, –116, and –120 dBm, respectively, and the corresponding noise figures are

$$NF_{GSM} = 174 - 106 - 10\log\left(182 \times 10^3\right) - 8 \cong 7.4 \text{ dB}$$
$$NF_{TDMA} = 174 - 114 - 10\log\left(27 \times 10^3\right) - 9 \cong 6.7 \text{ dB}$$

and

$$NF_{AMPS} = 174 - 120 - 10\log(27 \times 10^3) - 3 \cong 6.7 \text{ dB}.$$

In this typical case the noise figure of this multimode receiver shall be 6.7 dB or lower. The overall receiver noise figure is determined mainly by the noise figure and gain of the RF block from the antenna port to the down-converter output, and this block is shared between GSM and TDMA or AMPS. To ensure high enough sensitivity for all the operation modes, the lowest noise figure need be selected. The maximum receiver noise figure should be 9.2 dB or less, and thus the receiver sensitivity still has 1.5 dB margin in the worst case.

6.1.2.3. Linearity and Third-Order Intercept Point

In the superheterodyne receiver, the linearity of a receiver is usually measured by means of the third-order input intercept point or IIP_3 in short. The linearity requirement is more complicated to determine than the receiver noise figure. The requirement on the overall IIP_3 of a wireless mobile receiver is dominated by the allowed intermodulation distortion (IMD) or formally referred to as intermodulation spurious attenuation and the phase noise of UHF synthesizer LO.

The IMD performance requirements for different system receivers are defined in Table 6.2 to Table 6.4. They are, respectively,

GSM: desired signal –99 dBm, minimum interferers -49 dBm;
TDMA: desired signal –107 dBm, minimum interferers 62 dBc higher; and
AMPS: desired signal sensitivity level + 3 dB, minimum close-spaced interferers 65 dBc higher, and minimum wide spaced interferers 70 dBc higher.

We can also notice that the tone (CW) and modulated interferers for the intermodulation performance test are set at different offset frequencies from the signal carrier and with different frequency separation for GSM, TDMA and AMPS mobile receivers. The offset frequencies of intermodulation test interferers are as follows:

GSM: tone/modulated interferer, offset frequency ±800/±1600 kHz;
TDMA: tone/tone interferer, offset frequency ±120/±240 kHz; and
AMPS: close-spaced tone/tone interferer, offset ±120/±240 kHz, and wide-spaced tone/tone interferer, offset ±330/±660 kHz.

If ignoring other factor influence to the intermodulation performance, the third-order IMD product level, IM_3, is directly related to the receiver linearity or IIP_3 by (4.3.45), or

$$IIP_{3,\min} = S_{d,i} + \frac{1}{2}\left[3\left(I_{in,\min} - S_{d,i}\right) + CNR_{\min}\right]. \tag{4.3.45}$$

Using the IMD specifications data discussed above, we obtain the minimum receiver IIP_3 requirements for different mode operations to be

$$IIP_{3,\min|GSM} = -99 + \frac{1}{2}\left[3 \times 50 + 8\right] = -20 \text{ dBm}$$

$$IIP_{3,\min|TDMA} = -107 + \frac{1}{2}\left[3 \times 62 + 9\right] = -9.5 \text{ dBm}$$

and

$$IIP_{3,\min|AMPS} = -117 + \frac{1}{2}\left[3 \times 65 + 3\right] = -18 \text{ dBm for close-spaced}$$

$$IIP_{3,\min|AMPS} = -117 + \frac{1}{2}\left[3 \times 70 + 3\right] = -10.5 \text{ dBm for wide-spaced.}$$

In the IIP_3 estimation of the AMPS receiver, -120 dBm receiver sensitivity is assumed.

The above estimated IIP_3 is actually the minimum requirement of the receiver third-order input intercept point resulting from only considering the intermodulation interferers. In reality, the phase noise and spurious of the synthesizer LOs and the receiver noise figure will also impact the IMD performance as shown in (4.3.51). The noise figure has been determined by the receiver sensitivity, and it is a given value here. The LO phase noise and spurious level, especially at the offset frequencies equal to those frequencies where the intermodulation test interferers are located, should be low enough to ensure that the requested receiver IIP_3 for certain IMD performance is reasonable and feasible. A lower IIP_3 usually needs less current consumption. The VHF LO phase noise and spurious generally have a negligible impact on the IMD performance if the IF channel filter has good rejection to the interferers. This statement is indirectly proved by (4.5.4) and (4.5.5). The contribution of the VHF LO phase noise and spurious to the degradation of the IMD performance will be insignificant if the ΔR_{IF} in these formulas is large. In the following IIP_3 evaluation, we consider only the UFH LO influence to the IMD performance.

Indeed, the UHF LO phase noise and spurious not only affect the intermodulation spurious response attenuation performance but also determine the adjacent/alternate channel selectivity. Based the present synthesizer technology, in the typical the phase noise and spurious level of a synthesizer LO for Cellular and PCS band mobile receiver use are presented in Table 6.6. The phase noise may vary a couple of dB with temperature from room to hot (60°C) or to cold (-30°C). The UHF LO phase-noise and spurious requirements are also related to the selectivity of the IF channel filters, and the phase noise and spurious specifications will be looser if the rejection of the channel filters to interferers is higher.

In different systems, the demand for UHF LO phase noise and spurious levels to meet IMD and adjacent channel selectivity performance is different. The phase noise and spurious performance of the UHF LOs presented in Table 6.6 are derived based on the IMD and adjacent channel selectivity performance of the TDMA and AMPS receiver. In order to obtain a 3 dB margin on the IMD performance, the phase noise and spurious requirements of an UHF synthesizer LO for a GSM receiver can be less stringent than those given in Table 6.6 and they are presented in Table 6.7. This means that the UHF VCO in the GSM mode of operation can use less

Table 6.6 Phase noise and spurious specifications of UHF LOs

	Cellular Band Synthesizer		PCS Band Synthesizer	
Frequecy Offset (kHz)	Phase Noise dBc/Hz	Spurs (dBc)	Phase Noise dBc/Hz	Spurs (dBc)
± 30 kHz	≤ -105	≤ -60	≤-103	≤-60
± 60 kHz	≤ -117	≤ -85	≤-114	≤-85
±120 kHz	≤ -125	≤ -90	≤-122	≤-90
±240 kHz	≤ -131	≤ -95	≤-128	≤-95
±330 kHz	≤ -134	≤ -95	≤-131	≤-95
±660 kHz	≤ -140	≤ -95	≤-137	≤-95
±3000 kHz	≤ -144	≤ -95	≤-142	≤-95

Table 6.7. Phase noise and spurious requirements of UHF LO for GSM receiver

	800 MHz Band		1900 MHz Band	
Frequecy Offset (kHz)	Phase Noise dBc/Hz	Spurs (dBc)	Phase Noise dBc/Hz	Spurs (dBc)
±200 kHz	≤ -118	≤ -60	≤ -114	≤ -60
±400 kHz	≤ -124	≤ -65	≤ -120	≤ -65
±600 kHz	≤ -127	≤ -70	≤ -123	≤ -70
±800 kHz	≤ -130	≤ -78	≤ -126	≤ -78
±1600 kHz	≤ -136	≤ -85	≤ -132	≤ -85
> |3200| kHz	≤ -141	≤ -90	≤ -137	≤ -90

current than when operating in the TDMA or AMPS mode. If current saving is important in our design, we can let UHF VCO run with a different bias current for the GSM mode, or otherwise the UHF LO of the multi-mode receiver should have its phase noise and spurious performance based on the TDMA/AMPS mode requirements — i.e., the performance given in Table 6.6.

Using Tables 6.6 and 6.7 and formula (4.3.51), we can estimate the receiver IIP_3 for achieving the IMD performance with a 3 dB margin over minimum specifications defined in Tables 6.2 to 6.4. For example, the IIP_3 of 800 MHz band GSM receiver is calculated by using phase noise in Table 6.7 as

$$IIP_{3,GSM\_800} = \frac{1}{2}\left[3(-49+3) - 10\log\left(10^{\frac{-99-8}{10}} - 10^{\frac{-174+7.4+10\log 182\cdot 10^3}{10}} - 10^{\frac{-130+10\log 182\cdot 10^3-49+3}{10}} - 10^{\frac{-136+10\log 182\cdot 10^3-49+3}{10}} - 10^{\frac{-78-49+3}{10}} - 10^{\frac{-85-49+3}{10}}\right)\right] = -15 \text{ dBm},$$

and the required IIP_3 of 1900 MHz band GSM receiver is

$$IIP_{3,GSM\_1900} = -14.8 \text{ dBm}.$$

In a similar way, we can obtain IIP_3 for other modes and bands based on Table 6.6 phase noise and spurious as follows.

$$IIP_{3,TDMA\_800} = -3.4 \text{ dBm}, \quad IIP_{3,TDMA\_1900} = -2.8 \text{ dBm};$$

and

$$IIP_{3,AMPS\_close} = -12.6 \text{ dBm}, \quad IIP_{3,AMPS\_wide} = -9.3 \text{ dBm}.$$

From these results we can conclude that the TDMA receiver requires the highest linearity among these three receiver operation modes, and the receiver operating in the 1900 MHz band needs higher IIP_3 than when it is running in the 800 MHz band since the phase noise of the PCS band LO is worse than that of the cellular band LO. The linearity design of the receiver common path for different modes should be based on the TDMA requirement, but in the circuit design we should also consider adjustable bias circuitry to change the device bias based on operation modes to save the current consumption.

6.1.2.4. Selectivity and Blocking Performance

The receiver selectivity and blocking performance are mainly determined by channel filters and LO phase noise and spurious. The LO phase noise/spurious requirements is also partially determined by the IMD performance as described in the previous section, and therefore we have already had a basic idea what level phase noise/spurious can be used in our receiver system design. The channel filter characteristics affect not only the receiver selectivity and blocking performance but also the IMD performance since no matter what the adjacent/alternate channel interferers are, distance blocking signals, or intermodulation interference tones/modulated signals will be significantly attenuated when they pass through the channel filters. The channel filtering performance will indirectly impact the current consumption of the receiver too. The way to specify characteristics of channel filters is similar to defining the LO phase noise requirements. It is a trial and error procedure to make a tradeoff between filtering requirements and feasibility of implementation. Examples of channel filter characteristics for the GSM receiver and for the TDMA or the AMPS receiver are presented in Table 6.8.

Table 6.8. Rejection characteristics of channel filters for GSM, TDMA, and AMPS receiver

GSM Channel Filter			TDMA/AMPS Channel Filter		
	Insertion Loss (dB)			Insertion Loss (dB)	
	Typical	Worst		Typical	Worst
In-band	≤ 4.5	5.5	In-band	≤ 3.5	4.5
	Rejection (dB)			Rejection (dB)	
Offset frequency	Typical	Worst	Offset frequency	Typical	Worst
± 200 kHz	≥ 4	0	± 30 kHz	≥ 3	0
± 400 kHz	≥ 17	12	± 60 kHz	≥ 24	20
± 600 kHz	≥ 27	22	± 120 kHz	≥ 40	35
± 800 kHz	≥ 31	25	± 240 kHz	≥ 50	40
± 1600 kHz	≥ 40	30	± 330 kHz	≥ 45	40
± 3000 kHz	≥ 40	30	± 660 kHz	≥ 43	35

Utilizing (4.5.8) and Tables 6.6, 6.7, and Table 6.8, we are able to calculate selectivity and blocking characteristics. Only in calculating adjacent channel selectivity the phase noise and spurious impacts of the VHF LO may not be able to be ignored since the channel filter rejection is generally quite low at the adjacent channel frequency, such as ±200 kHz offset from the desired signal carrier for GSM and ±30 kHz offset for AMPS. For examples, the GSM adjacent selectivity is evaluated as

$$\Delta S_{adj} = 10\log\left(\frac{10^{\frac{-99-8}{10}} - 10^{\frac{-174+10\log 182\cdot 10^3+7.4}{10}}}{10^{\frac{-118+10\log 182\cdot 10^3}{10}} + 10^{\frac{-108+10\log 182\cdot 10^3-4}{10}} + 10^{\frac{-60}{10}} + 10^{\frac{-55-4}{10}}}\right) + 99,$$
$$= 46.8 \text{ dB},$$

and the AMPS adjacent channel selectivity is calculated in a similar way:

$$\Delta S_{adj} = 10\log\left(\frac{10^{\frac{-117-3}{10}} - 10^{\frac{-174+10\log 27\cdot 10^3+6.7}{10}}}{10^{\frac{-105+10\log 27\cdot 10^3}{10}} + 10^{\frac{-95+10\log 27\cdot 10^3-3}{10}} + 10^{\frac{-60}{10}} + 10^{\frac{-55-3}{10}}}\right) + 120$$
$$= 45.1 \text{ dB},$$

In the above selectivity calculations, it is assumed that the VHF LO phase noise at the corresponding adjacent channel is 10 dB worse than the UHF LO phase noise and the spurious is 5 dB worse. The results show margins over 37 and 29 dB, respectively, for GSM and AMPS cases. The adjacent/alternate/blocking performance evaluation results of GSM, TDMA, and AMPS receivers in 800 and 1900 MHz bands are summarized in Tables 6.9 and 6.10. The adjacent or alternate channel selectivity has a decisive margin to its specification, and the minimum margin is 6 dB, which is the margin of the alternate channel selectivity performance of the AMPS receiver. The margin of the GSM blocking characteristic reduces with increasing the offset frequency since the defined magnitude of the blocking interferer is raised with the frequency offset but the phase noise of synthesizer LO drops to its floor level at offset frequency beyond 1.5 MHz. In this case, it is better to check the worst case situation if the corresponding selectivity is still within the specification. The most critical blocking performance is the 1900 GSM receiver with a LO phase noise given in Table 6.7. It can be found that the blocking still has more than 0.5 dB margin in the worst case — i.e., the LO phase noise increases approximately 2 to 3 dB from its typical value at 60°C.

Table 6.9. Adjacent/alternate/blocking performance of GSM receiver

	800 MHz Band				1900 MHz Band			
GSM mobile receiver	Table 6.7 LO	Margin	Table 6.6 LO	Margin	Table 6.7 LO	Margin	Table 6.6 LO	Margin
Adjacent channel (dBc)	45.8	36.8	49.4	40.4	45.7	36.7	50.3	41.3
Alternate channel (dBc)	54.0	13.0	56.2	15.2	54.0	13.0	55.9	14.9
Block 0.6-1.6 MHz (dB)	59.9	3.9	61.1	5.1	58.3	2.3	60.9	4.9
Block 1.6-3.0 MHz (dB)	72.4	6.4	74.9	8.9	69.5	3.5	74.2	8.2
Blocking > 3MHz (dB)	79.4	3.4	82.0	6.0	75.4	2.4	80.1	4.1

Table 6.10. Adjacent/alternate performance of TDMA and AMPS receivers

	800 MHz Band		1900 MHz Band	
TDMA mobile receiver	Table 6.6 LO	Margin	Table 6.6 LO	Margin
Adjacent channel (dBc)	41.2	28.2	40.9	27.9
Alternate channel (dBc)	62.4	20.4	59.6	17.6
AMPS mobile receiver				
Adjacent channel (dBc)	45.1	29.1	NA	NA
Alternate channel (dBc)	66.2	6.2	NA	NA

Another blocking characteristic in the GSM system is referred to as AM suppression characteristics. The interferer is a GSM modulated burst signal with a level of –31 dBm, offset from the desired signal carrier by 6 MHz or higher in an integer multiple of 200 kHz. The burst interferer may block a direct conversion receiver due to the second-order distortion of the direct conversion receiver creating this problem, but it will not cause any trouble to a superheterodyne receiver. Therefore, we are not going to further discuss this specification here.

6.1.2.5. ADC Dynamic Range

The analog-to-digital converters (ADC) in the multi-mode receiver are shared among GSM, TDMA, and AMPS modes. In this case, the ADC dynamic range is determined by the operation mode, which must accommodate high CNR, fading margin, and AGC error. Estimations of the ADC dynamic range for the GSM and TDMA are listed in Table 6.11. The GSM mobile receiver with GPRS function needs a dynamic range approximate 60 dB or a 10 bit ADC, and the TDMA receiver requires a 9.5

bit ADC. Actually, the dynamic range of the ADC for the TDMA/AMPS receiver can be much higher dynamic range than needed since the ADCs in the I and Q channels are shared with the GSM receiver. On the other hand, the ADC for the TDMA/AMPS receiver can run at a sampling rate much higher than the signal bandwidth, but it is still lower than the sampling rate used for the GSM receiver ADC. This means that the dynamic range of the same ADC applied in the TDMA/AMPS receiver will provide even higher than 60 dB effective dynamic range.

Table 6.11. ADC dynamic range estimations

GSM		TDMA	
CNR (GPRS-TU50/CS-4)	27 dB	CNR	10 dB
Fading margin	12 dB	Fading margin	24 dB
Quantization noise floor	12 dB	Quantization noise floor	12 dB
Crest Factor	0 dB	Crest factor	3 dB
AGC Error	6 dB	AGC error	6 dB
DC offset	3 dB	DC offset	3 dB
Total	60 dB	Total	58 dB

A ΣΔ ADC has the maximum signal-to-(noise+distortion) ratio, $S/(N+D)$, at approximate 6 dB back-off from the maximum peak-to-peak voltage swing. For example, if the allowed maximum peak-to-peak voltage swing is 1.5 V, the ΣΔ ADC has the maximum $S/(N+D)$ when the signal peak-to-peak voltage swing is close to 0.75 V. Considering some head room for constructive fading, positive AGC error, and crest factor, the rms voltage applied to the ADC input will be between 75 mV and 150 mV determined by the AGC optimization.

6.1.2.6. System Line-Up Analysis and Design

We have the multimode superheterodyne receiver block diagram as shown in Fig. 6.1, and we have derived the receiver noise figure, IIP_3, UHF synthesizer phase noise performance and channel filter selectivity for achieving electrical performance of the receiver in different operation modes. Now, we should properly assign the gain, noise figure, and IIP_3 of individual stages to make the composite noise figure and IIP_3 equal to the results obtained in Section 6.1.2.2 to 6.1.2.4 or even better. An efficient tool to do this kind of analysis and design is the excel spreadsheet. To create a proper receiver system lineup — i.e., gain, noise figure and IIP_3 distribution in the receiver chain — we need have certain basic knowledge of the performance of key blocks or stages, such as band-pass filters, LNA, down-converter, IF VGA, I/Q demodulator, BB filter, and BBA, etc.

The wireless mobile station is basically built on silicon or more precisely on one or more IC chips or a chip. Based on our knowledge of possible performance of different function integrated circuits, off chip RF band-pass filters including the duplexer, and UHF synthesizer LO, we are able to layout a preliminary lineup of the designed receiver. Utilizing the calculation capability of the Excel spreadsheet and formulas (4.2.12)/(4.2.23) and (4.3.34), the cascaded noise figure and the cascaded IIP_3 of the receiver can be calculated. The gain, noise figure, and IIP_3 distribution in the receiver chain needs be adjusted until the overall noise figure and IIP_3 are at least equal to the values determined in Section 6.1.2.2 to 6.1.2.3 or even better. It is apparent that the receiver system lineup to meet the requirements is not unique. The system lineups for different mode receivers are presented in Table 6.12 to Table 6.16*. The detail gain, noise figure and IIP_3 distributions of 800 MHz GSM, TDMA, and AMPS receivers, and 1900 MHz GSM and TDMA receivers are described in these tables, respectively.

In these receiver operation modes and bands, some blocks and/or devices are shared between different modes and bands:

1. The whole block from IF VGA and the I/Q quadrature demodulator to the analog BBA in the I and Q channels is shared by all the operation modes and bands, but the bandwidth of the BB LPF should be adjustable to approximate 100 kHz for GSM and GPRS operation or to around 15 kHz for TDMA and AMPS operation.
2. The GSM receiver including the GPRS mode operating either in Cellular band or PCS band uses the same SAW filter with a characteristics as shown in Table 6.8 for their channel filtering. The 800 MHz and 1900 MHz TDMA receivers and the 800 MHz AMPS receiver share a 27 kHz channel filter, which has a frequency response also presented in Table 6.8.
3. In this multimode receiver architecture, there are two sets of RF blocks (one for each frequency band), and each set consists of a duplexer, LNA, RF BPF, and RF down-converter. The 800 MHz RF block is common for the GSM, TDMA, and AMPS receiver modes operating in this band, and another RF block is for the GSM and TDMA receivers running in the 1900 MHz band.

---

* In these tables, the performance results, such as sensitivity, etc., are calculated in terms of the corresponding formulas given in Chapter 4 instead of the simplified formulas presented in these tables, which are just symbolic expressions.

4. Similar to the RF blocks, there exist two UHF synthesizer LOs operating at 1 GHz band and 2GHz band in this multi-mode transceiver. Each of them is not only used for all the receiver modes working in the corresponding frequency band but also shared between the receiver and the transmitter.
5. By looking at Tables 6.12 to 6.16 carefully, we find that the required linearity or IIP_3 of the LNA and the RF down-converter for the GSM mode operation is much lower than that of the TDMA and AMPS modes in the each frequency band. On the other hand, the GSM receiver in both bands has looser specifications on the phase noise of the synthesizer LOs than the other modes. This means that we may be able to change the bias current of the LNA, RF down-converter, and the UHF VCO based on the operation mode to save the current consumption in the GSM mode operation since different modes share the RF block and the UHF synthesizer LO. These shared blocks and devices should otherwise operate at relatively high current to ensure that the LNA and RF down-converter have a high linearity (or IIP_3) and the UHF LO has a low phase noise as demanded by the TDMA mode for all the operation modes.
6. We know from Section 6.1.1 that the IF of the GSM receiver is slightly different from that of the TDMA and the AMPS receivers. This results from different channelization between the GSM system and the TDMA / AMPS system, but the UHF synthesizers in the GSM and TDMA/AMPS receivers use the same 19.2 MHz reference clock (VCTCXO).
7. The receiver system lineup analyses in Tables 6.12 to 6.16 present the receiver performance in typical case. In practical system design, we must do the worst-case analysis as well make sure the receiver performance is still within the specification.
8. From the typical- and worst-case system lineup analyses, we are able to develop the performance specifications of individual stages or blocks in the receiver chain. For the amplifiers and down-converters, values of gain, noise figure, IIP_3 and current consumption must be specified not only regarding their nominal values but also their tolerances under different conditions, such as over certain temperature and applied voltage ranges. For the filters, the main specifications are the insertion loss, in-band ripple, out-of-band rejection, and group delay and its distortion. It is apparent that the phase noise, spurious level, acquisition time, and current consumption are main concerns for the synthesizers.

Table 6.12. 800 MHz GSM receiver lineup analysis and typical performance evaluation

		Diplxr	Duplxr		RF BPF	MIXER	IF BPF	IF VGA	I/Q Demd							Total
Current mA																25.5
Power Gain dB		-0.4	-2.6	15.0	-2.5											
Voltage Gain dB		-0.4	-2.6	15.0	5.5	8.0	-4.5	27.6	30.0	0.0	20.0	0.0				98.60
Filter	@ Tx band		100		20	@ 200 kH	4.0			5.0						
Rejection(dB)	@ Rx band		100			400 kH	17.0			35.0						
						600 kHz	27.0			45.0						
						800 kHz	31.0			50.0						
						1.6 MHz	40.0			50.0						
						3.2 MHz	39.0			50.0						
'Pout dBm	-107.0	-107.4	-110.0	-95.0	-97.5											
Vout mV	0.0010	0.0010	0.0007	0.0040	0.0075	0.0188	0.0112	0.2688	8.5000	8.5000	85	85				
NF (dB)	0.0	0.4	2.6	1.5	2.5	8.0	4.5	8.5	20.0	50.0	45.0	0.0				
Cascaded NF dB	**5.82**	5.82	5.42	2.82	12.27	9.77	13.24	8.74	23.65	51.19	45.00	0.00				
NF Contribution	100.0%	28.80%	23.61%	21.62%	1.29%	15.65%	0.85%	8.00%	0.07%	0.07%	0.02%	0.00%				
IIP3 (dBm)	100.00	100.00	100.00	-3.00	100.00	0.00	100.00	-24.00	-10.00	31.60	16.60	100.00				
Cascaded IIP3 dBm	**-11.11**	-11.11	-11.51	-14.11	1.25	-1.25	12.75	-42.75	-15.09	31.60	16.60	96.99				
IIP3 Contribution	100.0%	0.00%	0.00%	7.75%	0.00%	69.10%	0.00%	0.31%	22.35%	0.49%	0.00%	0.00%				
Receiver Digital BB Perfomace, Test Conditions, and Synthesizer LO Phase Noise and Spurious																
Rx :	Rx BER =	2.0%	UHF PLL: Sp200kHz	-60	Sp600kHz	-70	Sp1.6MH	-85	dBc	VHF PLL: Sp200kHz =	-60	Sp600kHz	-100	Sp1.6MH	-100	dBc
	CNR(dB)	8.00	Nphs200=	-118	Nphs600=	-127	Nphs160(	-136	dBc/Hz	Nphs200=	-108	Nphs600=	-116	Nphs1.6M	-122	dBc/Hz
	BW(Hz)=	182000	Sp,400kHz =	-65	Sp,800kH	-78	Sp3Mhz=	-100	dBc	Sp,400kHz =	-90	Sp800kHz	-100	Sp3MHz=	-100	dBc
	Sd(dBm)	-99	Nphs400=	-124	Nphs800=	-130	Nphs3Mh	-141	dBc/Hz	Nphs400=	-112	Nphs800=	-117	Nphs3M=	-128	dBc/Hz

Receiver Main Performance Evaluation

				Margin (dBc)	Spec	
1. Rx Sensitivity						
MDS =	-174 + 10 log(BW) + CNR + NF =	**-107.6**	dBm	5.6	-102.0	dBm
2. Intermodulation Rejection						
IMD, =	[2*IIP3 + Sd - CNR]/3 -Sd =	**55.6**	dBc	5.6	50.0	dBc
3. Adj. Ch. Selectivity						
adj,200	10log(10^(Sd-CNR)/10-Pnf)/(Pn,spu+Pn,phs))-Sd =	**46.8**	dBc	37.8	9.0	dBc
adj,400	10log(10^(Sd-CNR)/10-Pnf)/(Pn,spu+Pn,phs))-Sd =	**55.2**	dBc	14.2	41.0	dBc
4. Blocking Characteristics						
Blk,0.6-1.6 MHz	10log(10^(Sd-CNR)/10-Pnf)/(Pn,spu+Pn,phs))-Sd =	**59.9**	dBc	3.9	56.0	dBc
Blk,1.6-3MHz	10log(10^(Sd-CNR)/10-Pnf)/(Pn,spu+Pn,phs))-Sd =	**72.5**	dBc	6.5	66.0	dBc
Blk,> 3MHz	10log(10^(Sd-CNR)/10-Pnf)/(Pn,spu+Pn,phs))-Sd =	**79.4**	dBc	3.4	76.0	dBc

Table 6.13. 800 MHz TDMA receiver lineup analysis and typical performance evaluation

		Diplxr	Duplxr	LNA	RF BPF	MIXER	IF BPF	IF VGA		LPF	BBA	ADC				Total
Current mA					4.0	12.0		8.0	5.0	1.5	1.0					31.5
Power Gain dB		-0.4	-2.6	15	-2.5											
Voltage Gain dB		-0.4	-2.6	15	5.5	8.0	-3.5	23.6	30.0	0.0	30.0					105.60
Filter	@ Tx band		100							24						
Rejection(dB)	Rx Band		100			60 kHz	24			40						
						120 kHz	40			50						
						240 kHz	50			60						
						330 kHz	45			50						
						660 kHz	43			50						
Pout dBm	-114	-114.4	-117	-102	-104.5											
Vout mV	0.0004	0.0004	0.0003	0.0018	0.0033	0.0084	0.0056	0.0850	2.6879	2.5879	85	85				
NF (dB)	0.0	0.4	2.6	1.5	2.5	10.0	3.5	8.5	20.0	50.0	45.0	0.0				
Cascaded NF dB	**6.21**	6.21	5.81	3.21	13.54	11.04	12.58	9.08	23.65	5[illegible].19	45.00	0.00				
NF Contribution	100.0%	26.35%	21.60%	19.78%	1.18%	24.26%	0.53%	5.82%	0.21%	0.21%	0.07%	0.00%				
IIP3 (dBm)	100.00	100.00	100.00	3.00	100.00	7.00	100.00	-24.00	-15.00	31.60	16.60	100.00				
Cascaded IIP3 dBm	**-3.20**	-3.20	-3.60	-6.20	9.35	6.85	29.66	-38.84	-15.10	31.54	-30.00	0.00				
IIP3 Contribution	100.0%	0.00%	0.00%	12.02%	0.00%	85.07%	0.00%	0.10%	2.75%	0.06%	0.00%	0.00%				
Receiver Digital BB Perfomace, Test Conditions, and Synthesizer LO Phase Noise and Spuri																
Rx :	Rx BER =	3.0%	UHF PLL	-60	Sp120kHz	-90	Sp330kHz=	-90	dBc	VHF PLL: Sp30kHz =	-60	Sp120kHz	-75	Sp330kHz=	-90	dBc
		9.00	Nphs30=	-105		-125		-134	dBc/Hz	Nphs30=	-94		-120	Nphs330kHz	-130	dBc/Hz
	BW(Hz)=	27000	Sp,60kHz =	-85	Sp,240kHz	-90	Sp660kHz=	-100	dBc	Sp60kHz =	-65	Sp240kHz=	-85	Sp660kHz=	-90	dBc
	Sd(dBm) =	-107	Nphs60=	-117		-131	Nphs660kHz	-140	dBc/Hz	Nphs60=	-114		-126	Nphs660kHz	-136	dBc/Hz

Receiver Main Performance Evaluation

					Margin (dBc)	Spec	
1. Rx Sensitivity							
	MDS =	-174 + 10 log(BW) + CNR + NF =	**-114.5**	dBm	4.5	-110	dBm
2. Intermodulation Rejection							
	IMD, =	[2*IIP3 + Sd - CNR] /3 -Sd =	**65.2**	dBc	3.2	62	dBc
3. Adj. Ch. Selectivity							
	adj,30 =	10log(10^(Sd-CNR)/10-Pnf)/(Pn,spu+Pn,phs))-Sd	**41.3**	dBc	28.3	13.0	dBc
		10log(10^(Sd-CNR)/10-Pnf)/(Pn,spu+Pn,phs))-Sd	**62.5**	dBc	20.5	42	dBc
4. Blocking Characteristics							
	Blk,0.6-1.6 MHz	10log(10^(Sd-CNR)/10-Pnf)/(Pn,spu+Pn,phs))-Sd	**59.9**	dBc	3.9	56	dBc
	Blk,1.6-3MHz	10log(10^(Sd-CNR)/10-Pnf)/(Pn,spu+Pn,phs))-Sd	**72.5**	dBc	6.5	66	dBc
	Blk,> 3MHz	10log(10^(Sd-CNR)/10-Pnf)/(Pn,spu+Pn,phs))-Sd	**79.4**	dBc	3.4	76	dBc

Table 6.14. AMPS receiver lineup analysis and typical performance evaluation

		Diplxr	Duplxr	LNA	RF BPF	MIXER	IF BPF	IF VGA	I/Q Demo	LPF	BBA	ADC				Total
Current mA				4.0		6.0		8.0	5.0	1.5	1.0					25.5
Power Gain dB		-0.4	-2.6	15	-2.5											
Voltage Gain dB		-0.4	-2.6	15	5.5	8.0	-3.5	30.1	30.0	0.0	30.0	0.0				112.10
Filter	@ Tx band		50			@ 30 kHz				24						
Rejection(dB)	Rx Band		42			60 kHz	24			40						
						120 kHz	40			50						
						240 kHz	50			60						
						330 kHz	45			60						
						660 kHz	43			50						
Pout dBm	-120	-120.4	-123	-108	-110.5											
Vout mV	0.0002	0.0002	0.0002	0.0009	0.0017	0.0042	0.0028	0.0850	2.6879	2.6879	85	85				
NF (dB)	0.0	0.4	2.6	2.0	2.5	8.0	3.5	8.5	20.0	50.0	45.0	0.0				
Cascaded NF dB	**5.72**	5.72	5.32	2.72	11.92	9.42	12.14	8.64	23.65	51.19	45.00	0.00				
NF Contribution	100.0%	29.35%	24.06%	22.03%	1.31%	15.94%	0.59%	6.48%	0.10%	0.10%	0.03%	0.00%				
IIP3 (Close) dBm	100.00	100.00	100.00	0.00	100.00	0.00	100.00	-24.40	-15.00	31.60	16.60	100.00				
Cascad IIP3 dBm	**-9.78**	-9.78	-10.18	-12.78	2.46	-0.04	28.27	-45.23	-15.09	31.60	16.60	96.99				
IIP3 Contritbution	100.0%	0.00%	0.00%	5.30%	0.00%	94.20%	0.00%	0.00%	0.48%	0.01%	0.00%	0.00%				
IIP3 (Wide) dBm	100.00	100.00	100.00	0.00	100.00	0.50	100.00	-24.40	-15.00	31.60	16.60	100.00				
Cascad IIP3 dBm	**-9.36**	-9.36	-9.76	-12.36	2.90	0.40	24.77	-45.23	-15.09	31.60	16.60	96.99				
IIP3 Contritbution	100.0%	0.00%	0.00%	5.81%	0.00%	92.02%	0.00%	0.02%	2.11%	0.05%	0.00%	0.00%				

Receiver Digital BB Perfomace, Test Conditions, and Synthesizer LO Phase Noise and Spurious

Rx :	Rx SINAD(dB)=	12.0	UHF PLL SP30kHz	-60	Sp120kHz	-90	Sp330kHz	-90	dBc	VHF PLL Sp30kHz	-60	Sp120kHz	-75	Sp330kHz	-90	dBc
	CNR(dB)=	3.00	Nphs30=	-105	Nphs120=	-125	Nphs330=	-134	dBc/Hz	Nphs30=	-94	Nphs120=	-120	Nphs330k	-130	dBc/Hz
	BW(Hz)=	27000	Sp,60kHz	-85	Sp,240kHz	-90	Sp660kHz	-100	dBc	Sp60kHz =	-65	Sp240kHz	-85	Sp660kHz	-90	dBc
	Sd(dBm) =	-99	Nphs60=	-117	Nphs240=	-131	Nphs660k	-140	dBc/Hz	Nphs60=	-114	Nphs240=	-126	Nphs660k	-136	dBc/Hz

Receiver Main Performance Evaluation

				Margin (dBc)	Spec	
1. Rx Sensitivity						
MDS =	-174 + 10 log(BW) + CNR + NF =	**-120.6**	dBm	4.6	-116	dBm
2. Intermodulation Rejection						
IMD, =	[2*IIP3 + Sd - CNR] /3 -Sd =	**68.8**	dBc	3.8	65	dBc
IMD,330kHz =	[2*IIP3 + Sd - CNR] /3 -Sd =	**73.3**	dBc	3.3	70	
3. Adj. Ch. Selectivity						
adj,30	10log(10^(Sd-CNR)/10-Pnf)/(Pn,spu+Pn,phs))-Sd =	**45.1**	dBc	29.1	16	dBc
adj,60	10log(10^(Sd-CNR)/10-Pnf)/(Pn,spu+Pn,phs))-Sd =	**66.2**	dBc	6.2	60	dBc

Table 6.15. 1900 MHz GSM receiver lineup analysis and typical performance evaluation

		Diplxr	Duplxr	LNA	RF BPF	MIXER	IF BPF	IF VGA	I/Q Demo	LPF	BBA	ADC				Total
Current mA				5.0		9.0		8.0	5.0	1.5	1.0	0				29.5
Power Gain dB	0	-0.4	-3	14.5	-2.5											
Voltage Gain dB	0	-0.4	-3	14.5	5.5	8.0	-4.5	27.5	30.0	0.0	20.0	0.0				97.60
Filter	@ Tx band		100		20	@ 200 kHz	4.0			5.0						
Rejection(dB)	@ Rx band		100			400 kHz	17.0			35.0						
						600 kHz	27.0			45.0						
						800 kHz	31.0			50.0						
						1.6 MHz	40.0			50.0						
						3.2 MHz	39.0			50.0						
'Pout dBm	-106.0	-106.4	-109.4	-94.9	-97.4											
Vout mV	0.0011	0.0011	0.0008	0.0040	0.0076	0.0190	0.0113	0.2688	8.5000	8.5000	85	85				
NF (dB)	0.0	0.4	3.0	1.5	2.5	10.0	4.5	8.5	20.0	50.0	46.0	0.0				
Cascaded NF dB	**6.84**	6.84	6.44	3.44	13.70	11.20	13.25	8.75	23.80	51.46	46.00	0.00				
NF Contribution	100.0%	22.68%	22.58%	18.67%	1.25%	25.70%	0.82%	7.75%	0.22%	0.23%	0.09%	0.00%				
IIP3 (dBm)	100.00	100.00	100.00	-3.00	100.00	0.00	100.00	-24.00	-15.00	31.60	16.60	100.00				
Cascaded IIP3 dBm	**-10.22**	-10.22	-10.62	-13.62	1.27	-1.23	12.85	-42.65	-15.09	31.60	16.60	100.00				
IIP3 Contribution	100.0%	0.00%	0.00%	8.66%	0.00%	68.80%	0.00%	0.31%	21.75%	0.48%	0.00%	0.00%				
Receiver Digital BB Perfomace, Test Conditions, and Synthesizer LO Phase Noise and Spurious																
Rx :	Rx BER =	2.0%	UHF PLL: Sp200kHz	-60	Sp600kHz =	-70	Sp1.6MHz=	-85	dBc	VHF PLL: Sp200kHz =	-80	Sp600kHz =	-100	Sp1.6MHz=	-100	dBc
	CNR(dB)=	8.00	Nphs200=	-114	Nphs600=	-123	Nphs1600=	-132	dBc/Hz	Nphs200=	-104	Nphs600=	-116	Nphs1.6MH	-122	dBc/Hz
	BW(Hz)=	182000	Sp,400kHz	-65	Sp,800kHz	-78	Sp3Mhz=	-100	dBc	Sp,400kHz	-90	Sp800kHz=	-100	Sp3MHz=	-100	dBc
	Sd(dBm) =	-99	Nphs400=	-120	Nphs800=	-126	Nphs3Mhz=	-137	dBc/Hz	Nphs400=	-110	Nphs800=	-117	Nphs3M=	-132	dBc/Hz

Receiver Main Performance Evaluation

				Margin (dBc)		Spec	
1. Rx Sensitivity							
MDS =	-174 + 10 log(BW) + CNR + NF =	**-106.6**	dBm	4.6	<	-102.0	dBm
2. Intermodulation Rejection							
$R_{IMD,}$ =	[2*IIP3 + Sd - CNR]/3 -Sd =	**55.9**	dBc	3.9	>	52.0	dBc
3. Adj. Ch. Selectivity							
$S_{adj,200}$ =	10log(10^(Sd-CNR)/10-Pnf)/(Pn,spu+Pn,phs))-Sd =	**44.5**	dBc	35.5	>	9.0	dBc
$S_{adj,400}$=	10log(10^(Sd-CNR)/10-Pnf)/(Pn,spu+Pn,phs))-Sd =	**53.9**	dBc	12.9	>	41.0	dBc
4. Blocking Characteristics							
$Blk_{,0.6\text{-}1.6\ MHz}$=	10log(10^(Sd-CNR)/10-Pnf)/(Pn,spu+Pn,phs))-Sd =	**58.3**	dBc	2.3	>	56.0	dBc
$Blk_{,1.6\text{-}3MHz}$=	10log(10^(Sd-CNR)/10-Pnf)/(Pn,spu+Pn,phs))-Sd =	**69.5**	dBc	3.5	>	66.0	dBc
$Blk_{,>3MHz}$=	10log(10^(Sd-CNR)/10-Pnf)/(Pn,spu+Pn,phs))-Sd =	**75.4**	dBc	2.4	>	73.0	dBc

Table 6.16. 1900 MHz TDMA receiver lineup analysis and typical performance evaluation

		Diplxr	Duplxr	LNA	RF BPF	MIXER	IF BPF	IF VGA	I/Q Demo	LPF	BBA	ADC				Total
Current mA				5.0		14.0		8.0	5.0	1.5	1.0					34.5
Power Gain dB		-0.4	-2.8	14.5	-2.5											
Voltage Gain dB		-0.4	-2.8	14.5	5.5	8.0	-3.5	24.3	30.0	0.0	30.0					105.60
Filter	@ Tx band		100			@ 30 kHz				24						
Rejection(dB)	Rx Band		100			60 kHz	24			40						
						120 kHz	40			50						
						240 kHz	50			60						
						330 kHz	45			50						
						660 kHz	43			50						
Pout dBm	-114	-114.4	-117.2	-102.7	-105.2											
Vout mV	0.0004	0.0004	0.0003	0.0016	0.0031	0.0078	0.0052	0.1191	3.7678	3.7678	150	150				
NF (dB)	0.0	0.4	2.8	1.5	2.5	10.0	3.5	8.5	20.0	50.0	45.0	0.0				
Cascaded NF dB	**6.58**	6.58	6.18	3.38	13.52	11.02	12.50	9.00	23.65	51.19	45.00	0.00				
NF Contribution	100.0%	24.12%	21.84%	18.96%	1.27%	26.10%	0.57%	6.26%	0.38%	0.38%	0.12%	0.00%				
IIP3 (dBm)	100.00	100.00	100.00	3.00	100.00	7.00	100.00	-24.00	-15.00	31.60	16.60	100.00				
Cascaded IIP3 dBm	**-2.59**	-2.59	-2.99	-5.79	9.33	6.83	28.98	-39.52	-15.09	31.60	16.60	100.00				
IIP3 Contribution	100.0%	0.00%	0.00%	13.22%	0.00%	83.44%	0.00%	0.09%	3.17%	0.07%	0.00%	0.00%				

Receiver Digital BB Perfomace, Test Conditions, and Synthesizer LO Phase Noise and Spurious

	Rx		UHF PLL							VHF						
Rx :	BER =	3.0%	SP30kHz=	-60	Sp120kHz	-90	Sp330kHz=	-95	dBc	PLL:	-60	Sp120kHz	-75	Sp330kHz=	-90	dBc
	CNR(dB)=	9.00	Nphs30=	-103	Nphs120=	-122	Nphs330=	-130	dBc/Hz	Nphs30=	-94	Nphs120=	-120	Nphs330kH	-130	dBc/Hz
	BW(Hz)=	27000	Sp,60kHz =	-85	Sp,240kHz	-95	Sp660kHz=	-100	dBc	Sp60kHz =	-65	Sp240kHz=	-85	Sp660kHz=	-90	dBc
	Sd(dBm) =	-107	Nphs60=	-114	Nphs240=	-128	Nphs660kh	-136	dBc/Hz	Nphs60=	-114	Nphs240=	-126	Nphs660kH	-136	dBc/Hz

Receiver Main Performance Evaluation

		Value	Unit	Margin (dBc)	Spec	Unit
1. Rx Sensitivity						
MDS =	-174 + 10 log(BW) + CNR + NF =	**-114.1**	dBm	4.1	-110	dBm
2. Intermodulation Rejection						
IMD, =	[2*IIP3 + Sd - CNR] /3 -Sd =	**64.7**	dBc	2.7	62	dBc
3. Adj. Ch. Selectivity						
adj,30 =	10log(10^(Sd-CNR)/10-Pnf)/(Pn,spu+Pn,phs))-Sd =	**40.9**	dBc	27.9	13.0	dBc
adj,60	10log(10^(Sd-CNR)/10-Pnf)/(Pn,spu+Pn,phs))-Sd =	**59.6**	dBc	17.6	42	dBc

6.1.2.6. Gain Control and RSSI Accuracy

The gain control range of the multi-mode receiver needs to cover the maximum dynamic range among these mode operations. From the minimum performance specifications of the GSM, TDMA, and AMPS mobile stations, Table 6.2 to Table 6.4, the 800 MHz GSM minimum dynamic range is 87 dB, the TDMA is 85 dB, and the AMPS is 91 dB. Assuming the margin of the receiver sensitivity plus dynamic range to be 10 dB, and gain change of the receiver chain over temperature and voltage to be 6dB, the overall gain control for the AMPS receiver needs approximately 108 dB. The LNA can be designed to obtain a 20 dB signal step gain control — say, from 15 dB to –5 dB or vice versa — and the rest of the 88 dB gain variation is implemented through the IF VGA (or VGA block). The adjustable gain stages in the receiver chain are listed in Table 6.17.

Table 6.17. Adjustable gain stages, adjustable ranges and approaches

	LNA	IF VGA	BBA
Adjustable range	20 dB (either 15 or –5 dB)	90 dB (+40 dB to –50 dB)	12 dB (either 32 or 20 dB)
Control Approach	Single step	Continuous or in 6 dB steps	GSM: 20 dB TDMA/AMPS: 32 dB

From Tables 6.12 to 6.17, we should notice that the BBA is set to a different gain in the GSM mode from the TDMA/AMPS mode — i.e., BBA gain equal to 20 dB in the GSM mode and equal to 30 dB in the TDMA/AMPS mode. The BBA gain in the GSM mode is set to 20 dB to make the CNR of the receiver high enough so that the receiver can still work properly after the LNA is switched to the low gain, -5 dB, operation. The LNA operates either at high gain, 15 dB, or in bypass mode with a –5 dB gain.

The LNA gain switching from high to low should take place as early as possible to save the current consumption. In this design example, the LNA gain in the GSM mode is switched down from high to low at –84 dBm, and switched back to high at –87 dBm, and in the TDMA/AMPS mode, the LNA two switching points are –92 dBm and –95 dBm, respectively, as shown in Fig. 6.2(a). Assuming that the IF VGA is continuously controlled, the gain variations versus received signal level for the GSM and TDMA/AMPS modes are depicted in Fig. 6.2(b). After calibration, the accuracy of the AGC in most of the dynamic region can be kept within ±3 dB and ±5 dB over temperature.

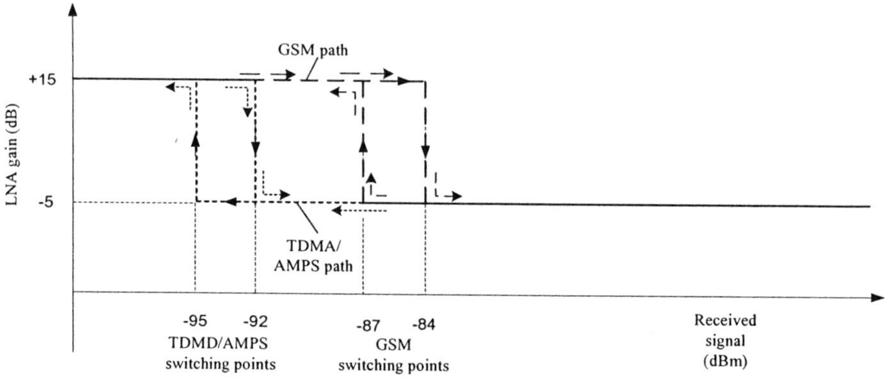

(a) LNA Gain Curve

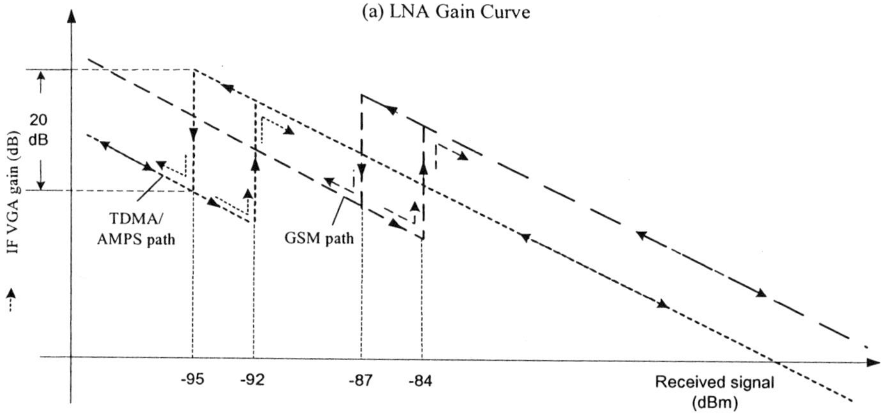

(b) IF VGA gain curve

Figure 6.2. AGC gain curves of LNA and IF VGA

The RSSI (received signal strength indicator) accuracy requirements for the GSM and the TDMA digital receivers are presented in Table 6.18.

Table 6.18. RSSI accuracy requirements of GSM and TDMA receivers

GSM RSSI		TDMA RSSI	
Measuring Range	-102 to -48 dBm	Measuring Range	-110 to -30 dBm
Range	Accuracy	Range	Accuracy
-102 to -70 dBm	± 4 dB	-110 to -85 dBm	±6 dB
-69 to -48 dBm	± 6 dB	-84 to -30 dBm	± 8 dB

6.1.3. Transmitter System Design

In the superheterodyne architecture transceiver, the UHF synthesizer LO for either the 800 MHz or the 1900 MHz frequency band is shared between the receiver and the transmitter. The corresponding IFs have been presented in Section 6.1.1. The superheterodyne mobile transmitters of different operation modes have the following things in common. The quadrature modulator converts the I and Q base-band signals into a single side band IF signal. No IF channel filter is used after the modulator. The frequency up-converter is an image rejection mixer. The I and Q BB amplifiers, I and Q BB LPFs, quadrature modulator, IF VGA, Up-converter, and RF driver are usually integrated on the same silicon die. The power amplifier is a separate gain block.

In the transmitter system design, the output power and control, the transmission spectrum and adjacent channel power (ACP) emission or ratio, and modulation accuracy will be mainly considered.

The minimum performance requirements for different modes of the transmitter are presented in Tables 6.19 to 6.21.

Table 6.19. Minimum performance specifications of GSM Mobile transmitter

GSM & GPRS 800/1900	**Specifications**	**Note**
Frequency band (MHz)	824 – 849 1850 – 1910	
Frequency accuracy	0.1 ppm	
Modulation	GSMK	
800 MHz maximum output power (dBm)	33 ± 2	
800 MHz output power dynamic range (dB)	28	In 2 dB step change
800 MHz minimum output power (dBm)	5 ± 5	
800 MHz output power tolerance (dB)	 3 5	Power Level: 31 to 13 dBm 11 to 5 dBm
1900 MHz maximum output power (dBm)	30 ±2	
1900 MHz output power dynamic range (dB)	30	In 2 dB step change
1900 MHz minimum output power (dBm)	0 ± 5	

Table 6.19. (Cont.)

1900 MHz output power tolerance (dB)	3 5	
Modulation Accuracy (degree)	< 5 rms < 20 peak	
Adjacent channel power ratio (dBc/30 kHz)	< -30	+/-200 kHz offset,
Atl. adjacent channel power ratio (dBc/30 kHz)	< -60	+/- 400 kHz offset
2nd Alt. adjacent channel power ratio (dBc/30 kHz)	< -60	+/- 600 kHz offset

Table 6.20. Minimum performance specifications of TDMA mobile transmitter

TDMA 800/1900	**Specifications**	**Note**
Frequency band (MHz)	824 – 849 1850 – 1910	
Frequency accuracy	200 Hz	
Modulation	$\pi/4$ DQPSK	
800 MHz maximum output power (dBm)	28 +2/-4	
800 MHz output power dynamic (dB)	35	4 dB step from 28 to 8 dBm 5 dB step from 8 to -7 dBm
800 MHz minimum output power (dBm)	-7 $\pm$ 5	
800 MHz output power tolerance (dB)	+2, -4 +5, -6	Power level: 28 to 8 dBm 3 to -7 dBm
1900 MHz maximum output power (dBm)	28 +2/-4	
1900 MHz output power dynamic (dB)	36	4 dB step from 28 to 8 dBm 6 dB step from 8 to 2 dBm 5 dB step from 2 to –8 dBm
1900 MHz minimum output power (dBm)	-8 $\pm$ 6	
1900 MHz Output Power Tolerance (dB)	+2, -4 +2, -6	Power level: 28 to 8 dBm 8 to -8 dBm
Frequency settling time (msec)	≤ 2	

Table 6.20. (Cont.)

Modulation accuracy (EVM)	< 12.5%	
Adjacent channel power ratio (dBc/30 kHz)	< -26	+/-30 kHz offset,
Atl. adjacent channel power ratio (dBc/30 kHz)	< -45	+/- 60 kHz offset
2nd alt. adjacent channel power ratio (dBc/30 kHz)	< -45	+/- 90 kHz offset

Table 6.21. Minimum performance specifications of AMPS mobile transmitter

AMPS	**Specifications**	**Note**
Frequency band (MHz)	824 - 849	
Frequency stability (ppm)	< 2.5	
Modulation	FM	
Maximum output power (dBm)	28 +2/-4	
Minimum output power (dBm)	8 +2/-4	
Output power switch step (dB)	4	
Power switch transition time between 2 levels (msec)	< 20	
Transmitter off RF power (dBm)	< -60	
Channel switch time (msec)	< 40	Carrier frequency settled within 1 kHz of its final value
Modulation peak deviation stability (%)	< |10|	
FM hum and noise (dB)	> 32	Ratio of audio outputs with FM and without modulation
Residual AM (%)	< 5	
Modulation distortion and noise (%)	< 5	
Spurious and Noise Emission Suppression (dBc/300 Hz)	> 26 > 63 +10x log(Tx power in W) < -80 dBm/30 kHz	20 to 45 kHz > 45 kHz Between 869 and 894 MHz

6.1.3.1. Transmission Power

The block diagram of this multimode transmitter is shown in Fig. 6.1. The overall gain of the transmitter chain is between 45 dB and 55 dB depending on the output level of the I and Q DAC and the output power. The PA power gain is usually around 30 dB, the power gain of the drive amplifier is generally not too high, approximate 8 to 10 dB, and the up-converter gain is in the 5 to 10 dB range. One possible gain distribution of the 800 MHz band transmitter operating in GSM and TDMA/AMPS modes is given in Table 6.22 (where it is assumed that different operation modes share the same transmitter path). The 1900 MHz transmitter has a similar gain distribution except the gain of the driver amplifier and/or the PA can be properly reduced since the maximum output power of the GSM transmitter is 3 dB lower than that in the 800 MHz band.

Table 6.22. A possible gain distribution of the 800 MHz transmitter

Function Block	DAC	I/Q Mixer	IF VGA	Up-Converter	Balun	Driver Amplifier	SAW Filter	PA	Duplexer	Diplexer	Antenna Port
800 MHZ GSM											
Gain (dB)	0.00	-1.00	12.50	6.00	-0.50	10.00	-3.50	30.00	-2.50	-0.50	
Cascad gain (dB)	0.00	-1.00	11.50	17.50	17.00	27.00	23.50	53.50	51.00	50.50	50.50
Pout (dBm)	-17.50	-18.50	-6.00	0.00	-0.50	9.50	6.00	36.00	33.50	33.00	33.00
Rejection in Rx band (dB)		20.00					31.00		38.00		
Rejection at L.O. frequency		20.00		15.00	0.00		24.00		29.00		
Rejection at image frequency		20.00		25.00	0.00		26.00		32.00		
NF (dB)	0.00	1.00	9.00	8.00	0.50	5.00	3.50	5.00	2.50	0.50	
Cascad NF (dB)	0.00	1.00	10.00	10.16	10.16	10.18	10.18	10.18	10.18	10.18	10.18
OIP3(dBm)	100.00	5.00	10.00	18.00	100.00	25.00	100.00	47.00	100.00	100.00	
Cascad OIP3	100.00	5.00	9.29	13.43	12.93	20.83	17.33	44.15	41.65	41.15	41.15
800 MHZ TDMA											
Gain (dB)	0.00	-1.00	10.50	6.00	-0.50	10.00	-3.50	30.00	-2.50	-0.50	
Cascad gain (dB)	0.00	-1.00	9.50	15.50	15.00	25.00	21.50	51.50	49.00	48.50	48.50
Pout (dBm)	-20.50	-21.50	-11.00	-5.00	-5.50	4.50	1.00	31.00	28.50	28.00	28.00

The maximum transmission output power of the mobile stations for GSM, TDMA, and AMPS systems is around 30 dBm — e.g., 30 to 33 dBm for the GSM and 26 to 28 dBm for the TDMA and AMPS. The output power change of this multimode transmitter is under command of the corresponding base stations and in steps. The control ranges are 20 dB (AMPS) and 36 dB (TDMA), respectively, as given in Tables 6.19 to 6.21. However, in practice the control rage should be 10 to 12 dB greater than 36 dB to compensate transmitter chain gain variation with temperature, frequency, operation mode, and control tolerance.

A power control loop as depicted in Fig. 6.3 is used in this transmitter for all operation modes to set and control their output power to an assigned level within the allowed tolerance. The power control loop consists of a power detector, temperature sensor, multiplexer, ADC, DSP, and DAC. The output of the power detector is a voltage, which is monotonically associated with the transmission power. The analog voltage is then converted to a digital signal after passing through the ADC that has a dynamic range determined by the power control range. In this application an 8 bit ADC is enough to cover control range, transmission gain variation, and control tolerance for this application. The DSP executes the power control algorithm, which is based on power detector calibration data obtained from different temperatures, frequencies, and operation modes. The output of the DSP is a digital control voltage value, and it is converted to an analog voltage through a 10 bit DAC. The proper functional range of the power control loop is usually around 25 dB. The control accuracy of the feedback loop is good when the output power relatively high, and the feedback loop will not provide additional accuracy when the output power is low. Generally the transmission power can be kept within ±1.5 dB by using this power control loop on the top 25 dB of the transmission power. The error contribution to the transmission power in the power control loop can be estimated as presented in Table 6.23.

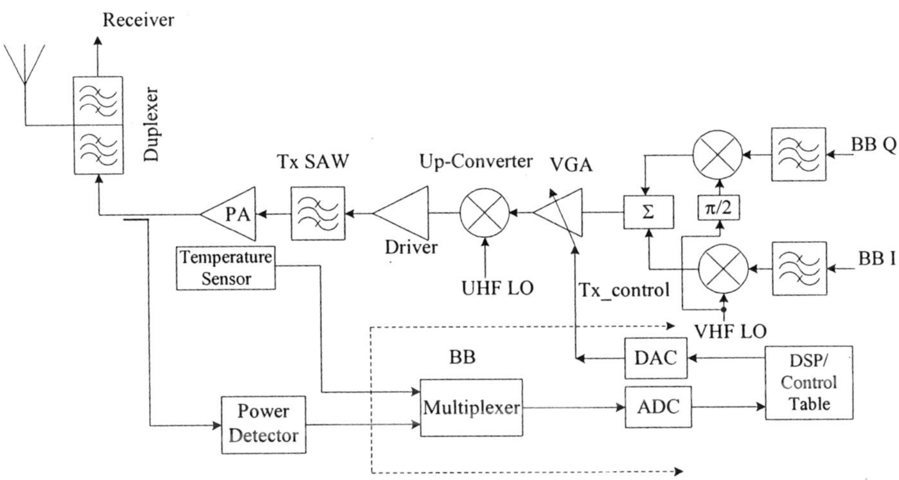

Figure 6.3. Transmission power control

Table 6.23. Error contribution to the feedback-controlled transmission power

Error Item	Close to Maximum Power	In Low Power Region	Notes
Tx_control	±0.25 dB	±0.75 dB	Tx control DAC step size error, nonlinearity, offset drift, and slop drift
Power detector accuracy	±0.3 dB	±3 dB	
Calibration accuracy	±0.5 dB	±0.5 dB	
I/Q error	~ 0 dB*	±1.1 dB**	I/Q DAC amplitude and I/Q channel gain variation with temperature
SAW attenuation variation	~ 0 dB*	±0.45 dB**	Due to full or partial suppression of feedback loop
PA gain change	~ 0 dB*	±0.7 dB**	Due to full or partial suppression of feedback loop
Total error	±1.05 dB	±6.5 dB	

* Error is efficiently suppressed by the power control feedback loop.

** Error is partially suppressed by the power control feedback loop.

The feedback loop is usually not used when the output power is lower than certain level since the transmission power error under the feedback loop control is high as seen from Table 6.23. The power level, when the feedback loop is disconnected, depends on the operation mode and frequency band. For instance, for the TDMA mobile transmitter these power levels are equal to or below 3 and 2 dBm when operating in the 800 and 1900 MHz, respectively, and for the GSM transmitter the corresponding power levels are 5 and 6 dBm, respectively. In these low output power regions, it is better to not use the feedback loop, and the transmission power is tuned based on a power control lookup table, which is measured as part of the transmitter tuning and calibration procedures and is installed in a memory for the DSP usage. The transmitter power accuracy in this case may be kept within ±2.5 dB.

6.1.3.2. Adjacent and Alternate Channel Power (ACP)

There is a mask to restrain the allowed transmission spectrum of the GSM transmitter. The mask restricts the power ratio of the integrated spectral power within 30 kHz at ±200 kHz and ±400 kHz offset frequencies to the GSM transmission power within the main spectral lobe are equal to or less than –30 dBc and –60 dBc, respectively. These are actually the adjacent and alternate channel power requirements for the GSM mobile transmitter.

The adjacent channel power is determined mainly by the spectral power density level of Gaussian pulse shaped GMSK modulation at a 200 kHz frequency offset from the carrier, phase noise, and spurious of the synthesized LOs at the same frequency offset from the LO frequencies, and the third-order nonlinearity of the transmitter chain, especially the PA. But the alternate channel power mainly results from the fifth-order nonlinearity of the transmitter chain instead of the third order. The maximum allowed contribution of the transmitter main blocks to the adjacent and the alternate channel powers are presented in Table 6.24. The adjacent and the alternate channel powers in the GSM case are measured in a 30 kHz bandwidth.

Table 6.24. Main block contributions of GSM mobile transmitter to adjacent and alternate channel powers

Blocks	±200 kHz	±400 kHz	Notes
Digital BB+I/Q DAC	-45 dBc	-70 dBc	Including pulse shaping, DAC noise and distortion
Modulator +IF VGA	-48 dBc	-72 dBc	
Up-converter	-45 dBc	-70 dBc	
LO phase noise + spur	-60 dBc	-65 dBc	Based on Table 6.12 LO spurious data
Driver amplifier	-40 dBc	-69 dBc	
Power amplifier	-36 dBc	-68 dBc	
Total ACPR	-30.5dBc	-60.6 dBc	
ACPR specification	≤ -30 dBc	≤ -60 dBc	

The total adjacent channel power ratio in Table 6.24 is approximately calculated by means of (5.4.22) with an assumption of all $\alpha_{k,j} = 0.5$, and the overall alternate channel power ratio in the same table is obtained from power summation. In fact, the adjacent channel power emission will not be an issue in the GSM mobile transmitter design, but the alternate channel (±400 kHz offset) power emission may be quite critical to meet –60 dBc specification. One main contribution to the alternate power emission is due to the LO spurious at ±400 kHz offset from its center frequency, which is too high if one is just using the level given in Table 6.12 that results from the receiver analysis. The GSM transmitter needs an UHF synthesizer LO to have a phase noise and spurious level, especially at ±400 kHz offsets, to meet that required by the TDMA mobile transceiver. When the LO spurious level reduces to –78 dBc from –65 dBc as estimated in

Table 6.24, the transmitter noise and spurious emission margin at ±400 kHz offset from the carrier can increase to 2.5 dB. The role of the PA nonlinearity in this specification is just secondary.

In the case of TDMA, the PA may dominate the contribution to the adjacent channel and alternate channel powers of its transmission since the waveform of pulse-shaped π/4 OQPSK modulation has large amplitude variation. The PA in the TDMA transmitter needs to operate in a more linear condition than in the GSM transmitter. However, the maximum output power of the TDMA transmitter is 2 to 5 dB lower than the power transmitted by the GSM. This means that the same PA can operate in the transmitter GSM or TDMA mode although the linearity requirement of these two modes on the PA may be significantly different.

The maximum allowed adjacent and alternate powers resulting from nonlinearity and noise of each block in the TDMA transmitter are presented in Table 6.24. The adjacent and alternate channel powers here are measured within a 300 Hz bandwidth centered at the assigned offset frequency. The cascade adjacent and alternate power ratios are approximately evaluated in terms of (5.4.22) with all $\alpha_{k,j} = 0.5$ and 0, respectively, as done in the GSM case.

Tables 6.24 and 6.25 also correspond to the worst case. The transmitter design should guarantee each block in the transmitter chain to have lower emissions than those given in these two tables. To achieve this, it is necessary for each block to have a proper linearity and noise figure.

Table 6.25. Main block contribution of TDMA mobile transmitter to adjacent and alternate channel powers

Block	±30 kHz	±60 kHz	Notes
Digital BB+I/Q DAC	-52 dB	-52 dB	Including square root raised pulse shaping, DAC noise and distortion
Modulator +IF VGA	-42 dB	-57 dB	
Up-converter	-40 dB	-55 dB	
LO phase noise + spur	-60 dB	-70 dB	
Driver amplifier	-36 dB	-53 dB	
Power amplifier	-32 dB	-49 dB	Dominating factor
Cascade ACPR	-26.3 dB	-45.4 dB	
ACPR specification	≤ -26 dB	≤ -45 dB	

The AMPS transmitter usually has no issues for the ACPR if the 800 MHz transmitter operating in TDMA mode is able to meet all the specifications. The AMPS has the same operation frequency band, the same 30 kHz channelization, and the same output power as the TDMA, but its RF transmission signal has a constant envelope — i.e., it imposes less stringent linearity on the transmitter chain. It is therefore unnecessary to further discuss the design related to the adjacent and alternate channel power emission of the AMPS transmitter.

6.1.3.3. Noise and Spurious Emission in a Receiver Band

In the transmitter wide-band noise and spurious emission specifications, the toughest one is usually the emission requirement in the corresponding receiver band. The minimum specifications of the GSM, TDMA, and AMPS transmitter emission in the receiver band are summarized in Table 6.26.

Table 6.26. Receiver band emission specifications of transmitters

System	Emission Requirement	Notes
800 MHz GSM	≤ -79 dBm/100 kHz	In 869 – 894 MHz
1900 MHz GSM	≤ -71 dBm/100 kHz	In 1930 – 1990 MHz
800 MHz TDMA	≤ -80 dBm/30 kHz	In 869 – 894 MHz
1900 MHz TDMA	≤ -80 dBm/30 kHz	In 1930 – 1990 MHz
AMPS	≤ -80 dBm/30 kHz	In 869 – 894 MHz

This specification for the transmitter with the architecture of Fig. 6.1 is not critical since the transmitter SAW and the duplexer can suppress the spurious and noise emission in the receiver band to lower than the minimum requirement. An example of the receiver band noise emission evaluation based on the 800 MHz GSM transmitter is shown in Table 6.27. Here it is assumed that the noise voltage from the base-band DAC is 18 nV/$\sqrt{Hz}$. Up to IF VGA output the noise is calculated in voltage in terms of (5.5.21) since the input impedances of these blocks are usually high. The noise power is then evaluated from the up-converter to the antenna port, and (5.5.10) and (5.5.12) are used. From the result presented in Table 6.27, the receiver band noise emission level is approximately –160.4 dBm/Hz, and the overall emission power in 100 kHz is –110.4 dBm. It is apparent that the emission power in the receiver band is 31 dB lower than the specification –79 dBm. The reference clock (19.2 MHz) harmonics in the receiver band may dominate the emission level.

Table 6.27. Example of receiver band noise emission evaluation

Function Block		I/Q Mixer	IF VGA	Up-Converter	Balun	Driver Amplifier	SAW Filter	PA	Duplexer	Diplexer	
Gain in Tx band (dB)	0.00	-1.00	12.50	6.00	-0.50	10.00	-3.50	30.00	-2.50	-0.50	
Rej in Rx band (dB)		20.00					31.00		38.00		
Cascad Gain in Rx band	0.00	-21.00	-8.50	-2.50	-3.00	7.00	-27.50	2.50	-38.00	-38.50	-38.50
NF (dB) in Rx Band	0.00	21.00	10.00	10.00	0.50	8.00	34.50	5.00	38.00	0.50	
Cascad NF in Rx	0.00	21.00	10.00	18.68	18.69	19.27	28.11	32.70	37.33	37.91	37.91
Noise in Rx band (dBm/Hz)	18.00	2.39	24.73	-132.93	-125.02	-115.02	-149.50	-119.47	-159.87	-160.35	-160.35

In a conventional GSM or TDMA transceiver, a transmitter/ receiver (T/R) switcher may be employed instead of using a duplexer. The noise and spurious emission in the receiver band can then become an issue if one does not properly control the emission, especially the reference clock harmonic spurious in the receiver band.

6.1.3.4. Spectrum of Burst Ramp-Up and -Down Transients

Both GSM and TDMA mobile transmitters operate in burst mode. In both systems, the time responses of bursts are well defined in [1] and [2], respectively. The burst ramp-up and -down transient for the GSM transmitter is specified in time domain and also in frequency domain. The maximum power level restraints at different offset frequencies from the carrier due to ramp transient for the GSM mobile transmitter are shown in Table 6.28.

Table 6.28. Maximum power due to burst ramp transient

Output Power	± 400 kHz	± 600 kHz	± 1200 kHz	± 1800 kHz
≤ 37 dBm	-23 dBm	-26 dBm	-32 dBm	-36 dBm

The burst ramp-up and -down transient is able to be properly controlled through the power control loop and the control voltage $Tx_{\_control}$ as depicted in Fig. 6.2. The transmission signal spectrum is equal to the convolution of the $Tx_{\_control}$ signal and the I/Q signal spectra. A right ramp transient of transmission bursts can be achieved by designing a proper $Tx_{\_control}$ voltage signal.

6.1.3.5. Residual Amplitude Modulation (AM)

The GSM and AMPS transmission waveforms should have a constant envelope. However, IQ imbalance, carrier feed-through, $T_{x\_control}$ control DAC noise, BB GMSK modulator and DAC noise, and PA AM-to-AM conversion will cause amplitude modulation. For the GSM transmission, the residual modulation shall be less than 1 dB or 12%, and it is specified equal to or less than 5% for the AMPS transmission.

In the GSM transmitter, the highest AM usually occurs at the low output power level where the AM suppression capability of the power control loop is poor. The worst-case AM in the GSM transmission is shown in Table 6.29. In the calculation of the total residual AM (*RAM*), the correlation terms are linearly added together first, and this subtotal and the uncorrelated terms are then summed up in the following way:

$$RAM_{GSM} = \sqrt{\left(\sum_k RAM_{cor\_k}\right)^2 + \sum_j RAM^2_{uncor\_j}} \, .$$

Table 6.29 Worst case residual AM of GSM transmission

Contributors	AM dB	AM %
BB modulator and DAC noise	0.2	2.3
DAC nonlinearity (correlated)	0.07	0.8
Amplitude imbalance (correlated)	0.25	2.9
Carrier feed-through (correlated)	0.18	2.1
Tx_control DAC noise	0.18	2.1
PA AM-AM conversion	0.45	5.3
Total	0.72	8.5

The residual AM of the AMPS transmission is mainly caused by the carrier feed-through, I/Q imbalance, and PA AM-to-AM conversion. The DAC noise contribution is small since its channel bandwidth is only around 25 kHz. To meet less than 5% residual AM requirement, the carrier leakage and the I/Q imbalance should be controlled to below –40 dBc and 0.2 dB, respectively, and the PA AM-to-AM conversion needs be less than 0.25 dB. The composite residual AM is approximately equal to

$$RAM_{AMPS} = \sqrt{10^{\frac{-40}{10}} + \left(10^{\frac{0.2}{20}} - 1\right)^2 + \left(10^{\frac{0.25}{20}} - 1\right)^2} = 4.1\% \, .$$

6.1.3.6. Modulation Accuracy

The modulation accuracy is measured in different way for GSM, TDMA and AMPS transmission signals. Phase error is used to measure the GMSK modulation accuracy. Error vector magnitude (*EVM*) and modulation distortion are employed as the measure of modulation accuracy for the TDMA and the AMPS system, respectively.

The GSM modulation accuracy requires the transmission signal phase error rms value and the peak phase error equal to or less than 5° and 20°, respectively. The main factors to the modulation accuracy degradation are the in-channel-bandwidth phase noise of the UHF and VHF LOs, base-band contribution including reconstruction filter, modulation approximation, and quantization noise, and the AM-to-PM conversion of the PA nonlinearity. Assuming that the integrated phase noise of the transmitter IC VHF LO is –32 dBc, and the integrated phase noise of the 1900 MHz UHF LO with an external VCO is –27 dBc, the total rms phase error caused by the LOs is

$$\Delta\vartheta_{N\_LO} = \sqrt{\Delta\vartheta_{N\_VHF}^2 + \Delta\vartheta_{N\_UHF}^2} = \frac{180}{\pi}\sqrt{10^{\frac{-32}{10}} + 10^{\frac{-27}{10}}} = 2.9^\circ.$$

Considering the base-band contribution to the modulation accuracy degradation is 1.4°, the error from the AM-to-PM conversion of the 1900 MHz PA is approximately 1°, and the carrier feed through is –35 dBc, the overall rms phase error is then

$$\Delta\vartheta_{MA} = \sqrt{\Delta\vartheta_{N\_LO}^2 + \Delta\vartheta_{BB}^2 + \Delta\vartheta_{AM/PM}^2 + \Delta\vartheta_{CFT}^2}$$
$$= \sqrt{2.9^2 + 1.4^2 + 1^2 + 1.8^2} = 3.8^\circ.$$

The peak phase error of 1900 MHz GSM GMSK modulation, which is usually worse than that of the 800 MHz GSM GMSK modulation, is estimated as in Table 6.30.

The modulation accuracy of the TDMA transmitter is measured in EVM and it is specified to be equal to or less than 12.5%. Assuming that the integrated VHF and 1900 MHz UHF LO phase noises are –32 and –26 dBc, respectively, the BB and DSP contribution is approximately 3%, the carrier feed-through is –35 dBc or less, and the PA nonlinearity degrades modulation accuracy approximate 4%, the rms EVM of the TDMA transmission modulation is evaluated as

$$EVM_{TDMA} = \sqrt{10^{\frac{-32}{10}} + 10^{\frac{-26}{10}} + 0.03^2 + 10^{\frac{-35}{10}} + 0.04^2} = 7.8\% .$$

The average EVM of the first 10 symbols is also defined in the TDMA transmitter specifications, and it shall be equal to or less than 25%. A possible EVM partition for the first 10 symbols of the 1900 MHz TDMA transmitter is presented in Table 6.31.

Table 6.30. Peak phase error partition of 1900 GMS modulation accuracy

Main Contributors	Peak Phase Error	Notes
Base-band	4°	Reconstruction filter, modulation approximation, quantization noise
LO phase noise	10°	
AM/PM conversion	2°	Derived from the peak AM-to-AM and AM-to-PM conversions
Carrier feed-through	1.8°	
Total peak phase error	17.8°	Linear summation of individual contribution

Table 6.31. The EVM partition for the first 10 symbols of 1900 MHz TDMA transmitter

Main Contributors		Estimated EVM
BB and DSP		3%
VHF LO integrated phase noise	-25 dBc	5.6%
I/Q imbalance	0.3 dB, 4°	3.2%
UHF LO integrated phase noise	-16 dBc	15.8%
Carrier feed-through	-28 dBc	4%
PA nonlinearity		5%
Total EMV		18.6%

The modulation accuracy of the 800 MHz TDMA is usually much better than the 1900 MHz one, and this is due to the fact that the in-channel-bandwidth phase noise of the UHF LO is lower.

The modulation distortion and noise of the AMPS transmission shall be within 5%. To meet this specification, the I and Q amplitude and phase imbalance need be less than 0.2 dB and 3°, and the integrated VHF and UHF phase noises should be lower than –35 and –30 dBc, respectively. The PA nonlinearity has insignificant influence on the FM distortion, and the carrier feed-through can be tuned down to below –40dBc. In this case, the total modulation distortion is thus approximately

$$MD_{AMPS} = \sqrt{10^{\frac{-30}{10}} + 10^{\frac{-35}{10}} + 10^{\frac{-32}{10}} + 10^{\frac{-40}{10}}} = 4.5\% .$$

The main contributors to the distortion are the I and Q imbalance and the UHF LO in-channel-bandwidth phase noise. The I and Q imbalance can be usually calibrated to be quite low. The phase locked loop bandwidth of the UHF synthesized LO can be designed as narrow as a couple of kHz since its channel frequency switching time is allowed to be as long as 40 msec. The integrated in-channel-bandwidth phase noise thus can be reduced to –36 dBc. It is possible to make the modulation distortion lower than 4% after reducing the UHF PLL loop bandwidth and the imbalance between the I and Q channel signals.

6.1.3.6. Radio Frequency Tolerance

In these three systems, the frequency tolerance of the mobile GSM transceiver is ±0.1 ppm, which is the toughest one among the specifications of these systems. The absolute minimum frequency error for the 800 MHz GSM is approximately 82 Hz, and 185 Hz for 1900 MHz.

Assuming that the DAC used in the AFC loop is 10 bits with a differential nonlinearity (DNL) ±0.6 LSB, and tuning voltage is from 0.2 V to 2.0 V, the tuning voltage per bit is

$$\frac{2.0 - 0.2}{10^{10} - 1} = \frac{1.8}{1023} = 1.76 \text{ mV/bit} .$$

If the DAC noise is 12 $\mu V/\sqrt{Hz}$ and the AFC loop noise bandwidth is 1 kHz, the noise voltage is equal to

$$12 \cdot 10^{-6} \cdot \sqrt{1000} = 379 \mu V .$$

The allowed maximum tuning slope is

$$\frac{0.1}{\left(1.6 \cdot 1.76 + 379 \cdot 10^{-3}\right) \cdot 10^{3} \cdot 1.76} = 18.2 \text{ ppm/V} .$$

Thus, the maximum tuning slope of the reference clock, VCTCXO, must be less than 18 ppm/V.

The frequency tolerances for the TDMA and AMPS mobile transceivers are looser than that demanded by the GSM transceiver. The VCTCO specification determined based on the GSM frequency tolerance requirement is also applicable to the TDMA and AMPS transceivers.

6.2. Direct Conversion Transceiver

In this section, a direct conversion CDMA transceiver design is discussed in detail. The receiver design is based on the configuration of Fig. 3.11, which employs the high dynamic range (or also called *high-resolution*) ADC. The traditional I/Q quadrature modulator incorporated with an up-converter to directly generate the modulated RF signal is used in the transmitter.

Many aspects of the transceiver design are common to all architectures. In this section, we discuss these common aspects and show how to handle the technical challenges of the direct conversion transceiver. The latter is emphasized in this design example.

The block diagram of a CDMA direct conversion transceiver is duplicated in Fig. 6.4. The receiver consists of a LNA with a two-step gain control, a single-step gain switching RFA, a quadrature demodulator directly down-converting an RF signal to I and Q channel base-band signals, a low-order BB LPF and a fixed gain BBA in I and Q each channel, respectively. In this CDMA receiver all the gain control is implemented in the RF block, and no gain variation takes place either in the BB I channel or in the BB Q channel. Since 10 bit ADCs are used here, it is possible to make the analog BB gain as low as only around 30 dB. The direct conversion receiver under such a gain distribution arrangement will possess many fewer issues on the DC offset and the I and Q mismatching. As used in the superheterodyne receiver, the receiver RF SAW is utilized here for suppressing the transmitter leakage. In the transmitter, the I and Q DACs interface with LPFs, and the filtered analog BB I and Q signals after amplitude adjustment then are directly converted to an RF transmission signal through a quadrature modulator. The transmission signal gains power when it passes through RF VGA, driver amplifier, and PA. The transmitter AGC is executed by means of the RF VGA, PA, incorporated with the step-control attenuator in the analog base-band. The transmitter RF SAW is adopted to control the out of band wide-band noise and spurious emissions. To connect the receiver input and the transmitter output into a common antenna, a duplexer is needed for the full-duplex CDMA transceiver.

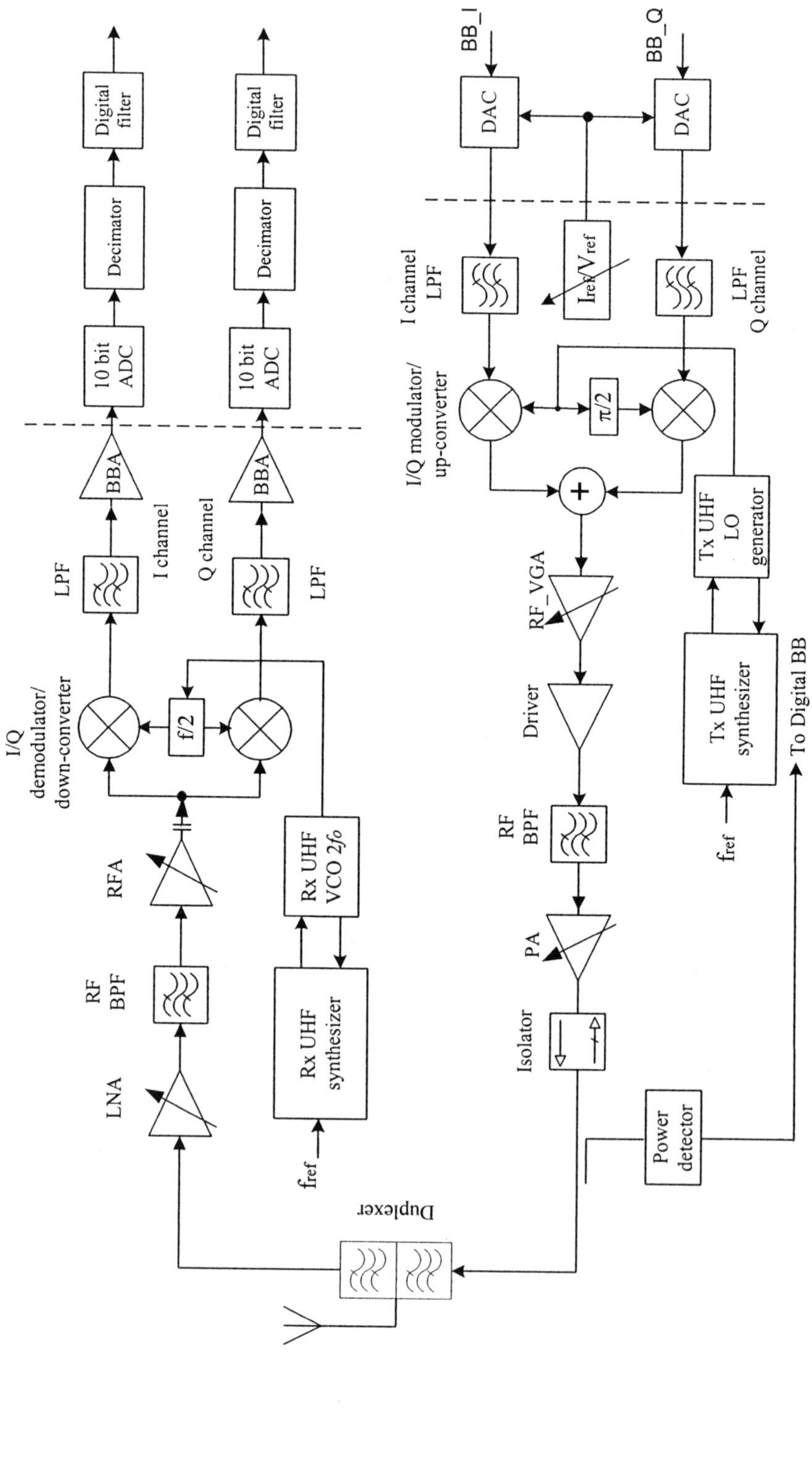

Figure 6.4. Block diagram of direct conversion transceiver with a high dynamic ADC

In this example, only a single frequency band, the 800 MHz cellular Band, CDMA transceiver design is discussed, but the same architecture and design approach can also be used for other frequency band transceiver. The data used in this example and the design results may not be optimized for a real transceiver.

6.2.1. Receiver System Design

The direct conversion receiver has the advantages of low cost, small size, lower part count, and simple configuration. However, in its implementation, many technical issues — such as DC offset, even-order distortion, I and Q channel amplitude and phase mismatching, flicker noise, etc. — must be resolved as described in Chapter 3. The receiver architecture of direct conversion presented in Fig. 6.4 is one of those that are able to properly handle these technical challenges.

The key to the direct conversion receiver architecture shown in Fig. 6.4 is that a high dynamic range (or high resolution) ADC makes it possible to have an AGC range much less than the receiver dynamic range and to significantly reduce the analog BB filter order and the BB amplifier gain as well. Thus the DC offset and I/Q channel mismatching issues become much easier to handle.

The minimum performance specification of an 800 MHz cellular band CDMA receiver is presented in Table 6.32. CDMA receiver performance specifications are measured based on the FER. In most cases, the minimum performance is determined while the FER reaches 1% except the receiver sensitivity uses 0.5%. The intermodulation spurious response attenuation, which is also referred to as *intermodulation distortion* (IMD), has three requirements corresponding to desired signal level — S_d, -101, -90, and –79 dBm, respectively. The offset frequencies of two intermodulation tones are defined as 900 kHz and 1.7 MHz lower or higher than the signal carrier frequency. The single-tone desensitization is a unique specification for the CDMA receiver, especially for the 800 MHz CDMA receiver. This is due to the fact that the AMPS system operates in the same 800 MHz band, and the CDMA receiver considers the AMPS interferer to be a CW tone.

Table 6.32. Minimum performance specifications of an 800 MHz CDMA receiver

Items	Min. Specification	Notes
Frequency band (MHz)	824 — 849	Band class 0
Spreading chip rate (MHz)	1.2288	Signal bandwidth
Sensitivity (dBm)	≤ -104	FER ≤ 0.5%
Dynamic (dBm)	≥ -25	FER ≤ 1%
Intermediation spurious response attenuation (dBm) Tone offset frequencies: @ 900 kHz & 1.7 MHz	≥ -43 ≥ -32 ≥ -21	*Sd* = -101 dBm *Sd* = -90 dBm *Sd* = -79 dBm FER ≤ 1%
Single tone desensitization (dBm) Tone offset: 900 kHz	≥ -30	Sd = -101 dBm TX Power ≥ 23 dBm FER ≤ 1%
Conducted noise and spurious emission (dBm)	≤ -76 ≤ -61 ≤ -43	In Rx band In TX band Other frequencies

6.2.1.1. Overall System Design Consideration

To achieve high receiver sensitivity, the RF block from the antenna port to the I/Q down-converter output must have low noise figure and high enough gain, 35 to 40 dB voltage gain. The reason for this is that a high gain RF block can effectively reduce the contribution of noisy later stages — i.e., the analog BB low-pass filter and the BBA — to the overall receiver noise figure. It will be equally important to maintain good receiver linearity (greater than –10 dBm, for instance) while increasing RF block gain to improve the receiver sensitivity.

As we know, the DC offset is one of the main issues of the direct conversion receiver. To resolve this, it is apparently necessary to decrease the analog base-band block gain. We may learn something from the BB block design of the superheterodyne receiver. The I/Q quadrature demodulator and the corresponding BB block in a superheterodyne receiver actually form a direct conversion receiver, which directly converts an IF signal down to a BB signal instead of converting an RF signal to a BB signal. We typically do not hear too much DC offset trouble from this IF direct conversion architecture. The reason for this is mainly due to the fact that the gain of the BB block of the IF direct conversion subsystem in the superheterodyne receiver is relatively low — say, 30 to 40 dB. Using local

DC offset compensation and digital base-band feedback DC offset correction can efficiently resolve the DC offset issue. In the RF direct conversion receiver, thus, an analog BB block with a similar gain and DC offset compensation strategy as used in the superheterodyne receiver can be employed to cope with the DC offset issue.

In a superheterodyne receiver, approximate 50 to 60% of the overall receiver gain comes from the IF block at the high gain mode. The IF block also dominates the AGC range — i.e., 80% of the total automatic gain control range is created by this block. It is apparent that there is no IF block in the direct conversion receiver. The missing IF block gain and gain control range in the direct conversion receiver are partially transferred to the RF block. The allowed maximum gain transfer is restricted by the receiver linearity requirement, and it is approximately 30 dB in the CDMA receiver case. The gain control range executed in the RF block can be 50 to 60 dB depending on the dynamic range (or resolution) of the ADC used in the direct conversion receiver.

The ADC used in the direct conversion receiver usually has higher dynamic range than that demanded by the superheterodyne receiver. The main reasons for this are that the gain control range is smaller than the received signal dynamic range; step gain control is used instead of continuous control; and filtering to interferers is loose because of not using the IF SAW channel filter. ADC resolution of 10 to 12 bits will be needed in a CDMA direct conversion receiver depending upon the missing gain and gain control range, and the loose rejection of the BB filter to the strongest and closest interferer. From the analysis and calculation in the following subsections, we shall see that the minimum voltage level of the desired signal appearing at the ADC input is only around 2.4 mV rms, and the strongest interference tone level in the receiver high gain mode may reach as high as 275 mV rms. Their ratio is 20log (275/2.4) $\cong$ 41 dB. Considering the dynamic range of the ADC used in a superheterodyne CDMA receiver to be approximately 24 dB, the total dynamic range of the ADC in this direct conversion receiver will be at least $D_{ADC} = 41 + 24 = 65$ dB. A pessimistic estimation of the resolution of the ADC is

$$b_{ADC\_ps} = \frac{D_{ADC} - 1.76}{6.02} = \frac{65 - 1.76}{6.02} \cong 10.5 . \tag{6.2.1}$$

An optimistic estimation of the ADC bit number can be obtained in terms of formula (3.4.13). Assuming that the sampling rate used in the ADC is 19.2 Msample/sec and considering the CDMA signal has a low-pass bandwidth of 615 kHz, then the ADC bit number is

$$b_{ADC\_op} = \frac{65 - 1.76 - 10\log(19.2/1.3)}{6.02} \cong 8.6. \tag{6.2.2}$$

It is probably fine to take the average of the results in (6.2.1) and (6.2.2), i.e., we should have a minimum 10 bit ADC for this direct conversion receiver:

$$b_{ADC} = \frac{b_{ADC\_ps} + b_{ADC\_op}}{2} = 9.5 \cong 10\,\text{bit}.$$

More accurately, we should use the ADC dynamic range instead of the number of bits since 5 dB per bit for the CDMA signal is employed usually. The minimum dynamic range of the ADC used for this direct conversion receiver is approximately 60 dB as described above.

It is assumed that the ADC used in this design has a ΔΣ type of ADC with a 1.5V maximum peak-to-peak voltage, sampling at a rate of 19.2 Msamples/sec, a bandwidth greater than the CDMA low-pass signal bandwidth 630 kHz, and an input impedance approximate 100 kΩ.

6.2.1.2. Evaluation of System Noise Figure and Linearity

In the CDMA mobile station minimum performance standards, IS – 95 and IS-2000_1x [4], the CNR for the receiver sensitivity, the intermodulation spurious response attenuation, and single-tone desensitization tests is defined as –1 dB at the data rate 9.6 kbps. This specification is unlike those defined in the other wireless systems, which usually need a positive CNR to achieve a certain BER or FER. The fact that the CDMA receiver is capable to operate at the negative CNR — i.e., the desired signal level lower than noise level, results from the processing gain of the CDMA, which is approximately 21 dB at the data rate 9.6 kHz. In practice, the CDMA receiver demodulator at digital base-band generally has at least 1 dB performance margin on the CNR — i.e., to reach 0.5% FER at 9.6 kbps CNR can be –2 dB or lower. However, in performance calculations CNR_{min} = -1 dB will be used in most cases.

The minimum sensitivity of the CDMA mobile station receiver is defined as –104 dBm for 0.5% FER at 9.6 kbps data rate as shown in Table 6.32. The maximum receiver noise figure can be calculated from (4.2.4). Considering the CDMA signal bandwidth, 1.23 MHz, the receiver noise figure in any condition must be better than

$$NF_{CDMA\_max} = 174 + S_d - 10\log\left(1.23 \times 10^6\right) - CNR_{min}$$
$$= 174 - 104 - 61 + 1 = 10\,\text{dB}.$$

In the CDMA receiver design, we would like to have 3 dB margin even in the worst case, and 4 dB margin or greater in the typical case. The corresponding receiver noise figure shall be approximately 7 dB in the worst case, and 6 dB or less in the typical case.

The linearity of the CDMA mobile station receiver using the direct conversion architecture is now not only measured by the third-order nonlinear distortion but also the second-order nonlinear distortion. From the minimum specification of the intermodulation spurious response attenuation or simply called intermodulation distortion (IMD) performance, we can determine the minimum third-order input intercept point (IIP_3) of the receiver as it has been done in the superheterodyne architecture receiver. The second-order nonlinear distortion of the direct conversion receiver will cause DC offset and low frequency interference products that are located in the receiver channel bandwidth. The main contributor to the DC component and the low frequency interference products is the I/Q down-converter or so-called quadrature demodulator in the direct conversion receiver. The second-order nonlinear products generated by the LNA will be blocked by the coupling capacitor and/or the receiver RF SAW filter, and the circuits after the I/Q down-converter have almost no response either to the RF interference tones or to the transmitter leakage. The second-order input intercept point of the I/Q down-converter, IIP_2, is determined by either some two-tone block specification or the allowed second-order product level due to the AM transmission leakage.

As seen in Table 6.32, there are three IMD performance specifications for three different desired signal levels, –101, –91, and –79 dBm, respectively. Using the same phase noise and spurious data of the UHF LO as given in the example of Section 4.3.3.3, and the receiver noise figure 10 dB, the minimum receiver IIP_3 for S_d = –101 dBm can be calculated by means of (4.3.51), and it is

$$IIP_{3,CDMA\_min} = \frac{1}{2}\left[3 \times (-43) + 10\log\left(10^{\frac{-101+1}{10}} - 10^{\frac{-174+10+61}{10}} - 2 \times 10^{\frac{-118.1}{10}} - 2 \times 10^{\frac{-118}{10}}\right)\right] = -12.7\,\text{dBm}.$$

In a similar way, we can obtain the minimum IIP_3 for $S_d = -90$ and -79 dBm to be

$$IIP_{3,-90\_\min} = \frac{1}{2}\left[3\times(-32)+\left(10^{\frac{-90+1}{10}}-10^{\frac{-174+19+61}{10}}\right)\right]=-2.7\ \text{dBm}$$

and

$$IIP_{3,-79\_\min} = \frac{1}{2}\left[3\times(-21)+10\log\left(10^{\frac{-79+1}{10}}-10^{\frac{-174+30+61}{10}}\right)\right]=8.3\ \text{dBm}.$$

where it is assumed that the receiver sensitivity is at least –95 dBm and –84 dBm or the receiver noise is $NF_{-90,\ max}$ = 19 dB and $NF_{-79,\ max}$ = 30 dB, respectively, when testing IMD performance at signal level –90 dBm and –79 dBm.

In the practical design, we would like to have 3 dB margin in the worst case IMD performance and 5 dB margin in the typical case when the desired signal is at a –101 dBm level. To reach these design goals, the overall IIP_3 in the LNA high gain mode operation needs to be –9.2 dBm or greater in the worst case, and –6.8 dBm or higher in the typical case, where the worst sensitivity –107 dBm and the typical sensitivity –108 dBm are respectively used. The IMD performance tests at desired signal –90 dBm and –79 dBm are usually carried out in the LNA middle gain mode. In this case, the receiver IIP_3 in these two tests will be the same, and thus the latter test is more critical due to the interference tone 11 dB higher than the former one's. The worst case IIP_3 should be determined based on the IMD performance with the desired signal –79 dBm, and it must be greater than 11 dBm.

The CDMA receiver has a special requirement on the IIP_3 of its LNA due to the single-tone desensitization issue, which results from the cross-modulation of the AM transmission leakage to a strong interference tone close to the desired signal through the third-order nonlinearity of the LNA. The reason why only the LNA IIP_3 impacts the single-tone desensitization instead of overall receiver IIP_3 is due to the RF SAW, which is able to efficiently suppress the transmission leakage power and thus the IIP_3 of the rest of the receiver chain after the RF SAW filter has negligible contribution to the cross-modulation product and the desensitization. From the single-tone desensitization performance we can determine the minimum LNA IIP_3, but it is not unique since it also relies on the rejection of the duplexer to the transmission, duplexer insertion loss and the output power of the transmitter. Assuming that the duplexer rejection to transmission is 48 dB, the insertion loss of the duplexer in receiver band is 3.5 dB, the transmission power 23 dBm, and the receiver noise figure is 10 dB, utilizing (4.4.14) we can calculate the LNA IIP_3 meeting the minimum single tone desensitization requirement — i.e., the interference level shall be –30 dBm as follows:

$$IIP_{3,LNA\_\min} = \frac{1}{2}(103.6 - 7 - 96 + 46 - 30 - 3.8) \cong 6.4 \text{ dBm}.$$

In the practical CDMA receiver design, we would like to have the LNA IIP_3 to be 8 dB or higher in the typical case. The single tone performance is usually 3 dB higher than –30 dBm or more.

For the CDMA direct conversion receiver, we need to also consider a special requirement on two-tone blocking performance. This specification results from the consideration that two strong AMPS signals with close enough carriers that may jam the cellular band CDMA direct conversion receiver due to the second-order nonlinear products generated by the quadrature down-converter. Assuming that two equal tones with a level –30 dBm and a frequency spacing of 615 kHz, we shall determine the minimum IIP_2 of the I/Q quadrature down-converter, which can keep the FER equal to 1% or less under the blocking tone attack when the desired signal S_d is at a level –101 dBm and the FER is 1% or less. If the designed receiver in the worst case has a noise figure NF_{Rx} = 8 dB, and the gain before the down-converter G_{before_Mxr} is 16.7 dB, the noise level at the down-converter input N_{Mxr_in} can be calculated by using either the method described in Subsection 3.2.3.2, or a modified method as follows:

$$\begin{aligned} N_{Mxr\_in} &= -174 + NF_{Rx} + 10\log(BW_{Rx}) + G_{before\_Mxr} \\ &= -174 + 8 + 61 + 16.7 = -88.3 \text{ dBm}. \end{aligned} \tag{6.2.3}$$

The allowed maximum noise/interference contribution at the quadrature down-converter input for 1% FER is equal to

$$\begin{aligned} D_{\max} &= S_d - CNR_{\min} + G_{before\_Mxr} \\ &= -101 + 1 + 16.7 = -83.3 \text{ dBm}, \end{aligned} \tag{6.2.4}$$

where CNR_{min} is the minimum carrier-to-noise/interference ratio, and it is –1 dB for 1% FER.

Comparing (6.2.4) and (6.2.3), we can find that the difference between the allowed maximum noise/interference contribution level and the actual noise level at the input of the quadrature down-converter is 5 dB. If the second-order intermodulation distortion product generated by the down-converter IMD_{2_Mxr} is equal to the noise level at the down-converter input N_{Mxr_in} — i.e., 3 out of 5 dB in $D_{max} - N_{Mxr\_in}$ is contributed by the IMD_{2_Mxr} and the other 2 dB is left for the LO phase noise and spurious contributions. Thus we have

$$IMD_{2\_Mxr} = N_{Mxr\_in} = -85.3\,\text{dBm}.$$

Using (3.2.11), we obtain the estimated minimum of the quadrature down-converter, IIP_{2_Mxr}, to be

$$IIP_{2\_Mx2} = 2(-30 + 16.7) + 85.3 = 58.7\,\text{dBm}.$$

The second-order nonlinear distortion of the quadrature down-converter may cause other issues, such as DC product, transmitter leakage self-mixing, etc., as discussed in Chapter 3. Among these issues caused by second order nonlinear distortion, the two-tone blocking demands the highest IIP2 of the down-converter, and therefore here we discuss only the two-tone blocking issue and the corresponding IIP_2 requirement.

6.2.1.3. System Lineup Analysis and Main Block Specifications

To achieve specified receiver system performance, a proper gain, noise figure, and the third-order input intercept point distribution along the whole receiver chain is necessary. Designing a proper distribution of key parameters in the receiver chain is usually a cut-and-try processing. Basic specifications of each stage in the receiver from the duplexer to the ADC need be first defined, and we then may appropriately adjust some parameters of each stage to trade off the overall receiver noise figure and the linearity for optimum performance. In the following, we are going to present main block specifications of this direct conversion receiver. It is necessary to consider the possibility of implementing the system in practical mass production with off the shelf products.

Let us begin with the cellular band duplexer specification. Two important parameters of a duplexer for the receiver system analysis and design are the insertion loss in the receiver band and the rejection (or suppression) of the transmitter signal. The former constraint will affect the receiver noise figure and sensitivity, and the later one will impact the single tone desensitization performance. A SAW duplexer used in 800 MHz Cellular band CDMA mobile station transceiver, usually has the specifications as shown in Table 6.33.

Table 6.33. Cellular band SAW duplexer specifications

	Minimum	Typical	Maximum
Antenna to receiver			
Frequency (MHz)	869		894
Insertion loss (dB)		2.5	3.5
Rejection* to Tx (dB)	47	50	
Transmitter to antenna			
Frequency band (MHz)	824		849
Insertion loss (dB)		1.5	2.5
Rejection* to Rx (dB)	40	44	

* Manufactories usually employ attenuation instead of rejection for convenience in their test. The attenuation is approximately equal to (rejection + insertion loss).

The low noise amplifier (LNA) follows the duplexer. It generally dominates the receiver sensitivity, and also the single-tone desensitization performance in the CDMA receiver. The main parameters are gain, noise figure, and IIP_3. In addition, for an active device we need also consider its power consumption. Typical specifications of an 800 to 900 MHz LNA used in the CDMA receiver are given in Table 6.34. To meet the IMD performance at –90 dBm and –79 dBm desired signal, the LNA will be switched to its mid-gain mode. In addition, to expand the gain control range of the receiver a low gain operation mode is added in this LNA. At mid- and low gain modes, the LNA is actually bypassed, and attenuators, which have very high linearity or IIP_3, may be used.

Table 6.34. LNA specifications for the direct conversion receiver

	Minimum	Typical	Maximum
Frequency (MHz)	869		894
High gain (dB)	15	16	16.5
Noise figure (dB)		1.5	2.5
IIP_3 (dBm)	6	8	
Current (mA)		5	
Mid-gain (dB)	-3	-2	-1.5
$IIP_{3\ mid\ gain}$ (dBm)	23	25	
Noise figure (dB)	1.5	2	3
Low gain (dB)	-21	-20	-19
$IIP_{3\ low\ gain}$ (dBm)	23	25	
Noise figure (dB)	19	20	21
Input/output impedance (Ω)		50	

An RF SAW band-pass filter is inserted in between the LNA and an RF amplifier. The function of this RF SAW is to further suppress the transmission leakage signal, and therefore the residual transmitter leakage will have no further impact on cross-modulation and single-tone desensitization performance, and the transmitter leakage self-mixing product generated in the UHF down-converter also becomes insignificant. The main parameters of this RF SAW BPF are similar to those of the duplexer, the insertion loss and the rejection to the transmitter leakage. The specifications provided by off the shelf RF SAW filter are presented in Table 6.35.

Table 6.35. Specifications of RF SAW BPF for CDMA receiver

	Minimum	Typical	Maximum
Frequency band (MHz)		849 - 894	
Insertion loss (dB)		2.5	3.0
Rejection* (dB)	27	32	
Input impedance (Ω)		50	
Differential output impedance (Ω)		200	

* See notes given in Table 6.33.

The RF amplifier (RFA) is a preamplifier of the I/Q quadrature down-converter. This amplifier can improve the receiver noise figure when the receiver operates with its LNA in high or mid-gain. On the other hand this amplifier also provides one-step gain control to further extend the receiver gain control range. The specifications of this RFA are shown in Table 6.36.

Table 6.36. Specifications of RF amplifier

	Minimum	Typical	Maximum
Frequency (MHz)	869		894
High gain (dB)	5	6	7
Noise figure (dB)		3	4
IIP_3 (dBm)	8	9	
Current (mA)		6	
Low gain (dB)	-13	-12	-11
$IIP_{3\ low\ gain}$ (dBm)	23	25	
Noise figure (dB)	11	12	13

The I and Q quadrature demodulator or down-converter is the key stage of the direct conversion receiver. It must have a high linearity, that is not only a high third-order intercept point IIP_3, but also a high second order intercept point since this stage dominates the contribution to the second-order nonlinear distortion products in the direct conversion receiver as discussed in the previous subsection. However, achieving the high IIP_3 usually needs spending more current. The high IIP_2 can be obtained based on symmetric circuit design and layout. The proposed specifications for the quadrature demodulator are presented in Table 6.37.

Table 6.37. Specifications of UHF quadrature demodulator

	Minimum	Typical	Maximum
Frequency (MHz)	869		894
Voltage gain (dB)	14	15	16
Noise figure (dB)		10	11.5
IIP_3 (dBm)	9.5	11	
IIP_2 (dBm)	60	65	
Current** (mA)		20	

** The current consumption includes currents of a frequency divider and a buffer amplifier.

The base-band (BB) low-pass filter in the direct conversion receiver plays the role of channel filter. It is assumed in this design that a fifth order elliptic filter is employed in the I and Q both channels, or otherwise the ADC will need more dynamic range or bits. Table 6.38 presents the basic requirements on this BB channel filter.

Table 6.38. Specifications of BB fifth-order low-pass filter

	Minimum	Typical	Maximum
3 dB Bandwidth (kHz)		630	
Voltage gain (dB)		0	
In-band ripple (dB)			0.2
Noise figure (dB)		25	30
IIP_3 (dBm)	30	35	
Rejection at 900 kHz offset (dB)	45		
Rejection at 1.7 MHz offset (dB)	50		
Group delay distortion (μsec)			4
Current consumption*** (mA)		4	

*** The current consumption for the BB filters in both channels

The ICI caused by the BB channel filter group delay distortion must be small enough to provide a carrier-to-interference ratio (CIR) is greater than 15 dB. This is necessary for the CDMA receiver to have a good multi-path fading performance and an expected high data rate operating capability.

The last stage before the ADC in the BB I and Q channels is a fixed gain BBA. This amplifier should be designed with a local DC compensation circuit (DC servo circuit). The main specifications of this amplifier are depicted in Table 6.39.

Table 6.39. Specifications of fixed gain BBA

	Minimum	Typical	Maximum
Frequency (kHz)	0		630
Voltage gain (dB)	33	34	35
Noise figure (dB)	32	28	
IIP_3 (dBm)	16	20	
Current*** (mA)		5	

*** The current consumption for the BB amplifiers in both channels

Finally, the characteristics of the synthesized LO for the CDMA mobile station receiver are in Table 6.40 where only the in-channel-bandwidth integrated phase noise, and phase noise and spurious near the IMD test tones are presented.

Table 6.40. Phase noise and spurious requirements of synthesized UHF LO

	Minimum	Typical	Maximum
Frequency (MHz)	1738		1788
Integrated phase noise within ± 615 kHz (dBc)			-30
Phase noise at 900 kHz offset (dBc/Hz)			-136
Phase noise at 1.7 MHz offset (dBc/Hz)			-138
Spurious near 900 kHz offset (dBc)			-80
Spurious near 900 kHz offset (dBc)			-85
Current consumption (mA)		12	

From these defined specifications of individual stages in the direct conversion receiver chain, we can evaluate the receiver performance. Either the Matlab or the Excel spreadsheet is a very useful tool to evaluate the transceiver performance. Here we are going to use the spreadsheet for our receiver system performance evaluation. In the performance evaluation, the formulas given in Sections 4.2.2 and 4.3.2 are used for the cascaded noise figure and IIP_3 calculations, and the formulas presented in Sections 4.2.1, 4.2.3, 4.3.3, and 4.4.2 need be employed for the receiver sensitivity, IMD performance, single tone desensitization, and two-tone blocking performance estimations. A performance evaluation of the designed CDMA direct conversion receiver in its high LNA gain mode is given in Table 6.41. From this table we can see that based on the specifications defined in Tables 6.33 – 6.40, the direct conversion receiver typically has more than 4 dB margin in the sensitivity, the IMD, and single tone desensitization, and 4 dB margin in the two-tone blocking performance.

Table 6.42 and Table 6.43 provide the performance evaluations at the desired signal level at S_d = –90 dBm and –79 dBm, respectively. These two tables mainly show the IMD performance at S_d = –90 dBm and –79 dBm. It is apparent that the IMD at S_d = –90 dBm has plenty of margin, 8 dB margin, but the IMD performance at S_d = –79 dBm is marginal because it is only 2 dB over spcification in the typical case. This situation can be improved if the IIP_3 of the quadrature demodulator can be increased a couple of dB, or by making the overall IIP_3 looking into the RFA close to 3 dB.

In addition, from Tables 6.42 and 6.43 we can also notice that the receiver sensitivity in the mid-LNA gain mode is just better than –96.5 dBm, and it will be even lower in the worst case. If the mid-gain sensitivity is close to –90 dBm, the hysteresis of the LNA gain switching will become narrow and thus the gain switching may not be reliable. The gain switching hysteresis needs to be greater than 2 dB to avoid unexpected switching, or otherwise the switching decision should be based on the average of multiple RSSI measurements. Increasing LNA or RFA gain will improve the mid gain sensitivity, but it will reduce the overall receiver IIP_3 and impact the IMD performance too. Proper tradeoff between the sensitivity and IMD performance is generally necessary when designing a receiver. On the other hand, the integrated circuit design will be the key to resolve this conflict between the receiver sensitivity and the IMD performance if we can obtain a quadrature demodulator with a high enough IIP_3.

In Table 6.41 to 6.43 and Table 6.44 as well, the sensitivity (MDS), IMD, single tone desensitization two-tone blocking are calculated by using the corresponding formulas given in Chapter 4 instead of the formulas on the left side of the table lower portion, which are just symbolic expressions. On the other hand, in these tables the synthesizer phase noise and spurious are those at ±900 kHz and ±1.7 MHz offset from the LO frequency, respectively, and the unit of the phase noise is dBc/Hz.

Table 6.41. Performance evaluation of CDMA direct conversion receiver in high LNA gain mode

		Line Loss					I/Q Mxr		BB-AMP	ADC_Input	Total
Current (mA)							20				40
Gain (dB)		-0.3	-2.5	16	-2.5	6.0					
V_Gain (dB)	0.0	-0.3	-2.5	16	3.5	6.0	15	0.0	34	0.0	71.70
Fltr Rej (dBc)		@Tx Band	50		25		@ 0.9 MH	45			
		@ Rx band	42				@1.7 MHz	50			
P_{out} dBm	-108.7	-109	-111.5	-95.5	-98	-92					
$V_{\_out}$ mV	0.0008	0.0008	0.0006	0.0038	0.0056	0.0112	0.0630	0.0630	3.16	3.16	
NF dB	0.0	0.3	2.5	1.5	2.5	3.0	10	25	28	40	
Cascade NF dB	**5.62**	5.62	5.32	2.82	13.22	10.72	16.02	29.78	28.03	0.00	
IIP3 dBm	100.00	100.00	100.00	8.00	100.00	9.00	11.00	35.00	20.00	20.00	
Cascade IIP3 (dBm)	**-7.60**	-7.60	-9.04	-12.04	4.01	1.51	10.36	34.98	-14.00	20.00	
Key Parameters for Receiver System Evaluation											
	RX	**FER =**	1.00E-03	**Sd(dBm) =**	-101	**Mixer**		**Synthesier Sp,adj (dB)=**	-85	Sp,atl (dB)=	-90
		CNR (dB) =	-1.5	**BW(Hz) =**	1.23E+06	**IIP2 (dBm)**	65.00	**Nphase=**	-138	Nphase=	-140
	TX	**P_{Tx}(dBm)=**	~ 25	PA Noise in Rx Band (dBm/Hz) =	-138						

			Margin (dBc)	Spec	
1. Receiver Sensitivity					
MDS = -174 + 10 log(BW) + CNR + NF =	**-108.69**	dBm	**4.69**	< -79	dBm
2. Intermodulation Rejection					
IMD = [2(IIP3 - Sd) - CNR] /3 =	**-38.61**	dBm	**4.39**	> -21	dBm
ST = Sd-CNR+2*IIP3-2(P_{Tx} - Rj_dplx) -C =	**-25.52**	dBm	**4.48**	NA	dBm
4. Two Tone Blocking					
$Blck_{2Tone}$ (Sd - CNR + IIP2) / 2 =	**-25.98**	dBm	**4.02**	NA	dBm

Table 6.42. Performance evaluation of CDMA direct conversion receiver in mid LNA gain mode with Sd = -90 dBm

		Line Loss		LNA		RFA	I/C Mxr	LPF	BB-AMP	ADC_Input	Total
Current (mA)		0.0	0.0	0.0	0.0	6.0	20	4.0	5.0		35
Gain (dB)	0.0	-0.3	-2.5	-2	-2.5	6.0					
V_Gain (dB)	0.0	-0.3	-2.5	-2	3.5	6.0	15	0.0	34	0.0	53.70
Fltr Rej (dBc)		@Tx Band	50		25		@ 0.9 MH	45			
		@ Rx band	42				@1.7 MHz	50			
P_{out} dBm	-90	-90.3	-92.8	-94.8	-97.3	-91.3					
$V_{\_out}$ mV	0.0071	0.0068	0.0051	0.0041	0.0061	0.0121	0.0683	0.0683	3.42	3.42	
NF dB	0.0	0.3	2.5	2.0	2.5	3.0	10	25	28	40	
Cascade NF dB	**18.02**	18.02	17.72	15.22	13.22	10.72	16.02	29.78	28.03	0.00	
IIP3 dBm	100.00	100.00	100.00	25.00	100.00	9.00	11.00	35.00	20.00	20.00	
Cascade IIP3 (dBm)	**10.39**	10.39	-9.04	-12.04	4.01	1.51	10.36	34.98	-14.00	20.00	
Key Parameters for Receiver System Evaluation											
	RX	**FER =**	1.00E-03	**Sd(dBm) =**	-90	**Mixer**		**Synthesier** **Sp,adj (dB)=**	-85	Sp,atl (dB) =	-90
		CNR(dB)=	-1.5	**BW(Hz) =**	1.23E+06	**IIP2 (dBm)**	65.00	**Nphase=**	-138	Nphase=	-140
	TX	P_{Tx}**(dBm)=**	~ 17	**PA Noise in Rx Band (dBm/Hz)**	-138						

Receiver Performance Evaluation

				Margin (cBc)	Spec	
MDS =	-174 + 10 log(BW) + CNR + NF =	**-96.57**	dBm	**6.57**	< -79	dBm
2. Intermodulation Rejection						
IMD =	[2(IIP3 - Sd) - CNR] /3 =	**-23.25**	dBm	**8.75**	> -21	dBm
3. Single Tone Desensitization						
ST =	Sd-CNR+2*IIP3-2(P_{Tx} - Rj_dplx) -C =	**-13.13**	dBm	**NA**	NA	dBm
4. Two Tone Blocking						
$Blck_{2Tone}$	(Sd - CNR + IIP2) / 2 =	**-11.64**	dBm	**NA**	NA	dBm

Table 6.43. Performance evaluation of CDMA direct conversion receiver in mid LNA gain mode with Sd = -79 dBm

		Line Loss		LNA		RFA	I/Q Mxr	LPF	BB-AMP	ADC_Input	Total
Current (mA)		0.0	0.0	0.0	0.0	6.0	20	4.0	5.0		35
Gain (dB)	0.0	-0.3	-2.5	-2	-2.5	6.0					
V_Gain (dB)	0.0	-0.3	-2.5	-2	3.5	6.0	15	0.0	34	0.0	53.70
Fltr Rej (dBc)		@Tx Band	50		25		@ 0.9 MH	45			
		@ Rx band	42				@1.7 MHz	50			
P_{out} dBm	-79	-79.3	-81.8	-83.8	-86.3	-80.3					
V_{out} mV	0.0251	0.0242	0.0182	0.0144	0.0216	0.0431	0.2424	0.2424	12.15	12.15	
NF dB	0.0	0.3	2.5	2.0	2.5	3.0	10	25	28	40	
Cascade NF dB	**18.02**	18.02	17.72	15.22	13.22	10.72	16.02	29.78	28.03	0.00	
IIP3 dBm	100.00	100.00	100.00	25.00	100.00	9.00	11.00	35.00	20.00	20.00	
Cascade IIP3 (dBm)	**10.39**	10.39	-9.04	-12.04	4.01	1.51	10.36	34.98	-14.00	20.00	

Key Parameters for Receiver System Evaluation

	RX	**FER =**	1.00E-03	**Sd(dBm) =**	-79	**Mixer**		**Synthesier Sp,adj (dB)=**	-85	Sp,atl (dB)=	-90
		CNR(dB)=	-1.5	**BW(Hz) =**	1.23E+06	**IIP2 (dBm)**	65.00	**Nphase=**	-138	Nphase=	-140
	TX	**P_{Tx}(dBm)=**	~ 6	**PA Noise in Rx Band (dBm/Hz)**	-138						

				Margin (dBc)	Spec	
1. Receiver Sensitivity						
MDS =	-174 + 10 log(BW) + CNR + NF =	**-96.57**	dBm	**17.57**	< -79	dBm
2. Intermodulation Rejection						
IMD =	[2(IIP3 - Sd) - CNR] /3 =	**-18.98**	dBm	**2.02**	> -21	dBm
ST =	Sd-CNR+2*IIP3-2(P_{Tx}	**-1.13**	dBm	**NA**	NA	dBm
4. Two Tone Blocking						
$Blck_{2Tone}$	(Sd - CNR + IIP2) / 2 =	**-5.64**	dBm	**NA**	NA	dBm

6.2.1.4. Dynamic Range, AGC, and DC Compensation

The dynamic range for the CDMA mobile station receiver is specified such that from –104 dBm to –25 dBm. The receiver shall properly operate at –25 dBm with a FER less than 1%, and therefore, the total minimum dynamic range is 79 dB. If we consider a 5 dB margin on the sensitivity and on the maximum input signal, the actual dynamic range is 89 dB or higher. In normal case, the receiver AGC range should cover not only this dynamic range, but also the receiver chain gain variation due to temperature and supply voltage changes (even after assuming the gain variation caused by the IC processing can be calibrated out). Then the AGC covering range may go up to 100 dB. However, this direct conversion receiver has only gain control range of 54 dB resulting from the LNA 2 gain-steps and the RFA one gain step, 18 dB each step. The shortage of the gain control range is made up by using a greater than 60 dB dynamic range or approximately 10 effective bit ADC as discussed in Section 6.2.1.1.

The receiver performance at –25 dBm CDMA signal input is evaluated and it is shown in Table 6.44. Operating in such a high input signal level, the LNA of the receiver is running at its –20 dB low gain mode and the RFA is also operating at –12 dB low gain mode. From Table 6.44, we can see that the receiver sensitivity at low LNA and RFA gain mode operating condition is approximately –61 dBm, and this means that we can switch to this mode as long as the received signal level is greater than –55 dBm. It is apparent that staying in LNA and RFA low gain modes time longer we can save more power consumption since the receiver spends the lowest current in this gain mode. However from fading performance point of view, we would like to make this gain switch taking place as late as possible and thus the range in which the receiver operates at the LNA mid gain mode will be wider. The RFA and LNA both gains in low is better to happen at the received signal level close to –35 dBm.

The AGC in this direct conversion receiver is digital in the sense that the gain control in analog domain is a discrete step control instead of continuous control and the fine gain control is performed in the digital domain through a digital variable gain amplifier (DVGA). An AGC loop incorporating DC offset compensation is proposed in [5], and its simplified block diagram is depicted in Fig. 6.5.

Table 6.44. Performance evaluation of CDMA direct conversion receiver in low LNA gain mode with Sd = -25 dBm

		Line Loss	Duplxr	LNA	RF SAW	RFA	I/Q Mxr	LPF	BB-AMP	ADC_Input	Total
Current (mA)		0.0	0.0	0.0	0.0	0.0	20	4.0	5.0		29
Gain (dB)		-0.3	-2.5	-20	-2.5	-12.0					
V_Gain (dB)		-0.3	-2.5	-20	3.5	-12	15	0.0	34	0.0	17.70
Fltr Rej (dBc)		@Tx Band	50		25		@ 0.9 MH	45			
		@ Rx band	42				@1.7 MHz	50			
Pout dBm	-25	-25.3	-27.8	-47.8	-50.3	-62.3					
V_out mV	12.57	12.15	9.11	0.91	1.36	0.34	1.93	1.93	96.49	96.49	
NF dB		0.3	2.5	22	2.5	12.0	10	25	28	40	
Cascade NF dB	**53.32**	53.32	53.02	50.52	30.52	28.02	16.02	29.78	28.03	0.00	
IIP3 dBm	100.00	100.00	100.00	25.00	100.00	25.00	11.00	35.00	20.00	20.00	
Cascade IIP3 (dBm)	**27.73**	27.73	-9.04	-12.04	4.01	1.51	10.36	34.98	-14.00	20.00	
Key Parameters for Receiver System Evaluation											
	RX	**FER =**	1.00E-03	**Sd(dBm) =**	-25	**Mixer**		**Synthesier Sp,adj (dB)=**	-85	Sp,atl (dB)=	-90
		CNR (dB)=	-1.5	**BW(Hz) =**	1.23E+06	**IIP2 (dBm)**	65.00	**Nphase=**	-138	Nphase=	-140
	TX	**PTx (dBm)=**	~ -50	**PA Noise in Rx Band (dBm/Hz) =**	-138						
Receiver Performance Evaluation											
1. Receiver Sensitivity							Margin (dBc)		Spec		
	MDS = -174 + 10 log(BW) + CNR + NF =				**-61.28**	dBm	**36.28**		< -25	dBm	
2. Intermodulation Rejection											
	IMD = [2(IIP3 - Sd) - CNR] /3 =				**10.65**	dBm	**NA**		NA	dBm	
3. Single Tone Desensitization											
	ST = Sd-CNR+2*IIP3-2(PTx-Rj_dplx) -C =				**52.95**	dBm	**NA**		NA	dBm	
4. Two Tone Blocking											
	$Blck_{2Tone}$ (Sd - CNR + IIP2) / 2 =				**39.40**	dBm	**NA**		NA	dBm	

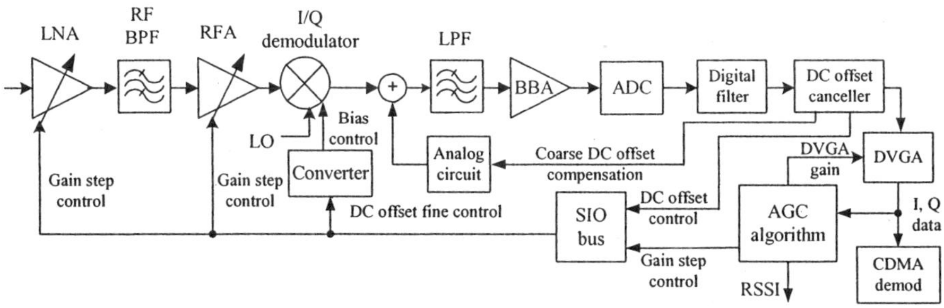

Figure 6.5. Digital AGC loop with DC offset compensation

In Fig. 6.5, the AGC loop is located after the digital filter and the DC offset canceller. The direct conversion receiver has no IF SAW channel filter and low-order low-pass filters are usually used in the analog base-band to save receiver IC die size and to reduce its cost. Thus the input to the ADC may still contain a certain amount of strong interferer power. The digital filter provides enough suppression to the out of channel interferers. After the digital filter, the received signal strength (RSS) can be accurately measured with negligible impact from interferers, and the AGC is then ruled by the in-channel-bandwidth signal strength only. On the other hand, we also do not like to see that the DC offset of the direct conversion receiver affects the AGC loop function, and thus it becomes a natural arrangement that the AGC loop is located behind the DC canceller.

There are two types of gain control in this AGC loop. The step gain control at the LNA and RFA stages is referred to as the coarse gain control, and the DVGA in the digital domain performs the fine gain control with a resolution close to 0.1 dB or even lower. In this example, only three gain steps are used for simplicity as described below in detail, but it is possible to employ more steps to cover the same control range or greater in the practical receiver design for reducing step size and therefore decreasing possible transient voltage amplitude overshoot or DC offset variation due to the gain step change.

In this direct conversion receiver, in order to meet the IMD performance requirements defined at the desired signal levels of –90 dBm and –79 dBm, the LNA gain from its high to mid or vice versa must take place at the desired signal level below –90 dBm if we would like to keep the receiver current consumption as specified in the previous section. The following gain step change is the RFA gain switched from its high to low when the desired signal increases equal to –54 dBm and the RFA gain is switched back to its high gain as the signal reduces to –57 dBm. These gain switches immediately cut down the receiver current consumption since the

LNA and the RFA are bypassed and they are sequentially turned off. When the desired signal reaches at –36 dBm, the LNA gain further steps down from its mid-gain to the low gain, and the LNA gain will be back to its mid-gain while the signal power decreases to –39 dBm. At the later two gain switch points (i.e., at -54 dBm and –38 dBm), the CDMA signal rms voltage at the input of the ΔΣ ADC should be slightly less than 200 mV because of the considerations of the CDMA signal crest factor and the constructive fading. The LNA and RFA gain step changes and the corresponding hysteresis are summarized in Fig. 6.6. An altenative of using the power hysteresis for preventing gain switching back and forth between two adjacent gain steps is to employ a temporal hysteresis instead.

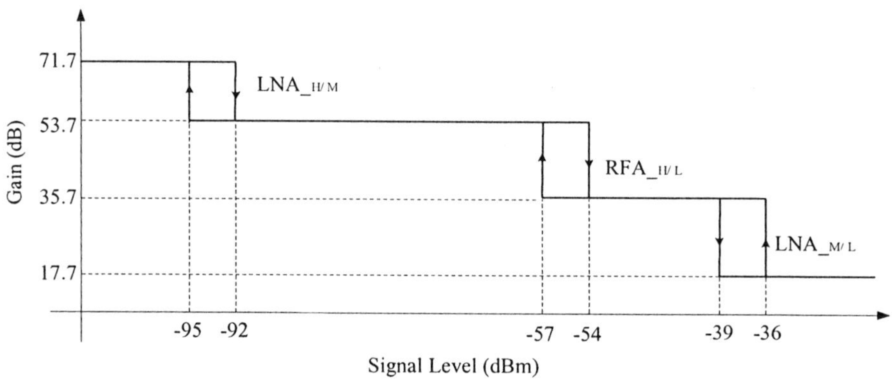

Figure 6.6. The LNA and RFA gain steps and hysteresis

In the DC offset control loop, the coarse compensation is carried out by providing a DC component opposite to the DC offset to the BB filter input via an analog circuitry in the analog base-band, and the fine compensation is achieved by adjusting the quadrature demodulator bias through the SIO control data generated by the DC offset canceller as shown in Fig. 6.5. The DC offset control loop has two operation modes, acquisition and tracking modes. The acquisition mode is used to quickly remove large DC offset as a result of a step change in the gains of the LNA and /or RFA, or the overall DC loop performing a periodic DC update, or other reasons. The tracking mode is used to perform DC offset correction in a normal manner, and its response is slower than that of the acquisition mode. However, for CDMA mode only receiver the complicated DC offset correction loop may not be necessary, and it can be simply replaced by employing AC coupling or high-pass circuitry design to block the DC and to solve DC offset problem as described in Section 3.2.

The details of the receiver AGC loop and DC offset correction loop designs are out of the scope of this book. Readers can refer to [5] if interested in these loop designs.

The example of the cellular band CDMA direct conversion receiver system design has been presented. It is just an exercise of direct conversion receiver system design, but it is definitely not optimized for practical application.

6.2.2. Transmitter System Design

The direct conversion transmitter does not have so severe technical challenges as those in the direct conversion receiver. First, the transmission signal is usually a controllable and deterministic signal. Second, the I and Q channels in the analog base-band of the direct conversion transmitter are much simpler than those of the direct conversion receiver, and they only contain a buffer amplifier with no gain and a low order passive LPF in each channel. The I and Q channel amplitude and phase imbalances, and the DC offset, thus, can be generally calibrated out or to an insignificant level in the initial tuning of individual mobile station. Third, the phase noise requirements of the synthesized LO for the CDMA transmitter are much lower than those required by the receiver. The transmitter block diagram applied for this design example has been presented in Fig. 6.4.

The minimum performance requirements for the CDMA direct conversion transmitter are exactly the same as defined in [4], and the main specifications for cellular band CDMA transmitter are presented in Table 6.45. The transmission power here is defined as effective radiated power (ERP), which has been explained in Chapter 5.

Table 6.45. Main specifications of CDMA mobile transmitter

Items	Min. Specification	Notes
Frequency band (MHz)	824 — 849	Band cass 0
Frequency accuracy (Hz)	≤ 300	
Maximum output power (dBm)	≥ 23	ERP
Minimum output power (dBm)	≤ -50	ERP
Standby output power (dBm/1 MHz)	≤ -61	
Gated output power change (dB)	≥ 20	With Tx FIR gate off
Adjacent channel power ratio (ACPR1) within 885 to 1.98 MHz frequency offset	≤ -42 or ≤ -54	dBc/30 kHz less stringent one dBm/1.23MHz

Table 6.45. (Cont.)

Alternate channel power ratio (ACPR2) within 1.98 to 4.0 MHz frequency offset	≤ -54 or ≤ -54	dBc/30 kHz less stringent one dBm/1.23MHz
Conducted spurious and noise emission beyond 4.0 MHz frequency Offset	< -13	dBm/1 kHz, 9 kHz ≤ f < 150 kHz dBm/10 kHz, 150 kHz ≤ f < 30 MHz dBm/100 kHz, 30 MHz ≤ f < 1 GHz dBm/1 MHz, 1 GHz ≤ f ≤ 12.5 GHz
Waveform quality factor, ρ	≥ 0.944	
Code channel power accuracy (dB)	± 0.25	At any data rate
Open Loop AGC accuracy (dB)	±9.5	
Closed loop AGC Range (dB) Accuracy (dB) Step size (dB) Response time (μsec)	 ±24 ±0.5 ±0.5 ≤ 200	

6.2.2.1. CDMA Transmission Signal and Output Power

The CDMA mobile station transmission signal not only has phase modulation, QPSK in IS-95 or HPSK in cdma2000_1x, but also amplitude modulation. The peak-to-average ratio (PAR)[†] of the IS-95 CDMA signal at 9.6 kbps data rate is approximately 3.85 dB. The PAR of cdma2000_1x signals is channel configuration dependent, and it can be as high as 5.4 dB in the case of the channel configuration: pilot + dedicate control channel (DCCH). A high PAR signal demands the transmitter to possess a high linearity to achieve the same adjacent channel power emission as obtained from a low PAR signal when the transmission power in both cases is the same. This means that for a high PAR signal the transmitter will operate in low power efficiency or high power consumption.

The spectrum of a cdma2000 channel configuration pilot + DCCH transmission signal passing through a power amplifier with a linear gain 27.4 dB and 1 dB compression point 31 dBm is shown in Fig. 6.7. The output power is only 26.8 dBm while the PA input is 0 dBm, and the adjacent channel power ratio is -42.2 dB just meeting the minimum performance requirement ≤-42 dB. The same PA is employed to amplify the cdma2000 voicc data with a PAR approximate 3.4 dB. The output power is

[†] The PAR is based on 99% CDF

now equal to 37.4 dBm when the input is still 0 dBm, and the worst ACPR1 becomes –48.4 dB — i.e., 6 dB better.

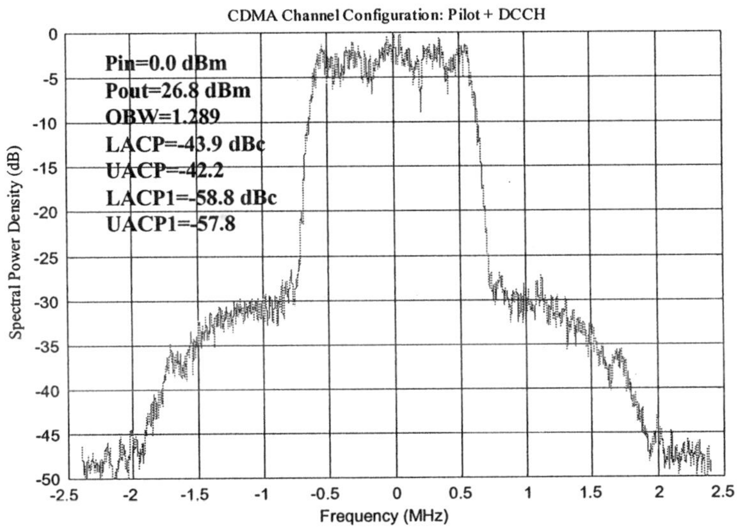

Figure 6.7. SPD of cdma2000 transmission configuration pilot+DCCH

Fortunately, in the minimum performance standard IS-98E a transmission power back-off 2 to 2.5 dB is allowed for high PAR signal transmissions as shown in Table 6.46. Therefore, it is not necessary to increase the transmitter linearity or to consume more power for the high PAR signal transmission if one takes the advantage of the power back-off. After the power back-off of high PAR transmissions, the PAR of the channel configuration [Pilot + FCH (9600bps) + SCH1 (153600bps)] dominates the

Table 6.46. Maximum transmission power back-off of high PAR signals

Cdma2000 Channel Configuration [4]	PAR (dB)	Maximum Power Back-Off (dB)
Pilot + DCCH(9.6k)*	5.4	2.5
Pilot+DCCH(9.6k)+ FCH(15k)**	5.2	2.0
Pilot+FCH(9.6k)+SCH1(9.6k)***	4.5	2.0
Pilot+FCH(9.6k) (Voice)	3.2	0
Pilot+FCH(9.6k)+SCH1(76.8k)	3.9	0
Pilot+FCH(9.6k)+SCH1(153.6k)	4.1	0
IS95	3.9	0

* DCCH: Dedicated Control Channel
** FCH: Fundamental Channel
*** SCH: Supplement Channel

requirement of the transmitter linearity. In a practical CDMA transmitter design, it may not be necessary to use the full amount of the allowed maximum power back-off. It has been verified that in most cases 0.5 to 1 dB back-off is enough to release the pressure on the PA linearity requirement.

The output power of the CDMA mobile station transmitter measured at the antenna port should be equal to or greater than 23 – ($G_{\_ant.}$ – 2.15) dBm, where $G_{\_ant}$ is the antenna of the mobile station at Cellular band, and 2.15 dB is the gain of a half wavelength dipole antenna. In this case, the transmitted ERP of the mobile station will be equal to or greater than 23 dBm. This output power level is not difficult to achieve as long as the overall transmitter chain has enough gain. However, it is more important that the reverse link transmission signal not only has proper power level, but also low ACPR and high waveform quality factor, which is discussed in the following sections.

A power detector is used to measure the transmission power, and its main function is to prevent the transmission power from going beyond a defined maximum level, which may violate the SAR specification or damage the power amplifier. When it detects that the output power is too high, the transmitter chain gain will be automatically reduced, or the transmitter will even be completely shut down.

6.2.2.2. Waveform Quality Factor (ρ) Evaluation

The waveform quality factor is another measurement of the modulation accuracy. It is related to error vector magnitude (EVM) by (5.3.13) in Section 5.3.1. The main contributions to the ρ degradation are the in-channel-band phase noise and spurious of synthesized LO, carrier leakage, image product caused by I and Q channel signal imbalance, reverse modulation due to the second harmonic current of the transmission signal reflected by the load of the drive amplifier and modulating the UHF VCO and the transmitter in-band noise and spurious dominating ρ in the low transmission power condition. A reasonable budget assigned to these contributors to ρ is presented in Table 6.47. From the budget data given in this table and using formula (5.3.47), we can calculate the typical and worst case ρ as follows:

$$\rho_{typical} = \frac{1}{1 + 10^{\frac{-27}{10}} + 10^{\frac{-34}{10}} + 10^{\frac{-32}{10}} + 10^{\frac{-33}{10}} + 10^{\frac{-12-25}{10}}} = 0.996$$

and

$$\rho_{worst} = \frac{1}{1+10^{\frac{-25}{10}}+10^{\frac{-27}{10}}+10^{\frac{-28}{10}}+10^{\frac{-30}{10}}+10^{\frac{-72+50}{10}}} = 0.986.$$

Actually, $\rho_{typical}$ represents the typical ρ in the high power case, and ρ_{worst} expresses the worst ρ in the –50 dBm output power case.

Table 6.47. Budget of contributors to waveform quality factor

Contributor	Typical	Worst Case
LO integrated phase noise and spurious (dBc)	-27	-25
Carrier leakage (dBc)	-34	-27
Image product (dBc)	-32	-28
Reverse modulation (dBc)	-33	-30
Transmitter in-band noise at maximum output power (dBm)	-12	-10
Transmitter in-band noise at –50 dBm output power (dBm)	-75	-72

6.2.2.3. Adjacent / Alternate Adjacent Channel Power and Linearity

The adjacent/alternate channel power level is a measure of the transmitter chain linearity. The adjacent channel power (ACP_{adj}) mainly results from third-order distortion, and the alternate channel power (ACP_{atl}) is predominantly contributed from fifth-order distortion products.

Assuming that the adjacent channel power ratio ($ACPR_{adj}$) at the maximum transmission power of 25 dBm for the signal with a PAR = 3.85 dB is designed to be equal to or less than –48 dBc/30 kHz, the minimum OIP_3 of the overall transmitter chain, OIP_{3_Tx}, can then be determined by means of (5.4.13)

$$OIP_{3\_Tx} = 25 - \left(10\log\left(10^{\frac{-48}{10}} - 10^{\frac{-60}{10}}\right) + 9 - 0.72 + 16\right) \bigg/ 2 = 37.0 \text{ dBm}.$$

A gain and OIP_3 partitioning of the transmitter to achieve $ACPR$ = -48 dBc performance is presented in Table 6.48. Here it is assumed that the digital BB DAC output is voltage, the maximum swing of the DAC output is 1.25 V peak-to-peak, and the minimum signal-to-(noise+distortion) ratio within the signal bandwidth, 615 kHz, is 35 dB. In this table, the function block 'Tx IC Chain' represents all the stages from the LPFs, I/Q modulator, to the driver amplifier in the transmitter block diagram of Fig. 6.4, and this function block is usually integrated on a silicon die referred to as transmitter chip, or simplified as Tx IC.

Table 6.48. Gain and OIP_3 partition and ACPR of the transmitter

Function Block	DAC Output	Tx IC Chain	Balun + SAW Fltr	PA	Coupler	Duplexer	Trace Loss	Antenna Port
Power Gain (dB)		23.6	-4.0	28.0	-0.4	-3.0	-0.2	
Cascade Gain (dB)		23.6	22.2	50.2	49.8	46.8	46.6	44.0
Pout (dBm)	-19.0	4.6	0.6	28.6	28.2	25.2	25.0	25.0
Vout (mVp-p)	1250							
NF in Tx Band (dB)		17.3	4.0	5.0	0.4	3.0	0.2	
Cascade NF (dB)		17.3	17.3	17.3	17.3	17.3	17.3	17.3
OIP3(dBm)		23.5	100.0	41.6	100.0	100.0	100.0	
Cascade OIP3 (dBm)		23.5	19.5	40.6	40.2	37.2	37.0	37.0
ACPRadj (dBc/30 kHz)	-60	-57.9	-57.9	-48.0	-48.0	-48.0	-48.0	-48.0

The alternate channel power ratio ($ACPR_{atl}$) at the maximum output power is mainly determined by the fifth-order nonlinearity of the PA since most of power amplifiers practically used in the CDMA mobile stations possess gain expansion near the maximum output power region and the high-order nonlinearity of the other stages in the transmitter chain is quite weak. In reality, the $ACPR_{atl}$ of the CDMA mobile transmission at the maximum output power is usually 10 dB or more below the $ACPR_{adj}$ — i.e., the $ACPR_{atl}$ will be –58 to –60 dBc if the $ACPR_{adj}$ is equal to –48 dBc. An accurate way to determine the $ACPR_{atl}$ is simulation by means of the PA amplitude and phase versus input signal level characteristics as described in Subsection 5.4.1. However, we may roughly estimate the fifth-order output intercept point (OIP_5) by using similar approach to the OIP_3 as discussed in Subsection 5.4.2. In this case, the $ACPR_{atl}$ is approximately expressed as

$$ACPR_{atl} \cong 4\cdot\left(P_{Tx} - OIP_5\right) - 15 - 10\log\left(30\times10^3/1.23\times10^6\right)\ \text{dBc}, \quad (6.2.5)$$

and the OIP_5 for a defined $ACPR_{atl}$ is roughly

$$OIP_5 \cong P_{Tx} - \frac{ACPR_{atl} + 31}{4}\ \text{dBm}. \quad (6.2.6)$$

If designing for $ACPR_{atl}$ = -60 dBc, the minimum OIP_5 is estimated as

$$OIP_5 \cong 25 - \frac{-60+31}{4} \cong 32.3 \text{ dBm}.$$

In the real case, the PA close to its maximum output power not only experiences fifth-order nonlinearity, but also seventh-order distortion. The latter one may also have a significant contribution to the $ACPR_{atl}$. Thus the OIP_5 should be greater than that given by (6.2.6).

6.2.2.4. Noise and Spurious Emission in Receiver Band

Another important emission specification is the noise/spurious emission in the CDMA mobile station receiver band. This emission may desensitize the receiver if its level is high. It should be noticed that only the excess noise — i.e., the noise portion beyond the thermal noise cases the receiver desensitization. The allowed desensitization is assumed to be 0.2 dB or less.

Table 6.49 presents the gain and noise figure of the transmitter in the receiver band, and the noise emission in the receive band as well. Where it is assumed that the DAC output noise level in the receiver band is 15 nV/√Hz, the source impedance is around 200 Ohm, the transmitter SAW has a 30 dB rejection in the receiver band, and the receiver band suppression of the duplexer is 42 dB. We can see that the conducted noise emission in the receiver band is –173.8 dBm/Hz, which is calculated by utilizing (5.5.5)

Table 6.49. Transmitter noise figure and noise emission in receiver band

Function Block	DAC Output	Tx IC Chain	Balun + SAW Fltr	PA	Directional Coupler	Duplx	Trace Loss	Antenna Port
Power Gain		23.6	-4.0	28.0	-0.4	-3.0	-0.2	
Cascad Gain		23.6	22.2	50.2	49.8	46.8	46.6	44.0
Pout (dBm)	-19.0	4.6	0.6	28.6	28.2	25.2	25.0	25.0
Vout (mVp-p)	1250							
Rej in Rx Band (dB)			30.0			42.0		
Cascad Gain in Rx Band (dB)		3.6	-30.4	-2.4	-2.8	-47.8	-48.0	-48.0
NF in Rx Band (dB)		37.7	34.0	5.0	0.4	45.0	0.2	
Cascad NF @ Rx		37.7	38.5	39.7	39.7	48.0	48.2	48.2
Noise in Rx band (dBm/Hz)	15 nV/Hz^(1/2)	-132.7	-165.9	-136.7	-137.1	-173.8	-173.8	-173.8

$$P_{N,Tx\_out} = 10^{\frac{-48}{10}} \left[\left(10^{\frac{48.2}{10}} - 1 \right) \cdot 10^{\frac{-174}{10}} + 10^{\frac{-174+18.5}{10}} \right] \cong 10^{\frac{-173.8}{10}} \text{ mW/Hz.}$$

In the above calculation, the DAC noise 15 nV/√Hz equivalent to 18.5 dB beyond the thermal is used.

The receiver desensitization due to the –173.8 dBm/Hz noise emission in the receiver band can be evaluated by means of (4.2.32) and (5.5.8) as follows. The emitted excess noise in the receiver band is equal to

$$P_{\Delta N\_Tx\_out} = P_{N,Tx\_out} - kT_o = 10^{\frac{-173.8}{10}} - 10^{\frac{-174}{10}} = 1.88 \times 10^{-19} \text{ dBm/Hz.}$$

Assuming that the receiver noise figure is 5.8 dB equivalent to –108.5 dBm sensitivity as given in Table 6.41, we obtain the receiver equivalent noise figure under the transmitter emission noise influence from (4.2.32) to be

$$NF_{Rx} = 10\log\left(\frac{1.88 \times 10^{-19}}{10^{-174/10}} + 10^{5.8/10} \right) = 10\log(0.047 + 3.8) \cong 5.85 \text{ dB.}$$

Comparing with the original receiver noise figure 5.8 dB, we can see the overall noise figure degradation is only 0.5 dB. It is definitely acceptable.

6.2.2.5. Automatic Gain Control and Power Consumption Management

In the CDMA mobile station, there exist two transmission power control loops — i.e., the open loop control and closed loop control [4]. The open loop power control is based mainly on the received signal strength. The mean transmission power is determined by the following power calculation formula for the 800 MHz band

$$P_{Tx\_OpenLoop} = -P_{Rx} - 73 + \text{Correction factors dBm.}$$

where P_{Rx} is the received signal power, the correction factors are defined in [4], and they are equal to 0 in normal operation. The open-loop power control accuracy is ±9.5 dB. The open-loop powers are calibrated at the received power levels as shown in Table 6.50.

Table 6.50. Calibration levels of open-loop power

P_{Rx} (dBm/1.23 MHz)	$P_{Tx\_OpenLoop}$ (dBm/1.23 MHz)
-25	-48 ± 9
-65	-8 ± 9
-93.5	20 ± 9

The CDMA mobile station provides a closed-loop adjustment to its open loop power estimation. Adjustments are made in response to valid received power control bits. The minimum closed-loop control range is ±24 dB around the open-loop power estimation. The step size of the closed-loop control can be 1 dB, 0.5 dB, or 0.25 dB. The accumulated error over 16 consecutive same direction steps for the 1 dB step control is ±3.2, or the average error is ±0.2 dB per step. The accumulated errors for the 0.5 dB and 0.25 dB step controls are ±4.0 dB and ±4.8 dB, respectively, over 32 and 64 consecutive same direction steps, or the average errors are ±0.125 dB and ±0.075 dB per step, respectively. The time response of single step control shall be less than 500 μsec following the reception of a valid closed-loop control bit to the mean output power reaching its final value within 0.3 dB for the 1 dB step control or 0.15 dB for the 0.5 and 0.25 dB step controls.

The dynamic range of the CDMA mobile station transmitter output power is approximately 75 dB from 25 dBm to –50 dBm. To cover this dynamic range, the RF VGA in the transmitter chain usually has a gain control range 90 dB and the PA has a gain adjustment around 6 to 10 dB when its reference current or voltage (depending on the type of PA) changes in order to reduce the current consumption. If the VGA has a linear gain control characteristic, an 8-bit or 10-bit DAC to control the 90 dB gain variation is enough for the 0.5 or the 0.25 dB step control. However, the VGA gain versus control voltage curve in most cases is nonlinear, and a 10-bit control DAC is needed to meet the control accuracy requirement for the 0.5 dB step control. Depending on the nonlinearity of the VGA control curve the 10-bit DAC may also be used for 0.25 dB step control. Usually, the DAC for controlling the PA reference current or voltage needs just a few bits in the case of step control, or otherwise a high resolution DAC may be adopted if a continuous control approach must be used for current consumption saving.

A simplified block diagram of the CDMA mobile station transmitter AGC is depicted in Fig. 6.8. The transmitter AGC, Tx_AGC, is determined by the received signal strength, Rx_Power, the closed-loop control bit, and code channel gain adjustment. The BB I and Q signal level from the data DAC is linearly controllable through adjusting the reference voltage V_{Ref} or reference current I_{Ref} based on the transmission power. The adjustment range of the BB signal can be up to 20 dB, and it also increases the same amount of the AGC dynamic range resulting from the RF VGA.

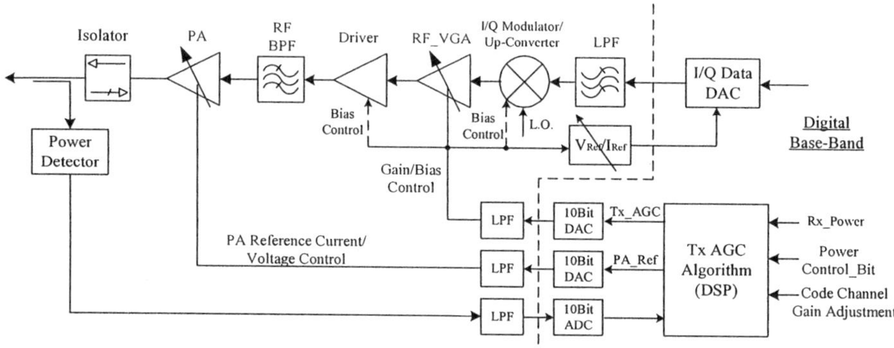

Figure 6.8. Simplified block diagram of transmitter AGC

Properly managing the power consumption of the transmitter chain plays a very important role in extending talk time of a mobile station. It is commonly used in the power management scheme that the bias current of the PA and the rest stages of the transmitter chain shall be adjustable with the transmission power but still keep the overall transmitter having high enough linearity. The bias control of the transmitter chain is usually implemented incorporating with the AGC. The current consumption of a mobile transmitter is often measured based on the CDG [8] statistical average.

The PA bias current is generally controlled by its reference current/voltage. Under acceptable linearity, the PA bias current should be adjusted down as fast as possible with the transmission power decreasing since the PA is one of the major power consumers in a mobile station. The PA reference current/voltage control, PA_Ref, can be either in steps or in a continuous way. In the step control case, the corresponding PA bias current is switched at certain power levels, which are selected based on the characteristics of the PA — i.e., its linearity and current consumption versus output power level. On the other hand the reference current/voltage will not only determine the PA current but also affect its gain. This gain change due to the reference current/voltage adjustment needs to be compensated by the VGA gain. At present, the CDG average current of a cellular band PA is around 55 mA.

It is similar to the PA case. The current consumption of the stages from the quadrature modulator up to the driver amplifier in the transmitter need be adjusted down as fast as and as low as possible with the transmission power decreasing, but the linearity of these stages should be still good enough to keep the ACPR of the output signal from them reasonable low — say, below –56 dBc. The adjustment of the DAC ouput BB signal level and therefore the input signal level to the quadrature (I/Q) modulator can further reduce the power consumption of the mobile station transmitter.

The design of the transmitter AGC is tightly associated with the detail circuitry design of the transmitter chain, and the PA performance. It is not practical to do further detailed discussions on this without necessary information.

6.2.2.6. Frequency Accuracy and Transmitter Synthesizer Performance

The frequency accuracy is mainly determined by the VCTCXO and the AFC system of the mobile station. The VCTCXO is used as a mobile station system reference frequency, and one of the commonly employed reference frequencies in CDMA is 19.2 MHz. The main specifications of a 19.2 MHz VCTCXO are presented in Table 6.51.

Table 6.51. Main specifications of 19.2 MHz VCTCXO

	Minimum	Typical	Maximum
Frequency (MHz)		19.2	
Preset accuracy (ppm)	-1		+1
Frequency temperature stability (ppm)	-2		+2
Aging stability (5 years) (ppm)	-4		+4
Output level (V)		1.0	
Second harmonic			-20
Control range (ppm)		± 12	
Settling time (msec)			5
Phase Noise (dBc/Hz) @ Offset Freq. 10 Hz 1 kHz 100 kHz			 -70 -130 -135

A simplified AFC loop is shown in Fig. 6.9. It consists of RF down-converter, LPF, frequency detector (FD), loop filter, integrator, and digital to frequency converter. The AFC loop is actually a digital frequency locked loop. Its design is out of this book's scope, and will not be further discussed here. Readers can refer to [6] and [7] if interested in details.

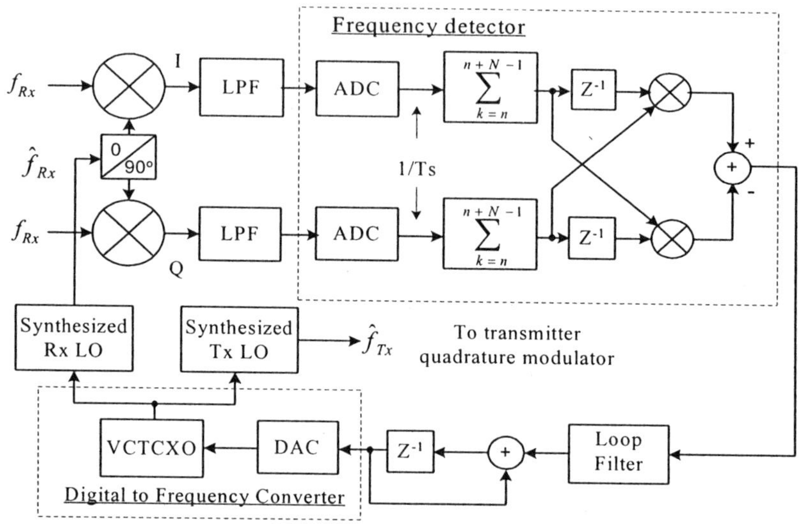

Figure 6.9. Simplified CDMA digital frequency locked loop

The receiver synthesized LO frequency $\hat{f}_{Rx}$ should be locked on the received signal carrier frequency f_{Rx} under the AFC loop control. The transmitter synthesized LO frequency $\hat{f}_{Tx}$ must be very close to the reverse link channel frequency within ±300 Hz tolerance. From the VCTCXO specification presented in Table 6.50, we can estimate that the frequency acquisition of the AFC loop needs approximately ±2.5 ppm or ±2.2 kHz.

In addition to the frequency accuracy of the synthesized LO in the CDMA transmitter, the most important specification is the integrated phase noise and spurs over the transmission signal bandwidth. The out of signal bandwidth phase noise and spurious requirements of the transmitter LO are much lower than those of the receiver LO. The minimum performance of the transmitter synthesized LO is summarized in Table 6.52.

Table 6.52. Minimum performance of transmitter synthesized LO

	Minimum	Maximum
Frequency (MHz)	824	849
Integrated phase noise in signal bandwidth (dBc/1.23 MHz)		-25
Phase noise at ± 885 kHz offset (dBc/Hz)		-120
Phase noise at ± 1980 kHz offset (dBc/Hz)		-128
Phase noise at ± 4.0 MHz offset (dBc/Hz)		-135
Spurious beyond 885 kHz offset (dBc)		< -80

Generally speaking, the minimum performance of the transmitter LO presented in Table 6.52, which represents actually the performance at the transmission carrier frequency, is not difficult to be implemented even if the VCO is integrated on the transmitter IC.

The LO signal can be generated by an LO generator, which consists of a by 4 divider, a single side band frequency converter, and a buffer amplifier as shown in Fig. 6.10. The left part of this figure is a simple block diagram of an integer N synthesizer with a comparison frequency of 48 kHz. The LO frequency is twice of the transmission carrier frequency. In this case, the π/2 phase shifter in the I/Q modulator circuitry of Fig. 6.4 transmitter side should be replaced by a by 2 frequency divider.

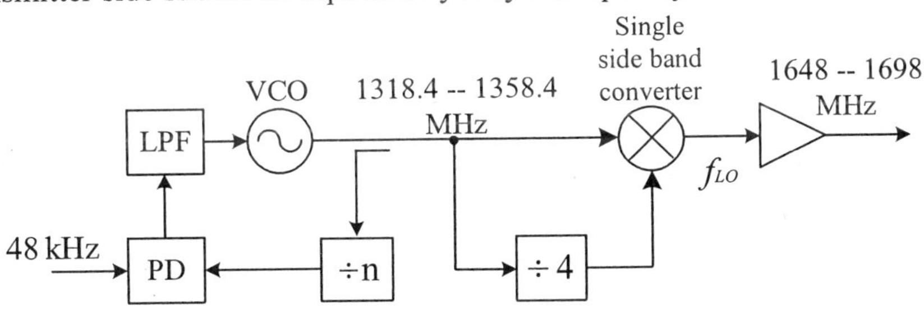

Figure 6.10. Block diagram of the LO generator.

In the previous sections, we have determined the most important parameters for a CDMA mobile station transmitter. The detail partitioning of the transmitter IC chain is not discussed here since it is much simpler than that of the receiver. The direct conversion transmitter system design for the CDMA mobile stations is ended here.

References

[1]. ETSI, *Digital Cellular Telecommunications System (Phase 2+); Radio Transmission and reception (GSM 05.05)*, July 1996.

[2]. TR45 TIA/EIA-136-270-A, *Mobile Stations Minimum Performance*, August 31, 1999.

[3]. TIA/EIA/IS-98-A, *Recommended Minimum Performance Standards for Dual-Mode Wideband Spread Spectrum Cellular Mobile Stations*.

[4]. TIA/EIA/IS-98-E, *Recommended Minimum Performance Standards for cdma2000 Spread Spectrum Cellular Mobile Stations*, Jan. 17, 2003.

[5]. Tao Li, C. Holenstein et al., "Direct Conversion Receiver Architecture," Patent WO 02/067420 A2

[6]. F. Lin, "Convergence and Output MSE of Digital Frequency-Locked Loop for Wireless Communications," *IEEE Trans. on Comm.*, vol. COM-44, no. 5, pp. 1215–1219 , May 1996.

[7]. A. D'Andreas and U. Mengali, "Design of Quadricorrelators for Automatic Frequency Control System," *IEEE Trans. on Comm.*, vol. COM-41, No. 6, pp. 988-997, June, 1993.

[8]. CDMA Development Group, *CDG System Performance Tests (optional)*, Rev. 3.0 draft, Apr. 9, 2003.

Associated References

[1]. G. G. E. Gieleg, "System _Level Design Tools for RF Communication ICs," *1998 URSI International Symposium on Signals, Systems, and Electronics*, pp. 422–426, Set. 1998.

[2]. B. A. Myers, J. B. Willingham et al., "Design Consideration for Minimal-Power Wireless Spread Spectrum Circuits and Systems," *Proceedings of the IEEE*, vol. 88, no. 10, pp. 1598–1612, Oct. 2000.

[3]. A. Matsuzawa, "RF-SoC: Expectations and Required Conditions," *IEEE Trans. on Microwave Theory and Techniques*, vol. 50, no. 1, pp. 245–253, Jan. 2002.

[4]. E. Sztein, "RF Design of a TDMA Cellular/PCS Handset, Part 1," *Communication Systems Design*, Feb. 2000.

[5]. E. Sztein, "RF Design of a TDMA Cellular/PCS Handset, Part 2," *Communication Systems Design*, March 2000.

[6]. S. K. Wong et al. "RF Transceiver Design and Performance Analysis for WCDMA User Equipment," *2002 IEEE International Conference on Communications*, vol. 1, pp. 507–511, May 2002.

[7]. P. Estabrook and B. B. Lusignan, "The Design of a Mobile Receiver Radio Using a Direct Conversion Architecture," *1989 IEEE 39th Vehicular Technology Conference*, pp. 63–72, May 1989.

[8]. A. M. Bada and M. Maddiotto, "Design and Realisation of Digital Radio Transceiver Using Software Radio Architecture," *Proceedings of 2000 IEEE 51th International Conference on Vehicular Technology*, vol. 3, pp. 1727–1737, May 2000.

[9]. A. Batra, J. Balakrishnan and A. Dabak, "Design Challenges for a Multi-Band OFDM UWB Communication System," *Wireless Design & Development*, pp. 24–36, Oct. 2004.

[10]. M. S. J. Steyaert et al., "Low-Voltage Low-Power CDMA-RF Transceiver Design," *IEEE Trans. on Microwave Theory and Techniques*, vol. 50, no. 1, pp. 281–287, Jan. 2002.

[11]. A. Springer, L. Maurer, and R. Weigel, "RF System Concepts for Highly Integrated RFICs for W-CDMA Mobile Radio Terminals," *IEEE Trans. on Microwave Theory and Techniques*, vol. 50, no. 1, pp. 254–267, Jan. 2002.

[12]. W. Krenik and J. Y. Yang, "Cellular Radio Integration Directions," *Proceedings of the Bipolar/BiCMOS Circuits and Technology Meeting*, pp. 25–30, Sept. 2003.

[13]. W. Krenik, D. Buss, and P. Rickert, "Cellular Handset Integration: SIP vs. SOC," *Proceedings of the IEEE 2004 Custom Integrated Circuits Conference*, pp. 63–70, 2004.

[14]. M. S. Heutmaker and K. Le, "An Architecture for Self-Test of a Wireless Communication System Using Sampled IQ Modulation and Boundary Scan," *IEEE Communications Magazine*, vol. 37, no. 6, pp. 98–102, June 1999.

[15]. A. A. Abidi, "Low-Power Radio-Frequency IC's for Portable Communications," *Proceedings of the IEEE*, vol. 83, no. 4, pp. 544–569, April 1995.

[16]. A. A. Abidi, "CMOS Wireless Transceivers: The New Wave," *IEEE Communications Magazine*, vol. 37, no. 8, pp. 119–124, Aug. 1999.

[17]. J. Loraine, "Counting the Cost of RF System-on-Chip," *IEE Electronics Systems and Software*, pp. 8–11, Sept. 2003.

[18]. A. Fernandez-Duran, "Application of Zero-IF Radio Architecture to Multistandard Compatible Radio Systems," *1995 IEE Conference on Radio Receivers and Associated Systems*, pp. 81–85, Sept. 1995.

[19]. W. J. McFarland, "WLAN System Trends and the Implications for WLAN RFICs," *2004 IEEE Radio Frequency Integrated Circuits Symposium*, pp. 141–144, 2004.

[20]. W. Thomann et al., "Fully Integrated W-CDMA IF Receiver and IF Transmitter Including IF Synthesizer and On-Chip VCO for UMTS

Mobiles," *IEEE Journal of Solid-State Circuits*, vol. 36, no. 9, pp. 1407–1419, Sept. 2001.

[21]. C. Bailet and C. Roberts, "Test Methodologies for Evaluating W-CDMA Receiver Designs," *RF Design*, pp. 60–68, June 2000.

[22]. P. Zhang, "Nonlinearity Test for a Fully Integrated Direct-Conversion Receiver," *Microwave Journal*, vol. 47, no. 10, pp. 94–112, Oct. 2004.

[23]. R. Magoon et al., "A Single-Chip Quad-Band (850/900/1800/1900 MHz) Direct Conversion GSM/GPRS RF Transceiver with Integrated VCOs and Fractional-N Synthesizer," *IEEE Journal of Solid-State Circuits*, vol. 37, no. 12, pp. 1710–1720, Dec. 2002.

[24]. M. Ugajin et al., "Design Techniques for a 1-V Operation Bluetooth RF Transceiver," *IEEE 2004 Custom Integrated Circuits Conference*, pp. 393–400, Oct. 2004.

[25]. M. Zargari et al., "A Single-Chip Dual-Band Tri-Mode CMOS Transceiver for IEEE 802.11 a/b/g Wireless LAN," *IEEE Journal of Solid-State Circuits*, vol. 39, no. 12, pp. 2239–2249, Dec. 2004.

[26]. Z. Shi and R. Rofougaran, "A Single-chip and Multi-mode 2.4/5 GHz Transceiver for IEEE 802.11 Wireless LAN," *Proceedings of 2002 3rd International Conference on Microwave and Millimeter Wave Technologies*, pp. 229–232, Aug. 2002.

[27]. W. J. McFarland, "WLAN System Trends and the Implications for WLAN RFICs," *Digest of 2004 IEEE International Symposium on Radio Frequency Integrated Circuits*, pp. 141–144, June 2004.

[28]. U. Karthaus et al., "Improved Four-Channel Direct-Conversion SiGe Receiver IC for UMTS Base Stations," *IEEE Microwave and Wireless Components Letters*, vol. 14, no. 8, pp. 377–379, Aug. 2004.

[29]. D. C. Bannister, C. A. Zelley and A. R. Barnes, "A 2–18 GHz Wideband High Dynamic Range Receiver MMIC," *2002 IEEE International Symposium on Radio Frequency Integrated Circuits*, pp. 147–149, June 2002.

[30]. C. L. Ko et al., "A CMOS Dual-Mode RF Front-End Receiver for GSM and WCDMA," *2004 IEEE Asia-Pacific Conference on Advanced System Integrated Circuits*, pp. 374–377, Aug. 2004.

[31]. F. Piazza and Q. Huang, " A 1.57-GHz RF Front-End for Triple Conversion GPS Receiver," *IEEE Journal of Solid-State Circuits*, vol. 33, no. 2, pp. 202–209, Feb. 1998.

[32]. L. Zhou et al., "S-Band Integrated Digital Broadband Receiver," *Proceedings of the IEEE Radar Conference*, pp. 535–540, April 2004.

[33]. P. U. Su ad C. M. Hsu, "A 0.25 μm CMOS OPLL Transmitter IC for GSM and DCS," *2004 IEEE International Symposium on Radio Frequency Integrated Circuits*, pp. 435–438, June 2004.

[34]. A. Hadjichristos, J. Walukas and N. Klemmer, "A Highly Integrated Quad Band Low EVM Polar Modulation Transmitter for GSM/EGDE Applications," *2004 IEEE International Conference on Custom Integrated Circuits*, pp. 565–568, Oct. 2004.

[35]. G. Brenna et al., "A 2-GHz Carrier Leakage Calibrated Direct-Conversion WCDMA Transmitter in 0.13-μm CDMA," *IEEE Journal of Solid-State Circuits*, vol. 39, no. 8, pp. 1253–1262, Aug. 2004.

[36]. T. Melly, et al., "An Ultralow-Power UHF Transceiver Integrated in a Standard Digital CMOS Process: Transmitter," *IEEE Journal of Solid-State Circuits*, vol. 36, no. 3, pp. 467–472, Match 2001.

[37]. B. Razavi, "A 900-MHz/1.8-GHz CMOS Transmitter for Dual-Band Applications," *IEEE Journal of Solid-State Circuits*, vol. 34, no. 575–579, May 1999.

[38]. D.S. Malhi et al., "SiGe W-CDMA Transmitter for Mobile Terminal Application," *IEEE Journal of Solid-State Circuits*, vol. 38, no. 9, pp. 1570–1574, Sept. 2003.

[39]. A. Italia et al., "A 5-GHz Silicon Bipolar Transmitter Front-End for Wireless LAN Applications," *2004 IEEE International Conference on Custom Integrated Circuits*, pp. 553–556, Oct. 2004.

[40]. M. Kosunen, et al., "Design of a 2.4 GHz CMOS Frequency-Hopped RF Transmitter IC," *Proceeding of the 1998 IEEE International Symposium on Circuits and Systems*, vol. 4, pp. IV-409–IV-412, May 1998.

[41]. A. Miller, "RF Exposure: SAR Standards and Test Methods," *Compliance Engineering Magazine*, Jan./Feb. 2003.

[42]. K. Fukunaga, S. Watanabe and Y. Yamanaka, "Dielectric Properties of tissue-equivalent liquids and their effects on Specific Absorption Rate," *IEEE Trans. On Electromagnetic Compatibility*, vol. 46, no. 1, pp. 126–129, Feb. 2004.

[43]. L. Pucker, "Paving Paths to Software Radio Design," *Communication Systems Design*, vol. 10, no. 6, June 2001.

Index

B

C

T